GENERAL TRADE MATHEMATICS

CONTENTS

APPENDIX

GENERAL TRADE MATHEMATICS

CHAPTER 1

WHOLE NUMBERS

INTRODUCTION

Before proceeding to the application of mathematics to mechanics, machines, and shop problems, we must ensure our thorough understanding of the arithmetic fundamentals. This means that we must be able to perform the four fundamental operations (addition, subtraction, multiplication, and division) with whole numbers, common fractions, decimal fractions, mixed numbers, and denominate numbers. We must also understand how to use percentages. Later, we must know how to use elementary algebra and how to perform simple geometric constructions. These tools are indispensable to any first-class workman.

Throughout this work, typical examples are given and solved. These solved examples, together with reasons for each step, are boxed so as to be readily found and should serve as guides in the solution of similar problems. The signs and symbols generally used in mathematics will be strictly adhered to.

It is an absolute necessity that the student know how to attack each problem he must solve. He must read carefully so that he understands the statements in the problem. He must be able to pick out the facts that are needed for the solution. When directions are given, they must be followed.

The rules and explained examples given should be carefully studied. In the actual solution of a problem, set the work down in logical order, and make all calculations on the sheet of paper you expect to keep as a record of your work. The result of each operation should be properly marked so that any part of the work can be referred to without confusion.

Approximate answers should be determined for comparison. For example: 45×65 equals approximately 50×60, or 3000. An answer greatly different from 3000 in this case is evidently an error.

Work carefully, and check your results as you go along.

REVIEW OF DEFINITIONS AND SYMBOLS

Arithmetic can be defined as the science and application of numbers.

The ten digits are 1, 2, 3, 4, 5, 6, 7, 8, 9, 0.

A **number** is one digit or a collection of digits, as 5, 17, and 343.

An **integer,** or **integral number,** is a whole number, as 6 and 73.

A **denominate number** is one that is applied to a particular thing as a unit of measure, as 3 ft. and 25 lb.

An **abstract number** is not applied to a particular object or quantity, as 6, 11, and 75.

A **prime number** is a number that can be divided exactly only by itself and the number 1, as 1, 2, 3, 5, and 7.

A **composite number** is one that can be divided exactly by other numbers as well as by itself and 1. For example, 6 can be divided exactly by 2 or 3, as well as by 6 and 1.

A **factor** of a composite number is any number that can be divided exactly into the composite number; for example, 2, 4, and 8 are factors of 16.

The **prime factors** of a given number are the prime numbers whose product is the given number.

Symbol	Represents	Indicates	Example
+	plus	addition	4 + 5 means 4 plus 5
−	minus	subtraction	9 − 4 means 9 minus 4
×	times	multiplication	5 × 6 means 5 times 6
÷	divided by	division	8 ÷ 4 means 8 divided by 4
/	divided by	division	$\frac{3}{4}$ means 3 divided by 4
=	equals	equality	2 + 1 = 3 means 2 plus 1 equals 3
:	is to	ratio	2:3 means 2 is to 3
: :	as	equality	2:4::3:6 means 2 is to 4 as 3 is to 6
%	per cent	per cent	6% means 6 per cent
$\sqrt{\ }$	radical sign	square root	$\sqrt{9}$ means the square root of 9
°	degrees	angular measure	30° means a 30-degree angle
°	degrees	temperature measure	80° means a temperature of 80 degrees
′	minutes	angular measure	29′ means 29 minutes
′	feet	linear measure	8′ means 8 feet
″	seconds	angular measure	35″ means 35 seconds
″	inches	linear measure	26″ means 26 inches

THE ARABIC DECIMAL-NUMBER SYSTEM

In English-speaking countries and in many others, the Arabic number system is used. In this system, there are ten symbols, or characters. These are 1, 2, 3, 4, 5, 6, 7, 8, 9, 0. These figures are sometimes called the alphabet of numbers; and, by combining them according to the principle of place value, any integral number can be written. The principle of place value is easily explained by the following chart:

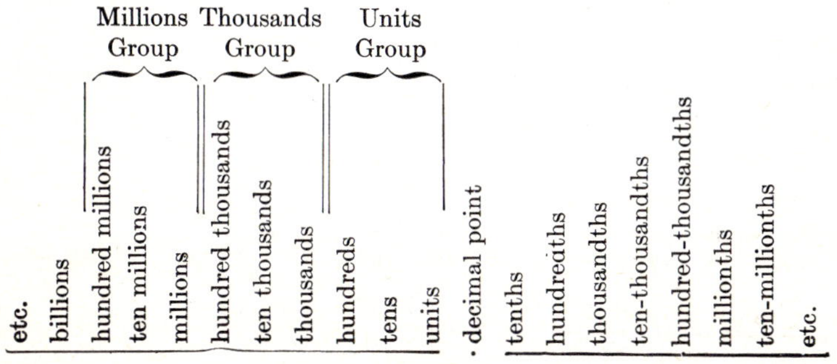

Millions Group: hundred millions, ten millions, millions
Thousands Group: hundred thousands, ten thousands, thousands
Units Group: hundreds, tens, units

etc. | billions | hundred millions | ten millions | millions | hundred thousands | ten thousands | thousands | hundreds | tens | units | · decimal point | tenths | hundredths | thousandths | ten-thousandths | hundred-thousandths | millionths | ten-millionths | etc.

The value of a figure is determined by its position with respect to the decimal point. Notice on the chart that all figures to the left of the decimal point represent whole number values, while figures to the right of the decimal point represent decimal values.

Example 1.—Read 763.

hundreds	tens	units
7	6	3

This number is read "seven hundred sixty-three."

Here the 3 represents 3 units, the 6 represents six tens, and the 7 represents 7 hundreds.

Example 2.—Read 5021.

thousands	hundreds	tens	units
5	0	2	1

This number is read "five thousand twenty-one."

Here the 5 represents 5 thousands, the 2 represents 2 tens, the 1 represents 1 unit. Note that since there are no hundreds a zero is used to "fill up" the hundreds' place.

Example 3.—Read .2315.

	tenths	hundredths	thousandths	ten-thousandths
.	2	3	1	5

This number is read "two thousand three hundred fifteen ten-thousandths."

In reading decimals (decimal fractions), read the number as if there were no decimal point and then give it the value of the place value for the **last** figure. Here the number is read "two thousand three hundred fifteen ten-thousandths" since the last figure, 5, is in the ten-thousandths' place.

Example 4.—Read 7.9

units	"and"	tenths
7	.	9

This number is read "seven and nine-tenths."

In reading a number made up of a whole number and a decimal, read the decimal point as "and"; then read the decimal.

Example 5.—Read $.83\frac{1}{3}$

.	8 tenths	$3\frac{1}{3}$ hundredths	This number is read "eighty-three and one-third hundredths." In reading complex decimals, note that the fraction has no place value of its own but is considered a part of the preceding place. Here we have three and one-third hundredths.

All numbers are read from left to right. In dictation, numbers are often read as they appear: 81.6 may be read "eight one point six." Telephone numbers are always read as they appear; for example, Central 4836-J is read "Central four eight three six jay."

In shopwork, numbers larger than billions and smaller than ten-thousandths are seldom used.

To write numbers from words, use the place-value chart. Where there is no figure for a place value, fill in the place with a zero: otherwise, you will change the value of the number.

EXERCISES

Read the following numbers:

1. 708	**2.** 68,754	**3.** 87,564	**4.** 358,743
5. 456,329	**6.** 22,222	**7.** 637,842	**8.** 5,437,895
9. 7854	**10.** 34,629	**11.** 456,783	**12.** 59,384
13. 4,567,893	**14.** 70,001	**15.** 150,051	**16.** 60,606,060
17. 77.6	**18.** 333.66	**19.** 5,493.621	**20.** 27.3946
21. 297,321.4	**22.** 5.832754	**23.** 217.4325	**24.** 9,001.002
25. 72.5480	**26.** 300.005	**27.** 66.0666	**28.** 0.00908
29. 10.1010	**30.** 501.501	**31.** 0.00006	**32.** 203.023
33. 4.100101	**34.** 0.003434	**35.** 56.69847	**36.** 789.3456
37. 0.000026	**38.** 20,000,045	**39.** 60,606.06	**40.** 200.002

Read the following telephone numbers:

41. 3633	**42.** 4365	**43.** 4382-W
44. 792-R2	**45.** Drexel 4843	**46.** Vermont 9427
47. VA-4536	**48.** 6-5236	**49.** 7464-2 green

EXERCISES

Write the following numbers in figures:

1. Seven hundred fifty-four.
2. Five thousand six hundred seventy-seven.
3. Two hundred forty-six thousand, nine hundred fifty-four.
4. Eight million, seven thousand, ninety-nine.
5. Five hundred forty-six million, nine hundred sixty thousand, sixty-one.
6. Twenty-one and five hundred seventy-six thousandths.
7. Six hundred forty-four and five hundred eighty-eight thousandths.
8. Forty-nine and forty-nine thousand forty-nine hundred-thousandths.
9. Eighty and fifty-nine thousand four hundred thirty three hundred-thousandths.
10. Four and one hundred one thousand one hundred one millionths.

REVIEW EXERCISES

Add the following numbers orally:

1. $7 + 8$ **2.** $3 + 7$

3. $8 + 4 + 6$ **4.** $17 + 8$

5. $7 + 9$ **6.** $18 + 5$

7.	**8.**	**9.**	**10.**
8	7	9	17
7	4	0	8
5	6	7	5
3	8	4	6
9	4	7	4

11.	**12.**	**13.**	**14.**
15	18	17	16
3	9	11	5
5	5	12	19
6	4	13	7
3	1	15	14

15. Starting with 5, count to 95 by adding 5 each time, as 5, 10, 15, etc.

16. Starting with 5, count to 96 by adding 7 each time.

17. Starting with 7, count to 97 by adding 9 each time.
18. Starting with 7, count to 95 by adding 8 each time.
19. Starting with 2, count to 90 by adding 11 each time.
20. Starting with 5, count to 97 by adding 4 each time.
21. Starting with 11, count to 95 by adding 7 each time.
22. Starting with 4, count to 100 by adding 12 each time.

ADDITION OF WHOLE NUMBERS

Addition is the process of finding the **sum** of two or more numbers called **addends.**

Example.—Add 643, 7264, 37, 9, and 1403.

```
  1 2
  6 4 3
1
7 2 6 4
    3 7
      9
1 4 0 3
-------
9 3 5 6
```

Write the numbers in vertical columns with units under units, tens under tens, etc.

Add the units' column: $3 + 4 + 7 + 9 + 3 = 26$.

Write the 6 below the units' column, and carry the 2 to the tens' column. Write the number 2 above the tens' column as shown.

Using the 2 carried over as the first number, add the tens' column: $2 + 4 + 6 + 3 + 0 = 15$. Write the 5 below the tens' column, and carry the 1 to the hundreds' column.

Add the hundreds' column: $1 + 6 + 2 + 4 = 13$. Write the 3 below the hundreds' column, and carry the 1 to the thousands' column.

Add the thousands' column: $1 + 7 + 1 = 9$. Write the 9 in the thousands' column.

Check by adding the column in the opposite direction.

EXERCISES AND PROBLEMS

1. $4{,}585{,}285 + 792 + 73{,}214 = ?$
2. $96 + 41 + 77 + 728 + 666 = ?$
3. $5003 + 304 + 50 + 25 = ?$
4. $75 + 36 + 942 + 1003 = ?$
5. \$21 + \$7 + \$95 + \$107 = ?
6. Add 4926, 7329, 9872, and 5280.
7. Combine 27, 41, 96, 728, and 777.
8. Find the sum of 143, 729, 896, and 148.

9. Add 720, 340, 680, and 500.

10. Add 1001, 101, 11, and 10,001.

11. Find the number of feet of fencing needed to enclose the lot shown.

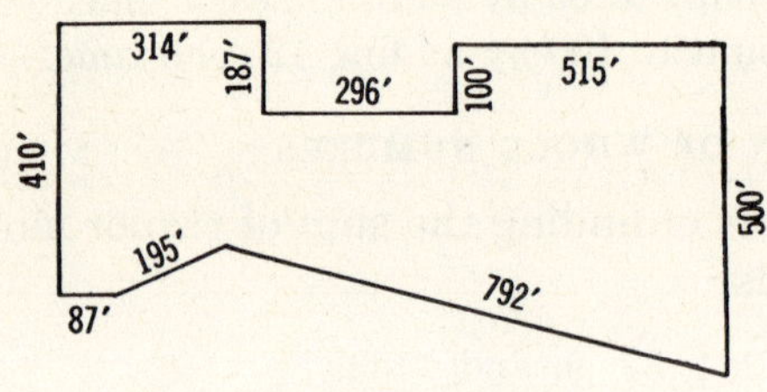

Prob. 11.

12. A weekly pay roll reads as follows: Rich, $65.00; Lane, $77.00; Smith, $88.00; Perkins, $80.00. Find the total pay roll.

13. A wiring job calls for the following quantities of No. 14 rubber-coated wire: kitchen, 57 ft.; dining room, 125 ft.; living room, 476 ft.; bath, 66 ft.; bedrooms, 146 ft.; hall, 127 ft. How many feet of wire are needed for the job?

14. Add 143, 35, 342, 98, 987, 34, and 642.

15. Truck loadings were 5674 lb., 9876 lb., 4532 lb., 4690 lb., and 8843 lb. Find the total weight in pounds.

16. Employees in three automobile-manufacturing plants numbered 99,743 men, 56,743 men, and 188,345 men. How many men were employed in the three plants?

17. Four consecutive years' production of wheat was 7,896,543 bu., 9,654,328 bu., 7,865,496 bu., and 7,865,432 bu. Find the total production in bushels for the 4 years.

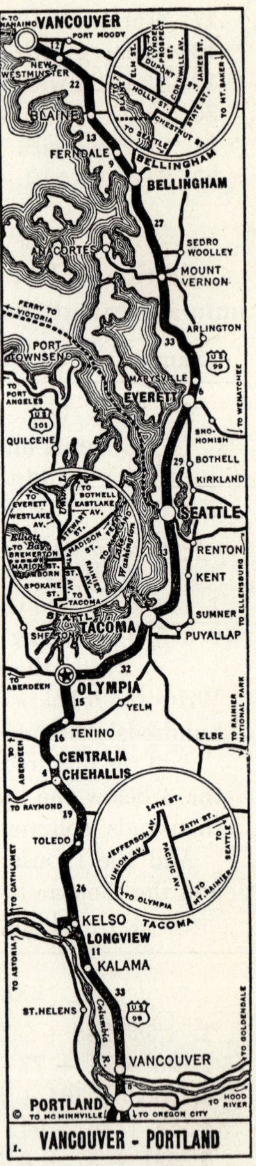

Prob. 18.

18. Use the strip road map shown to find the distance from Bellingham to Portland.

19. The table below shows the monthly expenditures of Mr. Brown and his family during the past year. Figure the total expenditure for each month and the total expenditure for each item. Copy the table in your notebook. Do not write in this book.

Month	Food	Water, light, and gas	Education	Clothing and linens	Doctors and dentists	Insurance	Taxes	Recreation	Total
Jan.	$ 78	$ 9	$18	$ 18	$ 7	...	$27	$ 3	$
Feb.	81	11	16	12	32	$36	36	12	
Mar.	92	14	11	24	...	83	84	9	
Apr.	86	12	10	6	11	...	27	15	
May	83	7	18	95	5	...	27	16	
June	80	8	4	16	...	83	27	20	
July	162	2	3	33	...	...	27	125	
Aug.	78	9	5	41	56	...	27	17	
Sept.	80	8	47	162	12	83	27	11	
Oct.	92	11	41	66	6	...	27	18	
Nov.	95	10	21	....	...	30	85	10	
Dec.	97	14	12	121	...	83	27	18	
Total	$	$	$	$	$	$	$	$	Grand total

Note.—By taking the sum of all the totals, you get the grand total. Since the grand totals obtained for the horizontal and vertical total columns should be the same, the problem is said to be self-checking.

SUBTRACTION OF WHOLE NUMBERS

Subtraction is the process of taking one number from another. The minus sign (−) indicates subtraction; 577 − 169 means "From 577 take 169."

The **minuend** is the larger number from which the smaller number is taken.

The **subtrahend** is the smaller number, to be taken from the minuend.

The **remainder, or difference,** is the result of a subtraction.

Example 1.—Subtract 385 from 976.

976 (minuend)
385 (subtrahend)
591 (difference)

Set the subtrahend below the minuend, with the units under units, tens under tens, etc.

In the units' column, think what number added to 5 gives 6.

Think "5 + 1 = 6." Write the 1 under the units' column.

Think "8 + 9 = 17." Write the 9 under the tens' column.

Think "1 (carried to the hundreds' column) + 3 = 4." Think "4 + ? = 9." Write 5 under the hundreds' column.

Check the result by adding the difference and the subtrahend. This is the "Austrian," or shop method.

Example 2.—From 577 subtract 169.

5̸77 (6 written above the crossed-out 7)
169
408

Set the subtrahend below the minuend with the units under the units, tens under tens, etc.

Subtract the right-hand figure. 9 cannot be subtracted ırom 7; but borrowing a 10 from the tens' digit and adding this to the 7 make 17; then

$$17 - 9 = 8,$$

remainder. In borrowing, always cross out the number borrowed from and write above it the next smaller number.

In the tens' column, 6 − 6 = 0. In the hundreds' column, 5 − 1 = 4.

Ans. 408, remainder.

Check the result by adding the remainder and the subtrahend.

Note that different methods are used in Examples 1 and 2. Use the one that is more familiar to you.

EXERCISES AND PROBLEMS

1. 91,543 − 2231 = ?
2. 343 − 277 = ?
3. 984 − 392 = ?
4. 10,001 − 777 = ?

5. 13,200 − 846 = ?

6. Subtract 3264 from 9371.

7. Take 746 from 1200.

8. How much is left when \$66 is taken from \$146?

9. How much is 9876 less 345?

10.

663	146	12000
−225	− 99	− 729

11. A casting weighed 526 lb. After machining, it weighed 498 lb. How much was machined off?

12. A tank contained 5266 gal. of oil. If 796 gal. were taken out, how many gallons remained?

13. Subtract 2,900,896 tons from 3,708,063 tons.

14. Take 287,969 bu. from 734,523 bu.

15. How many are 853,425 cars less 496,879 cars?

16. Take 5,070,908 yd. from 7,905,020 yd.

17. The minuend is 987,654. The subtrahend is 497,963. Find the remainder.

18. By how much is 77,728 less than 93,762?

19. How much is 2392 less 1644?

20. The table below gives the number and cost of motor vehicles manufactured in the United States over a period of years. Compute the increase or decrease for each succeeding year.

Year	Passenger cars	Cost	Motor trucks	Cost
1938	2,000,985	\$1,236,802,411	488,100	\$ 334,147,520
1939	2,866,796	1,765,189,067	710,496	494,829,231
1940	3,717,385	2,370,654,083	754,901	567,820,414
1941	3,779,682	2,567,205,996	1,060,820	1,069,799,855
1942	222,862	163,813,559	818,662	1,427,256,801
1943	139	101,799	699,689	1,451,794,475
1944	610	446,704	737,524	1,700,928,939
1945	69,532	57,254,655	655,683	1,181,955,532
1946	2,148,699	1,979,781,084	940,851	1,043,247,276
1947	3,558,178	3,963,896,000	1,239,744	1,708,622,000
1948	3,909,270	4,853,402,000	1,376,155	1,858,210,000
1949	5,119,466	6,768,418,000	1,134,136	1,407,435,000

REVIEW PROBLEMS ON ADDITION AND SUBTRACTION

1. Loaded cars of coal received by a dealer weighed 76,240 lb., 97,500 lb., and 68,054 lb. Find the total weight.

2. A trucking firm dispatched trucks with net loads as follows: 14,750 lb., 9329 lb., 4548 lb., 7479 lb., and 11,500 lb. Find the total weight.

(Courtesy of Tinius Olsen Testing Machine Co.)
PROB. 8.

3. A carload of lumber contained 17,280 bd. ft. If 5262 bd. ft. are sold, how many board feet remain?

4. A series of castings weigh 74 lb., 348 lb., 221 lb., 95 lb., 16 lb., and 58 lb. Find the total weight.

5. A dwelling requires 31,256 bd. ft. of lumber. The yard has only 24,094 bd. ft. on hand. Find the shortage.

6. A boxed piece of machinery weighed 458 lb. If the box weighed 95 lb., what did the machine weigh?

7. A casting made of iron weighs 325 lb.; if made of aluminum alloy, it weighs only 119 lb. What is the difference in weight?

8. On the testing machine shown, the tensile strength of steel was found to be 80,965 lb. per sq. in., while that of aluminum was only 31,256 lb. per sq. in. How much stronger is the steel than the aluminum?

9. Copper melts at 1981°F. and boils at 4172°F. How many degrees above the melting point is the boiling point?

10. The daily mileage on a trip was as follows: Monday, 325; Tuesday, 296; Wednesday, 98; Thursday, 405; Friday, 328; Saturday, 196. What is the total distance traveled?

11. Mr. Brown has an 88-acre farm. He sells 25 acres to Mr. Perkins and 16 acres to Mr. Jones. How much does he sell? How many acres has he left?

12. A machinist draws $88.00 the first week, $78.00 the second week, $72.00 the third week, and $96.00 the fourth week. What are his total earnings for the 4 weeks?

13. The total amount of gasoline used in a day by the trucks of a certain bakery was as follows: truck 1, 23 gal.; truck 2, 17 gal.; truck 3, 15 gal.; truck 4, 25 gal.; truck 5, 18 gal. Find the total amount consumed.

14. A contractor is to excavate a basement containing 50,000 cu. ft. During the first week he removed 375 cu. ft., 932 cu. ft., 458 cu. ft., 341 cu. ft., and 242 cu. ft. How many cubic feet remain?

15. A local freight agent checked in cars loaded as follows: 78,320 lb., 96,664 lb., 48,932 lb., and 21,968 lb. How many pounds were received in all?

16. A certain foundry used 5566 lb. of scrap iron, 3349 lb. of new pig iron, and 320 lb. of nickel in one heat. What is the total weight used in this heat?

17. In Prob. 16, how much more scrap iron than pig iron was used?

18. If the Packard automobile weighs 3960 lb. and the Deluxe Ford weighs 2878 lb., what is the difference in their weights?

19. A streamline train traveled 287 miles, while a steam locomotive traveled only 219 miles. Determine the difference in the distances.

20. Which is greater and by how much, 1,010,101 or 999,999?

MULTIPLICATION OF WHOLE NUMBERS

Multiplication is a process of increasing a number by its own value a certain number of times. It is really a short method of adding.

The number to be increased is called the **multiplicand.**

The number showing the number of times the multiplicand is to be taken as an addend is the **multiplier.**

The result of the multiplication is called the **product.**

The multiplication sign (×) is written between the numbers.

Example 1.—Multiply 746 by 425.

```
      746 (multiplicand)
      425 (multiplier)
     ----
     3730 (first partial product, 5 × 746)
    1492  (second partial product, 2 × 746)
   2984   (third partial product, 4 × 746)
   ------
   317050 (total product)
```

Place the multiplier below the multiplicand.

Begin by multiplying the multiplicand through by the figure in the units' place of the multiplier for the first partial product. 5 × 6 = 30. Put the zero in the units' column. The 3 represents 3 tens; therefore, we have 3 tens in addition to 5 × 4 tens, or 20 tens + 3 tens = 23 tens. Put the 3 under the tens' column, and carry the additional 2 hundreds. 5 × 7 = 35; adding in the 2 we carried gives us 7 in the hundreds' column and 3 in the thousands' column. This completes the first partial product, 3730.

Next multiply the multiplicand by the figure in the tens' place of the multiplier. Note that 2 (tens) × 6 = 12 (tens), and the 2 is set in the tens' column; the 1 is carried over to the hundreds' column. Proceed as for the first partial product.

In this way, multiply the multiplicand by the remaining figures of the multiplier.

Add the partial products to obtain the total product.

Check results by interchanging the multiplier and the multiplicand, and multiply again.

Example 2.—Multiply 545 by 500.

545 500 272500	Note here that the multiplier is set to the right. No multiplications are made by the zeros, for zero times any number is zero. Since there are two zeros at the end of the multiplier, the final product will have a zero in the units' place and a zero in the tens' place. Therefore, the units' place and the tens' place are filled by writing two zeros at the right of the product. Check as in Example 1.

Example 3.—Multiply 77 by 101.

77 101 77 770 7777	Note here that the tens' figure in the multiplier is zero. The tens' partial product will be zero; therefore, fill up the tens' place with a zero, and proceed to multiply by the figure in the next place. Check as in Example 1.

EXERCISES AND PROBLEMS

1. Multiply 7032 by 608.

2. $87 \times 100 = ?$

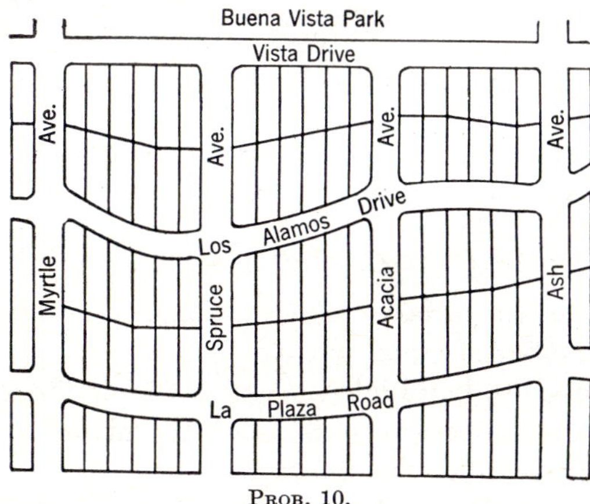

PROB. 10.

3. Find the product of 146 and 777.

4. Multiply 642 by 20.

5. How much is 532 × 366?

6. 666 × 10,000 = ?

7. 4800 × 99 = ?

8. 66 × 77 × 88 = ?

9. Find the product of 740 × 360.

10. On the map shown each lot is 50 ft. wide. The streets are 80 ft. wide. Get the total width of the subdivision from the west side of Myrtle Ave. to the east side of Ash Ave.

11. A mechanic earns $78.00 per week for 50 weeks. Compute his total wage.

12. A brickmason lays 3 bricks in 2 min. At this rate, how many bricks can he lay in a 7-hr. day?

13. A motorcar company loaded 3 automobiles to the freight car, at the rate of 5 freight cars per hour, for 24 hr. a day, counting 300 days in a year. Calculate the year's production.

14. In Prob. 13, figure the total weight of the iron in the year's production if the automobiles averaged 2116 lb. of iron each.

15. In a year, a coal company shipped 8744 cars of coal, averaging 40,200 lb. of coal per car. Find the total pounds shipped.

16. The cylinder block for a certain engine weighs 409 lb. Get the weight of 20,765 blocks.

17. If 1 cu. ft. of wood weighs 53 lb.. how heavy are 8059 cu. ft. of wood?

18. An empty dump truck weighs 9875 lb. Calculate the weight of 548 of these empty trucks.

19. Find the total weight of 7986 bags of cement if each bag weighs 96 lb.

20. If 5 bags of cement are required per lineal foot of highway, how many bags will be needed to build 120 miles of highway?

DIVISION OF WHOLE NUMBERS

Division is the process of separating a number into a required number of equal parts.

The number to be divided is called the **dividend.**

The number to divide by is called the **divisor.**

The result is called the **quotient.**

The sign of division (÷) is written between the numbers. The dividend is always written preceding the sign and the divisor following the sign; thus, 3 ÷ 4 means 3 divided by 4. The line of a fraction separating the two numbers also indicates division; thus, $\frac{7}{8}$ means 7 divided by 8.

Examples 1 and 2 show short division.

Example 1.—651 ÷ 7 = ?

7)651 Draw a frame around the dividend, and set the
93 divisor to the left of it. Write the quotient below the dividend.

Try the division of 7 into the first digit on the left. 7 into 6 is not possible.

Try 7 into 65. 65 ÷ 7 = 9, with 2 the remainder. Write the 9 under the last digit used, the 5, and carry the 2 (tens), making 21.

Then, 21 ÷ 7 = 3. Write the 3 at the right of the 9 in the quotient.

Check by multiplying the quotient by the divisor to obtain the dividend.

Example 2.—675 ÷ 4 = ?

4)675 or 4)675
$168\frac{3}{4}$ 168, R = 3

Check.

168
×4
672
+3
675

If in completing the division as in Example 1 there is a remainder, the remainder may be written in the answer over the divisor. In this example, remainder 3 over divisor 4, or $\frac{3}{4}$.

Or the remainder may be written after the letter R, as shown to the right above.

In checking, where the quotient contains a remainder, find the product of the divisor and the quotient, then add the remainder to obtain the dividend.

Example 3 shows long division. This is used when the divisor is a large number.

Example 3.—88,011 ÷ 127 = ?

```
      693
127)88011
     762    (127 × 6)
     1181
     1143   (127 × 9)
      381
      381   (127 × 3)
```

The process is the same as short division except that the partial products are set beneath the proper terms of the dividend. For example, the product of 127 × 6, which is 762, is written beneath 880. The 6 in the quotient is written above the last figure of the dividend used in this step.

Make the subtraction, writing the remainder directly below, and bring down the next number in the dividend, in this case, 1.

1181 ÷ 127 = 9, etc.

Check as in Example 1.

EXERCISES AND PROBLEMS

1. 175 ÷ 5 = ?

2. 362 ÷ 3 = ?

3. 18,550 ÷ 25 = ?

4. 10,153 ÷ 71 = ?

5. 86,952 ÷ 7246 = ?

6. Divide 625 by 5.

7. $\frac{1728}{12} = ?$

8. How much is 58,538 divided by 54?

9. Find the quotient of 6561 ÷ 81.

10. How much is 2405 divided by 37?

11. How many times is 177 contained in 44,035?

12. A $684 payroll is divided equally among 9 men. Find the earnings of each.

13. If 1296 people are enrolled in 24 classes, what is the average number of persons per class?

14. How much does each man receive if a weekly pay roll of $128,845 is divided equally among 1765 men?

15. A job that requires 1144 hr. was divided equally among 26 men. How many hours must each man work?

16. The struts in the bridge sketched here divide the length into 9 equal parts. How long is each part?

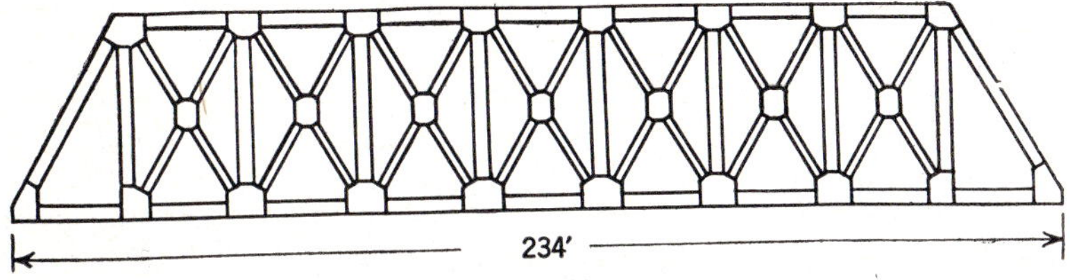

Prob. 16.

17. On a vacation trip, a family traveled 1296 miles in 6 days. What is the average mileage per day?

18. If average rate of travel per day is the same as in Prob. 17, how many days would it take to cover 3240 miles?

19. Divide an annual pay roll of $3,257,472 equally among 998 men.

Prob. 25.

20. If a punch-press operator finishes 7200 parts in 10 hr., how many seconds were required for each part?

21. Mr. Buckle earns $5700 in 12 months. What is his monthly salary?

22. If a milling-machine operator finishes 3 gears per day, how many days will it take him to deliver 240 gears? How many 5-day weeks is this?

23. A shop uses 96 kw.-hr. of electricity in a 24-day month. What is the average consumption per day?

24. A bar of iron 25 ft. long weighs 1575 lb. What is the weight per foot?

25. The photograph shows a modern streamline Santa Fe train on a double track. How many rails, each 30 ft. long, are there in a mile of double track?

26. If the rails in Prob. 25 each weigh 115 lb. per yd., find the weight of the rails for 1 mile of double track.

27. Find the price of the rails in Prob. 26 at $44 per ton.

28. A job allows for 252 hr. How many 7-hr. days is this?

29. There are 43,578 sq. ft. in a shop floor. If there are 9 sq. ft. in 1 sq. yd., how many square yards is this?

30. A truck is operated for 26 days at a total cost of $156.00. What is the daily cost of operation?

31. A certain boiler evaporates 3 lb. of water per hour for each square foot of heating surface. How many square feet of heating surface are there if 2769 lb. are evaporated in 1 hr.?

REVIEW PROBLEMS ON MULTIPLICATION AND DIVISION

1. If a train averages 48 m.p.h. for 9 hr., how far does it travel?

2. What is the average speed per hour of an airplane that travels from Chicago to Los Angeles, a distance of 1692 miles, in 5 hr.?

3. If Mr. Jones works 7 hr. a day, 5 days a week, for 45 weeks at $1.76 per hour, what does he earn?

4. Mr. Brown earns $95.00 per week for 48 weeks of each year. What is his total yearly income?

5. The daily attendance at U.H.S. for the 5 days last week was 4063, 4097, 4132, 4072, and 4116. What is the average daily attendance?

6. An automatic screw machine finishes a certain bolt every 3 sec. How many are finished in a continuous run of 4 hr.? Of 24 hr.?

7. Divide 4512 marbles equally among 47 boys.

8. The labor cost for assembling a certain kind of machine is estimated at $32.00 apiece. In 1 week, 20 machines are assembled by 8 men. If each man is paid the same amount, what does each get for his week's work?

9. If a shop uses an average of 15 kw.-hr. of electricity per day, how much is consumed in a month of 26 days?

10. A certain oil well flows at the rate of 2350 barrels per day. How much is produced in a month of 30 days?

PROB. 10.

11. A boiler requires 12 sq. ft. of heating surface for each horsepower. Find the horsepower of a boiler having 1716 sq. ft. of heating surface.

12. How many castings weighing 96 lb. each can be loaded on a truck that can carry 6 tons?

13. How many 8-in. pieces can be cut from a bar 54 ft. long?

14. A mechanic can assemble 4 machines per day. How many machines can 48 mechanics assemble in a month of 26 days?

15. How many feet are there in 19 miles? How many inches?

16. A train of 45 cars is loaded with 3 automobiles per car, each automobile weighing 3160 lb. Determine the total weight of the shipment.

17. A "joint," or length, of 2-in. pipe weighs 64 lb. How much will 125 joints weigh?

18. On a tank farm, there are 125 oil tanks of 50,000 bbl. each. What is the total storage capacity in barrels?

19. A company has 755 employees. The monthly pay roll is $199,335.00. What is the average wage per man?

20. How many pounds of coal will be required to run a 115-hp. boiler for 10 hr. if 3 lb. are needed per horsepower per hour?

CHANGING THE FORM OF DENOMINATE NUMBERS

Expressions like 12 gal. and 6 ft. are called **denominate numbers.**

The units in which denominate numbers are expressed can be changed from one form to another without actually changing the value. Tables of equivalents are given on page 523.

In changing denominate numbers from one form to another, it is necessary to keep only the following two rules in mind: (1) **In changing from a small unit to a larger unit, divide, since you must have in the result fewer of the large units than you originally had of the small units.** (2) **In changing from a large unit to a small unit, multiply, since you must have in the result more of the smaller units than you originally had of the large units.**

Example 1.—Change 12 gal. to cubic inches.

Note immediately that we are changing from a large unit to a small unit; therefore, we must multiply.

```
 231     There are 231 cu. in. in 1 gal.
  12     Hence, multiplying 231 by 12 gives the number of
 ---
 462     cubic inches in 12 gal.
231
----
2772
```

Ans. 2772 cu. in.

Example 2.—Change 17,325 cu. in. to gallons.

Note immediately that we are changing from a small unit to a large unit; therefore, we must divide.

```
       75     There are 231 cu. in. in 1 gal. Hence, the
231)17325     number of gallons in 17,325 cu. in. equals the
    1617      number of times 17,325 contains 231.
    ----
     1155
     1155
     ----
```

Ans. 75 gal.

Example 3.—Change 55 yd. to inches.

Note immediately that we are changing from a large unit to a smaller unit; therefore, we must multiply.

```
  55     1 yd. = 3 ft.
   3     Hence, 3 × 55 = 165, the number of feet in 55 yd.
 ---
 165     1 ft. = 12 in.
  12     Hence, 12 × 165 = 1980, the number of inches in
 ---     165 ft.
 330
165
----
1980
```

Ans. 1980 in.

Note.—This example can be solved in one operation by using this equivalent: 1 yd. = 36 in. 55 × 36 = 1980, the number of inches in 55 yd.

Example 4.—How many yards are there in 612 in.?

Note immediately that we are changing from a small unit to a larger unit; therefore, we must divide.

```
12)612        12 in. = 1 ft.
 3) 51        Hence, 612 ÷ 12 = 51, the number of feet
    17      in 612 in.
Ans. 17 yd.   3 ft. = 1 yd.
              Hence, 51 ÷ 3 = 17, the number of yards
            in 51 ft.
```

Note.—This example can be solved in one operation by using this equivalent: 36 in. = 1 yd. 612 ÷ 36 = 17, the number of yards in 612 in.

PROBLEMS

Change:

1. 41 gal. to cubic inches.
2. 56,200 tons to hundredweight.
3. 4620 cu. in. to gallons.
4. 78 miles to feet.
5. 1728 yd. to feet.
6. 242,682 ft. to yards.
7. 36,288 cu. in. to cubic feet.
8. 25 ft. to inches.
9. 47 doz. to units.
10. 48 doz. to gross.

11. 48 gross to dozen; to units.
12. 1152 sq. in. to square feet.
13. 108 sq. yd. to square inches.
14. 17,325 cu. in. to gallons.
15. 108 cu. yd. to cubic inches.
16. 50 gal. to pints.
17. 144 doz. to gross.
18. 279 degrees to seconds.
19. 144 doz. to units.
20. 250 bu. to quarts.
21. 162 cu. yd. to cubic feet.
22. 125 miles to rods.
23. 162 cu. ft. to cubic yards.
24. 48,000 lb. to tons.
25. 1,064,330 gal. to barrels of oil.
26. 1,000,000 lb. to long tons.
27. 546 fathoms to feet.
28. 15 tons to ounces.
29. 88 degrees to seconds.
30. 144 in. to feet.
31. 544,000 lb. to tons.
32. 621 cu. ft. to cubic yards.
33. 51,000 tons to pounds.
34. 2 cu. yd. to cubic inches.
35. 25,525 lb. of water to cubic feet.
36. 9 gal. to pints.
37. 96 pt. to gallons.
38. 766 sq. miles to acres.
39. 288 in. to yards.

CHAPTER 2

COMMON FRACTIONS

DEFINITIONS

A whole can be divided into equal parts.

On the rule shown, the inches are divided into 16 equal parts.

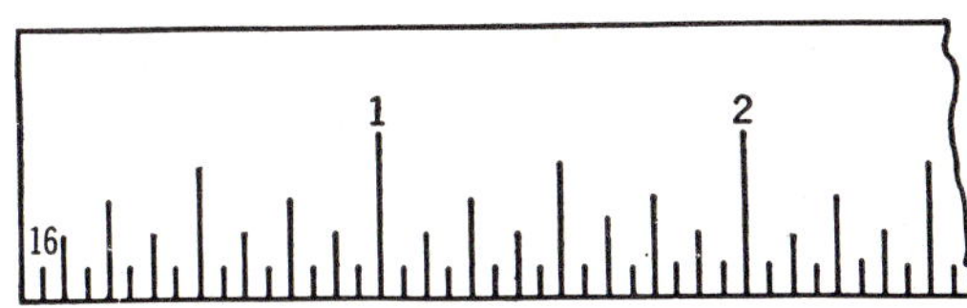

One part is equal to one-sixteenth of an inch, written $\frac{1}{16}$ in. or $\frac{1}{16}''$. Three parts are equal to three-sixteenths of an inch, written $\frac{3}{16}$ in. or $\frac{3}{16}''$.

A **common fraction** indicates a part or parts of a whole.

The **numerator** is the number written above the line of a fraction. It indicates how many parts of the unit are taken.

The **denominator** is the number written below the line of a fraction. It indicates into how many equal parts the unit is divided.

The numerator and denominator are called the **terms** of a fraction.

The line between the numerator and the denominator indicates division; the numerator is divided by the denominator. For example, in the fraction $\frac{7}{16}$, 7 is the numerator and 16 is the denominator. The line of the fraction indicates that 7 is divided by 16.

A **proper fraction** is a fraction in which the numerator is smaller than the denominator, as $\frac{5}{16}$ and $\frac{7}{32}$.

An **improper fraction** is a fraction in which the numerator is equal to or larger than the denominator, as $\frac{6}{3}$ and $\frac{7}{2}$.

A **mixed number** is made up of a whole number and a fraction, as $7\frac{1}{2}$. It is read seven and one-half and means seven whole units plus one-half a unit.

A **complex fraction** is a fraction in which either the numerator or the denominator (or both) is a fraction or a mixed number, as $\frac{\frac{1}{4}}{\frac{2}{3}}$, $\frac{7\frac{1}{2}}{5}$ and $\frac{9}{\frac{3}{8}}$.

CHANGING IMPROPER FRACTIONS TO WHOLE OR MIXED NUMBERS

An improper fraction can be changed to a whole number or a mixed number.

Example 1.—Change $\frac{8}{8}$ to a whole number or to a mixed number.

$8\overline{)8}$	$\frac{8}{8}$ means $8 \div 8$.
1	
$\frac{8}{8} = 1$	

Example 2.—Change $\frac{8}{5}$ to a whole number or to a mixed number.

$5\overline{)8}$	$\frac{8}{5}$ means $8 \div 5$.
$1\frac{3}{5}$	
$\frac{8}{5} = 1\frac{3}{5}$	

RAISING FRACTIONS TO HIGHER TERMS

When the numerator and the denominator of a fraction are *both* multiplied by the *same* number, the value of the fraction is not changed.

Example.—Multiply the terms of $\frac{3}{8}$ by 4.

$\frac{3 \times 4}{8 \times 4} = \frac{12}{32}$	For the numerator, $3 \times 4 = 12$. For the denominator, $8 \times 4 = 32$.
$\frac{3}{8} = \frac{12}{32}$	

Note.—This process is sometimes called **stepping up** the value of a fraction.

CHANGING THE DENOMINATOR OF A FRACTION TO A REQUIRED DENOMINATOR

The process just explained is used in shopwork when it is necessary to change a given common fraction to one with a required higher denominator.

Rule.—To change the denominator of a given fraction to a fraction having the required denominator, divide the denominator of the given fraction into the required denominator, then multiply both terms of the given fraction by this quotient.

Example.—Change $\frac{3}{4}$ to sixty-fourths.

$64 \div 4 = 16$	Divide 64, the required denominator, by 4, the given denominator.
$\frac{3 \times 16}{4 \times 16} = \frac{48}{64}$	4 must be multiplied by 16 to give 64.
$\frac{3}{4} = \frac{48}{64}$	Hence, 16 is the number used to multiply both terms of $\frac{3}{4}$.

EXERCISES

Change the following fractions to equivalent fractions having the required denominator:

1. $\frac{1}{2}$ to 8ths.
2. $\frac{3}{4}$ to 16ths.
3. $\frac{7}{10}$ to 50ths.
4. $\frac{1}{2}$ to 10ths.
5. $\frac{7}{12}$ to 72nds.
6. $\frac{3}{5}$ to 60ths.
7. $\frac{3}{13}$ to 39ths.
8. $\frac{7}{11}$ to 44ths.
9. $\frac{5}{8}$ to 144ths.
10. $\frac{7}{12}$ to 180ths.
11. $\frac{5}{7}$ to 42nds.
12. $\frac{3}{7}$ to 70ths.
13. $\frac{5}{8}$ to 64ths.
14. $\frac{7}{32}$ to 64ths.
15. $\frac{9}{16}$ to 64ths.
16. $\frac{7}{24}$ to 144ths.
17. $\frac{7}{24}$ to 360ths.
18. $\frac{3}{8}$ to 48ths.
19. $\frac{7}{18}$ to 144ths.
20. $\frac{5}{13}$ to 65ths.
21. $\frac{15}{32}$ to 128ths.
22. $\frac{13}{16}$ to 128ths.

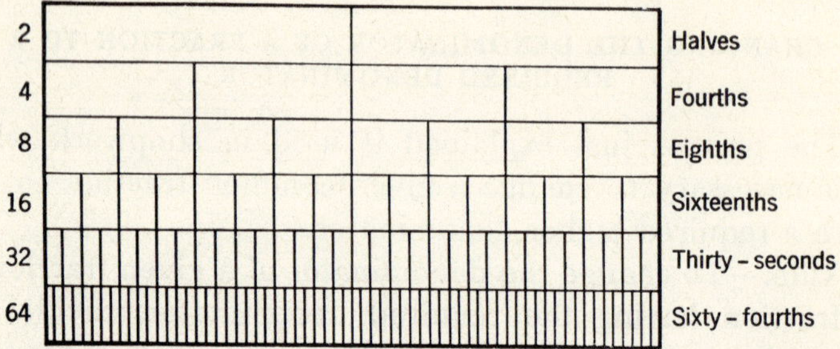

In practical shopwork the common fractions most used are halves, fourths, eighths, sixteenths, thirty-seconds, and sixty-fourths. By examining the ruler shown, which is divided into sixty-fourths, it is evident that

$$\frac{1}{4} = \frac{2}{8} = \frac{4}{16} = \frac{8}{32} = \frac{16}{64}$$

$$\frac{1}{2} = \frac{2}{4} = \frac{4}{8} = \frac{8}{16} = \frac{16}{32} = \frac{32}{64}$$

The student should become familiar with these and other equivalent fractions as he meets them in his work.

REDUCING FRACTIONS

When the numerator and the denominator of a fraction are *both* divided by the *same* number, the value of the fraction is not changed.

Example.—Divide both terms of $\frac{12}{32}$ by 4.

$$\frac{12 \div 4}{32 \div 4} = \frac{3}{8}$$

For the numerator, $12 \div 4 = 3$.
For the denominator, $32 \div 4 = 8$.

$$\frac{12}{32} = \frac{3}{8}$$

Note.—This process is referred to as **reducing the value** of a fraction.

REDUCING FRACTIONS TO LOWEST TERMS

Answers to problems should be expressed in lowest terms. To reduce a fraction to its lowest terms, the process is applied as follows:

Rule.—To reduce a fraction to its lowest terms, divide both terms of the fraction by a common factor, that is,

by any number that is a factor of each term. Continue dividing in the same manner until the terms of the fraction contain no common factors.

Example 1.—Reduce $\frac{28}{35}$ to lowest terms.

$$\frac{28 \div 7}{35 \div 7} = \frac{4}{5}$$
$$\frac{28}{35} = \frac{4}{5}$$

Both 28 and 35 can be divided evenly by 7. Since there is no integer that can be divided evenly into both 4 and 5. $\frac{4}{5}$ is in lowest terms.

Example 2.—Reduce $\frac{315}{455}$ to lowest terms.

$$\frac{315 \div 5}{455 \div 5} = \frac{63}{91}$$
$$\frac{63 \div 7}{91 \div 7} = \frac{9}{13}$$

Both 315 and 455 can be divided evenly by 5.

In the resulting fraction, both 63 and 91 can be divided evenly by 7.

Since 9 and 13 only contain the common factor 1, $\frac{9}{13}$ is in lowest terms.

EXERCISES

Reduce each of the following fractions to lowest terms:

1. $\frac{49}{84}$	**2.** $\frac{48}{84}$	**3.** $\frac{128}{144}$	**4.** $\frac{225}{625}$	**5.** $\frac{90}{135}$
6. $\frac{36}{48}$	**7.** $\frac{48}{36}$	**8.** $\frac{80}{600}$	**9.** $\frac{55}{330}$	**10.** $\frac{45}{135}$
11. $\frac{24}{82}$	**12.** $\frac{60}{105}$	**13.** $\frac{54}{98}$	**14.** $\frac{105}{120}$	**15.** $\frac{64}{128}$

FINDING THE LOWEST COMMON DENOMINATOR

In problems involving addition or subtraction of fractions, the denominators of the fractions must be the same. It is often necessary to change the given fractions to equivalent fractions. A common denominator can be obtained by multiplying all the denominators together. Or the **lowest common denominator** (L.C.D.), sometimes thought of as the **lowest common multiple** (L.C.M.), can be found.

Rule.—To find the lowest common denominator (L.C.D.),

1. Find the prime factors of each of the denominators.

2. Form a product of all the different prime factors that appear in step 1, using each the greatest number of times it occurs in any one of the given denominators.

Example.—Change the following fractions to equivalent fractions having the L.C.D.: $\frac{3}{4}$, $\frac{2}{3}$, and $\frac{5}{6}$.

$4 = 2 \times 2$ Prime factors of each denominator
$3 = 3 \times 1$ The L.C.D. must contain 2×2, since 2
$6 = 3 \times 2$ appears twice as a factor of 4. The 3 and 6 each contain the factor 3.

The 6 also contains the factor 2. Since 2 has already been taken care of as a factor of 4, it is not repeated. Therefore, the L.C.D. is $2 \times 2 \times 3$, or 12.

Divide the denominator of each of the fractions into the L.C.D. Then change each of the fractions to an equivalent fraction having the required denominator.

$$12 \div 4 = 3 \qquad \frac{3 \times 3}{4 \times 3} = \frac{9}{12}$$

$$12 \div 3 = 4 \qquad \frac{2 \times 4}{3 \times 4} = \frac{8}{12}$$

$$12 \div 6 = 2 \qquad \frac{5 \times 2}{6 \times 2} = \frac{10}{12}$$

ADDITION OF COMMON FRACTIONS

It is possible to add or subtract fractions only when they have a common denominator. When fractions to be added have different denominators, it is necessary to find a common denominator.

Rule.—To add fractions having different denominators, express each fraction as an equivalent fraction having a common denominator. Add the numerators, and put the sum over the common denominator. Reduce the resulting fraction to lowest terms.

Example 1.—Add $\frac{3}{8}$ and $\frac{1}{8}$.

$\frac{3}{8} + \frac{1}{8} = \frac{4}{8}$ Since $\frac{3}{8}$ and $\frac{1}{8}$ each have the same denominator, 8, add the numerators. $3 + 1 = 4$.
$\frac{4}{8} = \frac{1}{2}$ Place the sum, 4, over the common denominator, 8.

$\frac{4}{8}$ reduces to $\frac{1}{2}$.

Example 2.—Add $\frac{3}{4}$, $\frac{1}{8}$, and $\frac{9}{16}$.

> $\frac{3}{4} + \frac{1}{8} + \frac{9}{16} = \frac{12}{16} + \frac{2}{16} + \frac{9}{16} = \frac{23}{16}$
>
> $\frac{23}{16} = 1\frac{7}{16}$
>
> Here, the denominators are 4, 8, and 16. The L.C.D. = 16, $\frac{3}{4}$ is equivalent to $\frac{12}{16}$, $\frac{1}{8}$ is equivalent to $\frac{2}{16}$.
>
> Adding numerators, 12 + 2 + 9 = 23. Put 23 over 16. $\frac{23}{16}$ is an improper fraction and must be changed to a mixed number, $1\frac{7}{16}$.

EXERCISES AND PROBLEMS

1. $\frac{1}{2} + \frac{1}{4} = ?$
2. $\frac{1}{8} + \frac{1}{16} = ?$
3. $\frac{1}{4} + \frac{1}{32} = ?$
4. $\frac{1}{4} + \frac{5}{8} = ?$
5. $\frac{3}{4} + \frac{3}{8} = ?$
6. $\frac{3}{8} + \frac{7}{16} = ?$
7. $\frac{5}{8} + \frac{3}{4} = ?$
8. $\frac{1}{2} + \frac{9}{16} = ?$
9. $\frac{7}{16} + \frac{5}{32} = ?$
10. $\frac{3}{32} + \frac{11}{64} = ?$
11. $\frac{15}{64} + \frac{1}{8} = ?$
12. $\frac{13}{32} + \frac{9}{16} = ?$

13. $\frac{3}{8} + \frac{1}{4} + \frac{5}{16} = ?$
14. $\frac{1}{2} + \frac{9}{16} + \frac{7}{64} = ?$
15. $\frac{3}{8} + \frac{5}{16} + \frac{1}{4} = ?$
16. $\frac{15}{16} + \frac{7}{8} + \frac{10}{32} = ?$
17. $\frac{1}{2} + \frac{1}{4} + \frac{1}{8} + \frac{1}{16} = ?$
18. $\frac{21}{32} + \frac{5}{8} + \frac{9}{64} = ?$
19. $\frac{3}{8} + \frac{17}{64} + \frac{5}{32} = ?$
20. $\frac{1}{2} + \frac{3}{4} + \frac{9}{16} + \frac{1}{8} + \frac{9}{32} + \frac{5}{64} = ?$

21. Find the total length.

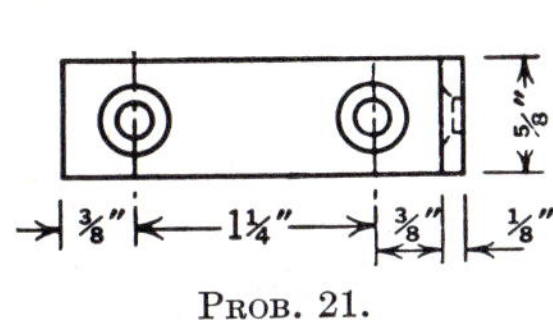

PROB. 21.

22. Find the total length.

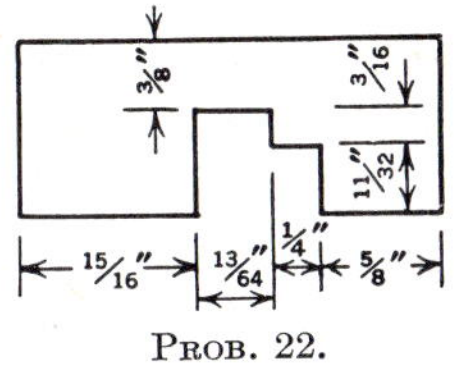

PROB. 22.

23. Find the pin length.

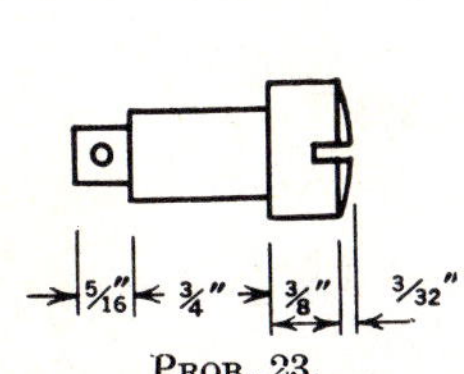

PROB. 23.

24. Find the width and the length.

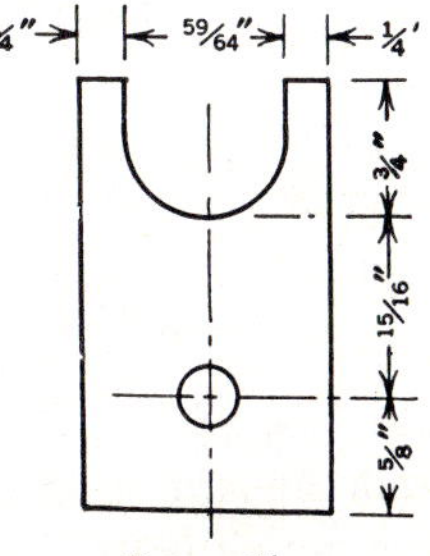

PROB. 24.

25. Find the pin length.

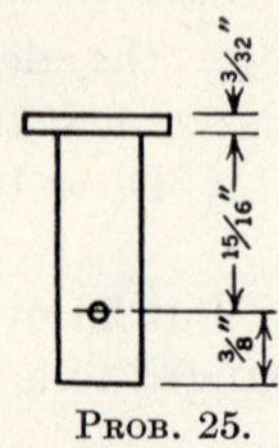

PROB. 25.

26. Find the over-all length.

1/2" 1/4" 3/8" 5/16" 3/8" 3/8" 3/8" 7/16" 3/8" 1/2" 3/8" 9/16" 3/8" 5/8" 3/8" 3/4" 1/2"

PROB. 26.

ADDITION OF MIXED NUMBERS

Rule.—To add mixed numbers, add the whole numbers and the fractions separately and combine the results.

Example.—Add $7\frac{3}{4}$, $9\frac{1}{2}$, $7\frac{5}{8}$, and $3\frac{9}{16}$.

	16ths
$7\frac{3}{4}$	12
$9\frac{1}{2}$	8
$7\frac{5}{8}$	10
$3\frac{9}{16}$	9
26	$\frac{39}{16} = 2\frac{7}{16}$
$2\frac{7}{16}$	
$28\frac{7}{16}$	

The L.C.D. is 16. Each fraction is changed to an equivalent fraction having 16 for its denominator.

The numerators are set in a column of sixteenths and added. The sum, 39, is put over 16, thus: $\frac{39}{16}$.

This improper fraction is reduced to a mixed number $2\frac{7}{16}$.

This is added to the sum of the units' column, 26, making the complete sum, $28\frac{7}{16}$.

EXERCISES AND PROBLEMS

1. Add $3\frac{7}{16}$, $9\frac{3}{4}$, $3\frac{5}{8}$, $26\frac{5}{8}$, $3\frac{3}{32}$, and 18.

2. Find the sum of $3\frac{1}{2}$, $7\frac{3}{8}$, $8\frac{7}{16}$, $\frac{9}{64}$, $\frac{3}{4}$, and $3\frac{7}{32}$.

3. How much is $9\frac{27}{32} + 5\frac{9}{16} + 8\frac{3}{8} + 5\frac{11}{64}$?

4. What length of stock is needed for the hammer shown?

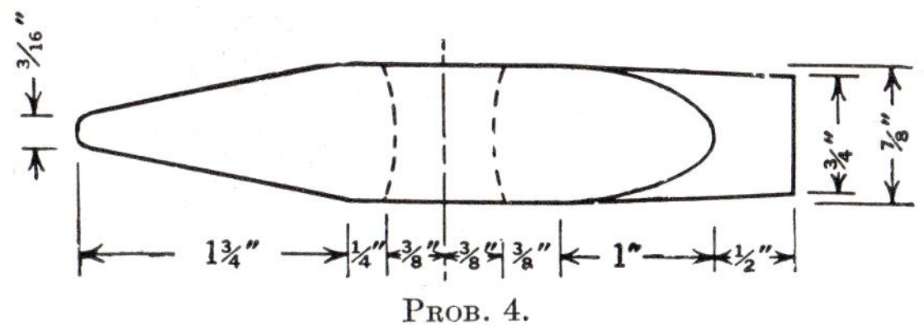

PROB. 4.

5. Find the sum of $3\frac{3}{32}$, $7\frac{1}{16}$, $9\frac{5}{64}$, and $21\frac{21}{32}$.

6. Add $4\frac{63}{64}$, $51\frac{11}{16}$, and $23\frac{23}{32}$.

7. How much is $3\frac{5}{8}$ plus $7\frac{3}{16}$ plus $5\frac{9}{64}$ plus $\frac{31}{32}$?

8. Total the following: $\frac{57}{64}$, $\frac{21}{32}$, $\frac{15}{16}$, $\frac{7}{8}$, $\frac{3}{4}$.

9. $21\frac{45}{64} + 34\frac{11}{16} + 59\frac{5}{8} + \frac{21}{32} + 5\frac{3}{4} = ?$

10. In the spark plug shown, find the lengths of A, B, and C.

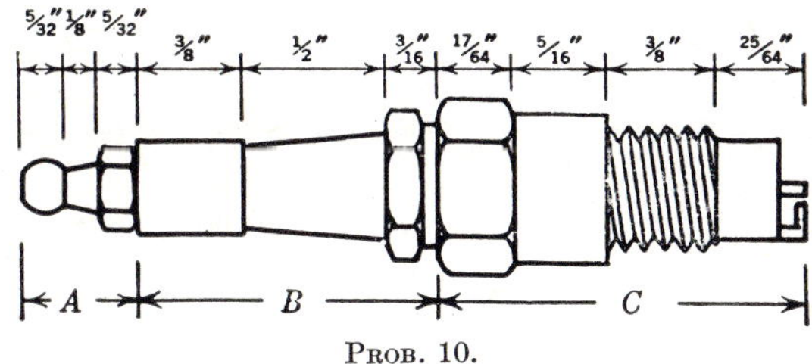

PROB. 10.

11. What is the total length of the spark plug?

12. In the diagram of the crankshaft, determine the length of A; of B; of D; of E.

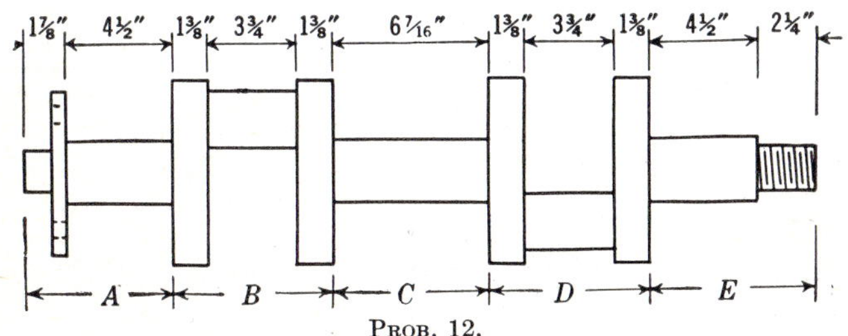

PROB. 12.

13. On the crankshaft of Exercise 12, find the length of A plus B.

14. On the same crankshaft, figure the length of D plus E plus C.

15. Find the total length of the same crankshaft, using the information obtained in Exercises 12 to 14.

SUBTRACTION OF COMMON FRACTIONS

Rule.—To subtract common fractions, express the fractions as fractions having a common denominator. Subtract the numerator of the fraction in the subtrahend from the numerator of the fraction in the minuend, and put the difference over the common denominator. Reduce results to lowest terms.

Example 1. $\frac{7}{8} - \frac{3}{8} = ?$

$\frac{7}{8} - \frac{3}{8} = \frac{4}{8}$	Subtract the numerator. $7 - 3 = 4$. Put 4 over 8.
$\frac{4}{8} = \frac{1}{2}$	Reduce $\frac{4}{8}$ to $\frac{1}{2}$.

Example 2. $\frac{3}{4} - \frac{5}{16} = ?$

$\frac{3}{4} - \frac{5}{16}$	L.C.D. = 16. $\frac{3}{4} = \frac{12}{16}$.
$\frac{12}{16} - \frac{5}{16} = \frac{7}{16}$	Subtract the numerator. $12 - 5 = 7$. Put 7 over 16.

Example 3.—From 8 subtract $\frac{3}{4}$.

$8 - \frac{3}{4}$	Take one unit from the integer, 8, and change it
$7\frac{4}{4}$	to fourths. $1 = \frac{4}{4}$.
$-\frac{3}{4}$	Find the difference between the fractions.
$7\frac{1}{4}$	$\frac{4}{4} - \frac{3}{4} = \frac{1}{4}$.

EXERCISES AND PROBLEMS

1. $\frac{9}{16} - \frac{3}{16} = ?$ **2.** $\frac{9}{16} - \frac{1}{16} = ?$ **3.** $\frac{15}{32} - \frac{3}{32} = ?$

4. $\frac{27}{64} - \frac{9}{64} = ?$ **5.** $\frac{7}{8} - \frac{3}{4} = ?$ **6.** $\frac{3}{4} - \frac{1}{8} = ?$

7. $\frac{7}{8} - \frac{1}{4} = ?$ **8.** $\frac{3}{4} - \frac{3}{16} = ?$ **9.** $\frac{9}{16} - \frac{3}{32} = ?$

10. $\frac{9}{64} - \frac{1}{8} = ?$ **11.** $\frac{27}{64} - \frac{1}{4} = ?$ **12.** $\frac{59}{64} - \frac{5}{8} = ?$

13. $16 - \frac{5}{16} = ?$ **14.** $19 - \frac{3}{8} = ?$ **15.** $8 - \frac{29}{32} = ?$

16. $11 - \frac{15}{64} = ?$ **17.** $3 - \frac{59}{64} = ?$ **18.** $1 - \frac{27}{32} = ?$

19. $5 - \frac{5}{16} = ?$ **20.** $5 - \frac{5}{64} = ?$ **21.** $3 - \frac{3}{16} = ?$

22. $9 - \frac{3}{8} = ?$ **23.** $\frac{7}{8} - \frac{5}{32} = ?$ **24.** $1 - \frac{9}{64} = ?$

25. A planer reduced the thickness of a board from $\frac{7}{8}''$ to $\frac{13}{16}''$. How much was removed?

26. How much of the bolt shown is not threaded?

27. A stud 4″ long is found to be $\frac{7}{16}''$ too long. What is the required length?

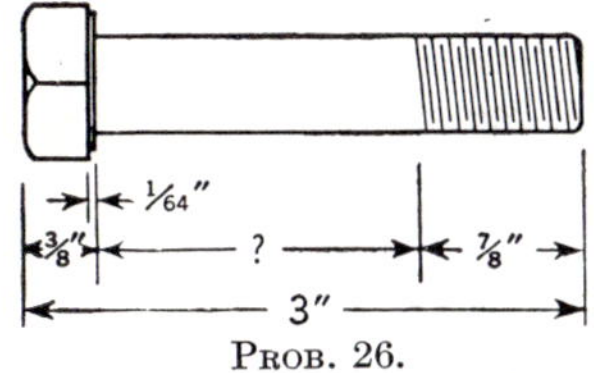

Prob. 26.

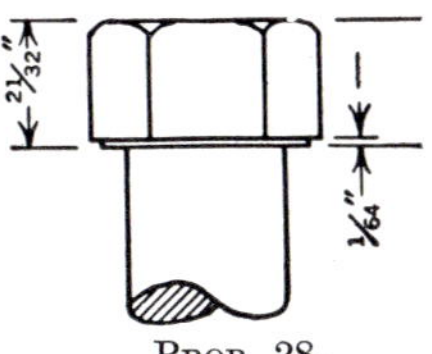

Prob. 28.

28. The head of the machine bolt shown, including the flat of $\frac{1}{64}''$, is $\frac{21}{32}''$ long. How long is the hexagonal part?

29. A 20″ casting shrinks $\frac{15}{64}''$ on cooling. Find the size when the casting is cold.

30. On the pipe nipple shown, how much is unthreaded?

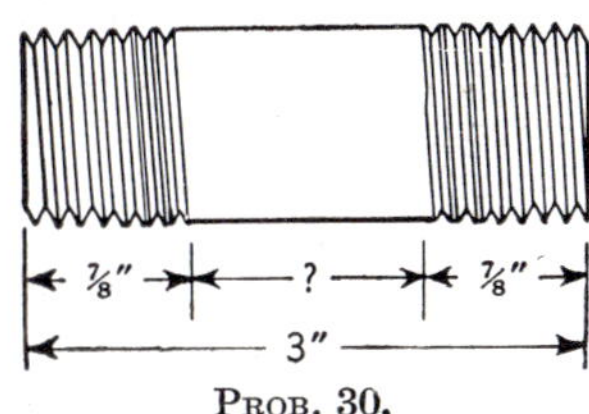

Prob. 30.

31. A 5″ iron billet is reduced in thickness $\frac{3}{16}''$ by each blow under a drop hammer. Find the thickness after the first blow. After the second blow. After the third blow.

32. What is the sheet metal width in next even inches needed to make a gutter like the cross-sectional shape shown?

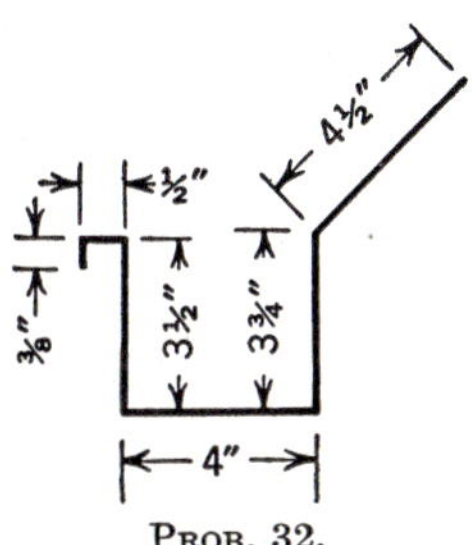

Prob. 32.

SUBTRACTION OF MIXED NUMBERS

Rule.—To subtract mixed numbers, change the fractions to equivalent fractions having a common denominator. Subtract the fractions; then subtract the whole numbers.

Example 1.—From $8\frac{15}{16}$ subtract $6\frac{3}{8}$.

	16ths
$8\frac{15}{16}$	15
$6\frac{3}{8}$	6
$2\frac{9}{16}$	$\frac{9}{16}$

L.C.D. = 16. Change the fractions to equivalent sixteenths.

Subtract the numerator. $15 - 6 = 9$. Put 9 over 16.

Subtract the whole number. $8 - 6 = 2$; remainder, $2\frac{9}{16}$.

Example 2.—Subtract $3\frac{15}{32}$ from $9\frac{1}{4}$.

	32ds	32ds
$\overset{8}{\cancel{9}}\frac{1}{4}$	$8 + 32$	$= 40$
$3\frac{15}{32}$	15	$= 15$
$5\frac{25}{32}$		$\frac{25}{32}$

L.C.D. = 32. Change the fractions to equivalent thirty-seconds.

Since 15 cannot be subtracted from 8, 1 unit (or $\frac{32}{32}$) is borrowed from the 9 units of the minuend and added to the $\frac{8}{32}$. $\frac{32}{32} + \frac{8}{32} = \frac{40}{32}$.

Subtracting numerator, $40 - 15 = 25$. Put 25 over 32.
Subtracting whole number, $8 - 3 = 5$.
The remainder is $5\frac{25}{32}$.

EXERCISES AND PROBLEMS

1. Subtract $5\frac{5}{8}$ from $11\frac{49}{64}$.
2. Subtract $7\frac{9}{16}$ from $12\frac{21}{32}$.
3. On the gauge shown, how much greater is D than C?

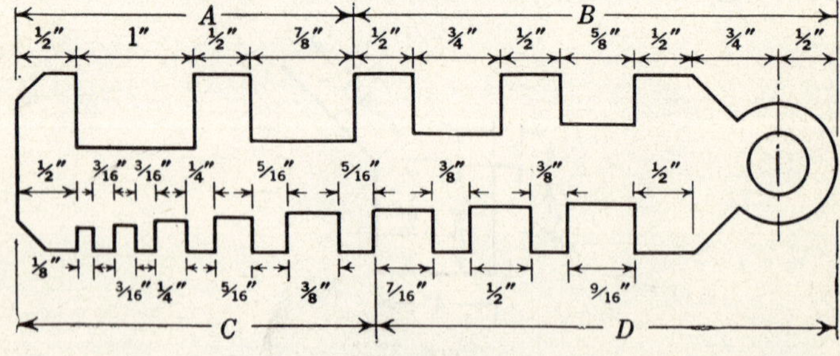

Probs. 3 to 5.

4. Find the difference between C and B.

5. How much is B minus A?

6. From 26 take $10\frac{9}{16}$.

7. How much greater is $14\frac{3}{16}$ than $10\frac{3}{8}$?

8. On the spindle shown, how much greater is B than A? Use the same diagram for Probs. 9 to 16.

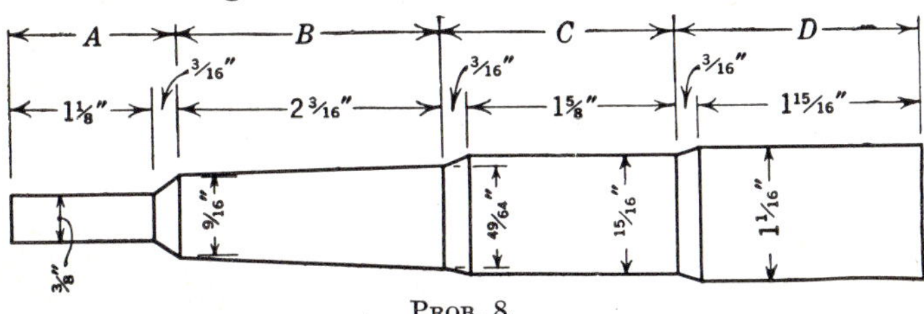

PROB. 8.

9. Find the difference between B and C.

10. Find the difference between A and C.

11. How much longer is B than D?

12. How much greater in diameter is D than the large end of B?

13. How much less in diameter is the large end of B than C?

14. How much less is the diameter at A than the small end of B?

15. How much greater in diameter is D than A?

16. How much larger is B at its large end than at its small end?

17. $\begin{array}{r} 54\frac{3}{16} \\ -39\frac{43}{64} \\ \hline \end{array}$

18. $\begin{array}{r} 27\frac{27}{32} \\ -\ 7\frac{17}{64} \\ \hline \end{array}$

19. $\begin{array}{r} 19\frac{19}{32} \\ -\ 9\frac{19}{64} \\ \hline \end{array}$

20. From a drum containing 52 gal. of oil, the toolroom boy filled four cans holding, respectively, $\frac{3}{8}$ gal., $1\frac{1}{2}$ gal., $\frac{1}{2}$ gal., and $4\frac{3}{4}$ gal. How many gallons of oil were taken out of the barrel? How many remained?

REVIEW PROBLEMS

1. A man spent $\frac{5}{8}$ of his income. What fraction of his income was saved?

2. John grew $\frac{7}{8}$ in. in a year, while Bill grew only $\frac{1}{4}$ in. How much more did John grow than Bill?

3. Mr. Jones can do one-half of a job in one day. His son can do one-fourth of it in a day. How much of the job can they do working together for one day?

4. Which has the greater diameter, a $\frac{3}{8}$-in. wire or a $\frac{5}{16}$-in. wire? How much greater?

5. Find the outside diameter of the hollow cylinder shown.

6. If a piece of steel $\frac{1}{64}$ in. thick is riveted to a plate $\frac{3}{8}$ in. thick, how thick is the combined unit?

7. Arrange the following drills according to size, starting with the largest: $\frac{1}{4}''$, $\frac{27}{64}''$, $\frac{5}{8}''$, $\frac{11}{16}''$, $\frac{3}{16}''$, $\frac{1}{2}''$.

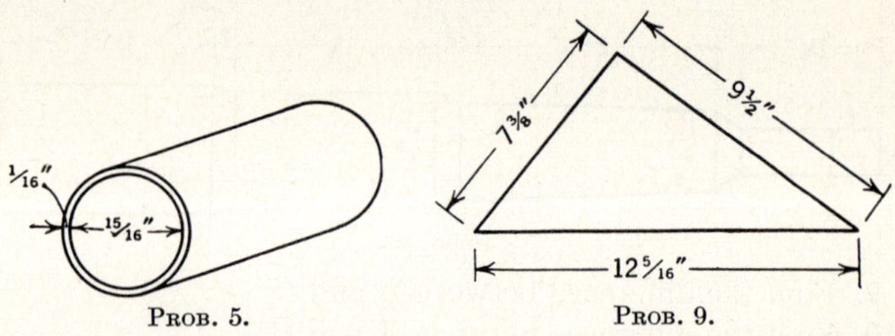

PROB. 5. PROB. 9.

8. A foreman ordered some brass $\frac{1}{4}$ in. thick. He received brass $\frac{7}{64}$ in. thick. Was it too thick or too thin? By how much?

9. Compute the perimeter of the triangle shown. (Perimeter is the distance around.)

10. A pile of wood contains 15 cords. What is left if Mr. Jones buys $2\frac{1}{2}$ cords; Mr. Brown, $3\frac{3}{8}$ cords; and Mr. Black, $2\frac{3}{4}$ cords?

11. During the week a truck used oil as follows: $1\frac{1}{4}$ gal., $\frac{1}{2}$ gal., $\frac{7}{8}$ gal., $\frac{9}{16}$ gal., 1 gal., $\frac{3}{8}$ gal. How many gallons were used?

12. If a man can do a piece of work in 16 days, how much of it can he do in 3 days? 5 days? 10 days? 12 days?

13. A bar of iron weighed $15\frac{1}{2}$ lb. It was machined down to $8\frac{7}{8}$ lb. What weight was removed?

14. The block shown is made up of several pieces bolted together. Find the total thickness.

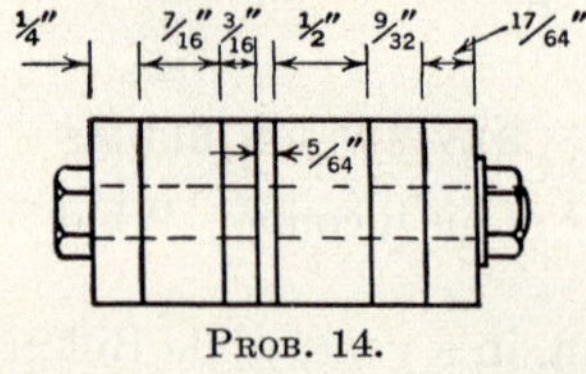

PROB. 14.

15. In one day the fleet of trucks of the Peerless Bakery used gasoline as follows: truck A, $9\frac{3}{8}$ gal.; truck B, $17\frac{3}{4}$ gal.; truck C, $18\frac{1}{2}$ gal.; truck D, $12\frac{5}{8}$ gal.; truck E, $9\frac{1}{2}$ gal. Compute the total consumption.

16. Four partly filled cans of oil containing $\frac{3}{8}$ gal., $\frac{7}{16}$ gal., $\frac{3}{4}$ gal., and $\frac{1}{2}$ gal., respectively, are poured into a large can. How many gallons are poured into the large can?

17. Bars of the following thicknesses are stacked together: $\frac{7}{16}''$, $\frac{3}{8}''$, $\frac{1}{2}''$, $\frac{5}{8}''$, $\frac{3}{32}''$. Find the thickness of the stack.

MULTIPLICATION OF COMMON FRACTIONS

Rule.—To multiply fractions, multiply the numerators of the fractions for the numerator of the product; multiply the denominators of the fractions for the denominator of the product. If possible, reduce the result.

Example 1.—$2 \times \frac{3}{8} = ?$

$\frac{2}{1} \times \frac{3}{8} = \frac{6}{8} = \frac{3}{4}$ Multiply the numerators. $2 \times 3 = 6$.
Multiply the denominators. $1 \times 8 = 8$.
All integers have the denominator 1.
Reduce $\frac{6}{8}$ to $\frac{3}{4}$.

Example 2.—$\frac{3}{4} \times \frac{5}{7} = ?$

$\frac{3}{4} \times \frac{5}{7} = \frac{15}{28}$ Multiply the numerators. $3 \times 5 = 15$.
Multiply the denominators. $4 \times 7 = 28$.

Example 3.—$\frac{8}{15} \times \frac{5}{12} \times \frac{2}{3} = ?$

$$\frac{\overset{2}{\cancel{8}}}{\underset{3}{\cancel{15}}} \times \frac{\overset{1}{\cancel{5}}}{\underset{3}{\cancel{12}}} \times \frac{2}{3} = \frac{4}{27}$$

When possible, reduce by canceling; that is, divide the same factor into both numerator and denominator.
Divide 4 into both 8 and 12.
Divide 5 into both 5 and 15.
Multiply the final numerators, $2 \times 1 \times 2 = 4$.
Multiply the final denominators, $3 \times 3 \times 3 = 27$.

Example 4.—Find $\frac{2}{3}$ of $\frac{9}{16}$.

$$\frac{\overset{1}{\cancel{2}}}{\underset{1}{\cancel{3}}} \times \frac{\overset{3}{\cancel{9}}}{\underset{8}{\cancel{16}}} = \frac{3}{8}$$

The word "of" when used in this way indicates multiplication.
Cancel as in Example 3.

EXERCISES

In the following exercises, cancel whenever possible. Fractional answers always should be reduced to lowest terms.

1. $5 \times \frac{3}{8} = ?$

2. $7 \times \frac{9}{16} = ?$

3. $3 \times \frac{5}{15} = ?$

4. $8 \times \frac{4}{5} = ?$

5. $6 \times \frac{3}{4} = ?$

6. $12 \times \frac{5}{16} = ?$

7. $\frac{3}{8} \times 9 = ?$

8. $\frac{11}{16} \times 4 = ?$

9. $\frac{2}{3} \times \frac{3}{4} = ?$

10. $\frac{1}{4} \times \frac{5}{8} = ?$

11. $\frac{4}{5} \times \frac{3}{8} = ?$

12. $\frac{1}{8} \times \frac{3}{4} = ?$

13. $\frac{1}{4}$ of $16 = ?$

14. $\frac{3}{8}$ of $\frac{11}{12} = ?$

15. $\frac{3}{8}$ of $12 = ?$

16. $\frac{5}{6}$ of $\frac{5}{9} = ?$

17. $\frac{3}{7} \times \frac{3}{5} = ?$

18. $\frac{5}{12} \times \frac{2}{5} = ?$

19. $\frac{3}{8} \times \frac{2}{3} = ?$

20. $\frac{15}{39} \times \frac{78}{95} = ?$

21. $\frac{15}{16} \times \frac{4}{45} = ?$

22. $\frac{1}{6}$ of $\frac{3}{8} = ?$

23. $\frac{9}{14} \times \frac{7}{18} = ?$

24. $\frac{5}{16} \times \frac{8}{15} = ?$

25. $\frac{3}{8} \times \frac{5}{9} \times \frac{7}{10} = ?$

26. $\frac{3}{5} \times \frac{4}{5} \times \frac{1}{2} \times 1 = ?$

27. $\frac{2}{15} \times \frac{4}{21} \times \frac{7}{8} \times 5 = ?$

28. $\frac{17}{32} \times \frac{64}{85} = ?$

29. $\frac{15}{16}$ of $\frac{7}{90} = ?$

30. $\frac{11}{17} \times \frac{85}{121} = ?$

31. $\frac{8}{27} \times \frac{9}{52} = ?$

32. $\frac{94}{133} \times \frac{19}{47} = ?$

33. $\frac{14}{45} \times \frac{9}{28} = ?$

34. $\frac{78}{95} \times \frac{15}{39} = ?$

35. $\frac{7}{27}$ of $\frac{9}{14} = ?$

36. $\frac{35}{19} \times \frac{76}{25} = ?$

37. $\frac{24}{27} \times \frac{18}{42} = ?$

38. $\frac{30}{75} \times \frac{15}{32} = ?$

39. $\frac{7}{15}$ of $\frac{5}{28} = ?$

40. $\frac{24}{55} \times \frac{13}{58} \times \frac{29}{39} \times 11 = ?$

CHANGING MIXED NUMBERS TO IMPROPER FRACTIONS

In shop problems, sometimes it is necessary to change a mixed number to an improper fraction.

Example 1.—Express $3\frac{5}{8}$ as an improper fraction.

$$3\frac{5}{8} = 3 + \frac{5}{8}$$

$$\frac{3 \times 8}{1 \times 8} = \frac{24}{8} \qquad \text{Changing 3 to eighths}$$

$$\frac{24}{8} + \frac{5}{8} = \frac{29}{8}$$

Rule for short cut.—**To change a mixed number to an improper fraction, multiply the whole number by the denominator of the fraction, and to this product add the numerator. Put the sum over the denominator of the fraction.**

Example 2.—Change $3\frac{5}{8}$ to an improper fraction.

$8 \times 3 = 24$	Multiply the whole number by the denominator of the fraction, and add the numerator.
$24 + 5 = 29$	
$3\frac{5}{8} = \frac{29}{8}$	Put the sum over the denominator of the fraction.

Example 3.—Express $2\frac{1}{3}$ as an improper fraction.

$$3 \times 2 = 6$$
$$6 + 1 = 7$$
$$2\frac{1}{3} = \frac{7}{3}$$

MULTIPLICATION OF MIXED NUMBERS

Rule.—**To multiply mixed numbers, change them to improper fractions and multiply. Cancel, if possible, and express results in lowest terms.**

Example.—$1\frac{5}{6} \times 5 \times 3\frac{1}{5} = ?$

$$\frac{11}{\underset{3}{\not{6}}} \times \frac{\overset{1}{\not{5}}}{1} \times \frac{\overset{8}{\not{16}}}{\underset{1}{\not{5}}} = \frac{88}{3} = 29\frac{1}{3}$$

$$1\frac{5}{6} = \frac{11}{6}; \qquad 5 = \frac{5}{1}; \qquad 3\frac{1}{5} = \frac{16}{5}$$

$$\frac{88}{3} = 3\overline{)88} \quad 29\frac{1}{3}$$

EXERCISES AND PROBLEMS

1. Find the product of $7\frac{1}{2} \times 3\frac{1}{4}$.
2. Multiply 10 by $\frac{3}{4} \times \frac{3}{8}$.
3. Find the product of $\frac{3}{8}$, $\frac{4}{9}$, and $\frac{7}{12}$.
4. How much is $10 \times 3\frac{1}{2} \times 4\frac{1}{5}$?
5. What is $\frac{3}{4}$ of $10\frac{1}{2}$?
6. What is $\frac{7}{8}$ of 33?

7. Find the length of A on the piston shown.

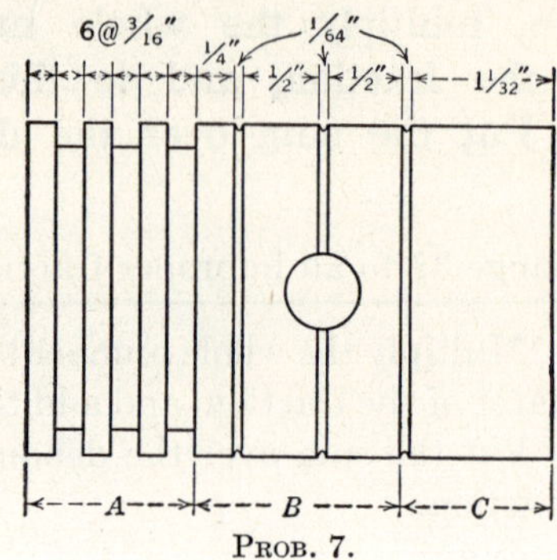

Prob. 7.

8. Using the piston shown, how long is B? C? $B - C$?

9. What is the whole length of the piston used in Probs. 7 and 8?

10. The distance between the centers of holes in a planer table is $1\frac{3}{8}''$. Find the distance from the 1st to the 10th hole.

(*Hint.*—Multiply the distance between centers by the 9 spaces between the 1st and the 10th hole.)

Find the distance from the 1st to the 25th hole. From the 1st to the 37th hole. From the 10th to the 25th hole. From the 25th to the 50th hole.

11. A bolt is $2\frac{1}{2}''$ long. Compute the length of another bolt $3\frac{5}{8}$ times as long.

12. The material allowance for a rivethead is $1\frac{1}{2}$ times the diameter of the rivet. Find the amount needed for a rivethead on $\frac{3}{8}''$ stock; on $\frac{7}{16}''$ stock; on $\frac{3}{4}''$ stock; on $1\frac{1}{8}''$ stock.

Prob. 12.

13. The height of an I beam is $2\frac{3}{8}$ times the width. Find the height for each of the following widths: 15″, $6\frac{1}{2}''$, $9\frac{3}{8}''$, $4\frac{3}{4}''$.

14. A jar holds $\frac{3}{4}$ gal. What part of a gallon does it contain when $\frac{3}{4}$ full? What part of a gallon does it contain when $\frac{4}{5}$ full? When $\frac{1}{8}$ full?

15. The width of a door is one-third the height. If the height is $6\frac{2}{3}$ ft., find the width. If the height is $8\frac{1}{4}$ ft. find the width.

16. Find the value of each of the following: $\frac{5}{8}$ of 1542; $\frac{7}{8}$ of 1728; $2\frac{5}{8} \times 3376$.

17. The circumference of a circle is $3\frac{1}{7}$ times the diameter. What is the circumference when the diameter is 21 ft.?

DIVISION OF COMMON FRACTIONS AND MIXED NUMBERS

Rule.—To divide fractions, invert the divisor and multiply. If possible, reduce the result.

A fraction is **inverted** when the numerator and denominator change places, as, $\frac{2}{5}$ inverted is $\frac{5}{2}$.

When choosing the fraction to be inverted, remember that the rule says, "Invert the divisor." The divisor always follows the division sign.

In each of the following examples, the dividend is multiplied by the reciprocal of the divisor. The **reciprocal** of any number is 1 divided by the number; thus the reciprocal of 3 is $\frac{1}{3}$. The reciprocal of a number is obtained by merely expressing the number as a fraction and inverting the fraction. The reciprocal of $\frac{3}{4}$ is $\frac{4}{3}$. The reciprocal of $3\frac{1}{2}$, or $\frac{7}{2}$, is $\frac{2}{7}$.

Example 1.—Divide $\frac{3}{4}$ by 3.

$\frac{3}{4} \div \frac{3}{1}$	$\frac{3}{1}$ is the divisor. When it is inverted, it becomes $\frac{1}{3}$.
$\frac{\overset{1}{\cancel{3}}}{4} \times \frac{1}{\underset{1}{\cancel{3}}} = \frac{1}{4}$	

Example 2.—$8 \div \frac{1}{2} = ?$

$8 \div \frac{1}{2}$	$\frac{1}{2}$ is the divisor. When it is inverted, it becomes $\frac{2}{1}$.
$\frac{8}{1} \times \frac{2}{1} = \frac{16}{1}$, or 16	

Example 3.—$9 \div 2\frac{1}{2} = ?$

$9 \div 2\frac{1}{2} = \frac{9}{1} \div \frac{5}{2}$	Change the divisor, $2\frac{1}{2}$, to an improper fraction, $\frac{5}{2}$.
$\frac{9}{1} \times \frac{2}{5} = \frac{18}{5} = 3\frac{3}{5}$	$\frac{5}{2}$ inverted is $\frac{2}{5}$.

Example 4.—$15\frac{1}{4} \div 2\frac{1}{2} = ?$

$15\frac{1}{4} \div 2\frac{1}{2} = \frac{61}{4} \div \frac{5}{2}$	Change both the dividend and the divisor to improper fractions.
$\frac{61}{\cancel{4}_{2}} \times \frac{\cancel{2}^{1}}{5} = \frac{61}{10} = 6\frac{1}{10}$	$\frac{5}{2}$, the divisor, inverted is $\frac{2}{5}$.

EXERCISES

1. $24 \div \frac{1}{8} = ?$

2. $32 \div \frac{1}{16} = ?$

3. $18 \div \frac{1}{12} = ?$

4. $\frac{3}{5} \div \frac{1}{15} = ?$

5. $3\frac{1}{2} \div \frac{1}{4} = ?$

6. $6\frac{3}{4} \div \frac{1}{4} = ?$

7. $4\frac{2}{5} \div \frac{1}{5} = ?$

8. $4\frac{5}{8} \div \frac{1}{8} = ?$

9. $6\frac{3}{4} \div \frac{1}{12} = ?$

10. $4\frac{1}{2} \div \frac{1}{24} = ?$

11. $\frac{2}{3} \div \frac{1}{24} = ?$

12. $3\frac{3}{8} \div \frac{1}{64} = ?$

13. $3\frac{5}{8} \div \frac{8}{29} = ?$

14. $1\frac{2}{3} \div 2\frac{1}{3} = ?$

15. $4\frac{3}{8} \div \frac{1}{3} = ?$

16. $7\frac{1}{2} \div \frac{3}{4} = ?$

17. $9\frac{3}{8} \div \frac{3}{32} = ?$

18. $23\frac{5}{8} \div \frac{7}{8} = ?$

19. $19\frac{1}{16} \div \frac{7}{16} = ?$

20. $27\frac{3}{4} \div 9\frac{1}{4} = ?$

EXERCISES AND PROBLEMS

1. Divide $3\frac{1}{2}$ by $\frac{3}{4}$.

2. Divide $19\frac{1}{2}$ by $\frac{1}{4}$.

3. Divide 172 by $2\frac{1}{2}$.

4. How much is $12\frac{1}{2}$ divided by $\frac{3}{4}$?

5. Divide $49\frac{1}{2}$ by 3.

6. Divide a 154″ strip of metal into pieces $5\frac{5}{8}$″ long.

7. Divide $25\frac{1}{2}$ by $\frac{7}{16}$.

8. Allowing $\frac{1}{8}$″ for cutting off each nut, how many nuts $\frac{3}{8}$″ thick can be cut from a piece of cold-rolled steel 12 ft. long?

9. How many sheets of metal, each $\frac{1}{32}$″ thick, are there in a pile $12\frac{7}{8}$″ high?

10. How many sheets of brass, each $\frac{1}{32}$″ thick, are there in a pile $25\frac{1}{2}$″ high?

11. $\frac{2}{3}$ is what part of 6?

12. $\frac{5}{8}$ is how many times as great as $\frac{1}{16}$?

13. Equally space the holes in the template shown.

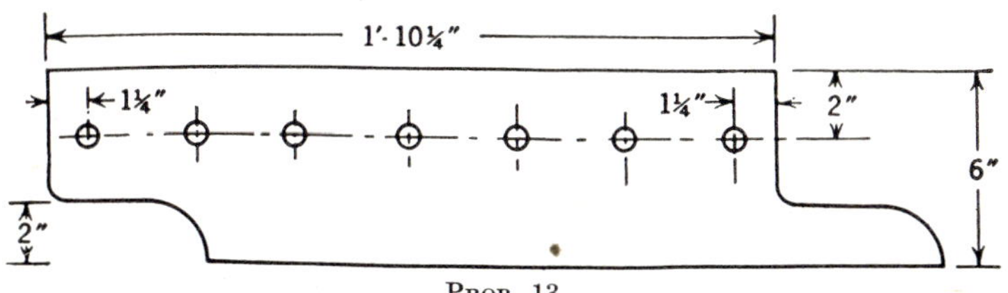

Prob. 13.

14. If a man chops $1\frac{1}{2}$ cords of wood per day, how long will it take him to chop $8\frac{3}{4}$ cords?

15. If an automobile tire is $7\frac{3}{4}$ ft. in circumference, how many revolutions (turns) will it make going 1 mile?

16. If the lead of the thread, that is, the distance the nut advances with one turn, on a bolt is $\frac{1}{4}''$, how many turns will a nut make to move $\frac{7}{8}''$?

17. A man drove his automobile 248 miles on $12\frac{1}{2}$ gal. of gasoline. What mileage did he get per gallon?

18. How many pieces $12\frac{1}{2}''$ long can be sheared from a piece $15'7\frac{1}{2}''$ long?

19. An airplane traveled 60 miles in $18\frac{3}{5}$ min. How many miles per minute did it average?

20. There are 75 lb. of bolts in a bin. If each bolt weighs $\frac{3}{8}$ lb., how many bolts are there in the bin?

21. How many bolts, each $3\frac{1}{2}''$ long, can be cut from a bar $9'8''$ long, if $\frac{1}{8}''$ is allowed for each bolt cut?

22. If a bar weighs $2\frac{1}{4}$ lb. per ft., how long must it be to weigh 25 lb.?

23. Sixteen trucks consume $133\frac{1}{3}$ gal. of gas in a day. What is the average consumption per truck?

24. How many flooring boards, each $3\frac{1}{8}''$ wide, are there in the width of a corridor $5'5''$ wide?

25. Find the number of threads of $\frac{1}{8}''$ pitch on a bolt threaded $3\frac{3}{4}''$.

26. If a book of 40 sheets is $\frac{1}{8}''$ thick, how many sheets are there in a book $1\frac{1}{4}''$ thick?

27. Divide a bar $7'6''$ long into 6 equal parts.

MATERIAL NEEDED FOR BENT PIECES

Calculating the length of material needed for bent metal pieces is of importance in the shop. Allowance must be made for the material needed in the bend itself. For square bends, that is, right-angle bends, such as are made in the vise by hammering, allow half the thickness of the stock for each bend.

Rule.—To find the length of material needed for square bends, find the sum of the inside dimensions, and to this length add one-half the stock thickness for each bend.

Example 1.—Calculate the stock needed for the angle piece shown.

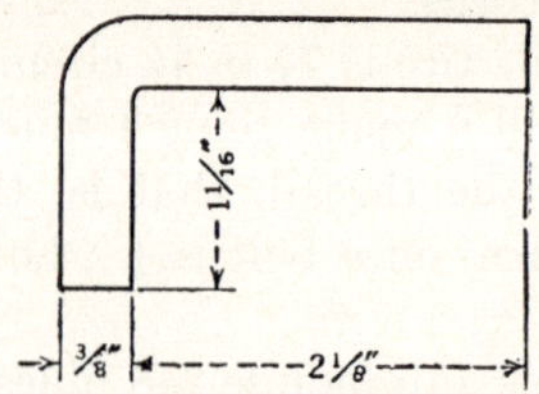

$2\frac{1}{8}'' + 1\frac{1}{16}'' = 3\frac{3}{16}''$	Sum of the inside dimensions
$\frac{1}{2} \times \frac{3}{8}'' = \frac{3}{16}''$	One-half the stock thickness
$3\frac{3}{16}'' + \frac{3}{16}'' = 3\frac{6}{16}'' = 3\frac{3}{8}''$	Total length

Example 2.—Find the length of material needed for the bracket shown.

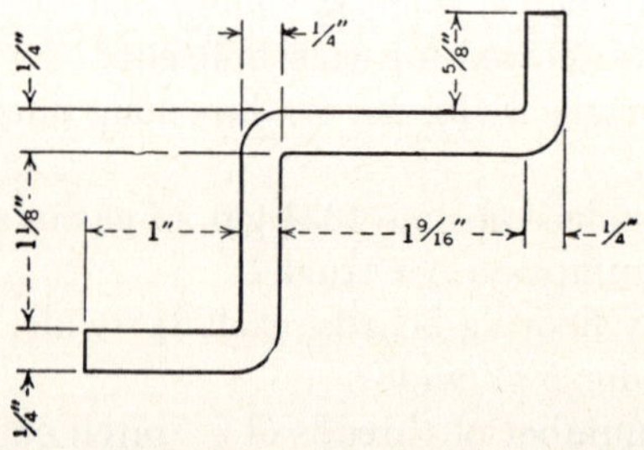

$1'' + 1\frac{1}{8}'' + 1\frac{9}{16}'' + \frac{5}{8}'' = 4\frac{5}{16}''$	Sum of the inside dimensions
$\frac{1}{2} \times \frac{1}{4}'' = \frac{1}{8}''$	One-half the stock thickness
$3 \times \frac{1}{8}'' = \frac{3}{8}''$	Allowance for the 3 bends
$4\frac{5}{16}'' + \frac{3}{8}'' + 4\frac{11}{16}'' = 4\frac{11}{16}''$	Total length

PROBLEMS

Calculate the length of stock needed for the bent pieces shown.

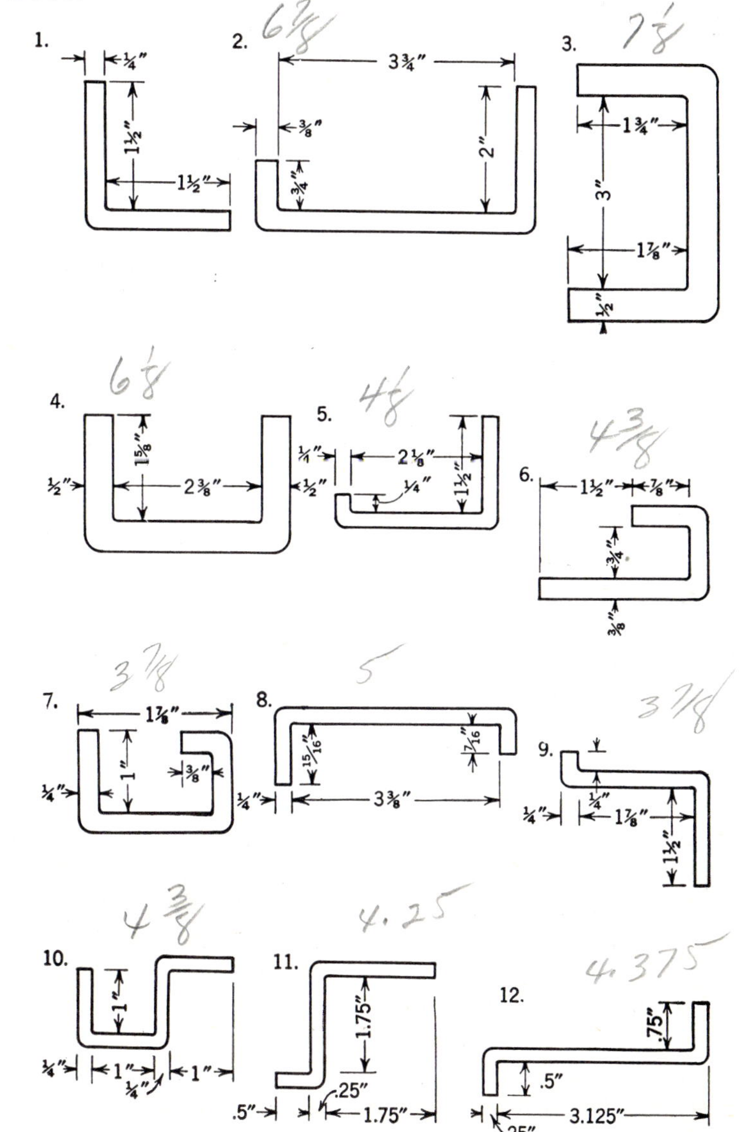

CHANGING DENOMINATIONS

Review the rules for denominate numbers in Chap. 1.

Example 1.—Change $\frac{7}{8}$ cu. yd. to cubic feet.

$\frac{7}{8} \times \frac{27}{1} = 23\frac{5}{8}$	There are 27 cu. ft. in 1 cu. yd. Hence, multiply $\frac{7}{8}$ by 27.
Ans. $23\frac{5}{8}$ cu. ft.	

Example 2.—Change 84 in. to yards.

$84 \div 36$	There are 36 in. in 1 yd. Hence, divide 84 by 36.
$\frac{\overset{7}{\cancel{84}}}{1} \times \frac{1}{\underset{3}{\cancel{36}}} = \frac{7}{3}$, or $2\frac{1}{3}$	
Ans. $2\frac{1}{3}$ yd.	

EXERCISES AND PROBLEMS

See page 549 for table of equivalents.

1. Change $3\frac{3}{4}$ ft. to inches.

2. Change 105 in. to feet.

3. Change 225 cu. ft. to cubic yards.

4. Change 1658 pieces to dozens.

5. How many pieces, each $2\frac{1}{2}$ ft. long, are there in $46\frac{1}{4}$ lin. ft.?

6. How many linear feet are there in 27 pieces, each $6\frac{3}{4}$ ft. long?

7. Change 513 pieces to gross.

8. Change $\frac{3}{4}$ cu. yd. to cubic inches.

9. Change 46,656 cu. in. to cubic yards.

10. In $16\frac{1}{2}$ ft., how many $4\frac{3}{4}$-in. pieces are there?

11. Change 75 in. to yards.

12. Find the total number of linear feet in 56 pieces each $16\frac{3}{4}''$ long.

13. Change $49\frac{1}{2}$ gross to pieces.

14. In $28\frac{1}{4}$ ft., how many pieces, each $3\frac{1}{4}$ in. long, are there?

15. Change 12 rd. to inches.

16. Change $5\frac{1}{2}$ tons to ounces.

17. Change $27\frac{3}{4}$ doz. to number of pieces.

GETTING HALF OF A MIXED NUMBER

Many occasions arise in the shop where it is necessary to divide a mixed number, usually a distance like $10\frac{3}{4}$ in., or $9\frac{7}{8}$ in., into two equal parts, as in finding the center line of a piece of stock. The following short-cut methods may be used to get half of a mixed number (1) when the integer part is an even number, as in $10\frac{3}{4}$, and (2) when the integer part is an odd number, as in $9\frac{7}{8}$.

Rule 1.—Combine half of the even integer with the numerator over the denominator doubled.

Example 1.—Get half of $10\frac{3}{4}$.

$10 \div 2 = 5$	Half of integer.
$4 \times 2 = 8$	The denominator doubled. Put numerator, 3, over $8 = \frac{3}{8}$
Ans. $5\frac{3}{8}$	Combined result.

Rule 2.—Subtract one from the uneven integer and take half of the remainder. Add the numerator to the denominator and put the sum over the denominator doubled. Combine results.

Example 2.—Get half of $9\frac{7}{8}$.

$9 - 1 = 8$	One subtracted from uneven integer.
$8 \div 2 = 4$	Half of remainder.
$7 + 8 = 15$	Numerator added to denominator.
$8 \times 2 = 16$	Denominator doubled. Put 15 over $16 = \frac{15}{16}$.
Ans. $4\frac{15}{16}$	Combined result.

EXERCISES AND PROBLEMS

1. Get half of $8\frac{1}{2}$. Of $6\frac{1}{4}$. Of $12\frac{3}{4}$. Of $32\frac{1}{2}$.

2. Get half of $9\frac{1}{2}$. Of $7\frac{1}{4}$. Of $13\frac{3}{4}$. Of $33\frac{1}{2}$.

3. Divide the following widths into two equal parts: $12\frac{7}{8}$ in.; $16\frac{15}{16}$ in.; $18\frac{21}{32}$ in.; $14\frac{9}{16}$ in.

4. Divide the following widths into two equal parts: $13\frac{7}{8}$ in.; $17\frac{15}{16}$ in.; $19\frac{21}{32}$ in.; $25\frac{9}{16}$ in.

5. Divide each of the following into four equal parts: 17 in.; $20\frac{1}{2}$ in.; $18\frac{3}{4}$ in.; $28\frac{3}{8}$ in.; $36\frac{1}{4}$ in.

6. Divide each of the following into four equal parts: $33\frac{1}{4}$ in.; $43\frac{1}{2}$ in.; $2\frac{3}{8}$ in.; $1\frac{9}{16}$ in.

7. Get half of $16\frac{1}{4}$ in. Get half of your first answer. Get half of your second answer. Get half of your third answer.

8. Get half of $33\frac{3}{8}$ in. Get half of your first answer. Get half of your second answer. Get half of your third answer.

REVIEW

1. $\frac{3}{8} + 5\frac{3}{4} + \frac{3}{16} + \frac{3}{16} = ?$

2. $\frac{7}{8} + 2\frac{5}{8} + 3\frac{7}{8} = ?$

3. $2\frac{9}{16} + 3\frac{5}{64} + 3\frac{29}{32} = ?$

4. $\frac{7}{32} + \frac{21}{64} + \frac{3}{8} + \frac{7}{16} = ?$

5. $3\frac{1}{2} + 5\frac{7}{8} + 3\frac{9}{16} = ?$

6. $8\frac{51}{64} + 8\frac{31}{32} + \frac{9}{16} = ?$

7. $11\frac{3}{4} - 6\frac{7}{16} = ?$

8. $8\frac{3}{8} - 4\frac{15}{16} = ?$

9. $9\frac{3}{16} - 5\frac{11}{32} = ?$

10. $3\frac{21}{32} - 2\frac{3}{4} = ?$

11. $16\frac{17}{32} - 8\frac{3}{4} = ?$

12. $21\frac{3}{8} - 11\frac{5}{16} = ?$

13. $\frac{3}{8} \times \frac{4}{5} = ?$

14. $\frac{5}{8} \times \frac{4}{25} = ?$

15. $\frac{5}{6} \times \frac{3}{5} = ?$

16. $\frac{5}{16} \times \frac{8}{15} = ?$

17. $\frac{5}{8} \times \frac{4}{15} \times \frac{6}{7} = ?$

18. $\frac{8}{9} \times \frac{3}{5} \times \frac{15}{16} = ?$

19. $\frac{5}{9} \times \frac{3}{7} \times \frac{21}{75} = ?$

Find the value of:

20. $\frac{3}{4} \times 1728$.

21. $\frac{5}{8}$ of 6408.

22. $\frac{5}{6}$ of 8032.

23. $\frac{7}{8}$ of 5665.

24. $\frac{3}{4}$ of 528.

25. $\frac{5}{6}$ of 7776.

26. $36 \div \frac{1}{9} = ?$

27. $24 \div \frac{1}{8} = ?$

28. $36 \div \frac{1}{12} = ?$

29. $5\frac{1}{2} \div \frac{1}{2} = ?$

30. $6\frac{3}{4} \div \frac{3}{8} = ?$

31. $7\frac{1}{2} \div \frac{3}{4} = ?$

32. $10\frac{5}{8} \div \frac{1}{2} = ?$

33. $5 \div 2\frac{2}{3} = ?$

34. $14 \div 3\frac{1}{2} = ?$

35. $60 \div 1\frac{7}{8} = ?$

36. $3\frac{5}{8} \div 1\frac{3}{4} = ?$

37. $1\frac{3}{8} \div \frac{3}{16} = ?$

38. $\frac{5}{2} \div \frac{2}{5} = ?$

39. $\frac{2}{5} \div \frac{5}{2} = ?$

40. $\frac{2}{5} \div \frac{5}{3} = ?$

41. $\frac{5}{2} \div \frac{5}{5} = ?$

42. $21\frac{5}{8} \div 7\frac{5}{8} = ?$

43. $48\frac{1}{4} \div 1\frac{3}{8} = ?$

44. $29\frac{1}{2} \div 4\frac{1}{2} = ?$

Change each of the following:

45. 90 yd. to inches.

46. 3.5 cu. ft. to cubic inches.

47. 3 ft. to eighths of an inch.

48. 9 miles to feet.

CHAPTER 3

DECIMAL FRACTIONS

REVIEW

A common fraction indicates a division. If the division is made, the result is a decimal fraction.

Any decimal fraction may be considered as a common fraction with a denominator of 10 or some power of ten. The power of ten indicates the number of times 10 is used as a factor, as 10 to the second power is 10×10, or 100; 10 to the third power is $10 \times 10 \times 10$, or 1000.

Decimals are read as indicated on page 4. Read .5 as five-tenths. The **tenths** indicate "Divide by 10." Therefore, $.5 = \frac{5}{10}$; .565 is read "five hundred sixty-five thousandths." The **thousandths** indicate "Divide by 1000." Therefore, $.565 = \frac{565}{1000}$. Decimal fractions are added, subtracted, multiplied, and divided, in the same way as whole numbers with additional rules for handling the decimal point.

ADDITION OF DECIMALS

Rule.—To add decimals, set the decimals in a vertical column, placing decimal points below decimal points. Add as for whole numbers, placing the decimal point in the sum below the column of decimal points.

Example 1.—$37.12 + 4.495 + 7.0 + .0013 = ?$

37.12 4.495 7.0 .0013 48.6163	Set the numbers down in a column with the decimal points below decimal points, the tenths below tenths, etc. Add as for whole numbers. Note that the decimal point in the sum is placed below the column of decimal points.

EXERCISES AND PROBLEMS

1. 7.341 + 3.1 + .076 + 3.1549 + .123 = ?

2. Add 731., 7.31, 73.1, .731, and .0731.

3. 9.65 + 6.695 + .8805 + 9.006 + 77.006 = ?

4. 77 plus .075 plus .00093 plus 44.333 plus 7896.3 equals what?

5. Add fifty-five hundredths and ninety-nine thousandths

6. Compute the length of the corner plate shown.

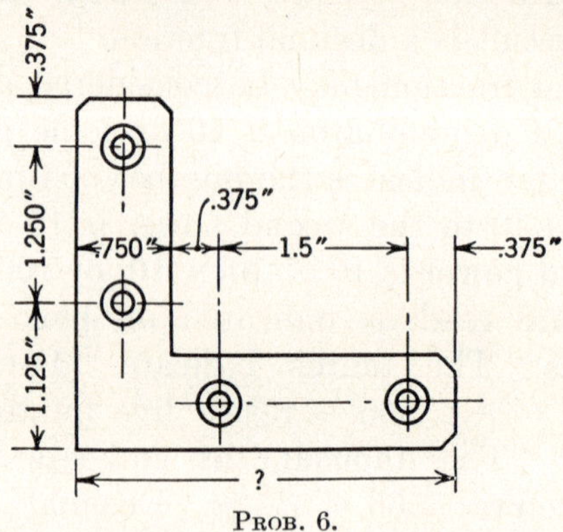

PROB. 6.

7. Find the width of the corner plate in Exercise 6.

8. Add 763.3, 55.576, 3456.78, 65.187, and 1.0098.

9. Add three tenths, seventy thousandths, and five millionths.

10. What is the sum of .009, 34.09, 123.25, 345.345, and 12.9876?

11. On the drawing of the spindle shown, find the length of A. (Use this diagram for Exercises 12 to 14.)

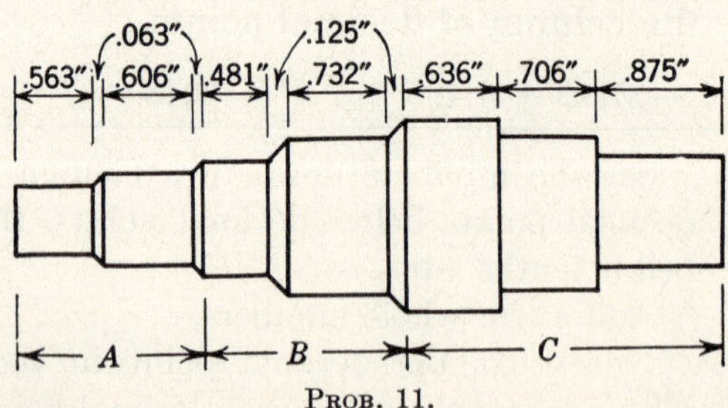

PROB. 11.

12. Find the length of B.

13. Find the length of C.

14. Find the total length of the spindle by adding the individual lengths. Check by adding A, B, and C.

15. On the drawing of the step block shown, find the length of A. (Use this diagram for Exercises 16 to 18.)

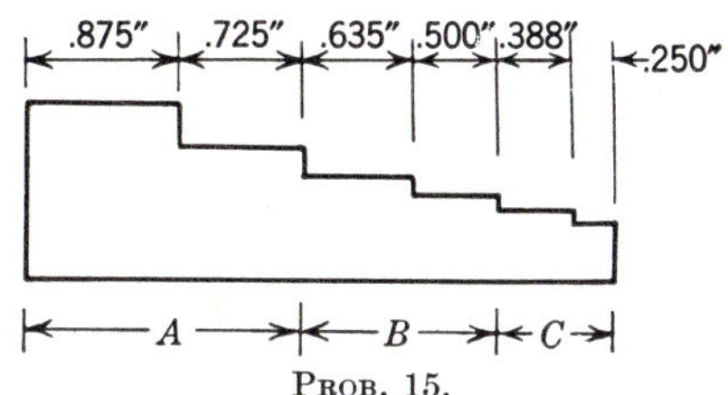

PROB. 15.

16. Find the length of B.

17. Find the length of C.

18. Find the total length of the step block.

19. Find the over-all dimensions on a drawing made up of the following detail dimensions: 3.17″, .625″, 2.125″, 4″, 2.625″, .750″, .4375″.

20. Write these fractions as decimal fractions, and add:

$$\frac{95}{100}, \frac{101}{10{,}000}, \frac{765}{1{,}000{,}000}, \frac{7}{10}, \frac{7654}{10{,}000}, \frac{1}{10{,}000}, \frac{1}{1000}.$$

21. Write each of the following as a decimal fraction and in words:

$$\frac{11}{1000}, \frac{101}{10{,}000}, \frac{4375}{100{,}000}, \frac{81}{1000}, \frac{15{,}432}{1{,}000{,}000}, \frac{3}{10}, \frac{566}{1000}, \frac{21}{10{,}000}$$

22. The inside diameter of a cast-iron cylinder is 3.875″. The wall thickness is .4375″. What is the outside diameter of the cylinder?

SUBTRACTION OF DECIMALS

Rule.—To subtract decimals, set the decimals in a vertical column, placing decimal point below decimal point.

Subtract as for whole numbers, placing the decimal point in the remainder below the column of decimal points.

Example.—Which is greater and by how much, .78321 or .25?

.78321 .25 .53321	To subtract one decimal from the other, place the subtrahend below the minuend with decimal point below decimal point, tenths below tenths, etc.

Subtract as for whole numbers.

Note that the decimal point in the remainder is placed below the column of decimal points.

PROBLEMS

1. Subtract 71.874 from 98.01.

2. How much is 99.66 less 66.99?

3. Take away $3.48 from $16.05.

4. Subtract .0987 from 1.001.

5. Which is greater and by how much, 998,765.9876 or 1,033,332.876?

6. Supply the missing dimensions of the bracket shown.

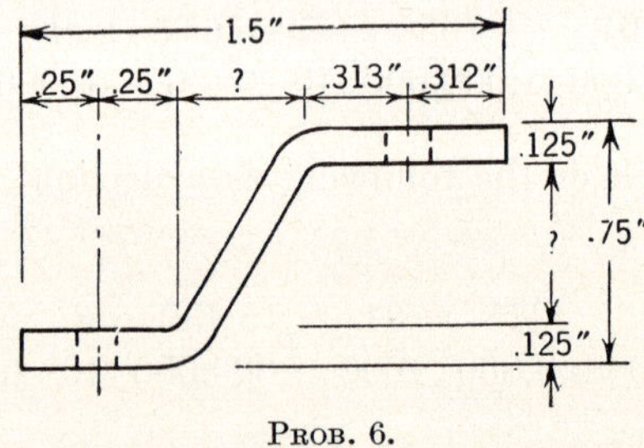

PROB. 6.

7. What is the remainder when 45.87 is subtracted from 99.09?

8. Take away 765,843.875 from 987,652.2.

9. How much is 8.875″ less 4.939″?

10. What remains if 66.5006 is taken from 79.3?

11.	**12.**	**13.**
90.48 −16.083	8.063 − .9876	78. −33.98

14.
```
  45.3
 -29.95
```

15.
```
  78.006
 - 7.9
```

16. 987.005 − 345.8 = ? **17.** 78.962 − 34.0097 = ?

18. 6 − .875 = ?

In the lathe spindle shown, find the difference in the length of the parts for each of the following:

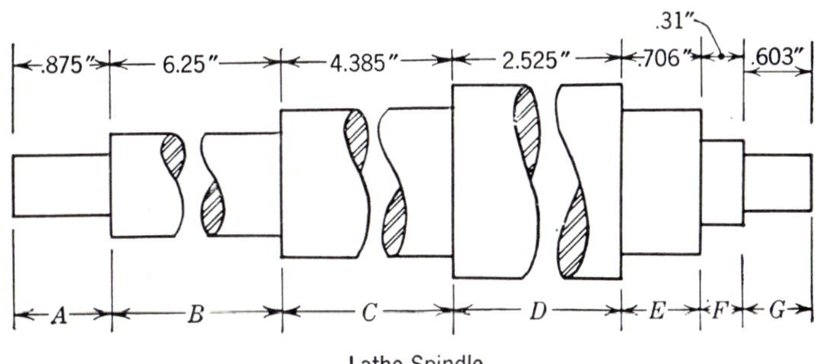

Lathe Spindle
PROBS. 19 to 27.

19. *A* and *D*. **20.** *B* and *A*. **21.** *B* and *C*.
22. *C* and *D*. **23.** *D* and *E*. **24.** *E* and *F*.
25. *A* and *G*. **26.** *B* and *D*. **27.** *F* and *G*.

MULTIPLICATION OF DECIMALS

Rule.—To multiply decimals, multiply as with whole numbers. Point off from the right of the product as many decimal places as there are in the sum of the decimal places of the numbers multiplied. Prefix zeros to the product, if necessary, to fill out the required number of decimal places.

Example 1.—Multiply .4375 by 8.

.4375 8 3.5000 *Ans.* 3.5	Set the numbers down, and multiply as with whole numbers. There are four decimal places in the multiplicand and none in the multiplier; hence, place the decimal point four places from the right in the product. The zeros at the end of the decimal may be dropped.

Example 2.—Multiply .0638 by .78.

.0638 .78 5104 4466 .049764	There are four decimal places in the multiplicand and two decimal places in the multiplier, or, altogether, six decimal places. Hence, place the decimal point six places from the right in the product. One zero is prefixed to the product, 49764, because the product must have six decimal places.

EXERCISES AND PROBLEMS

1. 17.5 × 6 = ?

2. 168 × .321 = ?

3. 4.8 × .067 = ?

4. 0.56 × 0.83 = ?

5. 62 × 37.91 = ?

6. 0.72 × 0.095 = ?

7. Multiply 6.85 by 81.2.

8. Multiply 0.2279 by 0.029.

9. How much is 306.693 × 2.61?

10. Find the product of 18.35 and 0.065.

11. How much is 70.79 multiplied by 20.36?

12. Find the product of 0.0695 and 0.0063.

13. Find each dimension for a shaft 1.25 times as large as the shaft shown in the diagram.

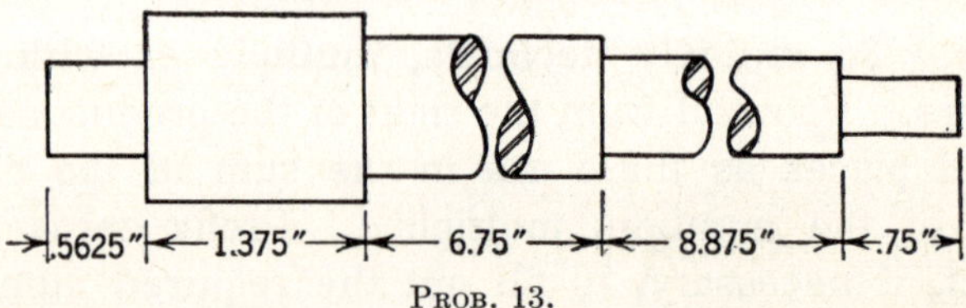

PROB. 13.

14. Find each dimension for a shaft 2.58 times as large as the shaft shown in the diagram.

15. Multiply 45.638 by each of the following: 0.1, 0.01, 0.001, 0.0001. State an easy rule for this type of problem.

16. For each of the rivetheads in the diagram shown, the dimensions of the rivethead are given in terms of the diameter *D* of the rivet. The dimensions of a rivethead are its diameter (sometimes top and bottom) and its thickness. Note that 1.71*D* means 1.71 × *D*. Find all the dimensions of the rivetheads shown.

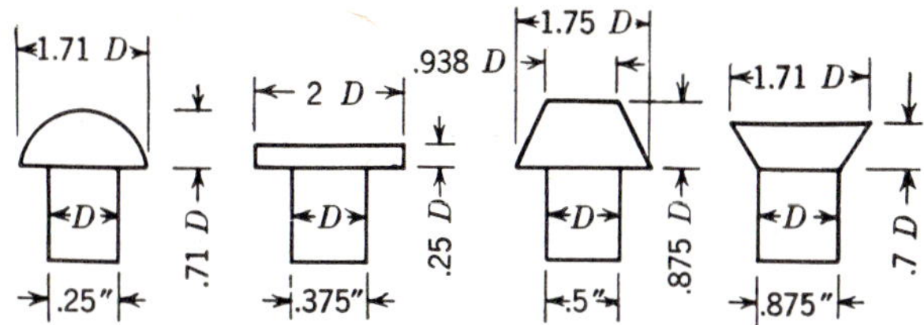

PROB. 16.

17. A man receives a wage of \$1.825 per hour. How much will he earn in 1 week if he works 38.5 hr.?

18. Scrap brass sells for 24.3¢ per pound. What is the price of 38.4 lb.?

19. An iron bar weighs 1.963 lb. per ft. Compute the cost of a 12.25-ft. iron bar when the price of iron is \$0.165 per pound.

20. What is the cost of 546 gal. of gasoline at 18.5¢ per gallon?

21. An oil well flows 1450 bbl. of oil per day. Find the value of a day's flow if the refinery pays \$1.925 per barrel.

DIVISION OF DECIMALS

Rule.—To divide a decimal by a whole number, divide as with whole numbers, placing the decimal point in the quotient directly above or below the decimal point in the dividend.

Example 1.—18.66 ÷ 6 = ?

```
6)18.66
   3.11
```

Divide as in short division of whole numbers.

Place the decimal point in the quotient directly below the decimal point in the dividend.

Check by multiplying the quotient by the divisor.

Example 2.—620.75 ÷ 25 = ?

```
    24.83
25)620.75
   50
   120
   100
    20 7
    20 0
       75
       75
```

Divide as in long division of whole numbers.

Place the decimal point in the quotient directly above the decimal point in the dividend.

Check by multiplying the quotient by the divisor.

Rule.—To divide a decimal by a decimal, move the decimal point of the divisor to the right until the divisor becomes a whole number; then move the decimal point of the dividend the same number of places to the right as you moved the decimal point in the divisor, annexing zeros if necessary. Divide as directed under the preceding rule.

Example 3.—18.66 ÷ .6 = ?

Solution 1.

$$\frac{18.66 \times 10}{.6 \times 10} = \frac{186.6}{6}$$

$$6\overline{)186.6}$$
$$31.1$$

Division of a decimal by a decimal is simplified if the divisor is made a whole number. The divisor, .6, can be made a whole number, 6, by multiplying it by 10. This necessitates multiplying the dividend by 10, making it 186.6.

Divide as in Example 1.

Check by multiplying the quotient by the divisor.

Solution 2, *by moving decimal points.*

.6.)18.6.6
3 1.1

The divisor's decimal point is always moved completely to the right, making the divisor a whole number. In this example, it will be moved one place.

The dividend's decimal point must be moved to the right exactly the same number of places the divisor's decimal point was moved. This is the same as multiplying by 10.

Note.—The loops beneath the numbers offer an easy way to keep track of the decimal point's movements. Never erase the decimal points to move them. Confusion may result, and it thus is difficult to check the problem.

Example 4.—620.75 ÷ .25 = ?

```
       24 83.
.25.)620.75.
     50
     120
     100
      20 7
      20 0
         75
         75
```

Move the decimal point two places to the right in both the divisor and the dividend.

Always keep the figures in the quotient in their proper positions.

Check by multiplying the quotient by the divisor.

EXERCISES

1. Divide 9801.9 by 0.9.

2. Divide 892.5 by 7.

3. Divide 58.32 by 18.

4. Divide 1728 by 0.12.

5. Divide 512 by 0.08.

6. Divide 343 by 0.7.

7. Divide 306.72 by 0.8.

8. Divide 793.1 by 0.07.

9. Divide 100.1 by 0.001.

10. $795.07 \div 43 = ?$

11. $42.875 \div 1225 = ?$

12. $19.36 \div 44 = ?$

13. $47.61 \div .69 = ?$

14. $132.651 \div 5.1 = ?$

15. $92.16 \div 0.096 = ?$

16. $0.6084 \div 0.078 = ?$

17. How much is $596.4 \div 0.007$?

18. Find the quotient when 912.673 is divided by 94.09.

CARRYING OUT AN ANSWER TO A DESIRED NUMBER OF DECIMAL PLACES

In each of the division problems given above, the quotient had no remainder. Often, division problems do not come out evenly; it is then customary to annex zeros to the right of the dividend in order to carry out the division to the accuracy desired, that is, to carry out the quotient to the number of decimal places desired.

Accuracy to more than four decimal places is seldom necessary. In order to express a quotient to a desired accuracy, the following rule should be applied:

Carry out the division to one more decimal place than is needed for the desired accuracy. If the figure in this place is 5 or more, drop it and add 1 to the figure in the preceding place. Put a minus sign after the quotient. If the figure is less than 5, drop it entirely, and put a plus sign after the quotient.

This process is often called **rounding off.** "Rounding off to three decimal places" means the same as "correct or accurate to three decimal places." Rounding off to three decimal places necessitates carrying out the division to four decimal places" and then following the rule given above.

Problems often require the answer expressed correct to the nearest hundredth or to the nearest thousandth.

"To the nearest hundredth," sometimes written "to the nearest 0.01," means that the answer should have two decimal places. Therefore, the quotient is carried out to three decimal places and rounded off to two decimal places.

The type of work being done must determine the degree of accuracy needed. In most cases the judgment of the operator must be relied upon in deciding when the result should be rounded off to a whole number or carried to one or more decimal places. Obviously, the number of teeth on a gear must be a whole number. The price of an article must be stated in the smallest coin, the cent, and costs should be rounded off to hundredths. In most of the following problems the degree of accuracy needed is indicated by directing how the result should be rounded off.

Example.—Divide 347 by 7.

7)347.000000
49.571428+

This quotient correct to four decimal places is 49.5714+.

Correct to five decimal places, it is 49.57143−.

Correct to three decimal places, it is 49.571+.

Correct to two decimal places, it is 49.57+.

Correct to one decimal place, it is 49.6−.

EXERCISES AND PROBLEMS

In Exercises 1 to 12, find the quotient correct to the nearest hundredth. This makes it necessary to divide to three decimal places and then round off to two decimal places.

1. 29.7 ÷ 4 = ?
2. 3.8 ÷ 51 = ?
3. 48.1 ÷ 23 = ?
4. 0.0734 ÷ 5 = ?
5. 0.0633 ÷ 21 = ?
6. 572.6 ÷ 921 = ?
7. 100 ÷ 21 = ?
8. 100 ÷ 3183 = ?
9. 56.43 ÷ 3.1 = ?
10. 345.9 ÷ 0.21 = ?
11. 234 ÷ 0.065 = ?
12. 1 ÷ 3 =

In Exercises 13 to 24, express the quotient to the nearest thousandth. Carry out the division to four decimal places, and then round off to three decimal places.

13. 457.1 ÷ 0.83 = ?

14. 63.75 ÷ 4.6 = ?

15. 82.3 ÷ 3.1416 = ?

16. 4 ÷ 0.3183 = ?

17. 80.70 ÷ 27.7 = ?

18. 807 ÷ 255 = ?

19. 236.6 ÷ 7.3 = ?

20. 500.9 ÷ 20.3 = ?

21. 45.843 ÷ 21 = ?

22. 66 ÷ 25 = ?

23. 0.48 ÷ 0.13 = ?

24. 1 ÷ 7 = ?

25. On the steel bridge shown, the struts are equally spaced. How long is each part to the nearest hundredth of a foot?

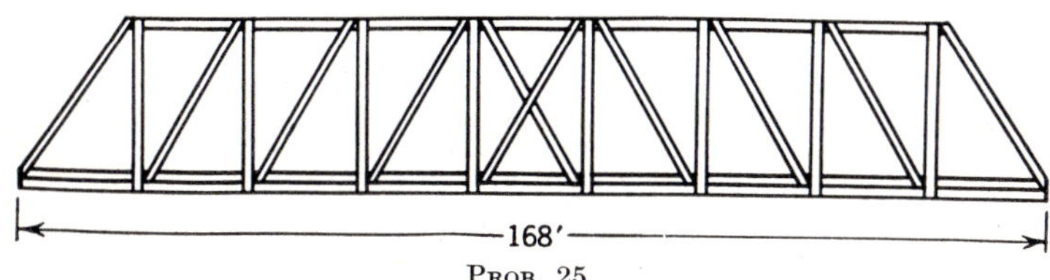

Prob. 25.

26. How many sheets of metal 0.1875″ thick can be placed in a box 4.5″ deep?

27. How many minutes will it take a railway train traveling 0.75 miles per min. to go 30 miles? To go 37 miles? To go 125.46 miles? Round off your results to two decimal places.

28. How long will it take an airplane traveling 2.3 miles per min. to go 394.6 miles? To go 2358 miles? Round off your results to one decimal place.

29. A 14-story building is 138.75 ft. high. What is the average height per story? Round off to two decimal places.

30. How many sheets of metal. each 0.0625 in. thick, are there in a pile 1 ft. 5.5 in. high?

AVERAGES

In finding the average of several quantities, it is customary to round off the result to the smallest part stated in the problem. If a problem is stated in whole numbers, the result is rounded off to a whole number. Likewise, if the problem is stated in decimal thousandths, the result should be rounded off to thousandths. The plus or minus sign should be put after the rounded-off result.

Rule.—To find the average of several quantities, divide their sum by the number of quantities.

Example.—Find the average attendance for last week at S.E.V.H.S., if the daily attendance was as follows: Monday, 4661; Tuesday, 4688; Wednesday, 4763; Thursday, 4752; and Friday, 4658.

$$\begin{array}{r} 4661 \\ 4688 \\ 4763 \\ 4752 \\ 4658 \\ \hline 23522 \end{array}$$

Add the quantities, and divide by the number of days, 5.

Rounding off to the nearest whole number gives 4704 as the average attendance.

$$\frac{23522}{5} = 4704.4, \text{ or } 4704+$$

PROBLEMS

1. Find the average of the following weights: 67 lb., 88 lb., 76 lb., 55 lb., 63 lb., and 59 lb.

2. Find the average of 7′6″, 3′8″, 8′2″, and 6′4″.

3. What is the average weekly wage of 6 men who received, respectively, $63.00, $55.70, $61.45, $64.65, $66.50, and $80.00?

4. Of the following dimensions, which is nearest the average: 0.325″, 0.363″, 0.346″, 0.352″, 0.328″?

5. What is the average output per hour if the screw machines in a shop finish bolts as follows: 360, 394, 448, 280, 424, and 400?

6. What should you consider the correct diameter of a wrist pin if you measured it three different times, with the readings as follows: 0.751″, 0.749″, and 0.750″?

7. The average weight of 16 boys is 114 lb. What is the combined weight of all 16 boys?

8. What is the mean (average) of the following dimensions: 1.733″, 1.750″, 1.668″, 1.700″, and 1.776″?

9. A student made the following grades on his daily work: 88%, 77%, 92%, 60%, 100%, 73%, and 82%. Compute his average grade.

10. What is the average distance per day if a car travels the following: Monday, 368 miles; Tuesday, 540 miles; Wednesday, 375 miles?

11. A man worked the following number of hours per day: Monday, 8 hr.; Tuesday, 7 hr. 45 min.; Thursday, 8 hr.; Friday,

9 hr. 30 min.; and Saturday, 5 hr. What is the average number of hours worked per day?

12. Noon temperatures for successive days last week were as follows: 77°, 82°, 65°, 79°, 89°, 72°, 56°. What was the mean daily temperature?

PRACTICE IN PLACING DECIMAL POINTS

Set A

In Exercises 1 to 54, the quotient figures are 343 in each case. Place the decimal point correctly, adding zeros as necessary. Do not write in this book. Try to finish Set *A* in 10 minutes.

1. 124)42.532 (quotient .343)
2. 124)425.32
3. 12.4)4253.2
4. 1.24)42,532
5. .124)425.32
6. 12.4).42532
7. 1.24)4253.2
8. 124).42532
9. .124).42532
10. 12.4)4.2532
11. 1.24).042532
12. 12.4)42,532
13. .0124)4253.2
14. .124).042532
15. .00124).0042532
16. .124).0042532
17. 1.24).0042532
18. 124).042532
19. 124)42532
20. .124)42,532
21. .0124)42,532
22. .0124)4.2532
23. .0124).042532
24. .0124).42532
25. .124).00042532
26. 12,400).42532
27. 12,400)42,532
28. 124).0042532
29. 1240).0042532
30. 1240)4.2532
31. 12,400)42,532
32. .124)42.532
33. 1.24)42.532
34. 12.4).0042532
35. .124)42.532
36. 12.4).00042532
37. 1.24)425.32
38. 12.4).042532
39. .0124).0042532
40. 124)4.2532
41. .0124).00042532
42. 1.24)42.532
43. 12,400)425.32
44. 1.24).42532
45. 1.24)4.2532
46. .124)4.2532

47. 12.4)$\overline{4.2532}$

48. .0124)$\overline{42.532}$

49. 1240)$\overline{42.532}$

50. .1240)$\overline{4253.2}$

51. 124)$\overline{4253.2}$

52. 12.4)$\overline{4253.2}$

53. 12.4)$\overline{4253.2}$

54. 1240)$\overline{4.2532}$

Try to finish Sets *B* to *F*, inclusive, in 10 minutes.

Set B

The quotient figures for Set *B* are 165. Proceed as in Set *A*

1. 15.5)$\overline{2.5575}$
2. 15.5)$\overline{25.575}$
3. 15.5)$\overline{255.75}$
4. 15.5)$\overline{2557.5}$
5. 15.5)$\overline{25,575}$
6. 1.55)$\overline{.25575}$
7. 1.55)$\overline{255.75}$
8. 15.5)$\overline{.25575}$
9. .155)$\overline{25.575}$
10. .155)$\overline{.25575}$
11. .155)$\overline{255.75}$

Set C

The quotient figures are 749.

1. 393)$\overline{294,357}$
2. 3.93)$\overline{294.357}$
3. .0393)$\overline{29,435.7}$
4. .393)$\overline{29.4357}$
5. .0393)$\overline{294.357}$
6. .00393)$\overline{294.357}$
7. 39.3)$\overline{2943.57}$
8. 3.93)$\overline{.294357}$
9. 393.)$\overline{.0294357}$
10. .393)$\overline{29.4357}$
11. 393)$\overline{.294357}$

Set D

The quotient figures are 625.

1. 625)$\overline{390,625}$
2. 625)$\overline{3.90625}$
3. 6.25)$\overline{390,625}$
4. .625)$\overline{3906.25}$
5. 625)$\overline{390.625}$
6. 6250)$\overline{39,062.5}$
7. 625)$\overline{.00390625}$
8. .0625)$\overline{3.90625}$
9. .00625)$\overline{3.90625}$
10. 6.25)$\overline{39.0625}$
11. 62.5)$\overline{.0390625}$

Set E

The quotient figures are 131.

1. 3.43)$\overline{4.4933}$
2. 34.3)$\overline{449.33}$
3. 3.43)$\overline{4493.3}$
4. .343)$\overline{44,933}$
5. .0343)$\overline{449.33}$
6. 343)$\overline{.0044933}$
7. .343)$\overline{.044933}$
8. .0343)$\overline{.44933}$
9. 3.43)$\overline{4.4933}$
10. 3.43)$\overline{44,933}$
11. 343)$\overline{.44933}$
12. .00343)$\overline{44,933}$
13. 343)$\overline{.44933}$

Set F

The quotient figures are 767.

1. 777)595,959 **2.** 7.77)59.5959 **3.** .777)59.5959
4. .0777).595959 **5.** 7.77).595959 **6.** .777).595959
7. 777)595.959 **8.** 7.77)5.95959 **9.** .777)5959.59
10. 7.77).0595959 **11.** 77.7)595.959 **12.** 77.7)595,959
13. .0777).595959 **14.** 77.7).595959

PRACTICE IN POINTING OFF ANSWERS

The results for the following exercises have not been pointed off. Write each product again, and place the decimal point where it belongs. Add zeros when they are needed.

1. 3.2 × 2.5 = 800
2. 56 × .75 = 4200
3. 5 × 8.3 = 415
4. .06 × 7.5 = 450
5. .70 × .05 = 350
6. .05 × .4 = 20
7. 42 × 4.2 = 1764
8. 12.5 × 12 = 1500
9. 3.5 × 3.5 = 1225
10. 4.5 × 45 = 2025
11. .05 × 700 = 3500
12. .01 × 10 = 10
13. 100 × .03 = 300
14. 40 × .40 = 1600
15. .50 × 6.0 = 3000
16. 1.6 × 0.3 = 48
17. 3.0 × 0.3 = 90
18. 4.0 × 2.00 = 800
19. 6.0 × 0.5 = 300
20. 0.7 × 7.0 = 490
21. 1735 × .4 = 6940
22. 3.75 × .5 = 1875
23. 1.74 × .4 = 696
24. 75.66 × .04 = 30,264
25. 1.6 × 7.5 = 1200
26. 28 × .25 = 700
27. .03 × 16.0 = 480
28. .60 × .75 = 45
29. .04 × .7 = 28
30. 15 × 1.5 = 225
31. 66.7 × 11 = 7337
32. 2.5 × 2.5 = 625
33. .375 × 400 = 1500
34. 8 × 400 = 32
35. .667 × .01 = 667
36. 3.1416 × 3 = 94,248

REVIEW EXERCISES

In the following division exercises, watch the decimal point:

1. 30)9 **2.** 3).09 **3.** .03).09
4. 3000)9 **5.** 30).9 **6.** .3)9

7. $.3\overline{).09}$ **8.** $.003\overline{)9}$ **9.** $.03\overline{)9}$

10. $3000\overline{).9}$ **11.** $3\overline{).9}$ **12.** $300\overline{)9}$

13. $3000\overline{).09}$ **14.** $.003\overline{).9}$ **15.** $.003\overline{).09}$

16. $.3\overline{).9}$ **17.** $300\overline{).9}$ **18.** $.03\overline{).9}$

19. $300\overline{).09}$ **20.** $.003\overline{)90}$

In the following exercises, place the decimal point where it belongs. Add zeros when they are needed.

21. $100 \times .04 = 400$

22. $20 \times .30 = 600$

23. $.50 \times 6.0 = 3000$

24. $\begin{array}{r} 546 \\ .132\overline{)72.072} \end{array}$

25. $\begin{array}{r} 546 \\ 13.2\overline{)72.072} \end{array}$

26. $\begin{array}{r} 5\ 46 \\ 13.2\overline{)720.72} \end{array}$

27. $1.8 \times 0.4 = 72$

28. $2.0 \times .3 = 60$

29. $2.0 \times 2.0 = 400$

30. $6.0 \times .5 = 300$

31. $0.5 \times 7.0 = 350$

32. $3735 \times .2 = 7470$

33. $\begin{array}{r} 5\ 46 \\ .132\overline{)720.72} \end{array}$

34. $\begin{array}{r} 546 \\ 132\overline{)72.072} \end{array}$

35. $\begin{array}{r} 546 \\ 1.32\overline{)7.2072} \end{array}$

36. $\begin{array}{r} 546 \\ 1.32\overline{)72.072} \end{array}$

37. $\begin{array}{r} 5\ 46 \\ 1.32\overline{)720.72} \end{array}$

38. $\begin{array}{r} 546 \\ 0.132\overline{)0.72072} \end{array}$

39. $\begin{array}{r} 546 \\ 0.0132\overline{)0.72072} \end{array}$

Add:

40. $37.56 + 9.063 + 126.37 + .0093 + .9736 = ?$

41. $2.004 + .026 + .939 + 700.00 + 9.32 = ?$

42. $179.163 + 1.087 + 183.1 + .0719 + 3.6007 = ?$

43. $96.99 + 969.9 + 9699 + 9.699 + .0096699 = ?$

Subtract:

44. $63.77 - 32.11 = ?$

45. $79 - 32.99 = ?$

46. $54 - 39.646 = ?$

47. $88.09 - 6.6 = ?$

48. $7.04 - 4.999 = ?$

49. $61.3 - 28.96 = ?$

50. $72.46 - 31.006 = ?$

51. $23.06 - 10.3 = ?$

52. $16.66 - 1.66 = ?$

Multiply:

53. 37.77 × 62.11 = ?

54. 79 × 32.99 = ?

55. 54 × 39.646 = ?

56. 88.09 × 6.6 = ?

57. 61.3 × 28.96 = ?

58. 7.04 × 4.999 = ?

59. 72.46 × 31.006 = ?

60. 23.06 × 10.3 = ?

61. 16.66 × 1.66 = ?

In these division exercises, round off the result to the nearest thousandth:

62. $41.1\overline{)6.236}$

63. $.95\overline{)425.1}$

64. $6.6\overline{)721.4}$

65. $.772\overline{)57.06}$

66. $5.4\overline{)63.96}$

67. $.41\overline{)21.111}$

68. $38.09\overline{)6.6}$

69. $6.6\overline{)88.09}$

70. $1.66\overline{)16.66}$

71. $16.66\overline{)1.66}$

72. $41\overline{)39}$

73. $32\overline{)63}$

In the following division exercises, round off the result to the nearest hundredth:

74. 364.09 ÷ 17 = ?

75. 297.4 ÷ 6.5 = ?

76. 11.11 ÷ 3.3 = ?

77. 428.8 ÷ 41 = ?

78. 37 ÷ 66 = ?

79. 64.96 ÷ 61.4 = ?

80. 66 ÷ 1.01 = ?

81. 440 ÷ 9.76 = ?

82. .72 ÷ 14.1 = ?

DIVIDING BY MULTIPLES OF TEN

Rule.—To divide a decimal by a multiple of ten, move the decimal point in the dividend as many places to the left as there are zeros in the divisor.

Example.—Divide 351.86 by 10; by 100; by 1000.

351.86 ÷ 10 = 35.186
351.86 ÷ 100 = 3.5186
351.86 ÷ 1000 = .35186

10 has one zero; move the decimal point one place to the left.

100 has two zeros; move the decimal point two places to the left.

1000 has three zeros; move the decimal point three places to the left.

PROBLEMS

Work mentally if possible.

1. Divide 175.6 by 10; by 100; by 1000; by 10,000.

2. Divide 217.68 by 10; by 1000; by 10,000.

3. State the quotient: 56.34 divided by 10; 356 divided by 1000; 587 divided by 100; .057 divided by 100; 563.4 divided by 100; 893.2 divided by 10; 68.3 divided by 1000; .003 divided by 10; 5634 divided by 1000; 8932 divided by 10,000; 78 divided by 10,000; 0.56 divided by 1000; .789 divided by 10; .0683 divided by 100; .06 divided by 100.

4. State a short method for dividing a number by 20; by 300; by 4000; by 5000; by any whole number ending in zeros.

5. State the quotient: 248 divided by 20; 67.5 divided by 500; 7593 divided by 5000; 89 divided by 400; 248 divided by 40; 246 divided by 600.

6. At 90¢ per 1000 cu. ft., what will 2468 cu. ft. of gas cost?

7. How many tons are there in 5240 lb.? In 137,801 lb.? In 756 lb.?

8. At $165 per 1000, what will 12,280 bd. ft. of lumber cost?

9. At $125.00 per thousand, what will 12,280 bd. ft. of lumber cost?

10. At $9.75 per ton, how much must be paid for 12,340 lb of coal?

11. A gallon of paint will cover 200 sq. ft. of any surface with two coats. State a rule for finding the number of gallons of paint required to give any surface two coats.

12. Using the rule from Prob. 11, find the number of gallons of paint required for two coats for 2578 sq. ft.; for 160 sq. ft. Round off to the nearest gallon.

13. A quart of grass seed is sufficient for seeding 300 sq. ft. of lawn. How many quarts will be required to seed a lawn of 2970 sq. ft.? Round off to the nearest quart.

TO CHANGE COMMON FRACTIONS TO DECIMALS

In the machine shop, dimensions are commonly expressed in thousandths, as .625 in. or 3.348 in. In the woodshop, common fractions are usually employed. Hence, it is necessary for the mechanic, especially the patternmaker, to be able to change from one form to the other.

Rule.—To change a common fraction to a decimal fraction, divide the numerator of the fraction by the denominator.

Example 1.—Express $\frac{3}{8}$ in. to the nearest thousandth of an inch

$\frac{3}{8} = 3 \div 8$ $\quad 8\overline{)3.000}$
$\quad .375$

Ans. .375 in.

Example 2.—Change $\frac{29}{64}$ in. to the nearest thousandth of an inch.

$64\overline{)29.0000}$ = .4531 *Ans.* .453 in.	As this division does not come out to even thousandths, it is necessary to carry the division one place beyond the three decimals desired and round off.

PROBLEMS

1. In the spark plug shown, express each dimension as a decimal to the nearest thousandth of an inch.

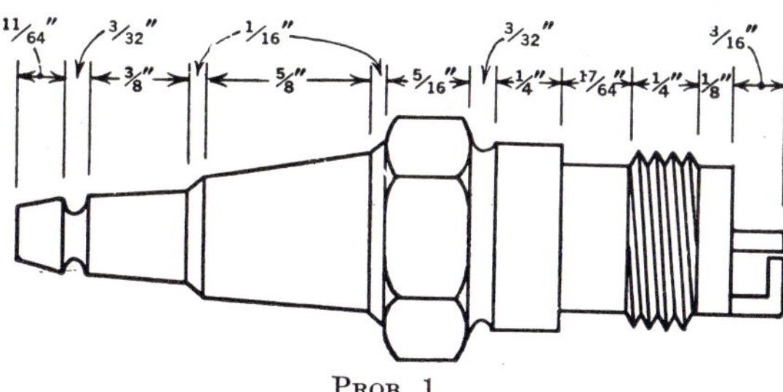

PROB. 1.

2. In the bolt shown, change each dimension to the nearest 0.001″.

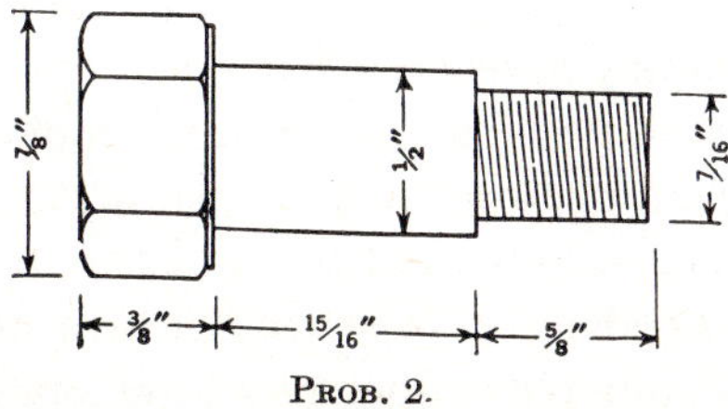

PROB. 2.

3. On the crankshaft shown, change the fractional part of each mixed number to the nearest thousandth.

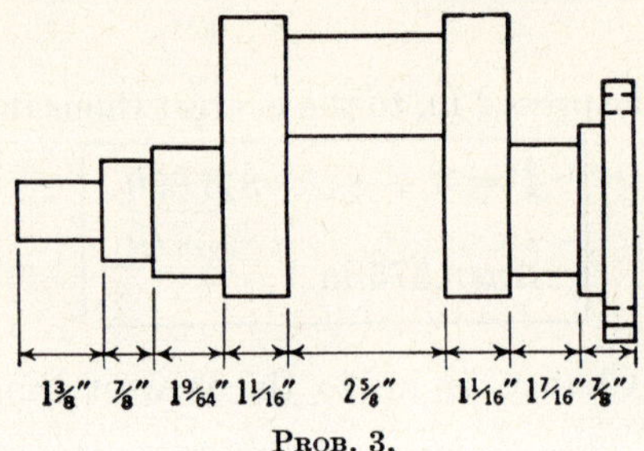

Prob. 3.

4. A wrist pin is $\frac{43}{64}''$ in diameter. Express this dimension to the nearest 0.001″.

5. How large should a pin boss be reamed if a $\frac{5}{8}''$ wrist pin is to be replaced with one 0.063″ oversize?

6. Find the decimal equivalent correct to three decimal places for each of the following: $\frac{25}{38}$, $\frac{224}{393}$, $\frac{17}{11}$, $\frac{6}{14}$, $\frac{22}{7}$.

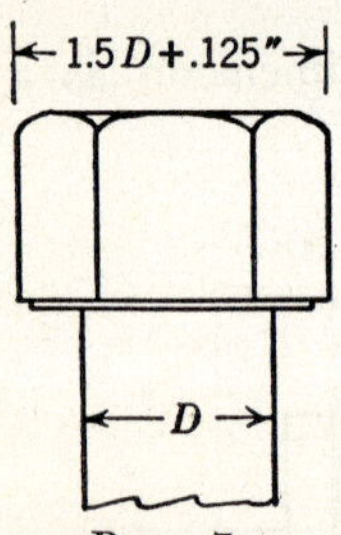

Prob. 7.

7. The distance across the flats of the hexagonal-head machine bolt shown is 1.5 times the diameter plus 0.125″. Get the distance across the flats of a $\frac{1}{2}''$ bolt; of a $\frac{3}{4}''$ bolt; of a $\frac{9}{16}''$ bolt.

8. How large (across flats) is the hexagonal stock needed to make a $\frac{3}{8}''$ bolt?

9. An engine cylinder is $3\frac{3}{8}''$ in diameter. The piston is to have 0.0035″ clearance. What is the diameter of the piston?

10. An engine cylinder $4\frac{5}{8}''$ in diameter is to be rebored to 0.062″ oversize. Find its size after reboring.

Many other occasions occur when it is necessary to change common fractions to decimals. The following table shows a typical example. In the table the first column gives the U.S. Standard sheet metal gauge number. The second column gives the thickness in common fractional form. It is possible to measure from No. 7-0 down to No. 14 with an ordinary rule. Below that, only a few thicknesses are found on the rule, and the decimal equivalents are then needed. These thicknesses are best measured with a micrometer caliper (see page 309).

Change each of the common fractions to an exact decimal or to a decimal correct to the nearest thousandth of an inch.

Gauge No.	Fraction inch thick	Decimal inch thick	Gauge No.	Fraction inch thick	Decimal inch thick	Gauge No.	Fraction inch thick	Decimal inch thick
7-0	$\frac{1}{2}$		9	$\frac{5}{32}$		24	$\frac{1}{40}$	
6-0	$\frac{15}{32}$		10	$\frac{9}{64}$		25	$\frac{7}{320}$	
5-0	$\frac{7}{16}$		11	$\frac{1}{8}$		26	$\frac{3}{160}$	
4-0	$\frac{13}{32}$		12	$\frac{7}{64}$		27	$\frac{11}{640}$	
3-0	$\frac{3}{8}$		13	$\frac{3}{32}$		28	$\frac{1}{64}$	
2-0	$\frac{11}{32}$		14	$\frac{5}{64}$		29	$\frac{9}{640}$	
0	$\frac{5}{16}$		15	$\frac{9}{128}$		30	$\frac{1}{80}$	
1	$\frac{9}{32}$		16	$\frac{1}{16}$		31	$\frac{7}{640}$	
2	$\frac{17}{64}$		17	$\frac{9}{160}$		32	$\frac{13}{1280}$	
3	$\frac{1}{4}$		18	$\frac{1}{20}$		33	$\frac{3}{320}$	
4	$\frac{15}{64}$		19	$\frac{7}{160}$		34	$\frac{11}{1280}$	
5	$\frac{7}{32}$		20	$\frac{3}{80}$		35	$\frac{5}{640}$	
6	$\frac{13}{64}$		21	$\frac{11}{320}$		36	$\frac{9}{1280}$	
7	$\frac{3}{16}$		22	$\frac{1}{32}$		37	$\frac{17}{2560}$	
8	$\frac{11}{64}$		23	$\frac{9}{320}$		38	$\frac{1}{160}$	

CHANGING THE FORM OF NUMBERS

The dimensions on shop drawings are often expressed in decimal form. The carpenter must change these decimals to common fractions, usually eighths, sixteenths, thirty-seconds, or sixty-fourths.

Rule.—To change a whole number, a mixed number, or a decimal to a fraction with a desired denominator, multiply by a fraction equal to 1 that has the desired denominator.

Example 1.—Change 6 to eighths.

$\frac{6}{1} \times \frac{8}{8} = \frac{48}{8}$ $\frac{8}{8}$ is 1 expressed in eighths. Do not cancel.

Example 2.—Change 6.125 to eighths.

$$\frac{6.125}{1} \times \frac{8}{8} = \frac{49.000}{8} = \frac{49}{8} = 6\tfrac{1}{8}.$$

Example 3.—Change .125 to eighths.

$$\frac{.125}{1} \times \frac{8}{8} = \frac{1.000}{8} = \frac{1}{8}$$

Example 4.—Change .489 to thirty-seconds.

$$\frac{.489}{1} \times \frac{32}{32} = \frac{15.648}{32}$$

$$\frac{15.648}{32} \text{ or } \frac{16}{32} = \frac{1}{2}$$

In this case the numerator 15.648 should be rounded off to 16, the nearest whole number.

PROBLEMS

1. In the size gauge shown, change each dimension to the nearest thirty-second of an inch.

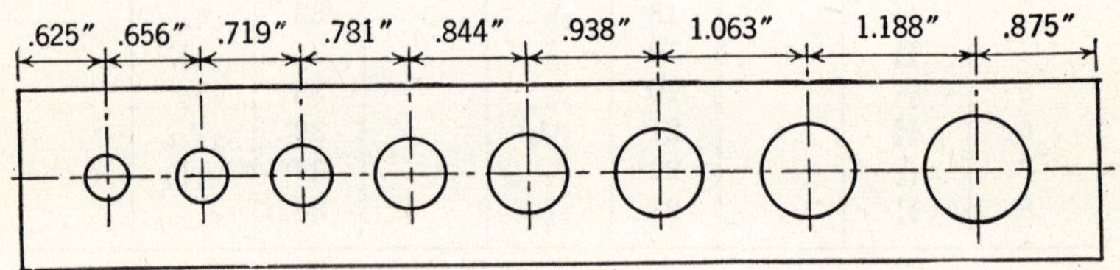

PROB. 1.

2. In the rivet set shown, reduce each dimension to the nearest sixty-fourth of an inch.

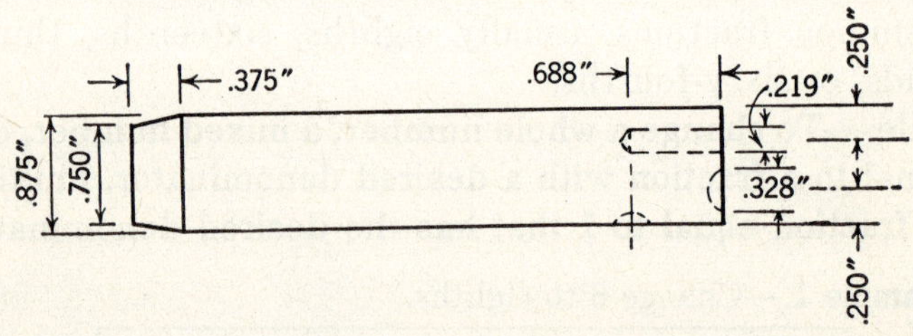

PROB. 2.

3. On charts showing drill sizes, many alphabetical drills are listed. Alphabetical drills are measured in decimal thousandths. If the mechanic has only fractional drills, what sizes to the nearest sixty-fourth of an inch will he substitute for each of the following alphabetical drills? Round off the numerators to the nearest whole numbers.

A—0.234″	B—0.238″	C—0.242″
D—0.246″	E—0.250″	F—0.257″
G—0.261″	H—0.266″	I—0.272″
J—0.277″	K—0.281″	L—0.290″
M—0.295″	N—0.302″	O—0.316″
P—0.323″	Q—0.332″	R—0.339″
S—0.348″	T—0.358″	U—0.368″
V—0.377″	W—0.386″	X—0.397″
Y—0.404″	Z—0.413″	

From a cabinet of fractional drills graduated in sixty-fourths of an inch, find the best drill to use instead of each of the following wire gauge drills. Round off the numerators to the nearest whole numbers.

No. 1—0.228 in.	No. 10—0.1935 in.	No. 50—0.070 in.
No. 2—0.221 in.	No. 20—0.161 in.	No. 60—0.040 in.
No. 3—0.213 in.	No. 30—0.128 in.	No. 70—0.028 in.
No. 6—0.204 in.	No. 40—0.098 in.	No. 80—0.0135 in.
No. 12—0.189 in.		

DENOMINATE NUMBERS INVOLVING DECIMALS

Example 1.—Change 2.72 tons to pounds.

```
2.72       There are 2000 lb. in a ton; hence, multiply
  2000   the given number of tons by 2000.
-------
5440.00
Ans. 5440 lb.
```

Example 2.—If 8″ flat files cost $16.85 per gross, determine the price of each.

```
       .117     There are 144 units per gross; hence, divide
144)16.850   the price per gross by 144.
     14 4
     ----
      2 45
      1 44
      -----
      1 010
      1 008
      -----
          2
Ans. 12¢ each     Round off to the nearest cent.
```

Example 3.—Find the cost per board foot when lumber sells at $32.50 per M. (M is the Roman numeral for 1000. It is customarily used in the trade.)

$32.50 \div 1000 = .0325$	Since the price per board foot
Ans. $3\frac{1}{4}$¢ per board foot	is asked for, do not round off.

EXERCISES

1. Change $126.50 per M board feet to price per board foot.
2. Change 45.9 gross to dozen.
3. Change 15¢ per piece to price per gross.
4. Change 721 fathoms to feet.
5. Change $0.098 per board foot to price per M board feet.
6. Change 2000 oz. to pounds.
7. Change $9.85 per dozen to price per piece.
8. Change 27.946 cu. in. to cubic feet.
9. Find the cost of 50 locks if one cost $87\frac{1}{2}$¢.
10. Change 14 quires to dozens.
11. Change 17.55 lb. to ounces.
12. Change 500 cu. ft. to cords.
13. Change 3.6 tons to ounces.
14. Change 5000 lb. of water to cubic feet.
15. Change 956,748 bd. ft. to thousands of board feet.
16. Change 270 miles to yards.
17. Change $145.55 per M board feet to cost per board foot.
18. Change 690 minutes to seconds.
19. Change 23,641 cu in. to gallons.

In the following, round off answers to the nearest 0.01.

20. Change 741.36 lb. of cast iron to cubic feet of cast iron.
21. Change 1,000,000 sq. ft. to acres.
22. Change 2.38 cu. yd. to cubic inches.
23. Change 75 gal. of water to pounds.
24. Change 58 lb. of mild steel to cubic inches of mild steel.
25. Change 17 gal. to cubic inches.
26. Change 48 gal. to cubic feet.
27. Change 210 days to hours.
28. If a box of 100 tool bits costs $9.36, find the cost of one tool bit.
29. Change 1 bu. to pints.

30. At $57\frac{1}{2}$¢ each, determine the cost per gross of $\frac{1}{2}''$ twist drills.

ADDITION OF DENOMINATE NUMBERS

Example.—Add $15'6\frac{1}{2}''$, $3'9\frac{3}{8}''$, and $6'5\frac{3}{4}''$.

	8ths
$15'6\frac{1}{2}''$	4
$3'9\frac{3}{8}''$	3
$6'5\frac{3}{4}''$	6
$24'20''$	$\frac{13}{8} = 1\frac{5}{8}''$
$1\frac{5}{8}''$	
$24'21\frac{5}{8}''$	
$21\frac{5}{8}'' = 1'9\frac{5}{8}''$	
$24'$	
$1'9\frac{5}{8}''$	
$25'9\frac{5}{8}''$	

Adding the column of fractions, as indicated on page 30, gives $1\frac{5}{8}''$.

Adding the column of the whole number of inches gives $20''$.

Adding the column of feet gives $24'$. Adding $1\frac{5}{8}''$ to $20''$, to get the total number of inches, gives $21\frac{5}{8}''$.

Since $21\frac{5}{8}''$ is greater than $1'$, dividing by 12 changes $21\frac{5}{8}''$ to $1'9\frac{5}{8}''$. This leaves $9\frac{5}{8}''$ in the inches' column. Add the $1'$ to $24'$, making $25'$.

EXERCISES AND PROBLEMS

1. Add $16'5\frac{3}{8}''$, $5'4\frac{1}{2}''$, and $3'8\frac{1}{2}''$.

2. Add 4 gal. 3 qt. 1 pt. and 7 gal. 3 qt. 1 pt.

3. Add 3 yd. 2 ft. 5 in., 9 yd. 1 ft. 8 in., and 4 yd. 2 ft. 8 in.

4. Add 4 bu. 2 pk. 1 qt., 2 bu. 3 pk. 3 qt., and 5 bu. 3 pk. 7 qt.

5. Add $9'3.75''$, $8'7.625''$, $5'9.5''$, and $3'8.375''$.

6. Add 5 bu. 3 pk. 2 qt. 1 pt. and 3 bu. 2 pk. 1 qt. 1 pt.

7. Add 7 tons 1350 lb., 1 ton 1540 lb., and 9 tons 1006 lb.

8. Add 16 lb. 14 oz., 10 lb. 9 oz., 11 lb. 12 oz., and 4 lb. 5 oz.

9. Add 4 gal. 3.5 qt., 5 gal. 2.7 qt., and 9 gal. 3.68 qt.

10. Add 7 yd. 2 ft. $6\frac{1}{2}$ in., 9 yd. 1 ft. $8\frac{3}{4}$ in., and 3 yd. 2 ft. $6\frac{3}{8}$ in.

11. 6 gal. 124 cu. in. + 7 gal. 131 cu. in. + 5 gal. 96 cu. in. = ?

12. 3 hr. 50 min. 25 sec. + 7 hr. 3 min. 48 sec. = ?

13. 5 gross 8 doz. 3 + 8 gross 3 doz. 10 + 11 gross 10 doz. = ?

14. 5 yd. 2 ft. 4.5 in. + 3 yd. 1 ft. 8.375 in. + 4 yd. 2 ft. 9.48 in. = ?

15. Find the sum of $3'7\frac{1}{2}''$, $4'9\frac{3}{8}''$, $5'8\frac{7}{8}''$, and $7'3\frac{1}{2}''$.

16. Add $70°58'11''$, $21°42'29''$, and $48'41''$.

17. 16 gal. 186 cu. in. + 12 gal. 200 cu. in. + 9 gal. 48 cu. in. = ?

18. Add 1′10$\frac{1}{2}$″, 7′8$\frac{3}{4}$″, 9′3$\frac{1}{8}$″, and 10′8$\frac{3}{16}$″.

19. The wing-flap pulley bracket shown is from a Lockheed airplane. Determine the over-all width. The dimensions are given in inches.

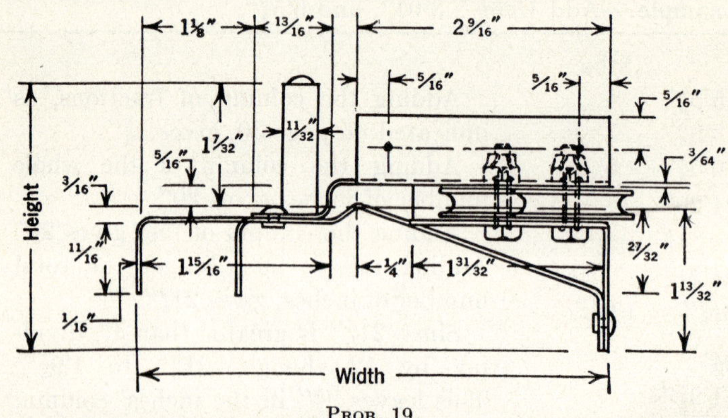

PROB. 19.

20. In the wing-flap pulley bracket of Prob. 19, determine the over-all height.

SUBTRACTION OF DENOMINATE NUMBERS

Example.—From 6 yd. 1 ft. 3 in. subtract 3 yd. 2 ft. 8 in.

5	3	15
~~6~~ yd.	~~1~~ ft.	~~3~~ in.
3 yd.	2 ft.	8 in.
2 yd.	1 ft.	7 in.

Start at the right. 8 in. cannot be subtracted from 3 in. Borrow 1 ft., or 12 in., from the number of feet in the minuend, 1 ft., and add it to the 3 in., making 15 in. Subtract 8 in. from 15 in., leaving 7 in.

2 ft. cannot be subtracted from 0 ft. Borrow 1 yd., or 3 ft., from the number of yards in the minuend, 6 yd., and add the 3 ft. to 0 ft., making 3 ft. Subtract 2 ft. from 3 ft., leaving 1 ft.

Subtract 3 yd. from 5 yd., leaving 2 yd.

EXERCISES AND PROBLEMS

1. From 9′8$\frac{7}{8}$″, subtract 6′5$\frac{1}{4}$″.

2. From 4′3$\frac{1}{4}$″, subtract 1′6$\frac{3}{8}$″.

3. Subtract 15′3$\frac{3}{4}$″ from 24′9$\frac{3}{8}$″.

4. Subtract 6 yd. 2 ft. 10 in. from 10 yd.

5. Subtract 3 tons 1640 lb. from 9 tons 200 lb.

6. Take 5 lb. 4 oz. from 8 lb. 2 oz.

7. Take 7 yd. 2 ft. $4\frac{1}{2}$ in. from 12 yd. 1 ft.

8. In the wing channel of the Lockheed airplane shown, the length of B is how much greater than that of A?

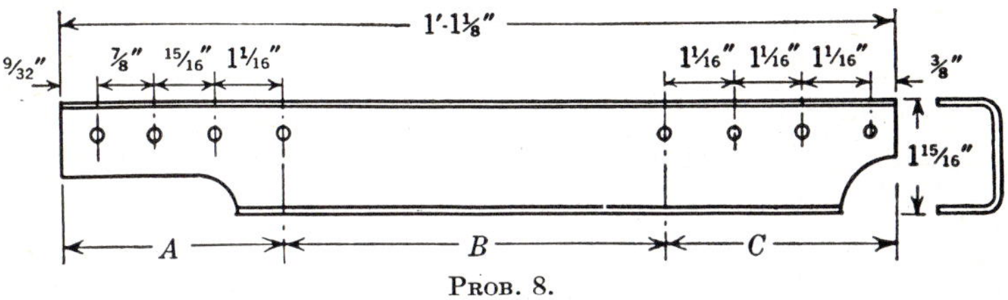

PROB. 8.

9. In the wing channel shown, C is how much greater than A?

10. In the wing channel shown, B is how much greater than C?

11. From 10 gal. take 4 gal. 3 qt. 1 pt.

12. From 10 gal. 148 cu. in. take 3 gal. 200 cu. in.

13. Take 33°50′10″ from 65°10′2″.

14. Take 3 gross 4 doz. from 10 gross 5 doz.

15. Take $48'9\frac{1}{2}''$ from $50'3\frac{1}{4}''$.

16. From 6 yd. subtract 2 yd. 2 ft. 9 in.

17. From 320° take 90°50′20″.

18. Subtract 15 yd. 2 ft. 8 in. from 27 yd.

19. From 5 yd. 2 ft. 4.375 in. take 1 yd. 1 ft. 2 in.

20. Take 16 gal. 2 qt. 1 pt. from 20 gal.

21. Subtract $9'5\frac{3}{8}''$ from 11 ft.

22. Subtract 5 gal. 3.5 qt. from 8 gal. 1 pt.

MULTIPLICATION OF DENOMINATE NUMBERS

Example.—Multiply 2 bu. 3 pk. 3 qt. 1 pt. by 5.

2 bu.	3 pk.	3 qt.	1 pt.
			× 5
10 bu.	15 pk.	15 qt.	5 pt.
+4 bu.	+2 pk.	+2 qt.	
	17 pk.	17 qt.	
14 bu.	1 pk.	1 qt.	1 pt.

Multiply each part of the multiplicand by 5.

5 pt. can be changed to 2 qt. 1 pt. Put the 1 pt. in the pints' column, and add the 2 qt. to the 15 qt.

17 qt. can be changed to 2 pk. 1 qt. Put the 1 qt. in the quarts' column, and add the 2 pk. to the 15 pk.

17 pk. can be changed to 4 bu. 1 pk. Put the 1 pk. in the pecks' columr and add the 4 bu. to the 10 bu.

EXERCISES AND PROBLEMS

1. 2×5 bu. 3 pk. 6 qt. = ?

2. Multiply 8′3½″ by 4.

3. Multiply 4 yd. 2 ft. 5 in. by 6.

4. Multiply 3 tons 900 lb. by 15.

5. Multiply 4 gal. 85 cu. in. by 6.

6. Find 6 times 16 lb. 14 oz.

7. Find 16 times 9′4.375″.

8. Find the dimensions of an I beam three times as large as the I beam shown.

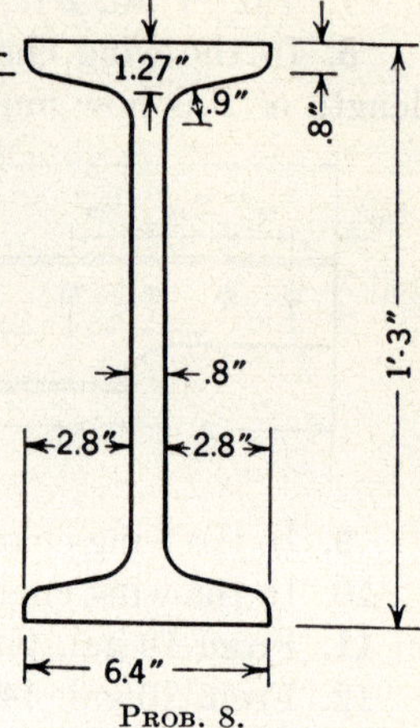

PROB. 8.

9. 14×7 lb. 4 oz. = ?

10. $12 \times 12'6.875''$ = ?

11. How many yards of felt will be needed to make 8 banners like the one shown?

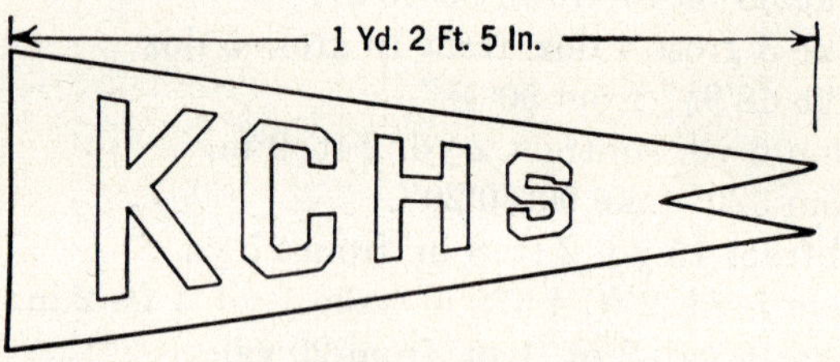

PROB. 11.

12. What is 8×7 bu. 3 pk. 2 qt.?

13. How much is $15°20'10'' \times 9$?

14. Find the product of 8 times 3 tons 750 lb.

15. 9×7 lb. 15 oz. = ?

16. 8×6 gross 8 doz. = ?

17. Multiply 5 gal. 16 cu. in. by 15½.

18. Multiply 3 hr. 5 min. 10 sec. by 8.7.

19. Multiply 12°8′15″ by 6.

20. Multiply 2 gal. 2 qt. 1 pt. 3 gills by 9.

DIVISION OF DENOMINATE NUMBERS

Example.—Divide 15′3½″ by 2.

12 2)15′ 3½″ 7′ 7¾″	2)15½″ 7¾″	Divide 15′ by 2, giving 7′ and 1′, or 12″, left over. Add the 12″ to the 3½″, making 15½″. Divide 15½″ by 2, giving 7¾″.

EXERCISES AND PROBLEMS

1. Divide 6′9$\frac{3}{8}$″ by 3.

2. Divide 12′7″ by 5.

3. On the airplane wing shown, the ribs, given as dotted lines, are equally spaced. Find the distance between the centers.

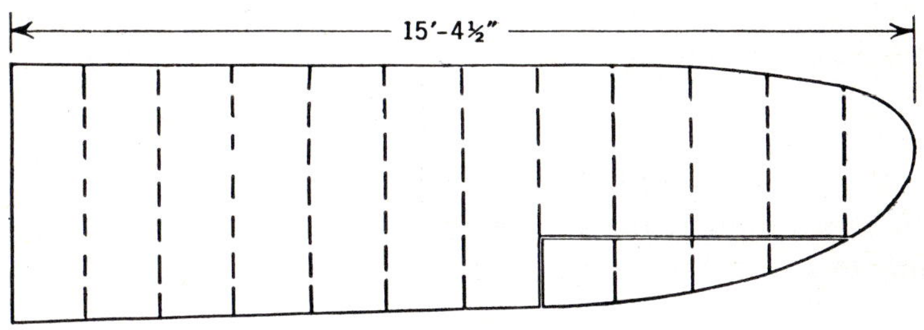

Prob. 3.

4. 4 tons 360 lb. ÷ 9 = ?

5. Find 11 yd. 2 ft. 4 in. divided by 5.

6. Find 5 bu. 3 pk. 1 qt. divided by 4.

7. 9 lb. 1 oz. ÷ 7 = ?

8. 9′5$\frac{3}{8}$″ ÷ 8 = ?

9. Divide 8 gal. 3 qt. 1 pt. by 6.

10. Divide 10 gross 8.5 doz. by 4.

11. What is the quotient when 48°10′50″ is divided by 6?

12. Find the quotient when 8 gal. 186 cu. in. is divided by 6.

13. In the template shown, the holes are equally spaced. Find the distance between centers.

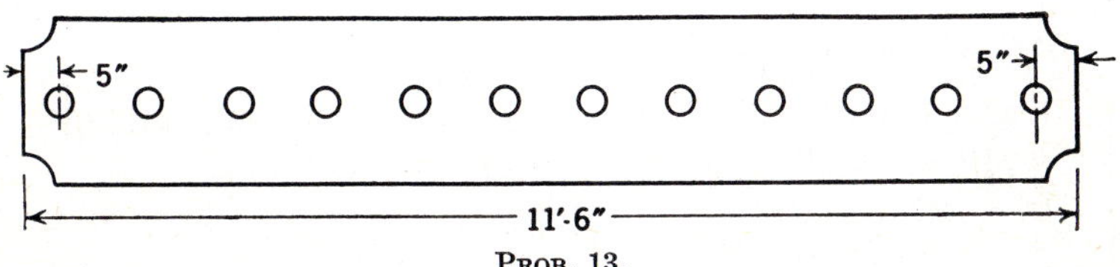

Prob. 13.

14. How much is 5 yd. 3 ft. 7 in. divided by 9?

15. Find the quotient when 3 hr. 50 min. 6 sec. is divided by 6.

16. Divide 6 gal. 124 cu. in. by 8.

17. Divide 9 tons 360 lb. 8 oz. by 4.

18. Divide 16 lb. 4$\frac{3}{4}$ oz. by 5.

19. Divide 6′4.565″ by 6.

20. How much is 3 yd. 2 ft. 10 in. divided by 15?

USING DENOMINATE NUMBERS IN PRACTICAL PROBLEMS

In practical problems the dimensions often are not all expressed in the same unit. Only denominate numbers expressed in the same unit can be multiplied or divided by each other. For example, inches times inches gives square inches, while inches times feet is impossible. The feet would have to be changed to inches to give square inches, or the inches would have to be changed to feet to give square feet.

Rule.—To multiply or divide denominate numbers, change all denominate numbers to the same unit, and then proceed with the multiplication or division.

Example 1.—Multiply $3'6\frac{1}{2}''$ by $4'5\frac{1}{4}''$.

$3'6\frac{1}{2}'' = 42\frac{1}{2}'' = \frac{85}{2}''$ Changing each compound dimension to inches
$4'5\frac{1}{4}'' = 53\frac{1}{4}'' = \frac{213}{4}''$

$$\frac{85}{2} \times \frac{213}{4} = \frac{18{,}105}{8} = 2263\tfrac{1}{8}, \text{ or } 2263\tfrac{1}{8} \text{ sq. in.}$$

$2263\frac{1}{8}$ sq. in. = 15 sq. ft. $103\frac{1}{8}$ sq. in.

Dividing by 144 sq. in.

15 sq. ft. $103\frac{1}{8}$ sq. in. = 1 sq. yd. 6 sq. ft. $103\frac{1}{8}$ sq. in.

Dividing by 9 sq. ft.

Ans. 1 sq. yd. 6 sq. ft. $103\frac{1}{8}$ sq. in.

Example 2.—Divide $10'4\frac{1}{4}''$ by $3'7\frac{1}{2}''$.

$10'4\frac{1}{4}'' = 124\frac{1}{4}'' = \frac{497}{4}''$ Changing each compound dimension to inches
$3'7\frac{1}{2}'' = 43\frac{1}{2}'' = \frac{87}{2}''$

$$\frac{497}{4} \div \frac{87}{2} =$$

$$\frac{497}{\underset{2}{\cancel{4}}} \times \frac{\overset{1}{\cancel{2}}}{87} = \frac{497}{174}$$

$$\frac{497}{174} = 2\frac{149}{174} = 2.856+, \text{ or } 2.86- \text{ times}$$

Ans. $3'7\frac{1}{2}''$ is contained in $10'4\frac{1}{4}''$ 2.86− times.

EXERCISES

1. Multiply $1'4\frac{1}{2}''$ by $2'3\frac{1}{4}''$.

2. Multiply $2'6\frac{1}{4}''$ by $2'6\frac{1}{4}''$.

3. Multiply $3'7\frac{1}{2}''$ by $2'6''$.

4. Multiply $8'5\frac{3}{8}''$ by $5'8\frac{7}{8}''$.

5. Divide $3'4\frac{1}{2}''$ by $6\frac{1}{2}''$.

6. Divide $11'8\frac{1}{2}''$ by $2'3\frac{1}{4}''$.

7. Divide $16'8\frac{1}{2}''$ by $4'2\frac{1}{8}''$.

8. Divide $7'2\frac{1}{2}''$ by $2'2\frac{1}{4}''$.

9. Multiply $9'3\frac{3}{4}'' \times 5'8\frac{3}{8}''$.

10. Multiply $40'5\frac{3}{8}''$ by $20'2\frac{1}{4}''$.

11. Divide $6'4''$ by $3'1''$.

12. Divide $3'5\frac{1}{2}''$ into $15'\frac{1}{2}''$.

13. Multiply $30'1\frac{1}{8}'' \times 10'5\frac{1}{2}''$.

14. Multiply $3'6''$ by $6'3''$.

15. Divide $6'3''$ into $18'3''$.

16. Divide $17'\frac{5}{8}''$ by $3'$.

17. Find the product of $8'5\frac{3}{4}''$ multiplied by $5'8\frac{1}{2}''$.

18. Find the quotient of $3'10''$ divided by $4\frac{1}{2}''$.

19. How much is $9'8\frac{1}{8}''$ multiplied by $8'9\frac{1}{2}''$?

20. How many pieces $4\frac{3}{8}''$ long can be cut from a bar 16 ft. long?

AN APPLICATION OF DENOMINATE NUMBERS

Many occasions arise for dividing a given length into a number of equal parts.

Example.—Divide $11'7''$ by 7.

Solution 1.

$7\overline{)11'7''}$

$1'$ with $4'$ left over

$4' = 48''$. Add to the $7''$ of the given length.

$48'' + 7'' = 55''$

$7\overline{)55.0000}$

7.8571

Ans. $1'7.857''$

The answer $1'7.857''$ would serve the purposes of a machinist, for he measures thousandths with a micrometer; but if a carpenter got this result, he would need to change the decimal to a common fraction with the denominator 16, for his rule is graduated in sixteenths.

$$\frac{.857}{1} \times \frac{16}{16} = \frac{13.712}{16}, \text{ or } \frac{14}{16} = \frac{7}{8}$$

The carpenter's answer would be $1'7\frac{7}{8}''$.

Solution 2.

$11 \times 12 = 132$ $132 + 7 = 139$	Changing the entire length to inches by multiplying the number of feet by 12 and adding the number of inches
7)139.000 19.857	Dividing the total number of inches by 7
19.857″ −12. ″ 7.857″	Obviously, 19.857″ is 1′ and some inches. Subtract 1′ (12″) from 19.857″ to find the number of inches left. This is the machinist's answer.
Ans. 1′7.857″	Find the carpenter's answer as in Solution 1.

EXERCISES AND PROBLEMS

Divide the following lengths, finding the part to the nearest thousandth of an inch for the machinist and then to the nearest sixteenth of an inch for the carpenter.

1. 9)11′7″	**2.** 5)11′7″	**3.** 3)11′7″
4. 5)19′5″	**5.** 7)19′5″	**6.** 9)19′5″
7. 11)19′5″	**8.** 13)19′5″	**9.** 15)19′5″
10. 3)23′7″	**11.** 5)23′7″	**12.** 7)23′7″
13. 9)23′7″	**14.** 11)23′7″	**15.** 13)23′7″
16. 5)5′2″	**17.** 7)5′2″	**18.** 9)5′2″
19. 5)11′3″	**20.** 7)11′3″	**21.** 9)11′3″
22. 11)11′3″	**23.** 3)12′5″	**24.** 5)12′5″
25. 7)12′5″	**26.** 9)12′5″	**27.** 5)25′7″
28. 7)25′7″	**29.** 9)25′5″	**30.** 11)25′5″
31. 3)33′8″	**32.** 4)33′8″	**33.** 5)33′8″
34. 6)33′8″	**35.** 11)21′6″	**36.** 13)21′6″
37. 15)21′6″	**38.** 17)39′5″	**39.** 19)39′5″
40. 21)39′5″	**41.** 10)21′10″	**42.** 11)21′10″

43. Divide a wall 19′4″ into 13 equal parts, getting the answer to the nearest sixteenth of an inch.

44. It is desired to drill 47 holes in a piece of sprinkler pipe 19′2″ long. How far apart should the holes be drilled if they are to be evenly spaced? Get the answer to the nearest sixty-fourth of an inch.

45. The reinforcing bars in a concrete wall a portion of which is shown, are to be equally spaced. The wall is 16′ long. There are 30 bars. What is the distance between centers to the closest sixteenth of an inch?

Prob. 45.

46. A man wishes to make 12 bolts from a piece of cold-finished steel that measures 6′3″, allowing $\frac{1}{4}$″ for cutting each bolt, including the last one. How long will the bolts be when finished?

47. Making no allowance for cutting, divide a board 16′ long into 11 equal parts. Get the answer to the nearest thirty-second of an inch.

48. The space under a desk top is 36″. Five drawers of equal height are to reach to the floor. Get the height of each drawer to the nearest sixty-fourth of an inch.

49. In a sprinkler pipe 16 ft. long, it is required to drill 150 holes, evenly spaced, with the first and last hole the same distance from the end as the holes are spaced. What is the distance between centers, expressed to the nearest sixty-fourth of an inch?

50. It is desired to panel a wall 27 ft. wide into 12 spaces. How wide is each space?

51. Divide a bar 16 ft. long into 45 parts. Find the quotient to the nearest sixty-fourth of an inch.

52. The footlights on a stage are evenly spaced. There are 47 lights. The distance from the first to the last light is 23 ft. Determine the distance between the centers.

53. A planer table is 10′ long. There are 17 holes, equally spaced. The end holes are 3″ from the ends. Find the distance between centers to the nearest sixty-fourth of an inch.

54. Eight rows of desks are equally spaced in a classroom 22 ft. wide. Get the distance between centers to the nearest fourth of an inch.

55. It is desired to divide the window sash shown, which is 7′4″ wide, into eight panes of equal width. Allowing 1″ for each vertical muntin bar and 3″ for the stiles at the sides, what is the width of each glass to the nearest eighth of an inch if the glass laps $\frac{1}{4}$″ on the wood?

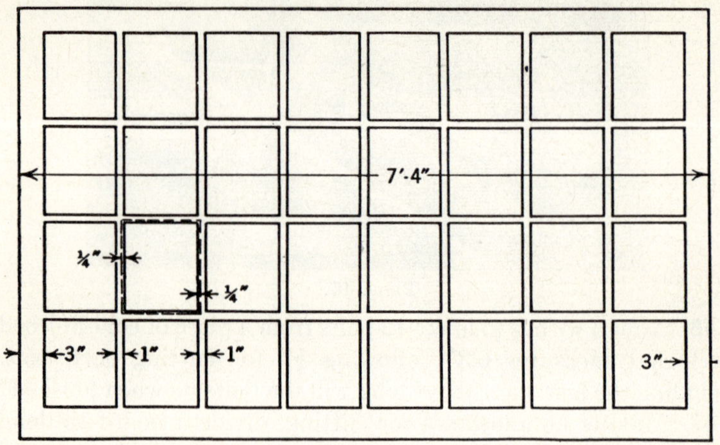

PROB. 55.

56. In a window 3′ wide, the stiles are $2\frac{1}{2}$″, two muntins are $\frac{7}{8}$″ each, and there are three panes of glass. Find the spacing of the muntins to the nearest fourth of an inch.

57. A locker room is 16′6″ wide. There are 15 lockers of equal width. Get the width of each to the nearest sixteenth of an inch.

58. Divide a sheet of paper $8\frac{1}{2}$″ wide into 11 equal columns. Find the answer to the nearest sixty-fourth of an inch.

59. Equally space 9 rods on a grill for a window 6 ft. wide. Find the result to the nearest sixteenth of an inch.

USE OF HANDBOOKS AND THE APPENDIX

One of the greatest aids to the student and mechanic, as a source of information and to shorten calculations, is a good handbook of tables, formulas, and general technical data.

In the Appendix (pages 523 to 544), are typical tables taken from various sources, to show their value and to give practice in their use. Other tables are given throughout the text.

In using a reference handbook or the Appendix, it will be necessary to study the explanations and the tables to learn how to get the information desired. In solving the problems on reduction (pages 23, 48, 74, etc.), tables in the Appendix should already have been used.

Every time it is possible to look up in the tables material that will aid in the computation at hand, it is intended that the proper table be used. If handbooks are available, they should be referred to, also.

A definite procedure should be followed in using tables.

1. Determine which table to use, either by searching through the tables available or, better, by looking in the index under some key word that appears in the problem being solved.

2. Study the table, and select the proper information.

Some tables include many types of information. For example, Table VI Size of Drills for Taps (page 529) has four columns of information for each size of tap. Column 1 gives the size, or tap diameter. Column 2 gives the number of threads per inch (pitch) for National Coarse (N.C.) threads. Column 3 gives the size of drill for N.C. taps. Column 4 gives the pitch for National Fine (N.F.) threads. Column 5 gives the size of drill for N.F. taps.

Some problems may be solved entirely from the tables. Others will require some calculation, based on values from the tables. The following examples are solved by use of the tables in the Appendix (pages 523 to 544).

Example 1.—Find the size of drill for a $\frac{7}{16}''$ N.F. tap. (See Table VI, page 529.)

1. Read down the first column, headed Tap diameter, to $\frac{7}{16}$. It is the fourteenth line.

2. In the fifth column, headed Tap-drill size, and opposite the $\frac{7}{16}$ in column 1, read $\frac{25}{64}$.

$\frac{25}{64}''$ is the size of drill for a $\frac{7}{16}''$ N.F. tap.

Example 2.—What is the pitch of a $\frac{5}{8}''$ N.C. thread? (See Table VI, page 529.)

1. Read down the first column, to $\frac{5}{8}''$. It is the seventeenth line.
2. In the second column, headed Pitch or threads per inch, and opposite the $\frac{5}{8}$, read 11.

11 threads per inch is the pitch of a $\frac{5}{8}''$ N.C. thread.

Example 3.—What is the proper number of revolutions per minute (r.p.m.) for a $\frac{3}{8}''$ twist drill when the cutting speed is 40 f.p.m. (feet per minute)? (See Table XI, page 533.)

Read down the first column, to $\frac{3}{8}''$. In the column under 40′, read opposite the $\frac{3}{8}$, 408.

408 r.p.m. is the correct speed.

Example 4.—Change 3.6 meters to feet. (See Table IV*a*, page 526.)

$$1 \text{ meter} = 3.281 \text{ ft.}$$
$$3.6 \times 3.281 = 11.8116 \text{ ft.}$$

Example 5.—Determine the next larger 64th inch drill than an M drill.

M drill is 0.295 in. (Table IX, page 531).
0.296875 is $\frac{19}{64}$ (Table II, page 524).
The next larger 64th drill is $\frac{19}{64}$.

Example 6.—Compute the end area of the outside of a $\frac{1}{4}$ in. diameter pipe.

Outside diameter of $\frac{1}{4}$-in. pipe = 0.54 in. (Table VII, page 530).

Area of a 54-in. circle = 2290.2 sq. in. (Table XV, page 539).

Area of pipe is 0.22902 sq. in. Place decimal point correctly.

PROBLEMS

Use the tables throughout the text and in the Appendix.

1. What is the pitch of a $\frac{3}{4}''$ N.C. tap?

2. What is the pitch of a No. 10 N.F. tap?

3. What is the size of drill for a $\frac{7}{8}''$ N.C. tap?

4. What is the proper number of r.p.m. for a $\frac{5}{8}''$ twist drill at 60 f.p.m.?

5. What is the next larger 64th inch drill than an R drill?

6. What is the next larger 64th inch drill than a No. 50 drill? How much larger is it?

7. What size drill should you turn at 340 r.p.m. when the number of f.p.m. is 100?

8. What is the diameter of No. 40 gauge steel wire?

9. What gauge is a copper wire 0.016″ diameter if Stubs gauge is used?

10. What is the decimal equivalent of $\frac{29}{64}''$?

11. Change 65 mm. to inches.

12. How many millimeters are there in $\frac{5}{8}''$?

13. How thick is No. 30 gauge U.S. Standard steel plate? Change this decimal to the lowest common fraction.

14. Which is larger and how much larger, No. 20 gauge wire, Stubs gauge, or No. 20 gauge wire, steel wire gauge?

15. What are the inside and outside diameters of $\frac{3}{4}''$ standard pipe?

16. What is the square root of 9.8?

17. What is the thickness of No. 18 gauge sheet zinc?

18. What diameter to the nearest thousandth of an inch should you turn a rod if the drawing calls for 55 mm?

19. What is the diameter of No. 38 gauge music wire?

20. At what r.p.m. should you run a $\frac{7}{8}''$ twist drill to give 30 f.p.m.?

21. What letter drills have exact fractional equivalents?

22. What pipe has the greater inside diameter, $\frac{3}{8}''$ standard pipe or $\frac{3}{4}''$ double-extra-strong pipe?

23. What is the taper per foot when the included angle is $14°15'$?

24. Find the size of drill for a taper tap and the pitch for $1\frac{1}{2}''$ pipe.

25. What is the included angle when the taper per foot is $\frac{7}{8}''$?

26. A drawing reads 0.36 by 0.24 meter. Change these dimensions to inches, and find the area in square inches.

27. What is the taper per inch from the center line if the taper per foot is 2″?

28. Convert 2500 kg. to pounds.

29. A portable electric drill runs at 1000 r.p.m. List all the drill sizes at each f.p.m. that would be within 10% of the values given in Table XI, page 533.

30. Look up the specific gravity of cast copper.

31. Find one-fourth the area of a circle 33″ in diameter.

32. Figure the weight of 1 cu. yd. of granite.

33. How much heavier is 10 cu. in. of lead than 10 cu. in. of iron?

34. What is the cross-sectional area of a $\frac{3}{8}$″ diameter rod?

35. Determine the weight of 8 cu. in. of mild steel.

36. Find the weight of 1000 bd. ft. of dry-pine lumber.

37. How much higher is the melting point of mild steel than that of gray cast iron?

38. Look up the circumference of a 4.8″ diameter shaft.

39. What weight will a 2-in.-diameter wrought-aluminum rod support?

40. What is the weight of a gray-iron casting 3″ × 4″ × 7″?

REVIEW OF DECIMALS AND FRACTIONS

Change each of the following fractions to the nearest sixteenths:

1. $\frac{1}{2}$	**2.** $\frac{1}{3}$	**3.** $\frac{1}{4}$
4. $\frac{1}{5}$	**5.** $\frac{1}{6}$	**6.** $\frac{1}{7}$
7. $\frac{1}{8}$	**8.** $\frac{2}{3}$	**9.** $\frac{1}{9}$
10. $\frac{3}{5}$	**11.** $\frac{3}{7}$	**12.** $\frac{3}{9}$
13. $\frac{3}{11}$	**14.** $\frac{3}{13}$	**15.** $\frac{3}{21}$

Change each of the following fractions to the nearest sixty-fourths:

16. $\frac{3}{11}$	**17.** $\frac{6}{11}$	**18.** $\frac{9}{11}$
19. $\frac{4}{15}$	**20.** $\frac{7}{15}$	**21.** $\frac{11}{15}$
22. $\frac{5}{13}$	**23.** $\frac{7}{13}$	**24.** $\frac{9}{13}$
25. $\frac{11}{13}$	**26.** $\frac{5}{12}$	**27.** $\frac{7}{12}$
28. $\frac{10}{12}$	**29.** $\frac{11}{12}$	

Change each of the following fractions to the nearest thirty-seconds:

30. $\frac{7}{11}$ **31.** $\frac{7}{9}$ **32.** $\frac{7}{13}$
33. $\frac{7}{15}$ **34.** $\frac{5}{11}$ **35.** $\frac{5}{9}$
36. $\frac{5}{13}$

Divide each of the following dimensions by 5, changing the decimal fractions to the nearest sixteenths:

37. 7″ **38.** 9″ **39.** 11″
40. 14″ **41.** 16″ **42.** 6″
43. 8″ **44.** 12″ **45.** 13″
46. 15″

Divide each of the following dimensions into 7 equal parts, changing the decimal fractions to the nearest thirty-seconds:

47. 13″ **48.** 25″ **49.** 26″
50. 18″ **51.** 16″

Divide each of the following dimensions into 9 equal parts, changing the decimal fractions to the nearest eighths:

52. 19″ **53.** 26″ **54.** 33″
55. 28″ **56.** 43″

57. Change each of the following decimals to the nearest sixty-fourth: 0.343, 0.739, 0.951, 0.014.

58. Change to the nearest thirty-second: 0.246, 0.734, 0.562, 0.295, 0.711, 0.799.

59. Change to sixty-fourths: 0.666, 0.734, 0.246, 0.719.

60. Change to thirty-seconds: 0.666, 0.734, 0.246, 0.719.

61. Change to sixteenths: 0.666, 0.734, 0.246, 0.719.

62. How do you account for the different common fractions for the same decimals in Probs. 59, 60, and 61?

63. In each of Probs. 59, 60, and 61, which common fraction is the closest approximation to the decimal fraction?

CHAPTER 4

POWERS AND ROOTS

DEFINITIONS

A number is **squared** when it is used as a factor twice. This is also called raising a number to its second **power.** For example, 5×5 is 5 raised to the second power, written 5^2, read "five squared." Likewise, $5 \times 5 \times 5$ is 5 raised to the third power, written 5^3, read "five cubed." 5^4 is read "five to the fourth power."

A root is the opposite of a power. It tells what number was raised to the power to give the number whose root is to be found. The cube root of 125 is 5, since $5 \times 5 \times 5 = 125$, or 5 is the number that was cubed to get 125. The square root of 25 is 5, since $5 \times 5 = 25$, or 5 was squared to get 25. The process of getting the equal factors to find the square root of a number is called extracting the square root.

The conventional sign indicating square root is $\sqrt{}$. The sign is placed over the number whose root is to be found, thus: $\sqrt{3}$, $\sqrt{25}$. This sign is called the square-root or **radical sign.**

To indicate a root other than square root, a small number is placed in the radical sign, as $\sqrt[3]{125}$, meaning "Find the cube root of 125."

Not all numbers are perfect squares. The square roots of numbers that are not perfect squares can be only approximate. Thus, $\sqrt{3} = 1.732$, when carried out to three decimal places. This is accurate enough for most purposes.

USING POWERS AND ROOTS

There are many uses for square roots. Several uses will be given in the problems following. Many other uses will present themselves in problems arising in the shop.

Tables giving the powers and roots of numbers are found in most handbooks. A simple table giving the powers and roots of numbers from 1 to 100 is given on page 528.

Example 1.—What is 17 cubed?

$17^3 = 4913$ In the table of powers and roots (page 528), read down the left-hand column marked Number to the number 17. Read the figure opposite the 17 in the column marked Cube.

Example 2.—What is 4.7 squared?

$4.7^2 = 22.09$ In the table of powers and roots (page 528), read down the left-hand column marked Number to 47. Opposite the 47, read the number 2209 in the column marked Square. Since 4.7×4.7 has two decimal places in the product, there must be two decimal places in the square.

Example 3.—Find $(\frac{2}{5})^2$.

$(\frac{2}{5})^2 = \frac{2}{5} \times \frac{2}{5} = \frac{4}{25}$ The parentheses indicate that the quantity inside the parentheses is to be used as a factor twice.

EXERCISES

Raise the following to the powers indicated. When you can, use the table on page 528.

1. 5^2 **2.** 9^2 **3.** 7^3 **4.** 20^2 **5.** 30^3 **6.** 2^6 **7.** 9^4
8. 25^2 **9.** 3^4 **10.** 6^4 **11.** 4^4 **12.** 11^3 **13.** 12^3 **14.** 3^5

15. 37^2 **16.** 75^3 **17.** 20^5 **18.** 68^2 **19.** 37^2 **20.** 11^5
21. $(\frac{5}{6})^3$ **22.** $(\frac{2}{3})^2$ **23.** $(\frac{1}{3})^3$ **24.** $(\frac{3}{5})^3$ **25.** $(\frac{3}{4})^2$ **26.** $(\frac{5}{8})^2$
27. $(\frac{4}{7})^3$ **28.** $(\frac{4}{11})^2$ **29.** $(\frac{1}{4})^4$ **30.** $(\frac{3}{8})^4$ **31.** $(\frac{1}{8})^5$ **32.** $(\frac{3}{7})^3$
33. $\frac{3^2}{5^2}$ **34.** $\frac{3^3}{8^2}$ **35.** $\frac{1^5}{4^2}$ **36.** $\frac{5^4}{9^3}$ **37.** $\frac{8^4}{25^2}$ **38.** $\frac{7^3}{11^3}$

39. $(1.3)^2$ **40.** $(2.5)^3$ **41.** $(0.7)^2$ **42.** $(0.05)^2$ **43.** $(0.22)^2$
44. $(2.02)^2$ **45.** $(1.1)^3$ **46.** $(1.5)^4$ **47.** $(0.363)^2$ **48.** $(0.006)^2$
49. $(0.004)^3$ **50.** $(0.04)^4$ **51.** $(0.05)^5$ **52.** $(11.1)^2$ **53.** $(1.11)^2$

Example 4.—$7^2 + 13^2 = ?$

$7^2 = 49$	Use the table of powers and roots (page 528)
$13^2 = 169$	to find each term separately
218	Adding.

Example 5.—$\sqrt{7921} = ?$

$\sqrt{7921} = 89$ You will not be able to find the square root of 7921 by looking first in the number column of the table, because that column lists numbers only up to 100. Find the column headed Square. Read down the column until you find 7921. The number of which 7921 is the square will be found in the column headed Number.

Example 6.—$\sqrt{\frac{33}{67}} = ?$

$\sqrt{\frac{33}{67}} = \frac{5.745}{8.185} = .701$ Find the square roots of the numerator and the denominator separately; then divide.

EXERCISES

Perform the indicated operations.

1. $6^2 + 4^2$ **2.** $9^2 + 11^2$ **3.** $16^2 + 12^2$ **4.** $8^2 + 3^2$
5. $11^2 - 7^2$ **6.** $7^3 - 4^2$ **7.** $18^2 - 5^3$ **8.** $11^3 + 15^2$
9. $3^2 \times 4^2$ **10.** $5^3 \times 6^2$ **11.** $9^2 \times 3^2$ **12.** $4^2 \times 2^2 \times 3^2$
13. $7^2 \div 3^2$ **14.** $15^2 \div 5^2$ **15.** $24^3 \div 4^2 \div 2^2$

16. $\sqrt{64}$ **17.** $\sqrt{144}$ **18.** $\sqrt{1}$ **19.** $\sqrt[3]{125}$ **20.** $\sqrt{16}$
21. $\sqrt{81}$ **22.** $\sqrt{625}$ **23.** $\sqrt[3]{16}$ **24.** $\sqrt[3]{14}$ **25.** $\sqrt[3]{35}$
26. $\sqrt[3]{88}$ **27.** $\sqrt{\frac{16}{81}}$ **28.** $\sqrt{\frac{25}{100}}$ **29.** $\sqrt{\frac{89}{64}}$ **30.** $\sqrt[3]{\frac{35}{13}}$
31. $\sqrt[3]{\frac{27}{15}}$ **32.** $\sqrt{\frac{19}{33}}$ **33.** $\sqrt[3]{\frac{19}{35}}$ **34.** $\sqrt{\frac{61}{72}}$ **35.** $\sqrt[3]{\frac{61}{72}}$
36. $\sqrt{\frac{29}{9}}$ **37.** $\sqrt{324}$ **38.** $\sqrt{529}$ **39.** $\sqrt{1600}$ **40.** $\sqrt{8836}$
41. $\sqrt{225}$ **42.** $\sqrt[3]{729}$ **43.** $\sqrt[3]{1728}$ **44.** $\sqrt[3]{24,389}$
45. $\sqrt{1369}$ **46.** $\sqrt{1225}$ **47.** $\sqrt{\frac{329}{1600}}$ **48.** $\sqrt{\frac{25}{225}}$ **49.** $\sqrt{\frac{441}{576}}$

50. $\sqrt{\frac{484}{5184}}$ **51.** $\sqrt{\frac{1296}{1444}}$ **52.** $\sqrt{\frac{2401}{10{,}000}}$ **53.** $\sqrt{\frac{9216}{256}}$ **54.** $\sqrt{\frac{8100}{2500}}$

55. $\sqrt[3]{\frac{27}{125}}$ **56.** $\sqrt[3]{\frac{216}{729}}$ **57.** $\sqrt[3]{\frac{12{,}167}{64}}$ **58.** $\sqrt[3]{\frac{74{,}088}{91{,}125}}$

59. $\sqrt{64} - \sqrt{25}$ **60.** $\sqrt{841} + \sqrt{225}$ **61.** $\sqrt{1089} + \sqrt{1681}$

62. $\sqrt[3]{512} - \sqrt{25}$ **63.** $\sqrt[3]{59{,}319} - \sqrt{441}$

64. $\sqrt{2401} \div \sqrt{121}$ **65.** $\sqrt{256} + \sqrt{7^2}$

66. $\sqrt{2025}$ **67.** $\sqrt[3]{1728} + \sqrt{3^2}$

68. $\sqrt{225} \times \sqrt{169}$ **69.** $\sqrt{576} \div \sqrt{64}$

70. $\sqrt[3]{1331} \times \sqrt[3]{343}$

FINDING SQUARE ROOTS BY ARITHMETIC

To find the square roots of decimals or numbers containing decimals, we shall use the arithmetical method. This method may be used when the square root of a number is not found easily or sufficiently accurately by use of the table.

The process will be explained by means of an example. Each numbered paragraph represents an operation to be performed before proceeding to the next operation.

Example 1.—Find the square root of 2937.64 by the arithmetical method.

```
        5    4. 2
     √29′ 37′.64

      25
100   4 37
 +4
104
 ×4   4 16
1080    21 64
 +2
1082
 ×2     21 64
```

(1) Beginning at the decimal point separate the number into groups of two figures, working both ways from the decimal point. For example, 2937.64 is separated as 29′37′.64. In the root, there will be one number for each group. The decimal point in the root will be directly above the decimal point in the number.

(2) Select the largest number that when squared is contained in the first, or left-hand, group. Write this number, in this case, 5, in the answer over the first group.

(3) Square the root. $5^2 = 25$.

(4) Subtract this first square, 25, from the first group, 29. Then bring down the next group, 37. The new dividend is 437.

(5) Multiply the root 5 by 2, getting 10. Since the next number in the root is one place over to the right, we must allow a zero for one more place, making the trial divisor 100. To find the number belonging in this place, we divide $100 + ?$ into 437, getting approximately 4. Hence the correct divisor is 104, and the 4 is the new number in the root. Place the 4 above its group, 37, and put the 4 also in its place in the divisor.

(6) Multiply the divisor, 104, by the new number in the root, 4. $104 \times 4 = 416$.

(7) Subtract.

(8) Repeat operations (4) to (7) to find each successive number in the root.

Repeating (4): $437 - 416 = 21$. Bringing down the next group of numbers, 64, gives 2164.

Repeating (5): $2 \times 54 = 108$. Filling the empty place with a zero gives the trial divisor, 1080. $2164 \div (1080 + ?)$ equals approximately 2. Place the 2 above its group, 64, and put the 2 also in its place in the divisor.

Repeating (6): Multiply the divisor by the last figure in the root.

Repeating (7): Subtract.

Since there is no remainder and there are no more groups of numbers to be brought down, the square root is 54.2.

Example 2.—Find the square root of 312.649 to two decimal places. (In the solution, the numbers in parentheses correspond to the numbers of the steps shown in Example 1.)

	1 7. 6 8
	$\sqrt{3'12'64'90}$
	1
20	2 12
+7	
27	
×7	1 89
340	23 64
+6	
346	
×6	20 76
3520	2 88 90
+8	
3528	
×8	2 82 24
	6 66

(1) Mark off groups of two figures each, and add a zero to complete the last group.

(2) Select the first root, in this case, 1.

(3) $1^2 = 1$.

(4) Subtract.

(5) $2 \times 1 = 2$. Allow a zero to fill up the last place in the trial divisor. $212 \div (20 + ?)$ equals approximately 7. Place the 7 above its group, 12, and put the 7 also in its place in the divisor.

(6) Multiply by the 7 just put in the root.

(7) Subtract.

Repeat operations (4) to (7) to find each successive number in the root.

In this problem, there is a remainder. Disregard the remainder if it is less than half the last divisor. If it is more than half the last divisor, increase the last number of the root by 1. Where there is a remainder, the square root is only approximate. The approximate root to two decimal places in this case is 17.68.

Study the changes in the roots as the decimal changes. Since the $\sqrt{100} = 10$, multiplying the number by 100 multiplies the square root by 10.

0. 2 5	2. 5	2 5	2 5 0
$\sqrt{0.06'25}$,	$\sqrt{6.25}$,	$\sqrt{6'25}$,	$\sqrt{6'25'00}$
0. 7 9	7. 9	7 9.	7 9 0. 6
$\sqrt{0.62'50}$,	$\sqrt{62.50}$,	$\sqrt{62'50.}$,	$\sqrt{62'50'00.00}$

EXERCISES

Find the square root of each of the following numbers. Notice that the same figures appear in each part of each example, but the decimal point is in a different place in each.

	a	*b*	*c*	*d*	*e*	*f*
1.	324	3.24	32.4	0.324	0.0324	32,400
2.	2304	23.04	0.2304	230.4	2.304	0.002304
3.	7569	756.9	7.569	75.69	0.7569	75,690
4.	4356	4.356	43.56	0.4356	0.04356	0.004356
5.	1225	12.25	1.225	.1225	.01225	.001225
6.	961	9.61	.961	96.1	9610	96,100
7.	1849	1.849	18.49	.1849	.01849	18,490
8.	441	.0441	.00441	44.1	4410	4.41
9.	11.56	1156	1.156	.1156	115.6	115,600
10.	8281	8.281	.8281	82.81	828.1	0.8281
11.	64	6400	640	6.4	.64	.0064
12.	196	1.96	19.6	.196	.0196	1960
13.	576	57.6	5.76	.576	.0576	5760
14.	289	.289	2.89	28.9	2890	.0289
15.	.0361	.361	3.61	36.1	361	3610
16.	256	25.6	2.56	.256	.0256	.00256
17.	324	3.24	3240	.324	3,240,000	
18.	42.25	.4225	422.5	42,250	0.04225	
19.	6084	60.84	.6084	0.06084	608.4	
20.	746	7.46	.746	.0746	74,600	

SQUARE ROOTS OF NUMBERS FROM TABLES

In Table XVII, pages 541 to 544, the first two figures of the numbers whose root we want are given in the left-hand column marked N, and the third figure is given at the top.

For example, on page 541 the square root of 2.0 is 1.414, this being opposite the number 2.0 and under 0. The square root of 2.18 is 1.476, this being opposite the number 2.1 and under 8, and so on. Likewise, on page 543 the square root of 35.6 is 5.967, this being opposite 35 and under 6.

The right-hand columns headed 1, 2, 3, . . . 9 show the number to be added to the root when the number whose root is desired contains a fourth figure. For example, we find $\sqrt{4.63} = 2.152$; but if we wish to find $\sqrt{4.637}$ we look further to the right-hand column under 7 and find 2. This means that we must increase the last number in the root by 2, that is, in this case we should add 0.002 to 2.152 giving $\sqrt{4.637} = 2.154$.

Consider the following examples:

Example 1.—$\sqrt{72.44} = 8.509 + 0.002 = 8.511$
Example 2.—$\sqrt{7'24.40} = 26.91 + 0.01 = 26.92$
Example 3.—$\sqrt{0.72'44} = 0.8509 + 0.0002 = 0.8511$

To find the square root of a number outside the range of the table separate the number into groups of two figures, as shown in Examples 2 and 3 above. As explained on page 95, moving the decimal two places in the number moves it only one place in the root.

EXERCISES

Using Table XVII on pages 541 to 544, find the square root of each of the following numbers:

	a	*b*	*c*	*d*	*e*	*f*
1.	729	7.29	72.9	0.729	0.0729	7290
2.	6	60	600	6000	0.06	0.006
3.	6561	6.561	65.61	656.1	0.6561	0.06561
4.	529	5290	52900	0.529	5.29	0.0529
5.	961	96.1	0.961	9.61	96,100	0.0961
6.	265	26.5	2650	0.265	0.0265	26,500
7.	6724	6.724	67.24	672.4	0.6724	0.06724
8.	792	7920	79200	79.20	0.792	0.0792
9.	12.96	129.6	1.296	12960	0.1296	0.01296
10.	3.55	35.5	3550	0.355	35,500	0.0355
11.	9801	9.801	98.01	980.1	0.9801	0.09801
12.	279	2790	2.79	27.9	27,900	0.279
13.	1.69	16.9	169	1690	0.169	0.0169
14.	54.2	542	5420	5.42	0.542	0.0542
15.	2198	2.198	21.98	219.8	0.2198	0.02198
16.	166	16.6	1.66	1660	16,600	0.166
17.	841	8410	8.41	84.1	0.841	0.0841
18.	7744	7.744	77.44	774.4	77,440	0.07744
19.	8500	850	8.5	0.85	85,000	0.085
20.	350	35	3500	3.5	35,000	0.035

CHAPTER 5

PERCENTAGE

DEFINITIONS

Percentage is a general term applied to arithmetic in which the whole, divided into 100 equal parts, is used as the standard of measure. **Per cent** is another name for **hundredths.**

The sign for per cent is %. It is easy to see that the per cent sign indicates division by 100, for it has the line of a fraction and two zeros.

When this sign is placed after a number, it means that that number of hundredth parts of the whole are to be taken. For example, 6% means that six hundredths of the whole are to be taken.

CHANGING THE FORM OF PER CENTS

In the actual working of a percentage problem the sign % cannot be used. Therefore, it is necessary to express the per cent either as a common fraction or as a decimal; for example, 5% may be written as $\frac{5}{100}$ or .05. You will observe that $\frac{5}{100}$ reduces to $\frac{1}{20}$.

The following is a list of fractional equivalents frequently used in percentage problems, where it is more convenient to work with the per cent expressed as a fraction than as a decimal. They were obtained by changing the per cent to a fraction and then reducing the fraction to lowest terms. These equivalents should be memorized.

$5\% = \frac{1}{20}$	$80\% = \frac{4}{5}$	$62\frac{1}{2}\% = \frac{5}{8}$
$10\% = \frac{1}{10}$	$90\% = \frac{9}{10}$	$87\frac{1}{2}\% = \frac{7}{8}$
$30\% = \frac{3}{10}$	$25\% = \frac{1}{4}$	$6\frac{1}{4}\% = \frac{1}{16}$
$70\% = \frac{7}{10}$	$50\% = \frac{1}{2}$	$33\frac{1}{3}\% = \frac{1}{3}$

$20\% = \frac{1}{5}$	$75\% = \frac{3}{4}$	$66\frac{2}{3}\% = \frac{2}{3}$
$40\% = \frac{2}{5}$	$12\frac{1}{2}\% = \frac{1}{8}$	$16\frac{2}{3}\% = \frac{1}{6}$
$60\% = \frac{3}{5}$	$37\frac{1}{2}\% = \frac{3}{8}$	$83\frac{1}{3}\% = \frac{5}{6}$

CHANGING A PER CENT TO A DECIMAL

Because per cent means hundredths, or division by 100, using the rule for dividing by multiples of ten results in the following rule for changing a per cent to a decimal.

Rule 1.—To change from a per cent to a decimal, remove the per cent sign and move the decimal point two places to the left.

Example 1.—Change 6.5% to a decimal.

6.5% = 0.065	To move the decimal point two places to the left, one zero must be prefixed.

Example 2.—Change 0.012% to a decimal.

0.012% = 0.00012	Here two zeros must be prefixed.

CHANGING A DECIMAL TO A PER CENT

In order to change from a decimal to a per cent, just the opposite operation is employed. Note that when the per cent sign is removed, it is replaced by its equivalent hundredths as a decimal; when the hundredths as a decimal disappear, the per cent sign appears.

Rule 2.—To change a decimal to a per cent, move the decimal point two places to the right and add a per cent sign.

Example 1.—Change 0.363 to a per cent.

0.363 = 36.3%	Moving the decimal point two places to the right and adding the per cent sign.

Example 2.—Change 0.0065 to a per cent.

0.0065 = 0.65%	Here the answer is less than 1%.

CHANGING A FRACTION TO A PER CENT

Rule 3.—To change a fraction to a per cent, change the fraction to a decimal containing two decimal places, then apply Rule 2.

Example.—Change $\frac{3}{7}$ to a per cent.

$\frac{3}{7} = 7\overline{)3.00}$ $.42\frac{6}{7} = 42\frac{6}{7}\%$	Applying Rule 3, then Rule 2.

Note.—In some problems the answer should be expressed with a decimal remainder. For example, the answer of the preceding example can be expressed exactly as $42\frac{6}{7}\%$ or rounded off to 42.9% or 42.86%, depending on the degree of accuracy needed for the work being done.

EXERCISES

Express each of the following per cents as a decimal:

1. $2\frac{1}{2}\%$	**2.** $5\frac{1}{2}\%$	**3.** $8\frac{1}{2}\%$	**4.** $12\frac{3}{8}\%$
5. $7\frac{1}{9}\%$	**6.** $6\frac{2}{3}\%$	**7.** $6\frac{1}{4}\%$	**8.** $5\frac{3}{4}\%$
9. $9\frac{1}{2}\%$	**10.** 125%	**11.** 190%	**12.** 216%

Express each of the following decimals as a per cent:

13. $0.33\frac{1}{3}$	**14.** 0.875	**15.** 0.3633	**16.** 0.5625
17. 0.987	**18.** 0.3125	**19.** $0.63\frac{1}{2}$	**20.** $0.43\frac{3}{4}$
21. $0.11\frac{1}{9}$	**22.** 1.25	**23.** 16.375	**24.** 3.635

Express each of the following fractions to the nearest tenth of a per cent:

25. $\frac{1}{8}$	**26.** $\frac{3}{8}$	**27.** $\frac{5}{8}$	**28.** $\frac{7}{8}$
29. $\frac{1}{7}$	**30.** $\frac{2}{7}$	**31.** $\frac{4}{7}$	**32.** $\frac{6}{7}$
33. $\frac{1}{9}$	**34.** $\frac{2}{9}$	**35.** $\frac{5}{9}$	**36.** $\frac{7}{9}$
37. $\frac{1}{11}$	**38.** $\frac{2}{11}$	**39.** $\frac{7}{11}$	**40.** $\frac{4}{11}$

Express each of the following per cents as a common fraction, and reduce to lowest terms:

41. 2%	**42.** 6%	**43.** 8%	**44.** 16%
45. 20%	**46.** 30%	**47.** 40%	**48.** 58%
49. 32%	**50.** 36%	**51.** 48%	**52.** 72%
53. 38.5%	**54.** 125%	**55.** 1.75%	**56.** $62\frac{1}{2}\%$

Express each of the following as (1) a per cent, (2) a decimal, and (3) a fraction or a mixed number:

57. 8%	**58.** 0.063	**59.** 7.5%	**60.** $\frac{3}{4}$
61. 0.925	**62.** $\frac{5}{8}$	**63.** $66\frac{2}{3}\%$	**64.** $0.02\frac{1}{2}$
65. $\frac{3}{7}$	**66.** $\frac{3}{4}\%$	**67.** $0.12\frac{1}{2}$	**68.** $\frac{64}{64}$
69. 650.00	**70.** 0.15%	**71.** $\frac{9}{16}$	**72.** $0.33\frac{1}{3}$
73. 125%	**74.** 0.375	**75.** $\frac{11}{42}$	**76.** $\frac{1}{2}\%$
77. 0.333	**78.** $3\frac{3}{8}$	**79.** $3\frac{5}{8}$	**80.** 0.5364
81. $1\frac{5}{9}$	**82.** 87.5	**83.** $3\frac{1}{4}\%$	

TERMS USED IN PERCENTAGE

The **rate** (R) is the number of hundredth parts taken; it is expressed as per cent. In 6% of 50 = 3, 6% is the rate.

The **percentage** (P) is the result of taking hundredth parts. In 6% of 50 = 3, 3 is the percentage, or the part taken.

The **base** (B) is the whole, or base, on which the per cent is reckoned. In 6% of 50 = 3, 50 is the base, or whole.

The **amount** (A) is the sum of the base and the percentage For 6% of 50 = 3, 50 + 3, or 53, is the amount.

The **difference** (D) is the remainder when the percentage is subtracted from the base. For 6% of 50 = 3, 50 − 3, or 47, is the difference.

THE THREE TYPES OF PERCENTAGE

There are three types of percentage problems, namely:

Type 1. Given the base and the rate, to find the percentage. This type makes it necessary to find the percentage, or part taken, of a given number.

Type 2. Given the base and the percentage, to find the rate. This type makes it necessary to find what per cent one number (the part) is of another (the whole).

Type 3. Given the rate and the percentage, to find the base. This type makes it necessary to find a whole number, when the percentage and the rate are known.

Note.—The percentage formulas given on the following pages make the solving of percentage problems easy after you have identified the known and unknown quantities. But these formulas are not automatic. In groups of mixed problems, you must decide which quantity is the base. which is the rate, and which is the percentage. Having done this, you must decide which type of percentage the problem involves. Then you are ready to use the right formula. No formula is a substitute for thinking a problem through.

In percentage formulas use the following abbreviations: P = percentage, B = base, R = rate, A = amount, D = difference.

PERCENTAGE: TYPE 1

Rule 4.—To find the percentage, multiply the base by the rate.

Formula. $P = BR.$

Rule.—To find the amount, add the base and the percentage.

Formula. $A = B + P.$

Rule.—To find the difference, subtract the percentage from the base.

Formula. $D = B - P.$

Example 1.—In a shop with 50 lathes, 6% are shut down. How many are idle?

$B = 50, \quad R = 6\%, \quad P = ?$

$P = BR$	Formula
$P = 50 \times 0.06$	Substituting. Change the 6% to a decimal before substituting.
$P = 3.00$	

Ans. There are 3 lathes idle.

Example 2.—A company employs 600 men. It is decided that 25% more men are needed. How many will be hired? How many men will be on the new pay roll?

$B = 600, \quad R = 25\%, \quad P = ? \quad A = ?$

$P = BR$	Here it is necessary to use two formulas. Find P first.
$P = 600 \times \frac{1}{4}$	Substituting
$P = 150$	

150 men are hired.

$A = B + P$	Formula
$A = 600 + 150$	Substituting
$A = 750$	

750 men will be on the new pay roll.

EXERCISES

Solve each of the following exercises by changing the per cent to a fraction and then multiplying. You should know the fractional equivalent of each of these per cents.

1. 5% of 1000
2. 50% of 1000
3. $6\frac{1}{4}\%$ of 800
4. 10% of 750
5. 25% of 1200
6. $16\frac{2}{3}\%$ of 180
7. 20% of 555
8. $12\frac{1}{2}\%$ of 480
9. $33\frac{1}{3}\%$ of 500
10. 20% of 500
11. $37\frac{1}{2}\%$ of 1200
12. 40% of 1200
13. 50% of 1200
14. 60% of 1200
15. $62\frac{1}{2}\%$ of 1200
16. $66\frac{2}{3}\%$ of 1200
17. 70% of 1200
18. 75% of 1200
19. 80% of 1200
20. $87\frac{1}{2}\%$ of 1200
21. $37\frac{1}{2}\%$ of 800
22. 40% of 2500
23. 70% of 2000
24. $62\frac{1}{2}\%$ of 800
25. $66\frac{2}{3}\%$ of 1500
26. $87\frac{1}{2}\%$ of 1600
27. 80% of 2000
28. 60% of 2000
29. 75% of 7.2
30. 90% of 90

In each of the following, change the per cent to a decimal and then find the percentage. Use $P = BR$.

31. 6% of 725
32. 24% of 1600
33. 16% of 750
34. 9% of 256
35. 12% of 729
36. 18% of 672
37. 11% of 585
38. 8% of 963
39. 46% of 562
40. 98% of 900
41. 15% of 785
42. 12.3% of 246
43. 3.3% of 789
44. 2.8% of 456
45. 7.5% of 759
46. 12.4% of 66
47. $1\frac{2}{3}\%$ of 48
48. $1\frac{1}{2}\%$ of 600
49. $\frac{1}{4}\%$ of 584
50. 0.005% of 648

51. 150% of 8500

52. 88% of 65.8

53. $6\frac{1}{4}$% of 77.77

54. $4\frac{3}{4}$% of 1.636

55. 125% of 1.6

56. 7.5% of 56.72

57. 6.3% of 5.432

58. 2.75% of 2.75

59. 276.5% of 700

60. 0.006% of 0.5625

PROBLEMS

Find the percentage.

1. Find 45% of 740 gal.

2. What is 2% of 250 lb.?

3. Find 3% of 8 bu.

4. Find $62\frac{1}{2}$% of 72 men.

5. Find one-half of 1% of 216 tons.

6. How much is 0.75% of 386 lb.?

7. A car contains a load of 59 tons of zinc ore, which is $3\frac{1}{2}$% zinc. Determine the zinc content in the load of ore.

8. A tank held 5000 gal. of oil from which $12\frac{1}{2}$% was drawn off. How many gallons were drawn off? How many remained in the tank?

9. A loaded freight car weighs 78,000 lb. If 85% of this represents the load, what is the weight of the load?

10. A shop has 43,560 sq. ft. of floor space. This is to be increased 25%. Find the amount to be added. Find the amount of floor space after the addition.

11. A wood pattern weighs 6% as much as the casting. If the casting weighs 725 lb., what is the weight of the pattern?

12. Out of a lot of 350 castings, 30% were rejected. Find the number rejected.

13. An engine listed at 87.5 hp. is only $87\frac{1}{2}$% efficient. What horsepower does it deliver?

14. A car salesman received 15% commission on his new-car sales. What is his commission on a car retailing at $2050.85?

15. With a screw jack the theoretical weight that can be raised is 10 tons. Because of friction, the jack is only 55% efficient. What actual weight can be raised?

16. A gasoline engine is found to be only 78% efficient. If it is rated at 120 hp., what actual horsepower does it develop?

17. A six-sheave pulley block is only 80% efficient. Find the actual load that can be raised if the theoretical load is 1320 lb.

18. The mechanical advantage of a machine was figured to be 12 to 1. The friction reduced this 5%. Find the actual mechanical advantage.

19. The shrinkage of molten iron when cooled is $1\frac{5}{8}$%. How many inches does a 4′6″ iron ingot shrink when cooled?

20. A bearing metal is composed of 58.5% copper and 41.5% zinc. Find the percentage of each in 750 lb. of the mixture.

21. A red brass is composed of 83% copper, 11% zinc, and 6% lead. Figure the percentage of each in 1 ton of red brass.

PROBLEMS

Find the percentage and the amount or the difference.

1. A line shaft revolves at 245 r.p.m. It is necessary to reduce this 20%. Determine the r.p.m. after the reduction. ($D = B - P$).

2. A pulley is 12.5 in. in diameter. The size is to be increased 15%. How large will it be after the increase? ($A = B + P$).

3. One plumber figured a job at $940.00. Another plumber bid 25% less. What was the second bid?

4. The list price of an automobile is $2756.60. The dealer will allow a shop owner a cash discount of $7\frac{1}{2}$%. How much is the discount? How much must a shop owner pay for the car?

5. If $12\frac{1}{2}$% of a sharp thread 0.532 in. high is removed at both the crest and the root to make the flats of a National thread, find the depth of the National thread.

6. Seventy-five pounds of brass contain 45% zinc and the balance copper. Figure the number of pounds of each.

7. One man did a job in $12\frac{1}{2}$ days. Two men working together can complete it in 40% as much time. How long will it require both men to do the job?

8. A gas engine develops 90 hp. at 2000 r.p.m. At 2400 r.p.m., its horsepower is increased 20%. What is the horsepower at 2400 r.p.m.?

9. A man's salary formerly was $4150 per year. Last year he received a raise of 10%. What was his salary last year?

This year his salary was cut 10% below what he received last year. What is his salary now?

10. A shipment of flooring was billed at $548.00, but it was damaged in transit. An allowance of 15% was made for the damage. What is the net amount due on the bill?

11. A bill of lumber amounted to $1248.60. If I get a discount of $16\frac{2}{3}$%, what sum must I pay?

12. A rough casting weighed 500 lb. The machinist removed 18% of this. What did the finished casting weigh?

13. An automobile-parts house buys brake lining at 69¢ per foot. If it is marked up 40% and sold at the nearest whole cent per foot, what is the selling price?

14. An automobile floodlight costs $2.40 to manufacture. The retail selling price is marked up to 300% of this. If the jobber gets 40% off the retail selling price, what does he pay? How much does the manufacturer make on this sale?

15. By using a lever, a man increases his lifting ability by $566\frac{2}{3}$%. If his normal ability is 185 lb., how many pounds can he lift with the lever?

16. A casting expands 0.003% when heated. How long does the casting become after heating if it is 3′3″ in length when normal?

17. An iron support 8′ long stretches one-hundredth of 1 per cent when loaded. What part of an inch is the stretch? How long is the support when loaded?

Discounts

A reduction from a price or an amount is called a **discount.** Discounts are expressed as a rate, or a per cent. The per cent of discount is called the rate of discount.

Rule.—To find the discount, multiply the list price, or base, by the rate of discount.

Formula. $P = BR$.

The **net price** is the price after the discount has been taken off the original, or *list*, price.

Rule.—To find the net price, subtract the amount of discount from the original, or list, price.

Formula. $D = B - P$.

Example 1.—A machine listed at \$625.00 was sold at a discount of 15%. Find the discount. Find the net price.

$B = 625, \quad R = 15\%, \quad P = ? \quad D = ?$	
$P = BR$	This is type 1 percentage, and two formulas will be needed.
$P = 625 \times .15$	Substituting
$P = 93.75$	
The discount is \$93.75.	
$D = B - P$	Second formula needed
$D = 625 - 93.75$	Substituting
$D = 531.25$	
The net price is \$531.25.	

Example 2.—An automobile tire is listed at \$22.50. Find the discount and the net price if the discount is 20%.

$B = 22.50, \quad R = 20\%, \quad P = ? \quad D = ?$	
$P = BR$	This is type 1 percentage. Use this formula first.
$P = \frac{\overset{4.50}{\cancel{22.50}}}{1} \times \frac{1}{\cancel{5}}$	Substituting (for ease in computation, $\frac{1}{5}$ is used instead of 0.20)
$P = 4.50$	
The discount is \$4.50	
$D = B - P$	Second formula needed
$D = 22.50 - 4.50$	Substituting
$D = 18.00$	
The net price is \$18.00.	

Chain Discount

In the wholesale business, two or more discounts are often allowed on the same purchase if cash is paid within a limited time, or for large quantity sales, or for a number of other reasons within the trade. This is sometimes called chain discount.

When a series of discounts is given, the discounts must be taken off in turn, separately. The net amount cannot be found by adding the rates of discount and then using the sum as a single rate of discount.

Example 1.—A lathe is listed at \$600.00. If discounts of 20% and 10% are allowed, what is the net amount?

Solution 1.

To find the first discount, $B = 600$, $R = 20\% = \frac{1}{5}$, $P = ?$ $D = ?$

$P = BR$	Four formulas are needed; this is the first.
$P = 600 \times \frac{1}{5}$	Substituting
$P = 120$	
$D = B - P$	Second formula needed
$D = 600 - 120$	Substituting
$D = 480$	

To find the second discount, $B = 480$, $R = 10\% = \frac{1}{10}$, $P = ?$ $D = ?$

$P = BR$	Third formula needed
$P = 480 \times \frac{1}{10}$	Substituting
$P = 48$	
$D = B - P$	Fourth formula needed
$D = 480 - 48$	Substituting
$D = 432$	

Net amount = \$432

Solution 2.

\$600 × .20 = \$120	First discount
\$600 − \$120 = \$480	First net amount
	The next discount, 10%, to be taken uses the first net amount, \$480, as base.
\$480 × .10 = \$48	Second discount. Note that the second discount is figured on the amount left after deducting the first discount.
\$480 − 48 = \$432	

The net amount is \$432.

Note.—In solving similar problems, use Solution 2 if it is easier for you.

PROBLEMS

1. Compute the net cost of a power grinder listed at \$115.00 if a discount of 15% is allowed.

2. An automobile is listed at $1960.00. Find the net cost if 25% discount is allowed.

3. A set of 8 new pistons is quoted at $3.40 each, with a discount of $12\frac{1}{2}$% allowed for cash payment. What is the cash price?

4. A drill press listed at $84.00 carried a discount of 20%. If an additional discount of 2% is allowed for cash, figure the cash price.

5. A machine listed at $746.00, less discounts of 20% and 10%, will net how much money?

6. A purchase of 6 vises, listed at $14.00 each, is discounted one-third. If an additional discount of 4% is allowed for cash, find the cash price.

7. A drafting machine is listed at $147.50. If school discounts of 20% and 10% are allowed, what will be the net price?

8. A truck was listed at $1000.00, less discounts of $\frac{1}{3}$ off and $\frac{1}{4}$ off. Determine the net price after taking advantage of these two discounts.

9. A washing machine was listed at $208.00. The price was discounted one-fourth. Later, a 10% discount was allowed. Find the net price.

10. A shear was listed at $75.50, less discounts of 20% and 20%. Find the net price if both discounts are taken.

11. A machine was listed at $600.00. One bidder offered a discount of 50%. Another offered discounts of 25% and 25%. Which was the better offer and by how much? Explain your results?

12. A house and lot were listed at $8000.00. The property was to be discounted 5% each day until sold. If it sold on the fourth day, what did the owner receive?

In each of the following, find the net amount after taking off all discounts:

13. $746.00 less 20% and 10%.

14. $221.00 less discounts of 25%, 10%, and 2%.

15. $748.00 less 25%, 10%, and 10%.

16. $1296.00 less discounts of 15%, 15%, and 15%.

17. $1000.00 less 50% and 50%.

18. $289.00 discounted 10%, 10%, and 10%.

19. $666.00 less one-half and one-third.

20. $595.00 less $33\frac{1}{3}$% and 10%.

Bills and Invoices

A typical wholesaler's bill, or **invoice**, is shown here.

LOS ANGELES
219 - 227 CENTRAL AVE.
PHONE TR. 0621

SAN FRANCISCO
656 - 676 TOWNSEND ST.
PHONE HE. 7550

DUCOMMUN
METALS and SUPPLY CO.

« INDUSTRIAL SUPPLIES
« TOOLS • METALS
« TUBULAR PRODUCTS

DUCOMMUN SINCE 1849
INCORPORATED 1907

LOS ANGELES
P. O. BOX 310 ARCADE STA.

PIONEER MERCANTILE COMPANY
BAKERSFIELD CALIFORNIA

INVOICE DATE July 12 1952

FOLLOWING AMOUNTS WILL APPEAR ON MONTHLY STATEMENTS AND ARE SUBJECT TO TERMS INDICATED ON ITEMS BELOW.

T	2% - 10th Prox.	X
E	½ of 1% - 10th Prox.	
R	1% - 10th Prox.	
M	Net Cash - 30 Days	
S		

Our Register No.	Your Order No.	Shipped Via
676	36457	Pacific Freight Lines

QUANTITY	DESCRIPTION	LIST OR NET PRICE	TOTAL
6 only 4" No 614 Athol Machinists vises @ $22.75 ea			$136.50
3 only 100 lb. Vulcan anvils @ 43¢ lb.			129.00
1 only Buffalo 10" bench drill @ $163.50			163.50
			429.00
Discount 25%			107.25
		net	321.75
less 2%			6.44
			$ 315.31

PAID
DATE July 20, 1952
BY B.A.A.
DUCOMMUN

We hereby warrant that there has been no violation on our part of any of the provisions of the Federal Fair Labor Standards Act insofar as the transaction represented by this invoice is concerned.

CLAIMS — For shortage, or defective material, must be made within 10 days.
CUT MATERIAL, when furnished correctly, cannot be returned for credit.

INTEREST will be charged on Past Due Accounts.
SALESMEN cannot authorize the return of merchandise, nor are they permitted to make any allowances.

Note the terms stated in the upper right-hand corner. The terms of this bill are "2%—10th Prox." This means 2% off if the bill is paid on or before the tenth of the following month. Since the bill was paid on the twentieth of the same month, the 2 per cent was deducted.

If the payment had not been paid by Aug. 10, the net amount of the bill would have been due.

PROBLEMS

Make out bills similar to the bill shown above for each of the following. Take off all discounts, and mark bills paid.

1. 6 Porter's No. 3 New-Easy bolt cutters @ $16.85 each.
4 Crescent No. 72 end nippers @ $3.85 each.
1 doz. Klein's pliers @ $2.90 each.
Discounts of 20% and 2%.

2. 1 No. 266 U.S. grinder complete @ $233.00.
1 No. 3 Scoggin's bench shear @ $155.00.
Discounts of 25%, 10%, and 2%.

3. 3 doz. No. 2 Champion ball peen hammers @ $24.50 per doz.
1 gross Simonds 10″ smooth mill files @ $9.65 per doz.
Discounts of 20% and 3%.

4. 1 doz. Starrett No. 94 combination square with 12″ blade @ $2.85 each, less 10%.
$\frac{1}{4}$ doz. Stanley No. 37-G 24″ levels @ $2.80 each, less $12\frac{1}{2}$%.
$\frac{1}{4}$ doz. Brown and Sharpe No. 59 micrometers @ $13.50 each, net. Terms: 2%—10th Prox.

5. 2 Little Giant No. 8 screw plates @ $55.60 each, less 10%.
3 Blackhawk No. 911L wrench kits @ $28.40 each, less $\frac{1}{3}$.
6 doz. Plomb No. 86 cold chisels @ $9.40 doz., less 25%.
Terms: 2%—10th Prox.

6. 1 Weber 8-in. bench grinder @ $145.00, less 10%.
1 Buffalo 10-in. Jr. bench drill with motor @ $163.50, less 5%.
1 Yankee No. 994 drill press vise @ $23.45, less 25%.
Additional discount of 5% for cash allowed on all items.

Interest

When money is lent, it is customary to require the borrower to pay for its use. The part paid, or the interest, is usually figured as a per cent of the money involved.

When a person buys anything and arranges to pay for it over a period of time, he is actually borrowing the money for the purchase. Because he does not pay the full price at the time the purchase is made, interest is included in the price he must pay. This is called installment buying. In installment buying, it is important to know the rate of interest being paid to ensure that not too large an amount of money is being spent for the privilege of extending the time of payment. High rates of interest are often hidden in what are apparently very reasonable terms of payment.

Terms Used in Interest

Interest (I) is the amount of money paid for the use of money.

In interest problems the base, that is, the money that earns interest, is called the **principal** (P).

The interest rate is called the **rate** (R). The rate is expressed as the per cent charged for 1 year's use of the money.

The **time** (T) in an interest problem is always expressed in terms of the years or parts of a year over which the money is being used.

The **amount** (A) in an interest problem is the sum of the principal and the interest. It is the amount of the loan plus the money paid for its use.

To Find the Interest for One Year

Rule.—To find the interest for 1 year, multiply the principal by the rate.

Formula. $I = PR$.

Example.—Determine the interest on $3600 for 1 year at 5%.

$P = 3600 \quad R = 5\%$	
$I = PR$	Formula
$I = 3600 \times 0.05$	Substituting
$I = 180$	
The interest is $180.00.	

To Find the Interest for Any Period of Time

Rule.—To find the interest for any period of time, multiply the principal by the rate by the time.

Note.—Except in transactions involving the Federal government, 30 days are considered a month and 360 days a year.

Formula. $I = PRT$.

Example.—Find the interest on $3600 for 2 years 7 months at 5%.

$P = 3600, \quad R = 5\% = \frac{1}{20} \quad T = 2$ yr. 7 mo.	
2 yr. 7 mo. $= 2\frac{7}{12}$ yr.	Changing the time to a fraction
$2\frac{7}{12} = \frac{31}{12}$	
$I = PRT$	Formula
$I = 3600 \times \frac{1}{20} \times \frac{31}{12}$	Substituting
$I = 465$	
The interest is $465.	

To Find the Amount

Rule.—To find the amount, add the principal and the interest.

Formula. $A = P + I$.

Example.—Find the interest and the amount when $300 are borrowed for 2 years 6 months 15 days at 4%.

Solution 1.		
$I = PRT$		Formula 2
$300 \times .04 \times 2$	$= 24.00$	Interest for 2 yr. is $24.
$300 \times .04 \times \frac{1}{2}$	$= 6.00$	Interest for 6 mo. is $6
$300 \times .04 \times \frac{15}{360}$	$= .50$	Interest for 15 da. is $.50.
	30.50	Total interest is $30.50.
$A = P + I$		Formula 3
$A = 300 + 30.50$		
$A = 330.50$		
The amount is $330.50.		

Note.—It is sometimes easier to change the time to the denomination of the smallest unit given and then solve the problem by one operation.

Solution 2.
2 yr. 6 mo. 15 da. = 720 da. + 180 da. + 15 da. = 915 da.
$= \frac{915}{360}$ yr.

$I = PRT$ Formula
$I = 300 \times .04 \times \frac{915}{360}$ Substituting
$I = 30.50$
The interest is $30.50.

EXERCISES

Find the interest and the amount.

Number	Principal	Rate, per cent	Time
1	$3000	4	3 yr.
2	60	6	1 yr.
3	1200	8	2 yr. 6 mo.
4	685	8	3 yr. 3 mo.
5	460	6	4 yr. 4 mo.
6	1800	5	5 yr. 2 mo.
7	8500	4	6 yr. 9 mo.
8	200	12	5 mo. 15 da.
9	3248	4	3 mo. 10 da.
10	1750	10	9 mo. 18 da.
11	1860	7	2 mo. 10 da.
12	1335	8	5 mo. 5 da.
13	1190	5	1 yr. 6 mo. 10 da.
14	1250	6	2 yr. 8 mo. 15 da.
15	2500	4	3 yr. 6 mo. 21 da.
16	1360	$5\frac{1}{2}$	6 yr. 7 mo. 17 da.
17	1600	$4\frac{1}{2}$	6 yr. 3 mo. 5 da.
18	750	$4\frac{1}{2}$	8 yr. 2 mo. 16 da.
19	2250	$4\frac{3}{4}$	9 yr. 3 mo. 10 da.
20	100	3.3	4 yr. 5 mo. 6 da.

PERCENTAGE: TYPE 2

Rule. To find the rate, divide the percentage by the base. Carry out the quotient to at least two decimal places, and then change to a per cent.

Formula. $R = \frac{P}{B}$.

This formula can be derived from the formula for type 1 percentage problems as follows:

$P = BR$ Formula for type 1 percentage

$\frac{P}{B} = R$ Solving the equation for R

or

$$R = \frac{P}{B}$$

Example 1.—16 is what per cent of 80?

	$P = 16, \quad B = 80, \quad R = ?$
	This is type 2 percentage.
$R = \frac{P}{B}$	Formula
$R = \frac{16}{80}$	Substituting
$R = 0.20$	Dividing and carrying the answer to two decimal places
$R = 20\%$	Changing the decimal to a per cent

Example 2.—What per cent of 16 is 4?

$P = 4, \quad B = 16, \quad R = ?$	
$R = \frac{P}{B}$	Formula
$R = \frac{4}{16}$	Substituting
$R = 0.25$	
$R = 25\%$	

Note.—If you recognize that $\frac{4}{16}$ equals $\frac{1}{4}$, which equals 25%, you will save yourself the division step in Example 2.

Example 3.—Pure gold is 24 carats fine. What per cent of pure gold is a 14-carat gold watch case?

$P = 14, \quad B = 24, \quad R = ?$	
$R = \frac{P}{B}$	Formula
$R = \frac{14}{24}$	Substituting
$R = 0.58\frac{1}{3}$	
$R = 58\frac{1}{3}\%$	
The watch case is $58\frac{1}{3}\%$ pure gold.	

Example 4.—A brass is composed of 165 lb. of copper and 70 lb. of zinc.

(*a*) The copper is what per cent of the mixture?

(*b*) The zinc is what per cent of the mixture?

(*c*) The zinc is what per cent of the copper?

(*d*) The copper is what per cent of the zinc?

Round off the quotients to three decimal places, and put plus, or minus after each as needed, as instructed on page 59.

The total mixture is 165 lb. + 70 lb., or 235 lb.

(*a*) $P = 165, \quad B = 235, \quad R = ?$

$R = \frac{P}{B}$ Formula

$R = \frac{165}{235}$, or 0.7021, or 70.2 + %

(*b*) $R = \frac{70}{235}$, or 0.2978, or 29.8 − %

(*c*) $R = \frac{70}{165}$, or 0.4242, or 42.4 + %

(*d*) $R = \frac{165}{70}$, or 2.3571, or 235.7 + %

ORAL EXERCISES

Using the formula $R = \frac{P}{B}$, express each of the following as a common fraction, reduce to lowest terms, and change to a per cent.

1. 4 is what per cent of 8? Of 16? Of 20? Of 36? Of 48?

2. 11 is what per cent of 22? Of 33? Of 66? Of 121? Of 4?

3. 15 is what per cent of 45? Of 90? Of 75? Of 10? Of 12?

4. 32 is what per cent of 64? Of 72? Of 24? Of 8? Of 4?

5. 10 is what per cent of 5? Of 30? Of 20? Of 60? Of 120?

6. What per cent is 24 of 12? Of 8? Of 4? Of 2? Of 3?

7. What per cent of 48 is 3? Is 6? Is 8? Is 12? Is 16?

8. What per cent of 60 is 5? Is 10? Is 15? Is 20? Is 30?

9. What per cent of 42 is 6? Is 21? Is 84? Is 7? Is 14?

10. 40 is what per cent of 60? Of 20? Of 80? Of 10? Of 200?

11. 35 is what per cent of 70? Of 45? Of 50? Of 5? Of 7?

WRITTEN EXERCISES AND PROBLEMS

Find results to the nearest tenth of a per cent.

1. 11 is what per cent of 27? Of 36? Of 95? Of 111? Of 7?

2. 13 is what per cent of 24? Of 38? Of 45? Of 50? Of 100?

3. 14 boys absent is what per cent of 89 boys enrolled?

4. 18 defective castings is what per cent of 142 in the run?

5. 9 parts spoiled out of every 250 finished is what per cent spoiled?

6. 3 gal. of gas used is what per cent out of a full 20-gal. tank?

7. 14 boys started a cross country run; 6 finished. What per cent finished?

8. If a helper earns 95¢ per hour and a machinist earns $1.85 per hour, the helper's wage is what per cent of the machinist's wage?

9. 41 machines were sold out of 1 month's quota of 50 machines. What per cent was sold?

10. If a 24-in. pulley is increased 1 in., what is the per cent increase?

11. If a man owns 10 lots in a subdivision of 45 lots, what per cent does he own?

12. Out of a population of 36,240 in Bakersfield, 1646 are foreign-born. What per cent are foreign-born? Find the answer to the nearest 0.1%.

13. A car manufacturer advised the owner to increase the pressure in his tires from 24 lb. to 30 lb. The increase is what per cent of the original?

14. 77 lb. of copper were obtained from 1 ton of ore. What per cent of the ore was copper?

PROBLEMS

1. A stock feed mixture calls for 20 lb. of mash and 70 lb. of grain. The mash is what per cent of the mixture? The grain is what per cent of the mixture?

2. A cubic inch of copper weighs 0.3211 lb. A cubic inch of lead weighs 0.4106 lb. The weight of the copper is what per cent of the weight of the lead?

3. A screw jack that is supposed to raise 25 tons will raise only 18 tons. What is the per cent efficiency?

4. With a lever a force of 25 lb. just balances a 245-lb. weight. The force is what per cent of the weight?

5. A cubic foot of wood weighs 50 lb., and a cubic foot of water weighs 62.4 lb. The weight of the wood is what per cent of the weight of the water?

6. A Boeing airplane has a top speed of 380 m.p.h. and a cruising speed of 310 m.p.h. The difference in speed is what per cent of the cruising speed? Of the top speed?

7. If your pay is raised from $1.37 per hour to $1.78 per hour, what is your per cent increase?

8. If your pay is cut from $1.78 per hour to $1.37 per hour, what is your per cent cut?

9. An automobile engine is rated at 120 hp. but delivers 98 hp. What is the per cent efficiency?

10. Molten cast iron on freezing shrinks $\frac{1}{8}$ in. per ft. What per cent shrinkage is this?

11. The pressure in a steam boiler was increased from 280 to 300 lb. per sq. in. What is the per cent increase in pressure?

12. Out of a shop run of 350 castings, 12 were defective and 8 were spoiled in machining. What per cent were defective? What per cent were spoiled?

13. By more efficient handling the production output was increased from 2548 machines to 2850 machines. What was the per cent of increase?

14. A cubic foot of steel weighs 485 lb.; a cubic foot of aluminum weighs 167 lb. The weight of the aluminum is what per cent of the weight of steel?

15. A contractor figured the cost of a dwelling at $15,250.00. He finished it for $12,866.00. How much did he gain? What per cent did he gain?

16. What per cent of a foot is 1 in.? Is $\frac{1}{4}$ in.? Is $\frac{1}{16}$ in.?

17. Mr. Jones can assemble 75 radios in a week, whereas Mr. Brown can assemble only 48 in the same time. If they are paid by the units assembled, Mr. Brown's wage is what per cent of Mr. Jones's wage?

18. Mild steel has a tensile strength of 65,000 lb. per sq. in. Brass has a tensile strength of 38,000 lb. per sq. in. The strength of brass is what per cent of the strength of steel?

PERCENTAGE: TYPE 3

Rule. To find the base, divide the percentage by the rate.

Formula. $B = \frac{P}{R}$.

This formula can be derived from the formula for type 1 percentage problems as follows:

$P = BR$ Formula for type 1 percentage

$\frac{P}{R} = B$ Solving the equation for B

or

$B = \frac{P}{R}$

Example 1.—The number 55 is 10% of what number?

$P = 55, \quad R = 10\% \quad B = ?$

This is type 3 percentage.

$B = \frac{P}{R}$ Formula

$B = \frac{55}{.10}$ Substituting

$B = 550$

Ans. 55 is 10% of 550.

Note.—In type 3 percentage, the whole number, or the original number (B in the formula) represents 100%. It is possible for the rate, R in the formula, to be larger than 100% (see Example 3). Since it is necessary to divide by the rate, the solution is usually easier if the rate is expressed as a decimal.

Example 2.—In a shop, 8% of the old machines were replaced by new ones. How many machines are there in the shop if 144 machines were replaced?

$P = 144, \quad R = 8\%, \quad B = ?$

$B = \frac{P}{R}$ Formula

$B = \frac{144}{.08}$ Substituting

$B = 1800$

There are 1800 machines in the shop.

Example 3.—There were 216 people at a picnic. This was 8% more than the number expected. How many people were expected?

Since the expected number is the base B and represents 100%, the number who attended represents 100% + 8%, or 108%.

$P = 216, \quad R = 108\%, \quad B = ?$

$B = \frac{P}{R}$ Formula

$B = \frac{216}{1.08}$ Substituting

$B = 200$

200 people were expected.

Example 4.—A machine was sold for $575. This amount was 8% below the marked price. Find the marked price.

The marked price is 100%. Then the selling price is 100% − 8%, or 92% of the selling price.

$P = 575, \quad R = 92\% \quad B = ?$

$B = \frac{P}{R}$ Formula

$B = \frac{575}{.92}$ Substituting

$B = 625$

The marked price of the machine was $625.

PROBLEMS

Do not forget to change the rate to a decimal.

1. 10 is 2% of what number? $2\frac{1}{2}$%? 25%? 1%?

2. 56 boys are 7% of what number of boys? 8%? 28%? 56%?

3. 33 machines are 3% of what number of machines? 11%? 33%? 25%?

4. 6% of what weight is 30 tons? 6% of what amount is $75.00?

5. 15% of what is 75 machines? $75.00? 75 tons? 75 boys?

6. Of how many bushels are 75 bushels, 10%?

7. Of what number of men are 84 men, 7%?

8. What number increased by 10% of itself is 1100? (1100 is 110% of what?)

9. What amount increased by 5% of itself is $210.00? (210 is 105% of what?)

10. What number of men increased by 11% of itself makes 333 men?

11. What horsepower decreased by 10% of itself is 81 hp.? (81 is 90% of what?)

12. What number of trucks decreased by 5% of itself is 475? (475 is 95% of what?)

13. A man lost $10.00. This was 25% of his earnings. Find his earnings.

14. 72 ft. is 90% of what length?

15. In a certain class, 7 boys were absent. This was 20% of the class. Find the enrollment.

16. An engine delivers 96 hp. at 85% efficiency. What is the horsepower at 100% efficiency?

17. What length diminished by 20% of itself is 480 ft.?

18. What number of machines increased by 5% of itself gives 525 machines?

19. If you increase the horsepower of an engine by 5% and get 525 hp., what was it before the increase?

20. How many cars diminished by 15% are then 340 cars?

21. If 30 men are added to a pay roll, the number on the original pay roll is increased 5% of itself. Find the number of men on the original pay roll.

22. $\frac{3}{16}''$ is 12% of what length?

23. 28.10 lb. are 11% of what weight?

24. By raising the compression in an automobile cylinder 10%, it was measured at 100 lb. What was the original compression?

25. A man saved $630.00 in a year. This was 11% of his income. Find his income.

26. A lathe, sold at a loss of 12%, brought $800.00. What should it have brought?

27. One ounce is one-half of one per cent of how many pounds?

28. There were 567 men on the pay roll this month. This is 5% more than last month. Find the number of men on the pay roll last month.

29. With a windlass, only 80% efficient, a man can raise a weight of 850 lb. What weight could he raise if there were no loss?

30. One ounce is 4% of how many ounces?

31. It was found that the volume of gasoline in a tank had expanded 3%. If the volume after expansion was 1146 gal., what was the original volume?

32. A dealer sold a car at a gain of 18% figured on the cost. He received $738.00. How much did the car cost him?

33. After being machined off 12%, a casting weighed 440 lb. What did it weigh before it was machined?

34. A contractor figured he had 8% of the bricks needed for a building. He had 12,000 bricks. How many bricks did the job require?

35. Of the castings in a run at the foundry, 9% were defective, 637 being satisfactory. What was the original run?

36. It was necessary to speed up a machine 135 r.p.m. This was found to be a 15% increase. What was the original r.p.m.?

37. The mechanical advantage of a lever was 6 to 1 when 95% efficient. What would it have been if there were no loss?

38. The turnover of men on a construction job averages 8% per month. If 48 men change monthly, how many men are on the pay roll?

39. An oil well producing 2500 bbl. of oil per day was found to be running 25% more than allowed under curtailment laws. To what must the production be cut?

40. The tensile strength of rope is 6000 lb. per sq. in. This is only $7\frac{1}{2}$% of the strength of steel cable. Find the strength of steel cable.

41. The actual weight raised by a hoist was 12,500 lb. From the power put in, friction consumed 18%. Had there been no loss, what weight could have been raised?

42. When heated, an iron pipe expanded $\frac{3}{16}$% of its original length. This expansion was exactly $\frac{3}{16}$ in. How long was the pipe before heating?

CHAPTER 6

VOCATIONAL FINANCE

BUDGETS

The plan of how to spend an income is called a **budget.** All branches of government (Federal, state, county, city) and each separate department of these branches make annual estimates of the financial needs for the year to come. All private corporations budget their income. It is also a wise policy for the family and for the individual to plan the spending of their incomes. Even a child should be taught to budget whatever money he receives or earns.

The budget should not be confused with an expense account. Everyone should make a record of expenditures, if for no other reason than to be able to plan a better future budget.

From figures gathered by the United States Bureau of Labor Statistics, the following budgets are conservative averages for the United States in general.

Budget for a Single Man with No Dependents

Item	Monthly income			
	$165	$185	$200	$250
	Per cent	Per cent	Per cent	Per cent
Savings and insurance	4	5	6	9
Taxes and social security	22	22	22	22
Board and room	50	45	40	32
Clothing and laundry	10	10	10	12
Charity, church, union dues	3	3	4	5
Education and advancement	6	7	8	9
Health and recreation	2	4	5	5
Auto and transportation	3	4	5	6

PROBLEMS

1. From the preceding table, determine for each item the monthly allowance and the yearly allowance for a single man with $165 monthly income.

2. Determine for each item the monthly allowance and the yearly allowance for a single man with $185 monthly income.

3. Determine for each item the monthly allowance and the yearly allowance for a single man with $200 monthly income.

4. Determine for each item the monthly allowance and the yearly allowance for a single man with $250 monthly income.

5. A single man who earned $225 in a month kept an expense account as follows: savings, $22.50; taxes, $45; board and room, $83; laundry and clothing, $20; education, $12; doctor bills, $11; automobile, $36; cash on hand at the end of the month $5.50. Figure the per cent of his income for each item, and determine whether or not he is keeping within the limits of a conservative budget.

Budget for a Family of Two, Three, or Four Persons

Item	Monthly income								
	$200			$250			$415		
	Number in family			Number in family			Number in family		
	2	3	4	2	3	4	2	3	4
	Per cent	Per cent	Per cent	Per cent	Per cent	Per cent	Per cent	Per cent	Per cent
Savings and insurance.......	6	5	3	8	5	3	9	6	5
Taxes and Social Security*..	12	6	3	12	8	4	15	12	9
Food........................	35	40	44	30	35	38	24	28	32
Rent, fuel, light...........	26	26	26	27	27	28	24	25	25
Clothing....................	10	12	13	10	12	14	12	13	13
Church, union dues, etc.....	2	2	2	2	2	2	3	3	3
Education...................	2	2	2	2	2	2	2	2	2
Health and recreation.......	2	2	2	3	3	3	4	4	4
Auto and transportation.....	5	5	5	6	6	6	7	7	7

* Decrease in taxes is due to lower income tax because of dependent exemption deductions

FAMILY BUDGETS

Family budgets are more complex. The table on page 124 shows the allowance for the various items for three different incomes. The per cents are determined for children under ten years of age. As the children grow older, the per cents allowed for clothing and education must be increased. Likewise, the per cent allowed for food may increase unless there is very wise planning.

PROBLEMS

Use the per cents given in the preceding table for each of the following problems.

1. Determine for each item the monthly allowance for a family of two adults with a $200 monthly income.

2. Determine for each item the monthly allowance for a family of two adults and one child when monthly income is $200.

3. Determine for each item the monthly allowance for a family of two adults and two children with $200 monthly income.

4. Determine for each item the monthly allowance for a family of two when the monthly income is $250.

5. Determine for each item the monthly allowance for a family of three when the monthly income is $250.

6. Determine for each item the monthly allowance for a family of four when the monthly income is $250.

7. Determine for each item the monthly allowance for a family of two when the monthly income is $415.

8. Determine for each item the monthly allowance for a family of three when the monthly income is $415.

9. Determine for each item the monthly allowance for a family of four when the monthly income is $415.

SOCIAL SECURITY

In the United States, over forty-five million workers, at least three-fourths of all employed persons and their families, are covered by insurance under the **Social Security Act** of 1935 and the **Federal Insurance Contributions Act** of 1950 (F.I.C.A.). Likewise, workers are covered against unemployment under the **Federal Unemployment Contribu-**

tion Act (F.U.C.A.), and in some states they are protected by disability insurance.

Social Security insurance must not be confused with State old-age pensions and relief payments, which are outright gifts to the needy from the state and Federal government. With Social Security, as with life insurance, you pay in advance for the benefits you will receive when you are eligible for retirement. Social Security "taxes" should not be considered taxes in the regular sense. The payments you make are just like those you make to a regular insurance company, with the addition that your employer contributes as much as you do toward your insurance. This makes your insurance cost you less and, more important, payments are made on your Social Security account only when you have income.

If your job is covered by Social Security (or F.I.C.A.), both you and your employer make regular payments to the **Old-age and Survivors Insurance Trust Fund.** This is the fund out of which your "benefits" will be paid when you retire at sixty-five years of age. When you die, either before or after you are sixty-five, benefits will be paid to your survivors, commencing as soon as application for survivors' benefits has been filed.

As an individual worker, you have a **Social Security account** and number under your own name in Baltimore, Md. When your account is opened, you are given a **Social Security card.** Every employer you have will want to see this card in order to have the information on it for his records.

Your employer takes the amount of your payments out of your wages and sends it with an equal amount that he must pay to the **U.S. Collector of Internal Revenue.** He makes the payments quarterly and will give you a receipt once a year for the payments you have made on your Social Security account. You may always check up on your Social Security account by writing the **Social Security Administration,** Baltimore, Md., or by calling at one of the local field offices in the principal cities.

The amount of taxes on wages for Social Security is as follows:

Years	Tax rate (per cent of wages) for	
	Employee	Employer
1937 through 1949	1	1
1950 through 1953	$1\frac{1}{2}$	$1\frac{1}{2}$
1954 through 1959	2	2
1960 through 1964	$2\frac{1}{2}$	$2\frac{1}{2}$
1965 through 1969	3	3
After Dec. 31, 1969	$3\frac{1}{4}$	$3\frac{1}{4}$

Beginning in 1951, many new occupations and most self-employed persons came under the Social Security laws for the first time. The tax rate for self-employment income is $1\frac{1}{2}$ times the rate for an employee. To 1951, payments were made on only the first $3000 of yearly income. Thereafter, the payments are made on the first $3600 of yearly income for both employees and self-employed. In self-employment the yearly earnings must be at least $900 before it is taxable for Social Security.

The 1950 amendments made the law retroactive to include some veterans of the Second World War. All servicemen who serve 90 days or more between Sept. 16, 1940 and Jan. 1, 1954, and who were not dishonorably released, are allowed $160 wage credit for each month of service.

TO CALCULATE THE TAXES

Since the calculations of the taxes are simple problems in percentage, explanation will be by example.

Example.—Mr. G. Smith's average wage was $69 per week during 1951. Determine the following:

1. Amount withheld weekly for Social Security.
2. Amount contributed weekly by Mr. Smith's employer.
3. Amount withheld yearly for Social Security.
4. Amount contributed yearly by Mr. Smith's employer.
5. Total yearly deposit with Collector of Internal Revenue.

$69 \times 0.015 =$	\$ 1.04	(1) Withheld weekly, $1\frac{1}{2}\%$ in 1951.
$69 \times 0.015 =$	\$ 1.04	(2) Contributed weekly by employer.
$1.04 \times 52 =$	\$ 54.08	(3) Withheld yearly.
$1.04 \times 52 =$	\$ 54.08	(4) Contributed yearly.
	\$108.16	(5) Total yearly deposit.

PROBLEMS

1. Make the necessary computations for the pay roll below.

Number	Name	Average weekly wage	Deduction for Social Security	Employer's weekly contribution	Total yearly deposit
Example	Smith, G.	\$69.00	\$1.04	\$1.04	\$108.16
1	Brown, J. E.	64.00			
2	Bush, Joe	60.00			
3	Case, F. E.	58.40			
4	Craig, C. S.	69.00			
5	Dort, Wm.	63.00			
6	Dow, L. W.	62.50			
7	Elkins, A.	68.50			
8	Elliot, C.	66.50			
9	Fox, W. W.	67.30			
10	Young, Al.	65.40			

2. For the following pay roll of salaried workers who receive their checks monthly, make the necessary computations for 1955. Only the first \$3600 per year is taxable.

Number	Name	Monthly salary	Deduction for Social Security	Employer's monthly contribution	Total yearly deposit
1	Allen, Sadie	\$210.00			
2	Bishop, Will	295.00			
3	Call, Frank	288.00			
4	Card, George	224.00			
5	Smith, Albert	268.50			
6	Towne, Fred	244.50			
7	Unis, W. J.	296.50			
8	Vance, Edward	344.00			
9	Wills, Jesse	525.50			
10	Woods, Ted	180.00			

The following example shows how to compute the total tax that will have been paid into a Social Security account by a workman and his employer during the worker's years of employment to the time of his retirement at sixty-five years of age.

Example.—Mr. A. Alder was forty-eight years old in 1948 when he was employed by a company that operated under the Social Security Act. His average annual income on this job was $3560. How much will have been paid into Mr. Alder's Social Security account when he retires at sixty-five years of age?

65 − 48 =	17 years before he is sixty-five.	
1948 + 17 =	1965	When he will retire.
$3000 × 0.01 × 2 =	$ 60.00	Paid by Alder for 2 years: 1948, '49.
$3000 × 0.015 × 1 =	$ 45.00	Paid by Alder for 1 year: 1950
$3560 × 0.015 × 3 =	$ 160.20	Paid by Alder for 3 years: 1951, '52, '53.
$3560 × 0.02 × 6 =	$ 427.20	Paid by Alder for 6 years: 1954, '55, '56, '57, '58, '59.
$3560 × 0.025 × 5 =	$ 445.00	Paid by Alder for 5 years: 1960, '61, '62, '63, '64.
	$1137.40	Total paid by Alder in 17 years.
	$1137.40	Total contributed by employer.
	$2274.80	Total deposited to Mr. Alder's Social Security account.

PROBLEM

Copy the following pay roll in your notebook and complete it. Remember, only the first $3600 per year is taxable.

Number	Name	Age in 1951	Average annual wage	1% to 1950	1.5% for 1950–1953	2% for 1954–1959	2.5% for 1960–1964	3% for 1965–1969	3.25% for 1970 and after	Total paid by worker and employer
Example	Alder, A.	51	$3560	$60	$205.20	$427.20	$445			$2274.80
1	Barker, J.	42	3220							
2	Cork, C. C.	51	2860							
3	Davis, J. H.	55	4850							
4	Evans, Leo	61	9510							
5	French, L.	22	3240							
6	Gage, Fred	29	3165							
7	Kelly, Wm.	54	3500							
8	Lowe, S. S.	53	2850							
9	Marks, Lee	21	3365							
10	Rapp, Larry	32	3000							

SOCIAL SECURITY BENEFITS

The money received by the retired worker from the Old-age and Survivors Insurance Trust Fund, formerly called Primary Insurance Benefit, or simply **benefit,** is now called **Primary Insurance Amount** (P.I.A.). In most cases benefit is paid monthly to the workman after he retires at sixty-five years of age, and his benefit is called **Old-age Insurance Benefit** (O.A.I.B.).

If the benefit, as figured, is less than $25, it will be raised to $25. Only the first $300 average monthly wage, or $3600 per year, may be used in figuring benefits.

There are two provisions of the Social Security Act that affect the payment of these benefits: (1) worker's insured status and (2) his earnings after retirement.

1. After 1950, the wage earner, at the time of death or retirement after reaching sixty-five, must have been **fully** or **currently insured** for benefit payments of any kind to be made.

To be **fully insured** the wage earner must have had at least (*a*) one quarter of coverage for each two calendar quarters elapsing after 1950, or after the quarter he attained age of twenty-one, whichever is later, and up to but excluding the quarter in which he attained the age of sixty-five or died, whichever comes first, but in any case not less than six quarters; or (*b*) have forty quarters of coverage.

To be **currently insured** a wage earner must have had at least six quarters of coverage during the 13-calendar quarter period ending with (*a*) the quarter in which he died or (*b*) the quarter in which he became eligible for old-age insurance benefits.

As defined by the Social Security Act, a quarter of coverage is a 3-months period ending Mar. 31, June 30, Sept. 30, and Dec. 31, in which the worker was paid at least $50 in a job covered by Social Security. The quarters of coverage may have been earned at any time after 1936, and the quarters need not be consecutive.

2. You may accept your benefit check even if you earn up to $75 a month in covered employment, instead of $15, as before. If you are seventy-five years old or over you may accept the payments regardless of the amount of your earnings.

The 1950 amendments increased the minimum benefit for those now receiving them and made it easier to become eligible for benefits. Only 1½ years of covered work will be necessary for a worker who reaches sixty-five or dies before the end of June, 1954. A wage of only $50 or more is necessary in each of six different calendar quarters.

Under the new law, for income after 1950, the amount of monthly benefit to a retired worker is 55 per cent of the first $100 of his average monthly wage, plus 15 per cent of the next $200. By the new rules, an eligible person will get at least $25, even if his average monthly wage is less than $40. Now, there is no provision made for adding a percentage to this amount for increment years, as was done before.

To get the **primary insurance amount,** or monthly benefit, for a retired worker, proceed as follows:

1. Take the total of the wages earned after Jan. 1, 1951, and divide by the months in the period when the worker

becomes sixty-five (but not less than 18 months). This gives the average monthly wage.

2. Take 55 per cent of the first $100.

3. Take 15 per cent of the balance up to $200.

4. Add items 2 and 3. This is the primary insurance amount, or monthly benefit for the retired worker.

Round off the average wage to the next lower dollar and the benefit to the next higher $0.10.

Example 1.—A man averaged $236 per month for the last 10 years before retiring at sixty-five, thus being fully insured. What is his primary insurance amount?

(1) $236		Average monthly wage.
(2) $100 × 0.55	= $55.00	55% of the first $100.
(3) $136 × 0.15	= $20.40	15% of balance, which is less than $200.
(4) add (2) and (3)	= $75.40	Primary insurance amount.

Example 2.—An employee earned in successive years after 1952, $2540, $3125, $3270, $3600, $3352, $3350, and $2690, when he reached the age of sixty-five in covered employment. What is his primary insurance amount?

(1) $2540 + 3125 + 3270 + 3600 + 3352 + 3350 + 2690	= $21,927	Total wages for 7 years, or 84 months.
$21,927 ÷ 84	= $261.03	Average monthly wage, which rounded off to next lower $1 is $261.
	$261.	Average monthly wage.
(2) $100 × 0.55	= $ 55.00	55% of the first $100.
(3) $161 × 0.15	= $ 24.15	15% of the remaining $161.
(4) Add (2) and (3)	= $ 79.15	Rounded off to next higher $0.10 is
	$ 79.20	Primary insurance amount.

PROBLEMS

Compute the monthly benefit for each of the men listed in the following pay roll who are eligible for retirement. Years of work are stated to determine if worker is fully or currently insured. In problems 2 and 8, remember that only the first $300 per month can be used in figuring monthly benefit.

Number	Name	Average monthly wage	Years of work	Monthly benefit
	Example 1	$236	10	$75.40
	Example 2	261	Last 7	79.20
1	Bird, Homer	212	12	
2	Brown, Wm. L.	356	15	
3	Duff, Edwin	184.50	Last 4	
4	Galt, W. G.	284	12	
5	James, J. J.	177.50	Last 6	
6	Ostler, Asa	224	18	
7	Perry, P. P.	296	33	
8	Quincy, Ross	524	44	
9	Younger, Edw.	255	Last 3	
10	Wilson, John	250	21	

BENEFITS FOR THE WAGE EARNER'S DEPENDENTS

Besides the wage earner's primary insurance amount, or old-age insurance benefit, there are additional monthly benefits payable to his dependents after his retirement and after his death. Those paid after retirement, called **life cases,** are determined as follows:

1. The **wife's insurance benefit** equals 50 per cent of the wage earner's primary insurance amount, if she is sixty-five years old or has in her care a child entitled to benefits. Wage earner must be fully insured for this benefit to be paid. If a wife is entitled to old-age insurance benefits from her own work record, she will receive an amount equal to the greater benefit, but not both.

2. A **husband's insurance benefit** equals 50 per cent of the woman worker's old-age insurance benefit, if he is sixty-five years old and was receiving at least half of his

support from her. For this benefit to be paid, she must be both fully and currently insured and thus entitled to her own old-age insurance benefit.

3. A **child's insurance benefit** equals 50 per cent of the worker's old-age insurance benefit for each unmarried child under eighteen years of age, and the worker must be fully insured. A lone child's benefit may not be less than $12.50 A child may be either natural, adopted, or a stepchild.

The maximum monthly benefit in life cases is $168.75 or 80 per cent of the average monthly wage, whichever is the lesser, but in no case less than $50.

BENEFITS FOR THE WAGE EARNER'S SURVIVORS

After the death of the wage earner or the retired wage earner, the family is entitled to survivor's monthly benefits. The most important of these are as follows:

1. A **widow's insurance benefit** equals 75 per cent of the primary insurance amount of her deceased husband, if she is sixty-five years old, and if he died fully insured. If she is entitled to old-age insurance benefits from her own work record, she will receive an amount equal to the greater benefit, but not both.

2. The title, **mother's insurance benefit,** which is equal to 75 per cent of the primary insurance amount, is now used instead of widow's current insurance benefit. The regulations for mother's insurance benefit makes it possible for a widow and every divorced former wife of the deceased worker to draw benefits at the same time on his work record, providing she and they qualify by being mothers of his unmarried children that are under eighteen years of age and who are in the care of the widow and a former wife divorced.

3. **A widower's insurance benefit** equals 75 per cent of the primary insurance amount of the deceased wife, if he is sixty-five years old, and providing she died fully and currently insured.

4. **A surviving child's insurance benefit** equals 75 per cent of the deceased worker's primary insurance amount, unless more than one child is entitled; then each child's benefit equals 50 per cent of the primary insurance amount plus one-fourth of the primary insurance amount divided by the number of entitled children. The worker may be fully or currently insured for a child's benefit in death cases, and the minimum benefit for one child is $18.75. If a child is entitled to benefits from two deceased parents, he will receive the benefit from the higher primary insurance amount.

5. A **parent's insurance benefit** equal to 75 per cent of the worker's primary insurance amount is payable to a parent who received one-half of his support from the worker at the time of a fully insured worker's death, but only an amount equal to the greater benefit if parent is entitled to old-age insurance benefit on his own work record. For a parent's benefit the worker must not be survived by an eligible widow, widower, children, or mother (divorced wife with dependent children of the deceased).

In death cases the maximum benefit payable to the surviving dependents is $168.75 or 80 per cent of the worker's average monthly wage, whichever is the lesser, but not less than $18.75 for one survivor.

6. A **lump sum** is payable to whoever paid the burial expense of the deceased worker or serviceman. The lump-sum amount equals three times the primary insurance amount of a worker who died either fully or currently insured.

Example 1.—When Mr. Brown is sixty-five years old, his wife is fifty years old and there are three children under eighteen years of age. Mr. Brown has worked on a job covered by Social Security for 20 years at an average monthly wage of \$216 and is, therefore, fully insured. Compute all of the possible benefits for Mr. Brown and his family.

$100 \times 0.55 = \$\ 55.00$	55% of the first \$100 of wages.
$116 \times 0.15 = \$\ 17.40$	15% of the next \$116 of wages.
$\$\ 72.40$	Mr. Brown's primary insurance amount, hence, his old-age insurance benefit.
$\$72.40 \times 0.50 = \$\ 36.20$	Wife's figured benefit, 50% of P.I.A.
$\$72.40 \times 0.50 = \$\ 36.20$	Each child's figured benefit.
$\$36.20 \times 3 = \$108.60$	Figured benefit for three children.
$\$108.60 + 36.20 + 72.40 = \$217.20$	Total figured benefit.
$\$216.00 \times 0.80 = \$172.80$	80% of the average monthly wage.
$\$168.75$	Maximum allowable benefit.

Of these three possible benefits, the maximum allowable benefit of \$168.75 will be paid, since it is the lesser, which rounded off to the next higher \$0.10 is \$168.80.

In a case like this, when the maximum allowable benefit is payable, it is necessary to reduce each dependent's benefit. Mr. Brown's old-age insurance benefit of $72.40 is not subject to reduction.

$\$168.80 - 72.40 = \$96.40$	Total for wife and three children.
$\$96.40 \div 4 = \$24.10$	Benefit for each dependent

Example 2.—Mr. Brown, of Example 1, died at the age of sixty-eight. Mrs. Brown and the three children who are nineteen, twelve, and six years old survive. Determine the survivor's benefits.

$\$72.40 \times 0.75 = \$\ 54.30$	75% of primary insurance amount is Widow's Insurance benefit.
$\$72.10 \times 0.50 - \$\ 36.20$	50% of primary insurance amount.
$\$72.40 \times \frac{1}{4} \div 2 = \$\ \ 9.05$	One-fourth of primary insurance amount divided by eligible children, 2.
$\$36.20 + 9.05 = \$\ 45.25$	Each child's figured benefit.
$\$\ 45.30$	Rounded off to the next higher $0.10.
$\$45.30 \times 2 = \$\ 90.60$	Benefits for two eligible children.
$\$90.60 + \$54.30 = \$144.90$	Survivor's monthly benefit, which is less than the maximum of $168.75.
$\$72.40 \times 3 = \$217.20$	Lump sum received by widow for burial expense is three times primary insurance amount.

PROBLEMS

In the following table all of the workers are fully insured and are being retired at the age of sixty-five, or they are deceased. as indicated. Determine the possible benefits. In problems 3. 7, and 8 remember that only the first $300 of average monthly wage may be used in figuring benefits.

Number	Name	Average monthly wage	Wife's age	Children under 18	Worker's primary insurance amount	Benefits: Wife or widow	Benefits: Children	Benefits: Total family
In the first six problems the workers are retired.								
Example 1	Brown, J. A.	$216	50	3	$72.40	$24.10	$72.30	$168.80
1	Meek, R. E.	175	46	5				
2	Nelson, L.	265	59	1				
3	Pratt, C. J.	490	66	0				
4	Snow, E. M.	146	45	4				
5	Wills, J. T.	275	55	2				
6	Worth, Geo.	296	51	4				
In the next six problems the workers are deceased								
Example 2	Brown, J. A.	$216	53	2	......	$54.30	$90.60	$144.90
7	Card, G. G.	330	51	2				
8	Davis, J. H.	420	67	1				
9	Ennis, Paul	244	48	4				
10	Gage, John	166	66	0				
11	Fike, A. L.	232	45	5				
12	Hare, Wm.	298	42	2				

UNEMPLOYMENT TAXES AND BENEFITS

The Social Security program in the United States, among other things, provides for a system of **unemployment insurance,** sometimes called **unemployment compensation,** to compensate workers, in part, for the loss of wages caused by unemployment. The system is a joint enterprise between the Federal government and the states and is usually referred to as the Federal-state system. The Federal laws impose a tax on most employers (certain types being exempt), but it is the states under their own laws which provide for the payment of benefits. For this reason the employers, who must also pay a pay-roll tax to the states, are excused from the payment of 90 per cent of the tax levied by the Federal government, the balance in the main being returned to the states to provide for the administrative costs, providing the state laws are approved by the Federal government.

Upon becoming unemployed, the worker should immediately go to the nearest **state employment office** to register for work and to make application for unemployment benefits. All but two states require a waiting period of 1 or more weeks before unemployment benefits are payable. Further, he can draw benefits for only a fixed period of time, depending on past earnings, usually from 6 to 26 weeks in a benefit year.

All states provide for certain eligibility requirements. The most important of these are: claimant must be able and available for any suitable work offered to him and usually he must seek employment. He may be held ineligible for limited periods of time for such reasons as voluntarily quiting a job without good cause, discharge for misconduct connected with his work, refusal of suitable employment, fraud in claiming benefits, and loss of employment due to a labor dispute.

Each state has its own formula for computing the amount of benefits payable. In most states the employee must have worked either a definite period of time or must have had a certain amount of wages in a past period of employment, called a **base period.** This usually means the previous year of employment, or the first four of the five quarters preceding the filing of the claim.

Under the state laws the weekly benefit amount varies with the worker's past wages within certain minimum and maximum limits. The maximum weekly benefit varies with the states from $15 to $26, most frequently it is $25 per week. Some states provide additional allowances for each dependent, which may increase the benefit to as high as $40. All states have a minimum, varying from $0.50 to $15 per week, most frequently it is from $5 to $6.

For **total unemployment** the benefit is usually taken as $\frac{1}{20}$ to $\frac{1}{26}$ of the "high-quarter" wage, that is, take $\frac{1}{26}$ of the total amount of wages earned in the quarter year of highest earnings in insured work during the base period. However, most states are adopting schedules, as the one following, giving the weekly benefit amount for the various wages earned. Inasmuch as each **worker's wage record** is on file in the state employment office, it is a simple process to determine his **weekly benefit amount** from the schedule.

For partial employment, that is, when a worker earns only a part of his regular wage, either on his regular job or at other work, his weekly benefit is reduced. In most states the **partial benefit amount** is usually the weekly benefit amount less the earnings in excess of $3 in the week of unemployment, with the amount rounded off to the next higher $1.

Schedule of Wages and Benefits

Wages in high quarter of base period	Weekly benefit amount	Wages in high quarter of base period	Weekly benefit amount
$ 75 to $199.99	$10	$340 to $359.99	$18
200 to 219.99	11	360 to 379.99	19
220 to 239.99	12	380 to 419.99	20
240 to 259.99	13	420 to 459.99	21
260 to 279.99	14	460 to 499.99	22
280 to 299.99	15	500 to 539.99	23
300 to 319.99	16	540 to 579.99	24
320 to 339.99	17	580 and over	25

In the following example and problems the schedule above may be used to determine the weekly benefit amount.

Example.—An eligible worker who had high-quarter wages of $564.50 is laid off. What will he draw for weekly benefit amount, or partial benefit amount, if he earns $16.40 per week?

\$564.50 is between \$540.00 and \$579.99, therefore,	
\$24.00 = weekly benefit amount.	
\$16.40 − 3.00 = \$13.40	Partial wage minus \$3.
\$24.00 − 13.40 = \$10.60	Weekly benefit amount minus wage in excess of \$3.
\$11.00 = partial benefit	Rounded off to next higher \$1.

PROBLEMS

Copy the following table in your notebook and complete it.

Number	Name	High-quarter wage	Weekly benefit amount	Partial wage earned	Partial benefit amount
	Example	\$ 564.50	\$24.00	\$16.40	\$11.00
1	Brown, H.	382.25		11.35	
2	Dow, James	425.00		6.10	
3	Rowe, Dick	378.50		10.45	
4	Steele, W.	542.80		18.00	
5	Strong, E.	444.68		12.50	
6	Tracy, J.	884.26		54.00	
7	Vance, W.	541.10		5.00	
8	Weeks, R.	578.80			
9	Wilson, J.	492.50		32.40	
10	York, E.	1249.50		22.00	

WORKMAN'S COMPENSATION DISABILITY INSURANCE

In some states and in many foreign countries, laws require employers to carry insurance for the protection of persons injured or meeting death in the course of their employment. The insurance covers any injury or disease arising out of the employment. Some states pay disability benefits for non-work connected sickness or injury.

Most states have organized a state agency, as a **state compensation insurance fund,** into which employers, and in some states employees, pay a certain per cent of their yearly pay roll. This fund is used to pay compensation to injured workmen. The employers are allowed to carry this insurance with private companies, or they may organize their own insurance department. All such organizations are subject to state rules and regulations.

There is no definite rate of tax for compensation insurance as there is for Social Security and unemployment. The state agencies have worked out rate schedules covering all of the occupations according to the hazards in each. States that pay liberal benefits naturally have higher rates than those paying low benefits. In one state that pays liberal benefits the lowest tax rate is $0.07 per $100 of pay roll for clerical workers, librarians, etc., while the highest rate is $17.66 per $100 of pay roll for cleaners and renovators of the outside of buildings.

DISABILITY COMPENSATION

Disability payments, called compensation or benefits, usually start 1 week after the workman leaves work as a result of the injury. Payments are made for 1 week in advance, as wages, on the eighth day, and thereafter on the employer's regular payday. In addition, all reasonable medical, surgical, and hospital treatment that is required to care for and relieve the injured is supplied without cost. Some states limit hospital and surgical expense.

Some states pay the same rate for disability benefits as for unemployment benefits, but generally a more liberal percentage is allowed. Most states allow 65 per cent of the average wage, some even using the highest wage earned in a base period, and some using the current wage at the time of injury. However, the maximum wage considered in any state is $50 per week; the lowest $10.

Weekly compensation is never less than $6 and generally not over $35, although in some states additional allowances are made for dependents.

Aggregate temporary-disability payments may not exceed 3 times the average annual earnings of the workman, nor may the payments extend beyond 240 weeks from the date of injury.

If the injury causes permanent disability, the number of weeks that compensation will be paid may be determined from the following table:

Per cent of permanent disability incurred	Number of weeks for which 65 % of average weekly wage is allowed	Per cent of average weekly wage allowed for remainder of life after first 240 weeks
1	4	0
10	40	0
20	80	0
30	120	0
40	160	0
50	200	0
60	240	0
70	240	10
80	240	20
90	240	30
100	240	40

If the injury causes death, **death benefits** are allowed to dependents. Where the workman was the only means of support, death benefits are figured at $3\frac{1}{2}$ times the average annual wage of the deceased workman. Death benefits are usually paid to the dependents in the same manner and amounts as if they were disability payments. In some states payment is made in a lump sum. Death benefits are not less than $2000 or more than $7000 where there is only a surviving widow, and from $7500 to $8750 where there are a surviving widow and dependent children.

Example.—Jesse Best normally earned $63 per week. While at work he received injuries that resulted in 10 per cent permanent disability. Compute the following:

1. His weekly disability compensation.
2. The number of weeks during which he would receive compensation and his total compensation.

3. Death benefit to his surviving wife and dependent children if the injury had been fatal.

Since his average weekly wage is higher than the $50 maximum, use this figure.

(1) $\$50 \times 0.65$	$= \$\ \ \ 32.50$	Weekly compensation at 65%.
(2) $\$32.50 \times 40$	$= \$\ 1300$	Total compensation is paid for 40 weeks on 10% disability.
(3) $\$63 \times 52 \times 3\frac{1}{2}$	$= \$11,466$	Since the figured amount, $11,466, is more than the $8750 maximum, only the latter will be allowed.
	$ 8750	Maximum death benefit.

PROBLEM

Copy the following table in your notebook and complete it. Use 65% of the average weekly wage to figure weekly disability compensation and $8750 as the maximum death benefit.

Number	Name	Average weekly wage	Weekly disability compensation	Per cent of permanent disability	Disability payments		Death benefit
					Weeks	Total	
Example	Best, Jesse	$63.00	$32.50	10	40	$1300	$8750
1	Cross, Ray	47.50		20			
2	Davis, Jake	71.50		30			
3	Ebb, Frank	48.75		1			
4	Elmer, T. J.	80.00		40			
5	Fields, Jess	50.00		50			
6	Grant, Ted	46.65		100			
7	Kuhn, Oscar	79.00		1			
8	Mann, C. E.	62.50		20			
9	Pratt, Elmo	44.00		10			
10	Rouse, Fred	48.00		20			

INCOME TAXES

In order to help finance the tremendous cost of operating our Federal government, **income taxes** must be paid by all employed persons. An employed person is one who receives payment in some form for work he does. These payments may be called wages, salaries, fees, bonuses, or commissions, and they may be based on the hour, day, week, month, year, or on a piecework or percentage plan. Likewise, some unemployed persons whose income is derived from taxable securities and all pensions, except Social Security and war-veteran pensions, are considered taxable income after the pensioner has recovered all he paid into the pension fund.

An employer is required by Federal law to withhold the income tax from each and every payment he makes to an employee. The employer remits the tax money he withholds to the Collector of Internal Revenue, as he does with Social Security payments. On or before Jan. 31, and at the end of employment, the employer must give each employee a **withholding statement** in duplicate showing (1) the total wages subject to income-tax withholding and the amount of income tax withheld, and (2) the amount of Social Security employee tax withheld and the amount of wages and employee tax deducted.

The employee must attach one copy of the withholding statement to his **Income-tax Return,** which he must file with the nearest local office of the Collector of Internal Revenue on or before Mar. 15 for the preceding year's income. Likewise, a self-employed person must file an income-tax return, and he must remit the amount of his income tax and Social Security tax. In some cases, self-employed persons must pay their income tax quarterly in advance. In some branches of employment, as some farming jobs, etc., the employer is not required to withhold the income tax from the employees. Nevertheless, the employee is required to file an income-tax return and to pay the taxes on his earnings just as a self-employed person does.

Income-tax deductions are uniform and fair to all and allow a 10 per cent standard deduction for such items as charitable contributions, property taxes, etc. However, they do not take into account many deductions, such as excessive outlay for medical expense, etc., that would be an unjust hardship on the worker, but these are adjusted when he files his income-tax return. If the return shows that he has paid more than he owes, the amount will be refunded without delay, after the tax examiner has determined that the correct figures have been submitted. Also, in fairness to the government, if he has additional income from other sources, such as interest on savings, etc., these additional earnings must be included and remittance made for them with the income-tax return.

A person with dependents pays less income tax than one without. So that the employer can determine the **income tax withholding amount,** each employee must give his employer a **Withholding-exemption Certificate** which shows the number of persons he claims as dependents. The employer can then figure the income-tax deduction by taking 20 per cent of the net earnings after exemption deductions, or he can use tables supplied by the government. These tables are based on 20 per cent deduction, but they are too extensive to include here. There may be other additional tax levies made when the income tax return is filed, but these will not be included here.

Each exemption provides deductions before figuring the tax as follows:

Pay-roll period	Amount of one withholding exemption
Daily or miscellaneous	$ 1.80
Weekly	13.00
Biweekly	26.00
Semimonthly	28.00
Monthly	56.00
Quarterly	167.00
Semiannually	333.00
Annually	667.00

In determining the amount of income tax to be deducted and withheld, the last digit of the wage amount may, at the election of the employer, be reduced to zero, or the wage amount may be computed to the nearest dollar. For example, if the weekly wage is $75.43, the employer may make this $75.40, or $75.00 from which to determine the deductions.

The steps in computing the income tax to be withheld are as follows:

1. Round off the total wage to the nearest dollar.
2. Multiply the amount of one withholding exemption (see preceding table) by the number of exemptions claimed.
3. Subtract the amount thus determined from the rounded-off wage.
4. Multiply the difference by 0.20 (for 20 per cent).
5. Subtract the percentage of step 4 from the actual wage payment. This gives the net after deduction.

Example.—J. Adams has a weekly wage of $75.43 and has in effect a Withholding-exemption Certificate claiming three exemptions. Determine his income withholding tax and the net after deduction.

(1)	Total wage, $75.43, rounded off to nearest dollar	= $75.00
(2)	Amount for three exemptions is 3 × $13.00	= $39.00
(3)	Subtract to get amount to tax	$36.00
(4)	Tax is 20% of $36.00, or 0.20 × $36.00 = $7.20, therefore, income-tax deduction	= $ 7.20
(5)	Net after deduction is $75.43 − 7.20	= $68.23

Note.—This is not the amount of take-home pay as the Social Security tax deduction has not been made.

PROBLEMS

In the following problems compute the amounts in the last three columns. Use 20% to figure the income-tax deduction and round off the weekly wage to the nearest dollar before figuring the amount to be withheld. Be sure to subtract the deduction from the weekly wage, not from the rounded-off wage.

Number	Name	Weekly wage	De-pend-ents	Amount subject to tax	Amount to with-hold	Net after deduc-tions
Example	Adams, J.	$75.43	3	$36.00	$7.20	$68.23
1	Black, R.	82.20	2			
2	Brown, E.	66.75	3			
3	Coe, C. C.	54.00	2			
4	Davis, J.	58.42	2			
5	Dons, F.	87.88	1			
6	Jacks, J.	96.92	4			
7	Kregs, C.	77.65	6			
8	Laret, A.	92.40	6			
9	Weil, L.	95.00	1			
10	Woe, R.	76.25	5			

TAKE-HOME PAY

The total wages of a workman are not all available to him for spending any more than are the total receipts of his employer. Wage deductions required by law must be made, and many others may be agreed upon between the employer, his employees, and the labor unions. Income-tax and Social Security (and in some states, disability and unemployment) deductions are required. Other deductions agreed upon may include union dues, welfare funds, etc. "Take-home pay" has become the common expression to denote the money actually received by the worker after the required deductions have been made, and only these will be made in the following examples and problems.

Example 1.—Determine the take-home pay for Frank Scott whose total wage is $195 per month in 1951 and who has an exemption certificate for one.

$195 − 56.00 =	$139.00	Taxable income after monthly exemption of $56.
$139 × 0.20 = $27.80		Income tax at 20%.
$195 × 0.015 = $ 2.93		Social Security tax at $1\frac{1}{2}$%.
$30.73		Total required deduction.
$195 − 30.73 =	$164.27	Monthly take-home pay.

PROBLEMS

Copy the following pay roll in your notebook and complete it. In problems 6, 7, and 8, remember that only the first $300 per month is subject to Social Security tax, whereas the full salary is subject to income-tax deduction.

Number	Name	Monthly pay	Exemptions	Income-tax deduction	Social Security deduction	Take-home pay
Example	Scott, Frank	$195	1	$27.80	$2.93	$164.27
1	Black, Mary	172	2			
2	Cook, James	196	4			
3	Jones, Ray	185	3			
4	Love, Cliff	200	2			
5	Brown, Bill	244	3			
6	Rowe, James	350	3			
7	Stone, Frank	480	5			
8	Varnes, F. F.	550	4			
9	Wirth, Gae	288	3			
10	Young, Fred	160	4			

Example 2.—Figure the take-home pay for John Doe whose rate is $1.85 per hr. for a 40-hr. week during 1951. He has an exemption certificate for two dependents.

$1.85 × 40		= $74.00	Total weekly wage.
$13.00 × 2		= $26.00	Exemption for 2 dependents.
		$48.00	Subject to income tax.
$48.00 × 0.20 =	$ 9.60		Income-tax deduction at 20%.
$74.00 × 0.015 =	$ 1.11		Social Security tax at $1\frac{1}{2}$%.
	$10.71		Total deduction.
$74.00 − 10.71		= $63.29	Take-home pay.

PROBLEMS

Copy the following weekly pay roll in your notebook and complete it. Consider 44 hr. a regular workweek. Use total wage rounded off to nearest dollar for figuring income-tax deduction. Use only $75 per week for figuring Social Security deduction.

Number	Name	Hourly rate	Hours worked	Exemptions	Total earned	Income-tax deductions	Social Security deductions	Take-home pay
Example 2	Doe, John	$1.85	40	2	$74.00	$9.60	$1.11	$63.29
1	Brown, J.	1.65	38	3				
2	Davis, N.	1.90	44	2				
3	Ellis, Joe	2.32	42	3				
4	Evers, C.	2.88	44	4				
5	Koons, F.	2.16	40	2				
6	Vance, Lee	1.95	44	3				
7	Wend, D.	1.65	44	5				
8	White, C.	2.32	42	1				
9	Young, L.	2.08	44	5				
10	Zener, T.	2.66	31	3				

WAGE COMPUTATIONS

Computing wages, keeping a record of the time worked, and determining the amount due should be of prime importance to the worker as well as to the employer. Every workman should keep his own record of work to compare with that of his employer.

Wages are no longer considered merely as the payment for services rendered. In addition, wages are a three-fold investment: (1) in unemployment compensation, (2) in disability compensation, and (3) in old-age security.

Many trade unions have secured additional benefits from their employers in the form of trust funds, often called **funded plans,** that will add to old-age pensions. Likewise, they have added to family benefits with plans called **health and welfare funds,** which may also provide free accident and health protection for the entire family. Often these benefits are paid for entirely by the employer, while in others both the employer and the employee contribute. These benefits are actually additions to the worker's wage and are often figured as such in wage agreements, but they are not paid in spendable money and are not generally included for tax purposes in the figured wage.

In the highly organized mechanical industries, efficiency experts use different methods to compute the wages due. Each workman should be familiar with the system used where he is employed. In addition, he should understand other systems so that, should he become an employer, he can apply this knowledge to his own best interests.

The simplest and most common system of wage payment is the hourly rate. With this system the wages due are determined by multiplying the number of hours worked by the rate per hour. The present tendency is for the worker to be able to expect a definite number of hours per day and per week. Any additional time worked is called **overtime.** Overtime is usually paid for at an increased rate. For example, double the regular rate may be paid for overtime on regular days, for all work on holidays, and in some cases for work on Saturday and Sunday. In calculating the amount due for overtime, many use the practice of figuring the rate per hour the same, but of multiplying the time spent in overtime by $1\frac{1}{2}$ or 2. Thus, "double time for overtime" means that in figuring the pay for the overtime the hours worked are doubled, but the rate per hour remains the same as for regular time.

The **Fair Labor Standards Act** of 1938 provided that in many specified industries, such as those producing goods for interstate commerce, the **maximum standard workweek** should be 40 hr. Work over this limit must be paid for at a rate at least $1\frac{1}{2}$ times the employee's regular hourly rate of pay. As defined by the act, a workweek is 7 consecutive 24-hr. days. There is no limit specified in the law as to the hours worked in any one day. However, many trade unions have agreements that limit work to 16 hr. in any 24-hr. period.

The minimum wage rate required by the act was 30 cents per hour. This was raised to 40 cents in 1940, and to 75 cents in 1950.

On many jobs, such as construction work, a timekeeper or the foreman usually keeps the record of the hours each employee works. In most industrial work, such as in factories, timecards are used to keep the record. When the workman starts his shift, he takes his card to a time clock and stamps on the card his starting time. When he finishes his shift, he stamps the time again. From this card a time clerk figures the workman's total time and the wages due him.

Example.—Lynn Abbot worked 8, 7, 8, 7, 6, and 4 hr. on the days of a week in 1951 at $1.65 per hr. Determine his total time, the amount due him, his income-tax deduction, Social Security deduction, and his take-home pay.

8 + 7 + 8 + 7 + 6 + 4 =	40 hr.	Total time.
$1.65 × 40 =	$66.00	Amount due him.
One exemption	$13.00	
	$53.00	Subject to income tax.
$53.00 × 0.20 =	$10.60	Income-tax deduction at 20%.
$66.00 × 0.015 =	$ 0.99	Social Security tax at $1\frac{1}{2}$%.
$10.60 + 0.99 =	$11.59	Total deductions.
$66.00 − 11.59 =	$54.41	Take-home pay.

PROBLEMS

1. Copy the following pay roll in your notebook and complete it. Make the necessary computations to show the hours worked, the amount due, the income-tax deduction at 20%, Social Security tax at $1\frac{1}{2}$%, and the take-home pay. Remember, use no more than $75 per week for figuring Social Security deduction, but use the full amount due for figuring the income-tax deduction after allowance for exemptions.

Number	Name	Hours per day						Total time	Wage per hour	Ex-emp-tions	Amount due	In-come tax, 20%	Social Se-curity tax 1½%	Take-home pay
		M	T	W	T	F	S							
Example	Abbott, L.	8	7	8	7	6	4	40	$1.65	1	$66.00	$10.60	$0.99	$54.41
1	Brown, J.	8	8	8	8	8	0		1.75	2				
2	Cole, F.	7	7	7	7	7	4		2.55	3				
3	Dennis, W.	8	7	6	8	7	3		1.90	2				
4	Thomas, R.	8	8	7	8	5	4		2.25	4				
5	Jonas, D.	8	8	6	8	6	4		1.88	1				
6	Kline, Ed.	7	8	7	8	7	2		2.05	3				
7	Love, Ned	9	9	9	9	0	0		1.95	5				
8	Wilkes, J.	8	8	8	8	4	4		1.88	2				
9	Wills, W.	8	8	7	7	8	0		0.95	1				
10	Yancey, C.	7	8	7	8	6	2		2.14	3				

2. On the following pay roll, the maximum standard workweek established by the Fair Labor Standards Act will apply, and 1½ times the overtime will be allowed for all time over 40 hr. per week.

Copy the table in your notebook and figure the total hours pay, the total weekly wage, exemption allowance at $13 each, income-tax deduction at 20% figured on the net wages when rounded off to the nearest dollar, Social Security tax at 1½% on wages up to $75 per week, and finally the take-home pay.

Number	Hours per day							Wage per hour	Regular hours worked	Overtime		Total hours pay	Total weekly wage	Ex-emp-tions	Ex-emp-tion allow-ance	In-come tax, 20%	Social Se-curity tax, 1½%	Take-home pay
	M	T	W	T	F	S	S			Hours worked	Hours pay							
Example	8	8	8	8	8	8	8	$2.25	40	16	24	64	$144	3	$39	$21	$1.13	$121.87
1	7	7	7	7	7	7	0	1.66						2				
2	9	9	9	9	9	0	0	1.88						4				
3	8	8	8	8	8	8	0	1.88						1				
4	9	8	7	9	8	7	0	1.50						2				
5	8	8	8	8	8	8	4	1.66						3				
6	7	8	8	7	8	7	0	1.75						6				
7	6	6	6	6	6	6	6	0.98						2				
8	8	8	0	8	8	0	0	1.42						1				
9	8	8	7	7	7	7	4	1.75						3				
10	7	7	7	7	7	7	7	1.32						4				

LIFE INSURANCE

In the tables illustrating the discussion of Budgets (pages 123 to 125), savings and insurance are the first items. These, like Social Security payments, are investments in future independence. Because Social Security benefits are not large, provision for additional future income should be made. One of the best ways to do this is to purchase life insurance from a reliable company.

Life insurance companies write life insurance contracts of many types. The contract itself is called the **policy.** The person who makes the contract, the insured, is called the policy holder. The amount the policy is written for is called the **face** of the policy. The money paid by the insured person to the company for his insurance is called the **premium.** Usually the premium is paid annually, but it can be paid semiannually, quarterly, and, in many cases, monthly or weekly. The amount of the premium depends on the age of the insured at the time the policy is started and the kind of policy. The **beneficiary** is the person to whom the face of the policy is paid at the death of the insured.

An insurance policy is said to **mature** at the time the face value is to be paid off by the insurance company. If at any time the policyholder decides to discontinue his policy, with some types he can collect a certain amount of what he had already paid into the policy. The amount is called the **cash surrender value.**

There are many types of life insurance policy, the most common of which are described below.

Ordinary Life Insurance.—In the case of an ordinary life policy, the insured pays a fixed premium till his death. The beneficiary then collects the face value of the policy.

Term Insurance.—In the case of a term policy, the insured pays a fixed premium for a stated interval of time. During this period of time, if the insured should die, his beneficiary would receive the face value of the policy.

When the term of the policy expires, however, the policy simply ceases to exist. There is no cash surrender value. Because people usually do not like to lose the money they have invested in a term policy, insurance companies have various ways in which term policies can be converted to other types of insurance.

Twenty-payment Life Insurance.—In the case of a 20-payment life policy, the insured pays a fixed premium for a period of 20 years. The policy is then said to be **paid up.** The face value will be paid to the beneficiary should the insured die during the period of payment or when the insured dies after the policy is paid up. This is one form of the **limited payment life policy.**

Twenty-year Endowment.—In the case of a 20-year endowment policy, the premiums are paid for 20 years. At the end of 20 years, the face value of the policy is paid to the policyholder himself, if he is living. If the policyholder should die before the 20 years are up, the face value of the policy is paid to his beneficiary. This is one form of the **endowment policy.**

Annual Premium in Advance for Each $1000 of Insurance

Age	Ordinary life	20-payment life	20-year endowment
15	$13.85	$21.26	$42.72
16	14.12	21.63	42.74
17	14.41	22.02	42.78
18	14.71	22.40	42.82
19	15.02	22.80	42.85
20	15.36	23.20	42.89
22	16.04	24 04	42.96
24	16.80	24.93	43.03
26	17.64	25.88	43.12
28	18.57	26.91	43.24
30	19.63	28.02	43.41
32	20.82	29.26	43.65

Semiannual premium will be 50.8 per cent of the annual premium.
Quarterly premium will be 25.7 per cent of the annual premium.

The table on page 155 is taken from the rate book of a large life insurance company. It is for nondividend policies.

Example.—Compute the (1) annual, (2) semiannual, and (3) quarterly premium for an ordinary life insurance policy for $3000.00 on a 20-year-old workman.

$15.36 × 3 = $46.08	1. Annual premium. This will be 3 times the amount shown in the table for each $1000.00 of insurance.
$46.08 × .508 = $23.41	2. The amount of each semiannual premium is 50.8% of the annual premium (see the table, page 155).
$46.08 × .257 = $11.84	3. Amount of each quarterly premium (see the table, page 155).

PROBLEM

Using the rate table given on page 155, compute the annual, semiannual, and quarterly premiums on each of the following. Copy the table below in your notebook.

Type of insurance	Age	Amount	Premium to be paid		
			Annually	Semi-annually	Quarterly
Ordinary life	16	$ 2,000.00			
20-payment life	18	2,500.00			
20-year endowment	20	3,000.00			
20-payment life	30	1,800.00			
20-year endowment	19	4,000.00			
Ordinary life	24	5,000.00			
20-payment life	28	10,000.00			
Ordinary life	32	1,500.00			
20-year endowment	22	1,000.00			
20-payment life	17	2,500.00			

Endowment Life Income Insurance.—This type of insurance is written by most companies. The plan works something like the Social Security plan of the Federal government. The person who takes out the insurance policy pays a stated amount regularly to the insurance company. Then, at a stated age, the insurance company pays back the face of the policy to him, either in a lump sum or in installments.

This type of policy offers the kind of insurance needed if the insured wishes to add to the income guaranteed by Social Security. It differs from Social Security in that all the premium must be paid by the insured person, for he is the one who has made the contract with the insurance company.

Annual Premiums Needed to Pay $10 per Month on Endowment Life Income Insurance per $1000

Age of insured	Maturing at age of		
	65	60	50
15	$16.70	$20.35	$34.15
16	17.18	21.01	35.63
17	17.68	21.61	36.53
18	18.23	22.79	38.93
19	18.90	23.57	40.77
20	19.39	24.40	42.74
22	20.65	26.73	47.15
24	22.07	28.28	52.32
26	23.68	30.63	58.43
28	25.48	33.33	65.73
30	27.87	36.44	75.52
32	30.19	40.10	85.21

If premiums are paid semiannually, take 50.8 per cent of the annual premium.

If premiums are paid quarterly, take 25.7 per cent of the annual premium.

Endowment income insurance offers many advantages, among them the following:

1. The policy can be written for any desired amount, to be paid back at any age selected.

2. The policy provides insurance at the death of the insured.

3. An endowment (lump sum) in cash at the maturity of the policy may be taken.

4. The policy has a cash value at any time after 3 years; this means that the person who has the policy can borrow some of the money he has put into his insurance.

5. Endowment income insurance will pay a fixed amount to the insured every month after the time stated in the policy till his death. After his death, his beneficiary will receive the money according to the terms stated in the policy.

Example.—John Smith, who is 20 years old, takes out an endowment life income policy for $3000.00 to mature when he is 65 and to pay him $30.00 per month until his death. What are his

1. Annual premium?
2. Semiannual premium?
3. Quarterly premium?
4. Daily savings to pay annual premium?

$19.39 × 3 = $58.17	1. Annual premium (see table on page 157)
$58.17 × .508 = $29.55	2. Semiannual premium (see table)
$58.17 × .257 = $14.95	3. Quarterly premium (see table)
$58.17 ÷ 365 = $ .16	4. The daily savings for the annual premium equals the total annual premium divided by the number of days in a year.

PROBLEM

Using the rate table given on page 157, compute the endowment life insurance premiums for each of the following. Copy the tabulation in your notebook.

Age when insured	Age to retire	Endowment per month	Annual premium	Semi-annual premium	Quarterly premium	**Daily savings for annual payment**
18	65	$20.00				
20	60	30.00				
30	65	40.00				
16	50	25.00				
24	60	40.00				
32	65	35.00				
17	50	30.00				
15	65	25.00				
19	65	50.00				
22	60	35.00				

REVIEW

Age when insured	Age when policy matures	Type of insurance	Amount or monthly endowment	Premium	
				Annual	Quarterly
32	65	Endowment life income	$ 45 mo.		
22	42	20-payment life	1500		
24	. .	Ordinary life	750		
28	50	Endowment life income	75 mo.		
32	52	20-yr. endowment	4500		
18	38	20-payment life	5000		
30	50	20-yr. endowment	5000		
30	60	Endowment life income	5000		

CHAPTER 7

RULES AND FORMULAS

DEFINITIONS

In the preceding chapters, you have learned rules, in words, for various operations. The words by which these rules were expressed were as few as possible. By the use of symbols, however, a rule can be expressed in an even briefer form, called a **formula.**

In a formula, the unknown quantities are represented by letters; for example, s is often used to stand for "sum," f is used for "feet," and b may be used for "brass." You will notice that each of these letters is the first letter of the word it stands for.

The letters representing unknown quantities are called **literal numbers.**

In a formula, the signs of operation are used in the following manner:

Wherever addition is implied, use the plus sign.

Wherever subtraction is implied, use the minus sign.

Wherever division is implied, use the line of fraction. The division sign is rarely employed.

Wherever multiplication is implied between abstract numbers or between the same literal number repeated, use the times sign or the raised dot; for example, $3 \times 2 \times 5$, $a \cdot a \cdot a$, $9 \cdot 8$. No sign is written to indicate multiplication between an abstract number and a literal number or between different literal numbers. For example, $3x$ means "3 times x"; LWT means "$L \times W \times T$."

An **equation** is a statement of equality. For example, $3b = 12$ states that 3 b's are equal to 12. An equation is said to have **sides** or **members;** in the equation above, $3b$ is the left side of the equation, and 12 is the right side.

A formula is sometimes called a literal equation.

Terms are made up of factors (numbers combined by multiplication or division); for example, $7ab$ is a term in which the factors are 7, a, and b; $n/2$ is a term with factors n and $\frac{1}{2}$.

In a term like $3b$, 3 is the numerical factor, and b is the literal factor. In writing a term with a literal and numerical factor, the numerical factor is written first, for example, $3b$.

An **expression** is a collection of terms combined by either addition or subtraction or both; for example, $5a + 8b$, $2y - 6x$, $3x + 7a - 5b$.

Substitution is the operation used when one thing is put in place of another. In writing formulas, literal numbers and symbols are substituted for words. In using a formula, a known value may be substituted for a letter representing an unknown value.

WRITING FORMULAS FROM RULES

Notice how the following word rules have been expressed as formulas:

Example 1.

Rule.—Cost equals the selling price minus the margin.

Formula. $C = s - m$, where C represents cost, s represents selling price, and m represents margin.

Example 2.

Rule.—To change a certain number of pounds to tons, divide by 2000

Formula. $T = \frac{P}{2000}$, where T represents tons and P represents pounds.

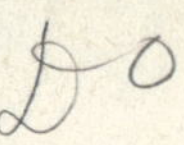

Example 3.

> *Rule.*—The circumference of a circle is equal to pi multiplied by the diameter.
>
> *Formula.* $C = \pi d$, where C stands for the circumference, π stands for pi, and d stands for the diameter.

Example 4.

> *Rule.*—The total weight of a truck loaded with gravel is equal to the weight of the truck plus the weight of the gravel.
>
> *Formula.* $W = T + G$, where W represents the total weight, T represents the weight of the truck and G represents the weight of the gravel.

EXERCISES

In the following rules and statements, substitute letters for words and write the formulas. See page 523 for tables of equivalents.

1. To change from inches to feet, divide the number of inches by 12. Write a formula for changing from feet to inches.

2. Dividing ounces by 16 gives pounds. Write a formula for pounds.

3. Gross multiplied by 144 equals units. Write a formula for units.

4. The number of cubic inches divided by 231 equals the number of gallons. Write a formula for obtaining the number of gallons from a number of cubic inches.

5. Write a formula for changing rods to feet. $F = ?$

6. Write a formula for changing feet to rods. $R = ?$

7. Write a formula for changing pounds to ounces. $O = ?$

8. Write a formula for changing ounces to pounds. $P = ?$

9. Write a formula for changing gross to dozen. $D = ?$

10. If there are 6 feet in 1 fathom, write a formula for changing a number of feet to fathoms. Write a formula for changing from fathoms to feet.

11. Write a formula for converting square yards to square feet.

12. There are 44 cu. ft. of coal per ton. Write a formula for changing from cubic feet to tons; for changing from tons to cubic feet.

13. The pay load equals the gross load minus the dead load. $P = ?$

14. The area of a parallelogram equals the width times the height. $A = ?$

15. The area of a triangle is equal to one-half the base times the height. $A = ?$

16. The area of a circle is equal to π times the radius squared. $A = ?$

Note.—To "square" a number means to use it as a factor twice, as 6 squared is 6×6, written 6^2.

17. The diagonal of a square is equal to 1.414 multiplied by the base. $D = ?$

18. The base of a square is equal to the diagonal divided by 1.414. $B = ?$

19. The surface of a sphere is equal to π times the diameter squared. $S = ?$

20. The volume of a rectangular solid is the product of its three dimensions, length, width, and height. $V = ?$

21. Interest on money is figured by multiplying the principal by the rate by the time. $I = ?$

22. To find the number of board feet, multiply the thickness in inches by the width in inches by the length in feet, and divide by 12. $B = ?$

23. The volume of a prism is equal to the area of the base times the height. $V = ?$

24. The diameter of a circle is found by dividing the circumference by π. $D = ?$

25. Twice the radius multiplied by π equals the circumference of a circle. $C = ?$

26. The radius of a circle is equal to one-half the diameter. $R = ?$

27. The area of the curved surface of a cylinder is equal to the product of π, the diameter, and the height of the cylinder. $A = ?$

28. The volume of a cylinder is equal to the product of π, the radius squared, and the height of the cylinder. $V = ?$

29. To get the horsepower of an electric motor multiply the number of volts by the number of amperes and divide by 746. H.P. = ?

30. The number of watts is equal to the product of the number of volts and number of amperes. $W = ?$

31. Brass is a mixture of copper plus zinc. $B = ?$

32. The size of a drill for a tap may be found by subtracting the double depth (D) from the tap diameter (T). $S = ?$

33. Liquid pressure equals the head multiplied by 0.433. $P = ?$

34. Mechanical work is the product of the weight and the height. $W = ?$

35. Energy is the product of the force and the distance. $E = ?$

36. In a right triangle the square of the hypotenuse is equal to the sum of the squares of the two sides. $h^2 = ?$

37. Volts is the product of the amperes and the ohms. $V = ?$

38. To find the number of amperes, divide the number of volts by the number of ohms. $A = ?$

39. To find the number of ohms, divide the number of volts by the number of amperes. $O = ?$

40. To find the number of volts, divide the number of watts by the number of amperes. $V = ?$

41. To find the number of amperes, divide the number of watts by the number of volts. $A = ?$

42. To find the number of kilowatts, divide the number of watts by 1000. $KW = ?$

43. To find the amount of horsepower, divide the number of kilowatts by 0.746. H.P. = ?

44. Divide the number of watts by 746 to get the amount of horsepower. H.P. = ?

SOLVING SIMPLE EQUATIONS

Some of the rules above, especially those for changing the form of denominate numbers, are familiar to you from your previous work in arithmetic. You then performed the conversion without thinking of a formula. Now you have expressed this process as a formula, so that you can solve any problem that comes under the particular formula by simply applying the formula directly.

When you solve simple formulas or equations, there are a few fundamentals to remember.

Because an equation says that two expressions are equal to each other, *whatever is done to one of the expressions must be done to the other* in order not to change the equality.

When an equation is solved for a quantity, only one of that quantity is wanted.

If the number 1 precedes a literal number, it is dropped, for m means $1m$.

When no sign is written before a term, it is understood that the term is added to the side of the equation on which it appears, or that a plus sign is understood preceding the term, as $5a + 6 = 7$ is understood to mean $+5a + 6 = +7$.

FOUR FUNDAMENTAL OPERATIONS

The four fundamental operations in mathematics are subtraction, addition, division, and multiplication. Whichever one of these four operations is applied to the left side of an equation must also be applied to the right side of the equation.

Operation 1. Subtraction

Operation 1.—The same number may be subtracted from each side of an equation without changing the equality.

Example 1.—$Y + 5 = 12$. Solve for Y.

Solution.

$$Y + 5 = 12$$
$$Y + 5 - 5 = 12 - 5$$
$$Y = 12 - 5$$
$$Y = 7$$

The plus sign indicates that there are 5 units more than Y on the left side of the equation. To find Y, 5 must be subtracted from each side of the equation.

Check.

$$7 + 5 = 12$$
$$12 = 12$$

EXERCISES

Solve for each letter.

1. $X + 3 = 9.$ **2.** $X + 4 = 7.$

3. $Y + 6 = 15.$ **4.** $7 = A + 3.$

5. $8 = A + 4.$ **6.** $12 = Z + 2.$

7. $11 = A + 4.$ **8.** $Z + 3 = Y.$

9. $Y + 8 = 21.$ **10.** $A = T + 4.$

11. $L = 3 + Y.$ **12.** $S = B + b.$

13. $Z = S + A.$ **14.** $Z = A + 6.$

15. $L = R + t.$ **16.** $s = t + r.$

17. $A + 6 = 14.$ **18.** $C + 13 = 16.$

19. $16 + B = 20.$ **20.** $5 + A = 9.$

21. $r = P + s.$ **22.** $R + T = L.$

23. $S = t + r.$ **24.** $Z + 8 = Q.$

25. $P = a + b + 9.$ **26.** $S = P + a + 6.$

27. $Y + 2 = S + B.$ **28.** $A + B = w + 3.$

29. $S + c + d = P.$ **30.** $L = W + T + B.$

31. $A = b + c + d + e.$ **32.** $S + 12 = C + H.$

Operation 2. Addition

Operation 2.—The same number may be added to each side of an equation without changing the equality.

Example 1.—Solve for x in the equation $x - 2 = 9.$

$x - 2 = 9$	Since the left side is 2 units less
$x - 2 + 2 = 9 + 2$	than x, 2 must be added to each side
$x = 9 + 2$	of the equation to find x.
$x = 11$	

Example 2.—Solve for C in the equation $C - 4 = A.$

$C - 4 = A$	
$C - 4 + 4 = A + 4$	Add 4 to both sides of equation
$C = A + 4$	

EXERCISES

Solve the following exercises by applying Operation 2.

1. $M - 5 = 32.$

2. $47 = S - 3.$

3. $H - 10 = 10.$

4. $12 = K - 12.$

5. $P - 3 = s.$

6. $F = A - 1.3.$

7. $12 = M - 13.$

8. $B = C - 25.$

9. $S = T - D.$

10. $A = W - c.$

11. $R = X - 3.$

12. $W - A = B.$

13. $T - C = K.$

14. $A = B - 3.$

15. $T - f = s.$

16. $B - 5 = 10.$

17. $L = P - 4.$

18. $S = T - D.$

19. $A = W - c.$

20. $F = P - 1.3.$

21. $R = S - d.$

22. $B = R - r.$

23. $U - M = Q.$

24. $T - f = c.$

25. $B - 12 - A.$

26. $R - 5 = 9.$

27. $U - N = K.$

28. $4 + S = 13.$

29. $T - C = S + A.$

30. $A + 7 = X - 3.$

31. $Y - 2 = B + 3.$

32. $M = n + v - p.$

Operation 3. Division

Operation 3.—Each side of an equation may be divided by the same number without changing the equality.

Example 1.—Solve for P in the equation $8P = 24$.

$$8P = 24$$

$$\frac{8P}{8} = \frac{24}{8}$$ Divide both sides by 8 by applying Operation 3.

$$P = \frac{24}{8}$$ Cancel $\frac{8}{8}$ on the left side and divide 24 by 8 on the right side.

$$P = 3$$

Check $8 \times 3 = 24$ Check by substituting 3 for P.

$24 = 24$

Example 2.—Solve for s in the equation $n = 9s$.

$$n = 9s$$

$$\frac{n}{9} = \frac{9s}{9} \quad \text{Divide both sides by 9.}$$

$$\frac{n}{9} = s \quad \text{By canceling } \frac{9}{9} \text{ on the right side.}$$

Notice that in each of these examples there was an unwanted number attached to the unknown letter by multiplication. The unwanted number is used as a divisor on both sides.

Example 3.—Solve the equation $A = bh$ for b and for h.

$$A = bh$$

$$\frac{A}{h} = \frac{bh}{h} \quad \text{To get } b \text{ divide both sides by } h.$$

$$\frac{A}{h} = b \quad \text{By canceling } \frac{h}{h} \text{ on the right side.}$$

$$\text{or } b = \frac{A}{h}$$ It is common practice to place the single letter to the left of the equal sign.

$$\frac{A}{b} = \frac{bh}{b} \quad \text{To get } h \text{ divide both sides by } b.$$

$$\frac{A}{b} = h, \text{ or } h = \frac{A}{b}$$

EXERCISES

In the following equations, solve for each letter by applying Operation 3.

1. $7x = 42.$	**2.** $5Y = 30.$	**3.** $10c = 24.$
4. $K = 5x.$	**5.** $m = 3n.$	**6.** $W = 4z.$
7. $7Y = T.$	**8.** $5f = c.$	**9.** $42g = A.$
10. $5r = 44.$	**11.** $55 = 7P.$	**12.** $RN = 40.$
13. $h = 0.866r.$	**14.** $W = 110A.$	**15.** $P = RS.$
16. $ND = S.$	**17.** $E = ir.$	**18.** $S = 60h.$
19. $P = 16z.$	**20.** $A = bh.$	**21.** $A = LW.$
22. $W = va.$	**23.** $E = IR.$	**24.** $C = 327G.$
25. $H = 2.3p.$	**26.** $G = 7.5f.$	**27.** $E = SN.$
28. $1.4b = d.$	**29.** $0.88d = s.$	**30.** $60H = S.$

Eliminating More than One Factor at a Time

In some problems more than one factor must be eliminated in order to solve for the unknown.

Example 1.—Solve for L, for W, and for H in formula $V = LWH$.

$$V = LWH$$

$$\frac{V}{WH} = L; \qquad W = \frac{V}{LH}; \qquad H = \frac{V}{LW}$$

Example 2.—Solve $pl = wh$ for each letter.

$$pl = wh$$

$$l = \frac{wh}{p}; \qquad p = \frac{wh}{l}; \qquad w = \frac{pl}{h}; \qquad h = \frac{pl}{w}$$

EXERCISES

In each of the following, solve in turn for each literal number.

1. $S = \pi DH$.	**2.** $A = 0.5bh$.	**3.** $I = PRT$.
4. $L = RNS$.	**5.** $5280 = \pi DN$.	**6.** $P = HAW$.
7. $PL = WT$.	**8.** $10F = RD$.	**9.** $L = RFA$.
10. $PD = rst$.	**11.** $nW = PRN$.	**12.** $Pp = Ws$.
13. $7L = 8R$.	**14.** $5rs = 4ay$.	**15.** $9LMR = TF$.
16. $PC = WL$.	**17.** $s = 5TSA$.	**18.** $F = 0.03WRN$.

Operation 4. Multiplication

Operation 4.—Each side of an equation may be multiplied by the same number without changing the equality.

Example 1.—Solve for V in the formula $g = \frac{V}{231}$

$g = \frac{231}{V}$	To solve for V, 231 must be eliminated. Since the line of a fraction indicates division, both sides multiplied by 231 will eliminate it from the dividend on the right side.
$\frac{231}{1} \times g = \frac{V}{231} \times \frac{231}{1}$	
$231g = V$	Found by canceling $\frac{\cancel{231}}{\cancel{231}}$ on right side.

EXERCISES

Solve the following formulas by applying Operation 4.

1. $W = \frac{A}{b}$. **2.** $V = \frac{W}{A}$. **3.** $d = \frac{C}{\pi}$. **4.** $r = \frac{d}{2}$.

5. $\frac{E}{i} = R$. **6.** $\frac{i}{12} = f$. **7.** $N = \frac{X}{a}$. **8.** $\frac{Y}{F} = 3$.

9. $d = \frac{G}{12}$. **10.** $\frac{G}{7.5} = f$. **11.** $\frac{h}{60} = s$. **12.** $N = \frac{W}{P}$.

13. $S = \frac{P}{R}$. **14.** $R = \frac{40}{N}$. **15.** $\frac{W}{F} = P$. **16.** $b = \frac{v}{a}$.

17. $F = \frac{Y}{3}$. **18.** $G = \frac{C}{231}$. **19.** $\frac{S}{0.88} = d$. **20.** $W = \frac{P}{f}$.

21. $H = \frac{W}{746}$. **22.** $\frac{W}{P} = S$. **23.** $R = \frac{P}{s}$. **24.** $X = \frac{N}{D}$.

25. $A = \frac{bh}{2}$. **26.** $R = \frac{C}{2\pi}$. **27.** $\frac{V}{WL} = H$. **28.** $\frac{i}{T} = PR$.

29. $P = \frac{i}{rt}$. **30.** $N = \frac{V}{\pi d}$. **31.** $\frac{P}{w} = \frac{W}{p}$. **32.** $\frac{P}{r} = \frac{W}{R}$.

33. $\frac{P}{L} = \frac{W}{H}$. **34.** $\frac{P}{d} = \frac{W}{R}$. **35.** $\frac{R}{10} = \frac{F}{D}$. **36.** $\frac{p}{t} = \frac{w}{L}$.

Changing Signs in Equations

All the signs of an equation may be changed without changing the equality.

Equations often require more than one step in their solutions. It is best to perform the operations in the following order. (1) Do all additions and subtractions. (2) Do all multiplications and divisions.

Example 1.—In $d = D - 2V$, solve for V.

$d = D - 2V$ $-d = -D + 2V$	Since the sign preceding V must be plus, change all signs on both sides.
$D - d = D - D + 2V$	Add D to both sides.
$D - d = 2V$	
$\frac{D - d}{2} = \frac{2V}{2}$	Divide both sides by 2.
$\frac{D - d}{2} = V$	By canceling $\frac{2}{2}$.

Example 2.—Solve $d = D - YL$ for Y and L.

$d = D - YL$	
$-d = -D + YL$	Applying Operation 5.
$D - d = D - D + YL$	Add D to both sides.
$D - d = YL$	
$\frac{D - d}{L} = \frac{YL}{L}$	Divide both sides by L. By canceling $\frac{L}{L}$.
$\frac{D - d}{L} = Y$	
$\frac{D - d}{Y} = \frac{YL}{Y}$	Divide both sides by Y.
$\frac{D - d}{Y} = L$	By canceling $\frac{Y}{Y}$.

EXERCISES

In the following, solve for each letter:

1. $S = A - 3B$.
2. $f = V + AH$.
3. $6A - 5 = 12$.
4. $2Y - 3 = 13$.
5. $2A + 5 = 15$.
6. $3A + 4 = 2B$.
7. $D = 2r + d$.
8. $2L = W + 2$.
9. $3S = 4D - 3$.
10. $3X = A + B$.
11. $P = 2b + 2h$.
12. $T = 2t + c$.
13. $d = D - YL$.
14. $F = WA + f$.
15. $AS = D - d$.
16. $2Y = 3T + s$.
17. $Y + 2 = QT - 3$.
18. $2A = 3X + Y$.
19. $16H = D - 12$.
20. $D = N - 3W$.
21. $YL = H + 4$.
22. $T = 0.7s + 0.005D$.

PARENTHESES, FRACTIONS, AND RADICAL SIGNS

When a number of quantities are grouped together, parentheses ( ), brackets [ ], braces { }, and the vinculum (or raised bar) $\overline{\quad}$ are used to indicate the desired grouping. These enclosing symbols indicate that the calculations within the enclosure are to be completed, clearing the innermost first, before the remaining calculations are commenced. For example,

1. $6 + [5 \times 4(3 + 7)] = 6 + [5 \times 4 \times 10] = 6 + 200 = 206$.

The line of a fraction acts the same way as parentheses.

2. $7 + \dfrac{2(15 + 9)}{3} = 7 + \dfrac{2 \times 24}{3} = 7 + \dfrac{48}{3} = 7 + 16 = 23$.

An expression under a radical sign must be treated as a whole unless it can be simplified.

3. $24 + \sqrt{9 + 16} = 24 + \sqrt{25} = 24 + 5 = 29$.

4. $24 + \sqrt{9 \times 16} = 24 + \sqrt{144} = 24 + 12 = 36$.

In equations and formulas with parentheses, fractional lines, or radical signs, treat the enclosed expression as a single unit, as above.

Example 1.—$t = a(D - c)$. Solve for a, D, and c.

$t = a(D - c)$	
$\frac{t}{D - c} = a$	Divide both sides of the equation by $D - c$ as a unit and omit the parentheses.
$\frac{t}{a} = D - c$	Divide both sides of the original equation by a and omit the parentheses.
$\frac{t}{a} + c = D$	Add c to both sides.
$c = D - \frac{t}{a}$	Subtract $\frac{t}{a}$ from both sides.

Example 2.—$A = \frac{h(B + b)}{2}$. Solve for h, B, and b.

$A = \frac{h(B + b)}{2}$	
(1) $2A = h(B + b)$	Multiply both sides by 2 to clear fraction.
$\frac{2A}{B + b} = h$	Divide both sides of the equation by $B + b$ and omit the parentheses.
(2) $\frac{2A}{h} = B + b$	Divide both sides in equation (1) by h, to get $B + b$.
$\frac{2A}{h} - b = B$	Subtract b from both sides of equation (2).
$\frac{2A}{h} - B = b$	Subtract B from both sides of equation (2).

Example 3. $c = \sqrt{a^2 + b^2}$. Solve for a and b.

$c = \sqrt{a^2 + b^2}$	
(1) $c^2 = a^2 + b^2$	Square both sides to remove radical sign.
$c^2 - b^2 = a^2$	Subtract b^2 from both sides of equation (1).
$\sqrt{c^2 - b^2} = a$	Extract square root of both sides.
$c^2 - a^2 = b^2$	Subtract a^2 from both sides of equation (1).
$\sqrt{c^2 - a^2} = b$	Extract square root of both sides.

EXERCISES

In the following formulas, solve for each letter.

1. $t = \frac{D - d}{L}$.
2. $s = \frac{D - d}{2}$.
3. $r = \sqrt{\frac{A}{n}}$.
4. $s = T - \frac{1.3}{N}$.
5. $L = \pi S(R + r)$.
6. $A = \pi(R^2 - r^2)$.
7. $T = \frac{12(D - d)}{l}$.
8. $L = \frac{s(P + p)}{2}$.
9. $b = \frac{8B}{9} - 0.4$.
10. $V = \frac{\pi r^2 h}{3}$.
11. $R_c = \sqrt{r^2 + R^2}$.
12. $D = \sqrt{\frac{S}{\pi}}$.
13. $V = \frac{4\pi r^3}{3}$.
14. $d = \sqrt{\frac{V}{0.7854h}}$.
15. $A = \frac{0.866s^2}{2}$.
16. $P = n\sqrt{2(a^2 + b^2)}$.
17. $h = \sqrt{s^2 - \frac{b^2}{2}}$.
18. $s = \frac{L(D - d)}{2l}$.
19. $L = \frac{\pi s(D + d)}{2}$.
20. $f = \frac{1}{8N}$.
21. $D = \sqrt{h^2 - b^2}$.
22. $V = 4.1888r^3$.
23. $A = 0.0087\pi r^2$.
24. $L = 1.57s(D + d)$.
25. $D_c = \sqrt{d^2 + D^2}$.
26. $S = T - \frac{0.975}{N}$.

CHAPTER 8

RATIO AND PROPORTION

RATIO

A ratio is the comparison of two like quantities. For example, if one gear has 50 teeth and another has 10 teeth, a comparison of the number of teeth in the two gears would be 50 teeth for the first gear to 10 teeth for the second gear. This ratio can be stated in four ways, as the ratio of 50 to 10, 50:10, $50 \div 10$, or $\frac{50}{10}$. All these expressions are read in the same way, the ratio of 50 to 10. Note that a ratio may be expressed as a fraction. Reducing the ratio $\frac{50}{10}$ to its simplest form gives us 5 to 1. This means that the first gear has 5 times the number of teeth on the second gear. We say that the number of teeth on the gears are in the ratio of 5 to 1.

A 50-tooth gear cannot be compared with 5 men. Both quantities must be of the same kind. Teeth must be compared with teeth or men with men.

The two numbers of a ratio are called the terms of the ratio.

The first term of a ratio is called the **antecedent.** In the ratio 50 teeth to 10 teeth, 50 teeth is the antecedent.

The second term of a ratio is called the **consequent.** In the ratio 50 teeth to 10 teeth, 10 teeth is the consequent.

An **inverse ratio** requires the reversing of the sequence of the terms. For example, for the ratio 50 teeth to 10 teeth, the inverse ratio would be 10 teeth to 50 teeth.

Ratios are usually reduced to their lowest terms.

Rule.—To reduce a ratio to its lowest terms, treat the ratio as a fraction, and reduce the fraction to its lowest terms.

Example 1.—Reduce the ratio 7 to 28 to its lowest terms.

> 7 to 28 = 7 ÷ 28 = $\frac{7}{28}$ = $\frac{1}{4}$
>
> Write the ratio as a common fraction, and reduce it. Hence, the ratio 7 to 28 is the same as the ratio 1 to 4.

Example 2.—Express the ratio $\frac{3}{4}$ to $\frac{5}{8}$ in its lowest terms.

$$\frac{3}{4} \text{ to } \frac{5}{8} = \frac{3}{4} \div \frac{5}{8} = \frac{3}{\cancel{4}_1} \times \frac{\cancel{8}^2}{5} = \frac{6}{5}$$

> Hence, the ratio $\frac{3}{4}$ to $\frac{5}{8}$ is the same as the ratio 6 to 5

Example 3.—Write the inverse ratio of $2\frac{1}{2}$ to $9\frac{1}{4}$, and reduce it to its lowest terms.

> $2\frac{1}{2}:9\frac{1}{4}$ Given ratio
>
> $9\frac{1}{4}:2\frac{1}{2}$ In the inverse ratio the order of the terms in the given ratio is reversed.

$$9\frac{1}{4} \div 2\frac{1}{2} = \frac{37}{4} \div \frac{5}{2} = \frac{37}{\cancel{4}_2} \times \frac{\cancel{2}^1}{5} = \frac{37}{10}, \text{ or 37 to 10}$$

Rule.—To separate a number according to a given ratio, add the terms to find the whole; then, find the fractional part each term is of the whole. Divide the number into parts corresponding to these fractional parts.

Example 4.—On a certain job, one man worked 7 hr., and another man worked 15 hr. If the job paid $44.00, divide the pay according to the number of hours each man worked.

> 7 + 15 = 22 Adding the terms
>
> One man will receive $\frac{7}{22}$ of the $44.00.
>
> $\frac{7}{22} \times 44 = 14$ This man will receive $14.00.
>
> The other man will receive $\frac{15}{22}$ of the $44.00.
>
> $\frac{15}{22} \times 44 = 30$ The other man will receive $30.00
>
> *Check.*
>
> The ratio of $14.00 to $30.00 = $\frac{14}{30}$ = $\frac{7}{15}$, or the ratio of the hours worked.

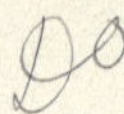

EXERCISES AND PROBLEMS

Express each of the following ratios in lowest terms:

1. Reduce each of the following ratios to lowest terms: 6 to 3; 9 to 27; $\frac{4}{5}$ to $\frac{3}{4}$; $\frac{1}{2}$ to 10; $2\frac{5}{8}$ to $5\frac{3}{4}$; 3.16 to 18.00.

2. Write the inverse ratios for each of the following: 8 to 3; $9\frac{1}{2}$ to 3; $\frac{1}{4}$ to 12; $6\frac{1}{2}$ to $2\frac{1}{4}$; $7\frac{1}{4}$ to $21\frac{3}{4}$. Reduce to lowest terms.

3. Two gears have 88 teeth and 14 teeth, respectively. Find the ratio of the number of teeth in the gear having more teeth to the number of teeth in the gear having fewer teeth.

4. One pulley running 540 r.p.m. is belted to another running 90 r.p.m. Compare the speed of the faster pulley with the speed of the slower pulley.

5. Divide $270.00 between two men in the ratio of 3 to 6.

6. Brass is composed of copper and zinc in the ratio of 66 parts copper to 24 parts zinc. Find the amount of each in 1600 lb. of brass.

7. The cutting speed of brass is 80 f.p.m., and that of steel is 35 f.p.m. Find the ratio of the cutting speed of brass to the cutting speed of steel.

8. The circumference of a circle is about 2512 ft., and the diameter is about 800 ft. Compare the circumference with the diameter.

9. A penny is composed of 19 parts copper and 1 part tin. Find the weight of each metal in 1000 lb. of pennies.

10. A machinist earns $95.00 a week, and his helper earns $55.00 a week. Find the ratio of the earnings of the helper to those of the machinist.

11. A bronze is composed of 8 parts copper and 2 parts tin. Calculate the number of pounds needed of each to make 1 ton of bronze.

12. The weight of a cubic foot of a substance is 500 lb. The weight of a cubic foot of water is 62.4 lb. Compare the weight of the substance with that of water.

13. Pulley diameters are in inverse ratio to their speeds. If two pulleys belted together are running 200 and 600 r.p.m., find the ratio of the diameter of the large pulley to the diameter of the small pulley.

Do

14. An airplane cruises at 210 m.p.h. and has a top speed of 280 m.p.h. What is the ratio of the top speed to the cruising speed?

15. Dowmetal F is composed of 96 parts magnesium and 4 parts aluminum. Determine the amount of each in 1000 lb. of Dowmetal. Find the ratio of magnesium to aluminum.

16. A concrete mixture is in the ratio of 1 part cement, 3 parts sand, and 5 parts gravel. Find the number of cubic feet of each in 4 cu. yd. of concrete.

17. A feed mixture calls for 1 part mash, 3 parts grain, and 2 parts greens. Compute the amount of each in 1000 lb. of the mixture.

18. A certain bell bronze is composed of 5 parts tin and 16 parts copper. Find the number of pounds of each needed to cast a bell weighing 840 lb.

19. Pewter has 100 parts tin and 17 parts antimony. What is the amount of each metal needed to make 1 ton of pewter?

20. Statuary bronze is made by alloying 2 parts tin, 90 parts copper, 5 parts zinc, and 9 parts lead. Find the amount of each in a ton of bronze.

21. Locomotive-bearing bronze is an alloy of 7 parts tin, 64 parts copper, and 1 part zinc. Find the amount of each in 360 lb. of bronze.

PROPORTION

A proportion is the expression of equality between two ratios. For example, 2 has the same ratio to 3 as 4 has to 6; or the ratio 2:3 = the ratio 4 to 6. This proportion may be written 2:3 = 4:6, 2:3::4:6, or $\frac{2}{3} = \frac{4}{6}$. Each of these proportions is read "two is to three as four is to six."

In a proportion the first and last terms (the outside terms) are called the **extremes,** and the second and third terms (the inside terms) are called the **means.** For example,

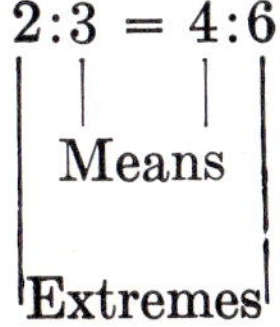

Writing the preceding example in fraction form,

$$\frac{2}{3} = \frac{4}{6}$$

Using this proportion as an equation and clearing all denominators:

$$2 \times 6 = 3 \times 4$$

In the original proportion, 2 and 6 were the extremes, and 3 and 4 were the means. Therefore, 2×6 is the product of the extremes, and 3×4 is the product of the means. This may be stated as a rule, as follows:

Rule.—In a proportion, the product of the means equals the product of the extremes.

This rule is used to solve a proportion when one of the terms is unknown.

Example 1.—In the proportion $2:3 = 4:x$, find the unknown term.

$2:3 = 4:x$	
$2x = 12$	Multiplying the extremes gives $2x$; multiplying the means gives 12. The rule states that these products are equal.
$x = 6$	Solving the equation for x
Check.	Check by substituting the value of x in
$2:3 = 4:6$	the original proportion. The product of
$2 \times 6 = 3 \times 4$	the extremes should equal the product of
$12 = 12$	the means.

Example 2.—In the proportion $\frac{3}{4}:\frac{1}{2} = \frac{5}{8}:x$, find the unknown term.

$$\frac{3}{4}:\frac{1}{2} = \frac{5}{8}:x$$

The product of the means equals the product of the extremes, by rule.

$$\frac{3}{4}x = \frac{1}{2} \times \frac{5}{8}$$

$$\frac{3}{4}x = \frac{5}{16}$$

$$x = \frac{5}{16} \div \frac{3}{4}$$

Solving the equation

$$x = \frac{5}{\overset{}{\cancel{16}}_{4}} \times \frac{\overset{1}{\cancel{4}}}{3}$$

$$x = \frac{5}{12}$$

Check.

$$\frac{3}{4}:\frac{1}{2} = \frac{5}{8}:\frac{5}{12}$$

$$\frac{3}{4} \times \frac{5}{12} = \frac{1}{2} \times \frac{5}{8}$$

$$\frac{15}{48} = \frac{5}{16}$$

$$\frac{5}{16} = \frac{5}{16}$$

PROBLEMS

Solve for the unknown in each of the following:

1. $2:5 = 8:x$.
2. $12:3 = x:4$.
3. $4:8 = a:10$.
4. $w:8 = 25:2$.
5. $1:b = 3:6$.
6. $10:1 = 48:p$.
7. $x:7 = 3:21$.
8. $15:x = 10:5$.
9. $5:9 = 4:y$.
10. $3:8 = 21:z$.
11. $\frac{1}{2}:\frac{1}{4} = \frac{3}{8}:m$.
12. $\frac{3}{4}:\frac{7}{8} = 4:n$.
13. $\frac{1}{8}:2 = x:5$.
14. $5:\frac{1}{4} = c:\frac{1}{2}$.
15. $3.3:6.5 = 9:a$.
16. $7.5:x = 2.8:11$.

USING PROPORTIONS IN PRACTICAL PROBLEMS

Many practical problems that come up in business or in the home, as well as in the shop, can be solved easily by the use of the proportion.

Example.—A bus traveled 259 miles in 7 hr. At the same average speed, how far would the bus go in 13 hr.?

Solution 1.

Let M = number of miles the bus would go in 13 hr.

$$\frac{259 \text{ miles}}{M \text{ miles}} = \frac{7 \text{ hr.}}{13 \text{ hr.}}$$ Comparing, miles is to miles as hours is to hours.

$$\frac{259}{M} = \frac{7}{13}$$

$$7M = 3367$$ Product of means = product of extremes.

$$M = 481$$

Ans. 481 miles.

Check.

$$\frac{259}{481} = \frac{7}{13}$$

$$481 \times 7 = 259 \times 13$$

$$3367 = 3367$$

Solution 2.

$$\frac{259 \text{ miles}}{7 \text{ hr.}} = \frac{M \text{ miles}}{13 \text{ hr.}}$$ Comparing, miles is to hours as miles is to hours.

$$7M = 3367$$

$$M = 481$$

Ans. 481 miles

The preceding example is an illustration of a *direct proportion.*

You will remember that a ratio is a comparison of two like quantities and a proportion is the expression of equality between two ratios. In a proportion, there are always two like and two unlike quantities.

In a direct proportion the unlike quantities change in the same order in each ratio. For example, in Solution 1 above, the first ratio compares the small number of hours, 7, to the large number of hours, 13; the second ratio keeps the same order and compares the small number of miles, 259 (corresponding here to the 7 hr.), to the unknown, or large, number of miles, M (corresponding to the 13 hr.).

Although in a ratio only like quantities can be compared, in a proportion each ratio can contain the unlike quantities, provided that the quantities are compared in the same order. For example, in Solution 2 above, the small number of miles, 259, is to the small number of hours, 7, as the unknown large number of miles, M, is to the large number of hours, 13.

PROBLEMS

In the following, omit the descriptive words while working the problem. Note that in a proportion the terms are not always of the same kind in each ratio, but the order of the comparison in the proportion is the same for each ratio.

1. 10 acres: 40 acres = x bu.: 200 bu.
2. 6 hammers: \$3.00 = x hammers: \$10.00.
3. 6 boys: 15 apples = 10 boys: a apples.
4. \$2.25: \$5.00 = 9 balls: b balls.
5. 16 men: 48 men = x machines: 180 machines.
6. 10 men: 20 machines = 50 men: x machines.
7. 14 cows: t tons of hay = 5 cows: 15 tons of hay.
8. d days: 12 days = 200 miles: 600 miles.
9. 5 days: 10 days = 1000 miles: M miles.
10. 9 gal: g gal. = 108 miles: 1200 miles.

WORD PROBLEMS

In each of the following problems, write the correct proportion, and solve.

1. An automobile runs 36 miles on 3 gal. of gas. How many miles can it run on 15 gal.?

2. If it takes a machinist 3 hr. to finish 9 gears, how many can he finish in an 8-hr. day?

3. If 4 automobile tires cost \$87.50, what is the cost of 10 tires?

4. A man can cut 5 bolts in 12 min. At the same rate, how long will it take him to cut 85 bolts?

5. If a 12-lb. casting costs 96¢, what will a 50-lb. casting cost?

6. If 12 men can dig 150 ft. of trench a day, how many feet of trench can 16 men dig per day?

7. If 1 in. on a map represents 125 miles, how many inches will represent 540 miles? Round off to three decimal places.

8. If 5 machines drill 240 holes in 1 hr., how many holes can 16 machines drill in 1 hr.?

9. If a truck uses 10 qt. of oil in 3 days, how long will 50 gal. last?

10. If a post 8 ft. high casts a shadow 10 ft. long, how high is an oil derrick that casts a shadow 40 ft. long (see the figure shown)?

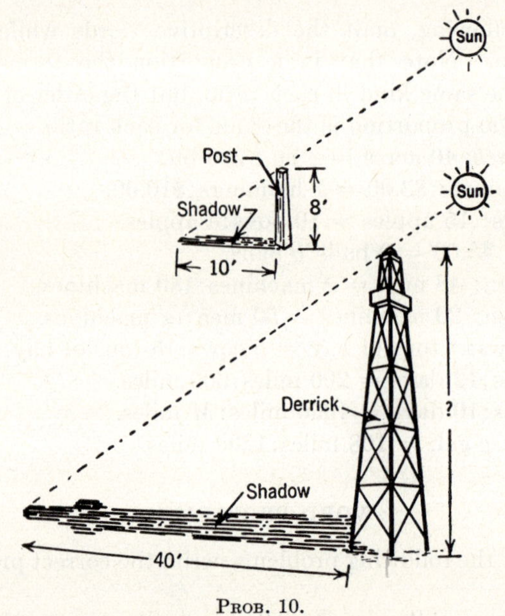

PROB. 10.

11. If 12 drills cost $8.75, how much will 20 drills cost?

12. When 6 mechanics can assemble 14 cars in 1 day, at the same rate how many cars can 150 mechanics assemble per day?

13. A company bought 12 lathes for $540. At the same rate, what would 9 lathes cost?

14. A shop operating 12 lathes finishing 90 drill stems a day must increase production to 150 stems a day. How many lathes must it operate daily?

15. A concrete floor containing 7500 sq. ft. cost $1875.00. At this rate, determine the cost of a floor of 1200 sq. ft.

16. If 1000 ft. of wire weighs 45 lb., calculate the weight of 1 mile of wire.

17. If 4 men can drill 260 holes per hour, how many holes can 10 men drill in a 7-hr. day?

18. If 36 bars of stock last 2 screw machines 3 hr., how many bars are needed for 10 machines for an 8-hr. day?

19. If 3 men lay 25 squares of roofing per day, how many men are needed to lay 375 squares in a day?

20. If an automobile runs 90 miles on $5\frac{1}{2}$ gal. of gas, how far will it run on a full 20-gal. tank?

21. If a man walks 9 miles in 2 hr., how long will it take him to walk 30 miles if the same average rate is maintained?

22. If a train runs 280 miles in 6 hr., how long will it require to run 3000 miles at the same average rate per hour?

23. How long will it take a man to lay 12,000 shingles if he can lay 625 shingles in 2.5 hr.?

24. If a mechanic assembles 7 machines in 2.5 days, at the same rate how long will it take him to assemble 50 machines?

25. Two gears in mesh have a speed ratio of 2 to 5. If the smaller gear makes 70 r.p.m., calculate the r.p.m. of the larger.

AUTOMOBILE AXLE RATIOS

In an automobile, the pinion gear on the drive shaft has fewer teeth than the ring gear on the axle. This fact is

shown on the accompanying illustration. Thus, the revolutions of the axle are fewer than the revolutions of the drive shaft. The ratio is expressed in this way:

$$\frac{\text{Number of teeth on the ring gear}}{\text{Number of teeth on the pinion gear}}$$

It is known as the rear axle ratio or the gear ratio.

In tables, the number of teeth on the pinion gear is expressed as 1. The first number in the gear ratio is usually rounded off to two decimal places.

Example 1.—The rear axle ratio of the Buick car is 4.44:1. Since there are 9 teeth on the pinion gear, how many teeth are there on the ring gear?

Let x = number of teeth on the ring gear.

$\frac{4.44}{1} = \frac{x}{9}$ Writing the proportion,

$1x = 9 \times 4.44$ Finding the product of the means and the product of the extremes.

$x = 39.96$ Since there cannot be a fraction of a tooth on a gear, rounding off to the nearest whole number, there are 40 teeth on the ring gear.

Example 2.—In the Cadillac car, the number of teeth on the ring gear is 59, and the number of teeth on the pinion gear is 12. What is the gear ratio?

Let x = first number in the gear ratio (or the numerator of the gear ratio)

$\frac{x}{1} = \frac{59}{12}$ Writing a proportion

$12x = 59 \times 1$ Finding the product of the extremes and

$x = 4.916$ the product of the means

Rounding off to the nearest hundredth, the first number in the gear ratio is 4.92.

PROBLEMS

In the following table, find the missing number in each line. In your answers, round off the number of teeth on a gear to the nearest whole number. Round off the missing number in a gear ratio to the nearest hundredth. The first two problems have been solved in the preceding illustrative examples.

Number	Car make	Gear ratio	Number of teeth on ring gear	Number of teeth on pinion gear
1	Buick	4.44:1	?	9
2	Cadillac	? :1	59	12
3	Chevrolet	? :1	38	9
4	Chrysler	4.30:1	43	?
5	De Soto	4.10:1	?	10
6	Dodge	? :1	49	11
7	Ford	? :1	40	9
8	Hudson	4.64:1	51	?
9	Mercury	4.44:1	40	?
10	Nash	4.10:1	47	?
11	Oldsmobile	? :1	53	12
12	Packard	4.09:1	?	10
13	Plymouth	3.73:1	?	11
14	Pontiac	? :1	39	10
15	Studebaker	? :1	53	11
16	Willys	? :1	46	11

17. In a certain car there are 40 teeth on the ring gear and 13 teeth on the pinion gear. Find the gear ratio.

18. How many teeth are there on the ring gear of a car if the pinion gear has 15 teeth and the gear ratio is 3.94 to 1?

CHAPTER 9

SURFACE MEASUREMENT

MENSURATION

In surface measurement, many formulas are developed and used to find the lengths, areas, and volumes of the geometric figures usually encountered in the various branches of shopwork. Finding these measurements is called **mensuration.** There are two important methods used in mensuration.

Direct measurement applies the unit of measurement directly to the object to be measured. For example, in the shop in order to find the length of some object, the workman may apply his rule directly to the job and read off the length from the rule. If the length is small, he may use a micrometer to get a more accurate reading. Or perhaps it is required to find the number of gallons of oil in a tank. This he may do by simply using a gallon container and drawing off the oil 1 gal. at a time, counting the number of gallons of oil.

Indirect measurement is used when it is inconvenient or impossible to get the measurements directly. Indirect measurement makes use of formulas. For example, the formula for the side of a square has been given as $S = \sqrt{A}$, when the area is known. If a workman is told that he is to make a square block having an area of 49 sq. in., by applying the formula, he knows that the side of the block will be 7 in. He did not do any actual measuring of square blocks till he found one that was 49 sq. in. in area.

The kind of unit used in mensuration depends upon what is to be measured.

For **lengths,** the common units of measure are the inch, foot, yard, rod, and mile. As we have already learned, the carpenter usually employs common fractions when parts of an inch are needed, whereas the machinist uses decimal fractions.

For **areas,** sometimes called surfaces, square units are used.

For **volumes,** or solids, cubic units are used.

Geometric figures are made up of points, lines, surfaces, and solids.

A **point** has position only. It has neither length, breadth, nor thickness.

A **line** has length only. It has neither breadth nor thickness.

A **surface** has length and breadth. It has no thickness. Surfaces are sometimes called **planes.**

A **solid** has length, breadth, and thickness.

ANGLE MEASURE

The opening between two meeting lines is called an **angle.** Two intersecting lines, that is, lines that cross each other, also form angles.

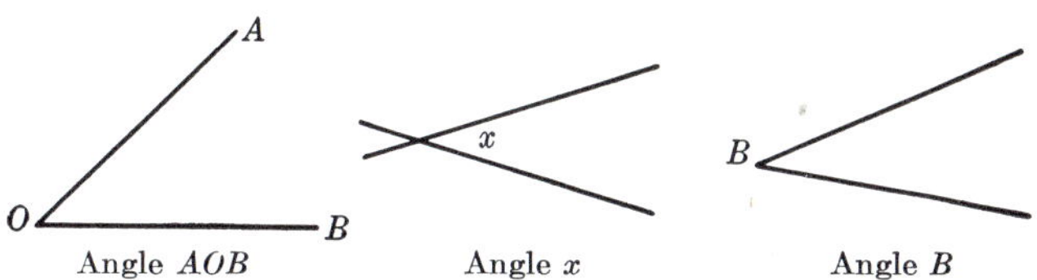

Angle *AOB* Angle *x* Angle *B*

The lines that form the angle are called the **sides.** In angle *AOB*, *AO* and *BO* are the sides. The point where the lines meet is called the **vertex.** Point *O* is the vertex in angle *AOB*. The symbol for an angle is a small angle, $\angle$.

Angles may be identified in three different ways, as follows: (1) With capital letters following the angle sign, as $\angle AOB$. The vertex letter is written between the others. (2) With a small letter placed in the angle sign, as $\angle x$. (3) With a capital letter, as *B*, outside the angle sign at the vertex.

Angles are measured by the opening between the sides, not by the lengths of the sides. The unit of measure for angles is the **degree.** One degree equals sixty minutes. One minute equals sixty seconds. An angle of seventy-nine degrees twenty-four minutes forty-four seconds is written 79°24′44″.

There are 360° around a point, as shown. A **straight angle** is an angle of 180°. A **right angle** is an angle of 90°. An **acute angle** is an angle less than 90°. An **obtuse angle** is an angle more than 90°. Acute and obtuse angles are also called **oblique angles.**

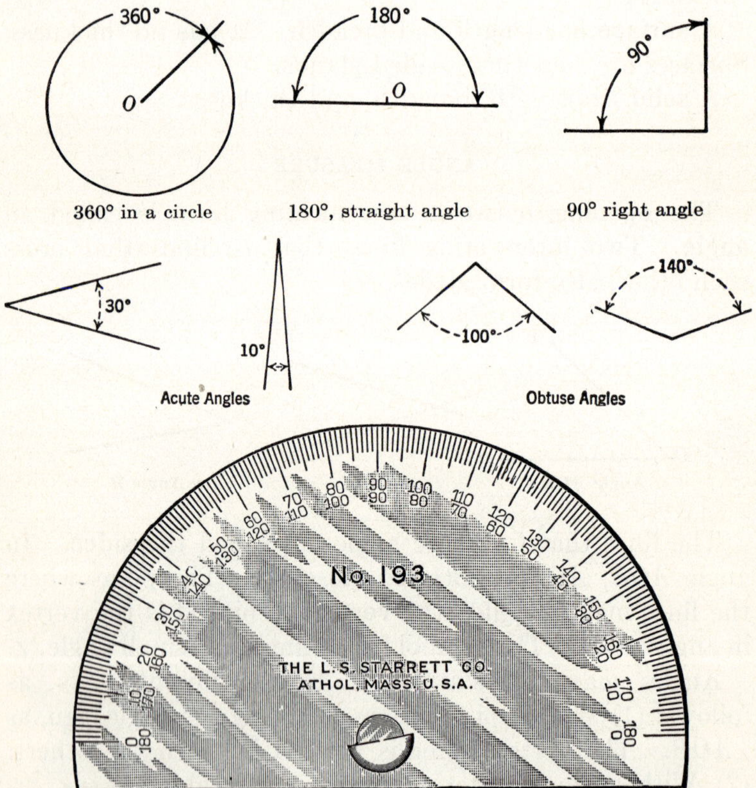

The protractor shown is used to measure and lay off angles.

Rule. To measure an angle with a protractor, place the center so that it coincides with the vertex of the angle and one 0° side coincides with one side of the angle and read the degrees where the other side of the angle crosses the graduated scale.

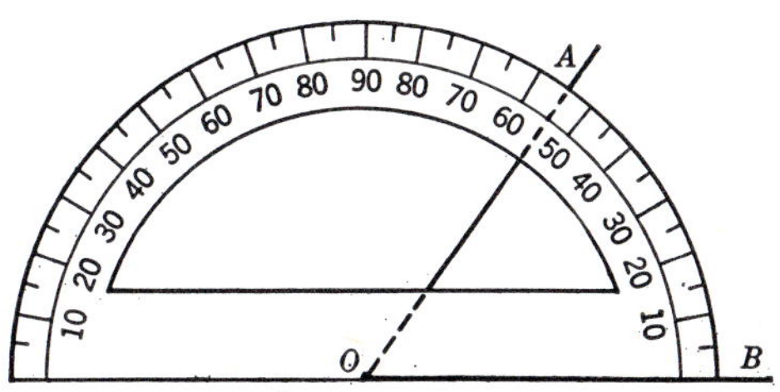

An angle of 55° measured with a protractor.

To draw an angle with a protractor, first draw a line for one side of the angle. Then place the center so that it coincides with the point that is to be the vertex of the angle. On the graduated scale, find the number that corresponds to the size of the angle to be drawn. Place a dot on your paper at this point. Connect this dot, or point, and the vertex of the angle.

PROBLEMS

1. Make a list of 10 objects or places that have right angles.

2. Make a list of 10 objects or places that show acute angles.

3. Make a list of 10 objects or places that show obtuse angles.

4. How many degrees in (*a*) one-half a right angle? (*b*) one-third a right angle? (*c*) One-fourth a right angle?

5. How many degrees are there in the angle formed by the hands of a clock at (*a*) 1 P.M.? (*b*) 2 P.M.? (*c*) 3 P.M.? (*d*) 4 P.M.? (*e*) 5 P.M.? (*f*) 6 P.M.?

6. How many degrees are there in the angles at the corners of a normal street intersection?

7. If two streets intersect at an acute angle, name the other angles formed. Make a drawing, and indicate which angles are larger than a right angle.

8. Using your protractor, draw the following angles: (*a*) $\angle AOB = 65°$; (*b*) $\angle STU = 32°$; (*c*) $\angle x = 111°$; (*d*) $\angle m = 38°$; (*e*) $\angle s = 144°$; (*f*) $\angle B = 77°$.

9. Measure the angles below with your protractor, and name each.

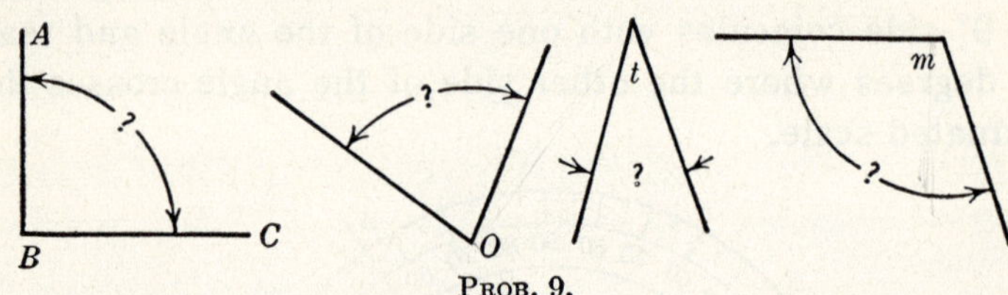

PROB. 9.

10. With your protractor, draw angles of the following sizes, and label each one: (*a*) 135°; (*b*) 90°; (*c*) 75°; (*d*) 135°; (*e*) 166°.

11. With your ruler and protractor, draw figures like those below, and measure the angles marked with question marks.

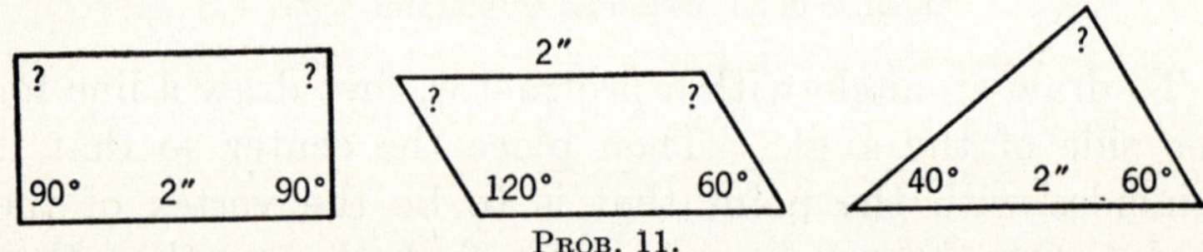

PROB. 11.

12. Draw several figures with three sides (triangles), and measure the interior angles at the corners with your protractor.

13. Measure the angles in the figures shown below.

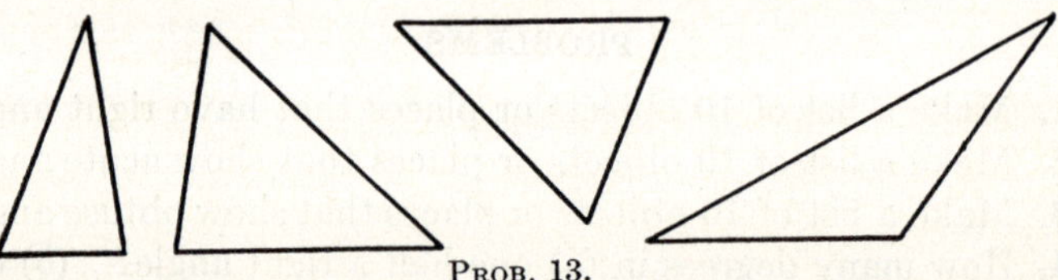

PROB. 13.

14. Add up the interior angles of each triangle in Prob. 13. What can you deduce about the sum of the angles of a triangle?

15. How many degrees are there in the third angle in each of the following triangles?

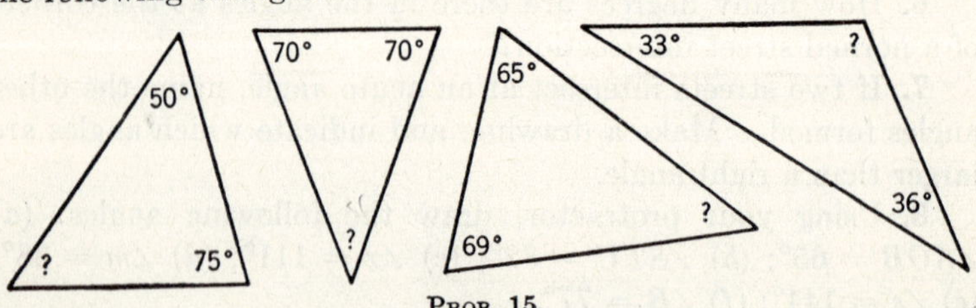

PROB. 15.

Draw triangles with two angles of the sizes given here. The sides can be of any convenient length. Determine the size of the angle not given.

	Angle *A*	Angle *B*	Angle *C*
16	65°	75°	?
17	90°	?	30°
18	110°	33°	?
19	?	60°	60°
20	?	88°	44°
21	70°	70°	?

POLYGONS

A **polygon** is a closed geometric figure bounded by straight-line segments, lying in a plane.

The figure *ABCDE* is a polygon. *AB*, *BC*, etc., are called the sides.

Points *A*, *B*, etc., are called the vertices. Any point, as *A*, is called a vertex. A **vertex** is a point where two lines meet.

The line *EC* is a **diagonal.** A diagonal joins two non-consecutive vertices.

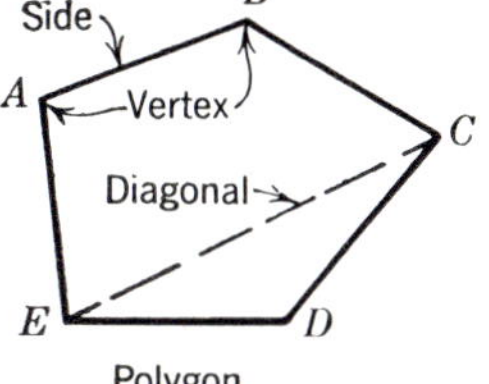

Polygon

The **perimeter** of a polygon is the sum of the lengths of the sides. It is the distance around the figure.

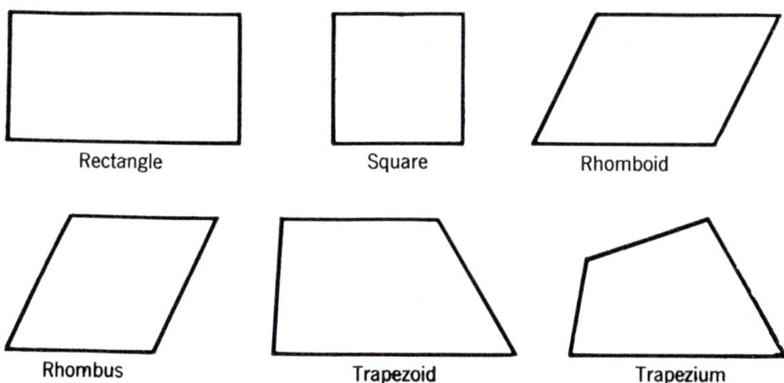

Polygons with four sides are called **quadrilaterals.** The different types of quadrilateral are shown at the foot of page 213.

The **parallelogram** is a quadrilateral in which the opposite sides are parallel.

The **rectangle** is a parallelogram that has four right angles. The opposite sides are equal. The rectangle is also called an oblong.

The **square** is a parallelogram in which the angles are all right angles and the sides are all equal. The square is also a rectangle.

The **rhomboid** has opposite sides parallel but no right angles.

The **rhombus** is a parallelogram in which all the sides are equal but none of the angles is a right angle. The rhombus is also called a **diamond** or **lozenge.**

The **trapezoid** is a quadrilateral with only two sides parallel.

The **trapezium** is a quadrilateral with no two sides parallel.

The **base** of a geometric figure is the side on which it seems to stand.

The **height** (altitude) of a parallelogram or a trapezoid is the length of the line drawn from the highest point to the base, forming a right angle with the base. The height is said to be **perpendicular** to the base.

The **area** of a geometric figure is the amount of surface it covers. In measuring area, square units are used. One square inch represents an area equivalent to the area of a square one linear inch on a side.

THE RECTANGLE

Rule.—The area of a rectangle equals the base multiplied by the height or the length multiplied by the width.

Formulas. $A = bh$, or $A = lw$.

Example 1.—Find the area of a rectangle with a height of 5 in. and a base of 6 in.

$A = bh$	Formula
$A = 6 \times 5$	Substituting
$A = 30$	
Ans. 30 sq. in.	

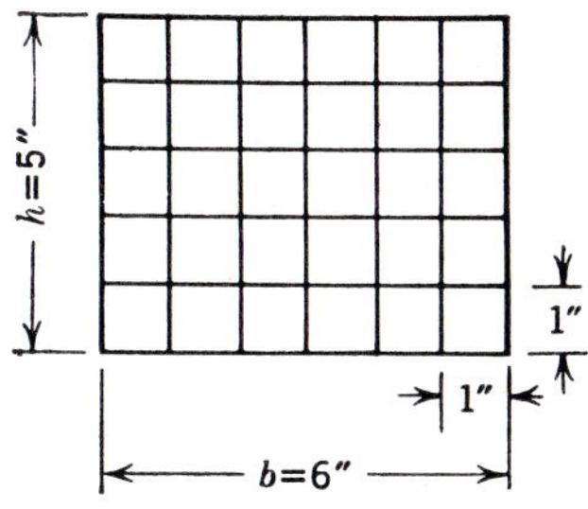

Note.—In the drawing, observe that the square inches in the area have been marked off. There are 5 rows with 6 square inch units in a row. The number of squares is obviously 30, and the area of the rectangle is 30 sq. in.

The formula $A = bh$ may be solved for b or h when either is unknown, as follows:

$$b = \frac{A}{h}, \qquad h = \frac{A}{b}$$

Example 2.—Find the base of a rectangle 5 in. high whose area is 30 sq. in.

$b = \frac{A}{h}$	Formula
$b = \frac{30}{5}$	
$b = 6$	
Ans. 6 in.	

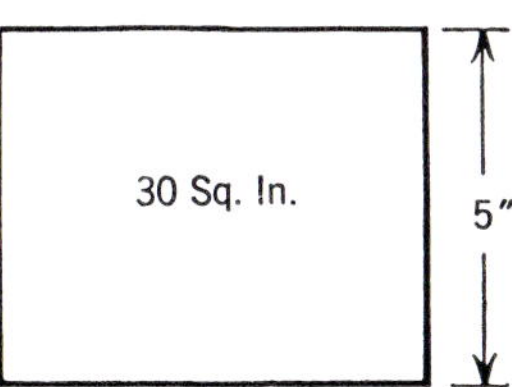

Example 3.—Find the height of a rectangle when the area is 30 sq. in. and the base is 6 in.

$h = \frac{A}{b}$	Formula
$h = \frac{30}{6}$	
$h = 5$	
Ans. 5 in.	

EXERCISES AND PROBLEMS

Find the area of each of the following rectangles. Draw a diagram for each problem.

Note.—3″ × 4″ means 3″ by 4″; that is, $b = 3''$ and $h = 4''$.

1. 16″ by 31″ **2.** 27″ by 36″ **3.** 13″ × 16″

4. 25″ × 38″ **5.** $12\frac{1}{2}'' \times 19''$ **6.** $5\frac{1}{4}' \times 3\frac{7}{8}'$

7. $9\frac{3}{8}' \times 5\frac{1}{4}'$ **8.** $8\frac{1}{2}' \times 8\frac{1}{2}'$ **9.** 12″ × 14.4″

10. 6.75′ × 3.5′ **11.** 3.86′ × 9.42′ **12.** 16′ × 0.001′

13. 16′4″ by 3′0″ **14.** 12′10″ by 5′6″ **15.** 9′3″ by 2′1″

16. 3 yd. 2 ft. 4 in. × 5 yd. 1 ft. 9 in.

17. 2 yd. 1 ft. 6 in. × 11 in.

18. 9 rd. 2 ft. 6 in. × 3 rd. 1 ft. 10 in.

19. 4 rd. by 3 ft. 6 in.

20. A field is 2 miles 80 rd. long by 20 rd. wide. Find the area in acres. Notice that an acre is a surface or square unit of measure.

THE PARALLELOGRAM

The square and the rectangle are parallelograms containing right angles. The rhombus and the rhomboid are also parallelograms.

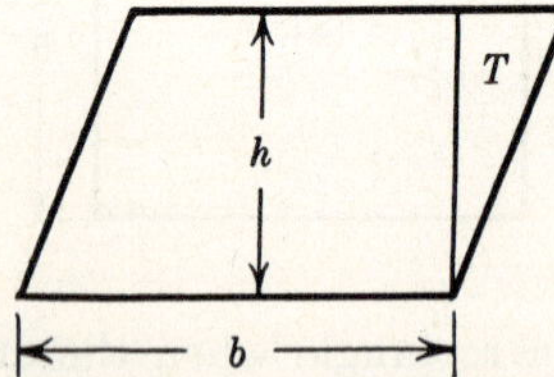

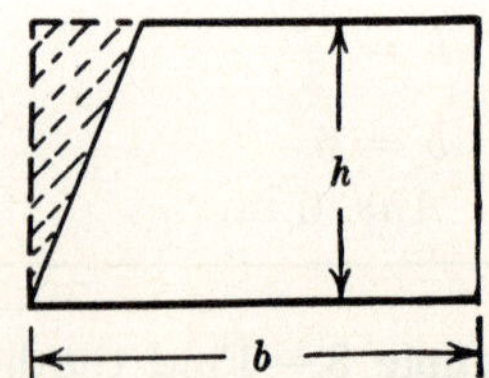

The rule and formula for finding the area of a parallelogram will be seen by studying the figures shown here. Each one has a base b and a height h. If the triangle T is cut off the right end of the figure at the left by a line perpendicular to the top and if then this triangle is placed at the left end of the figure, as shown on the right, the resulting figure is a rectangle. It is evident that this rectangle has the same area as the original parallelogram; hence, we use the same formula for finding the area of a parallelogram as for finding the area of a rectangle.

Rule.—The area of a parallelogram is equal to the product of the base and the height.

Formula. $A = bh$.

Example.—Find the area of a parallelogram with a base of 15 in. and a height of 6 in.

$A = bh$	Formula
$A = 15 \times 6$	Substituting
$A = 90$	
Ans. 90 sq. in.	

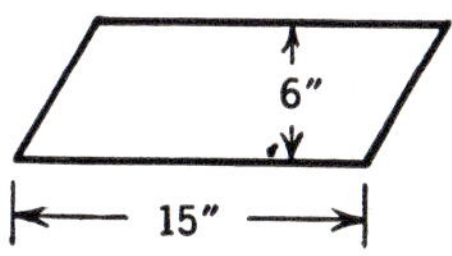

THE TRAPEZOID

The trapezoid has only two sides parallel. A rectangle can be made of the trapezoid by a line m, parallel to the two parallel sides and midway between them. The length of line m is the average of the lengths of the two

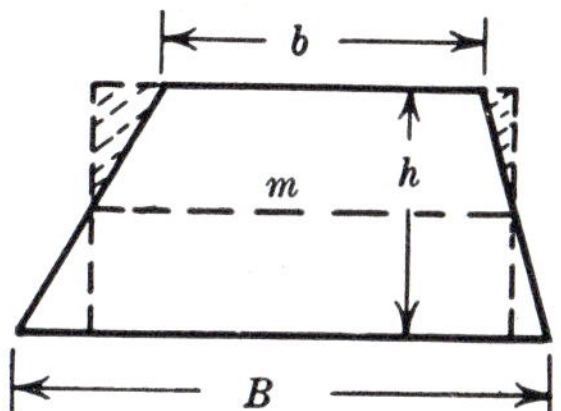

parallel sides b and B, or $m = \frac{1}{2}(b + B)$. Perpendiculars drawn at the ends of line m cut off triangles below m that are exactly equal to the triangles needed above m to form the rectangle; hence, the following may be stated:

Rule.—The area of a trapezoid equals one-half the sum of the two bases (parallel sides) multiplied by the height of the trapezoid.

Formula. $A = \frac{B + b}{2} \times h$, or $A = \frac{1}{2}(B + b)h$.

Note.—When terms are enclosed by parentheses, perform the operations indicated inside the parentheses first.

Example.—Find the area of the trapezoid whose bases are 15 in. and 31 in. and whose height is 9 in.

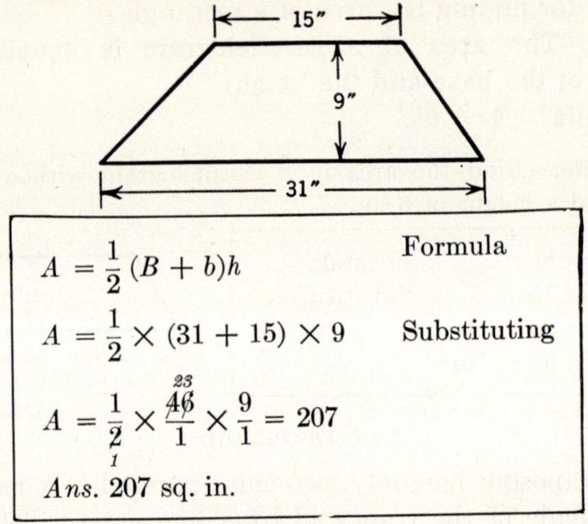

$$A = \frac{1}{2}(B + b)h \qquad \text{Formula}$$

$$A = \frac{1}{2} \times (31 + 15) \times 9 \qquad \text{Substituting}$$

$$A = \frac{1}{\cancel{2}_{1}} \times \frac{\cancel{46}^{23}}{1} \times \frac{9}{1} = 207$$

Ans. 207 sq. in.

PROBLEMS

Illustrate, give the formula, and solve for each of the following:

1. A parallelogram is 16 ft. wide and 9.6 ft. high. Find its area.

2. Determine the area of a rhomboid 17.6 yd. on the base and 9.3 yd. high.

3. What is the area of a rhombus 12.3 ft. on the base and $7\frac{1}{2}$ ft. high?

4. A parallelogram is 72′6″ on the base and 11′3″ high. Determine the area in square feet.

5. A rhomboid is 18′9″ on the base and 6′3″ high. Determine the area in square feet.

6. A rhombus is 24.6 ft. high and 94.3 ft. wide. Find its area.

7. The area of a parallelogram is 196 sq. ft. The base is 11 ft. Find the height.

8. The area of a rhomboid is 343 sq. ft. The height is 11 ft. Find the base. Express the answer to two decimal places.

9. The area of a rhombus is 996 sq. ft. If the base is 44 ft., what is the height? Round off the answer to one decimal place.

10. The altitude of a trapezoid is 7 ft., the lower base is 15 ft., and the upper base is 9 ft. What is the area?

11. On a trapezoid the parallel sides are 19 and 33 ft., respectively. If the altitude is 17 ft., find the area.

12. The top base of a trapezoid is 12.6″. The bottom base is 18.4″. The height is 11.9″. Determine the area. Round off the answer to one decimal place.

13. A city lot is 50 ft. wide. It has parallel sides 129 and 156 ft. long. Find the area of the lot.

14. The area of a trapezoid is 400 sq. ft. The bases are 20 and 30 ft. Find the height.

15. The parallel bases of a trapezoid are 5′9″ and 11′3″, respectively. The altitude is 6′6″. Determine the area.

16. The parallel sides of a trapezoid are 6′6″ and 9′5″, respectively. The altitude is 4′3″. Find the area in square feet and square inches.

17. A rhomboid is 77 ft. on the base. If the area is 847 sq. ft., what is the height?

18. A parallelogram has an area of 1000 sq. ft. The base is 66 ft. 6 in. Determine the height in feet and inches.

19. List two practical needs for knowing the length of all the sides of a parallelogram.

20. List two practical needs for knowing the length of all the sides of a trapezoid.

EXERCISES

The rules and formulas given on preceding pages can be used to compute the areas of some of the figures below. For each of these figures, write down its name and find its area.

Each of the other figures can be divided so that the formulas you already know can be used for the parts. Copy each of these figures. Divide each one into simple figures for which you can find the area. Write down the name of each figure you have made, and find the total area of each numbered figure shown here and on page 200.

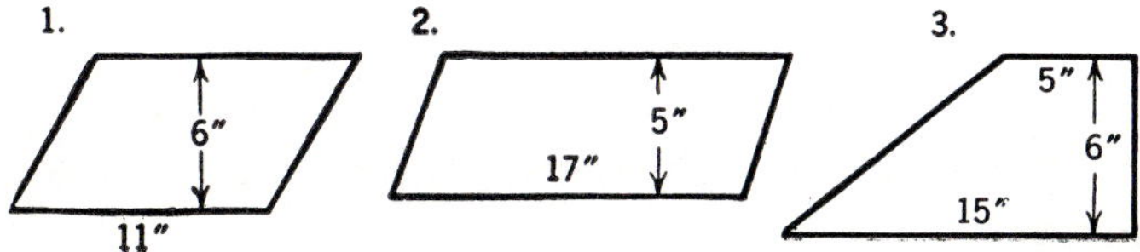

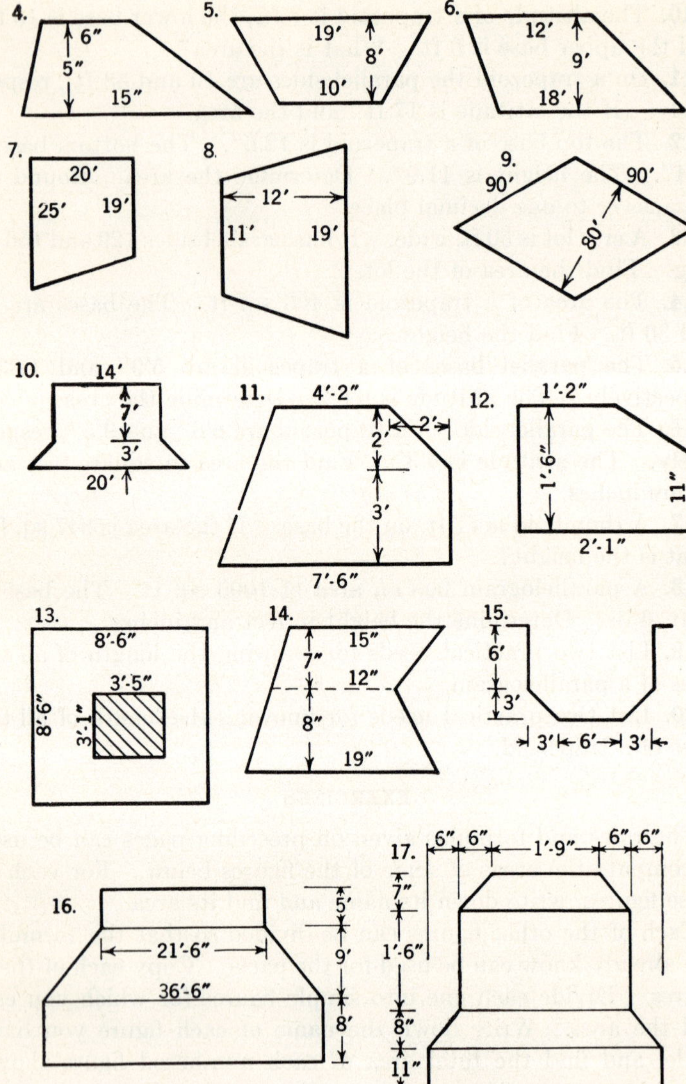

FINDING THE AREA OF A SQUARE

Rule 1.—The area of a square equals the length of the side of the square multiplied by itself, that is, the side squared.

Formula. $A = s \times s$, or $A = s^2$.

Example.—Find the area of a square 4 in. on a side.

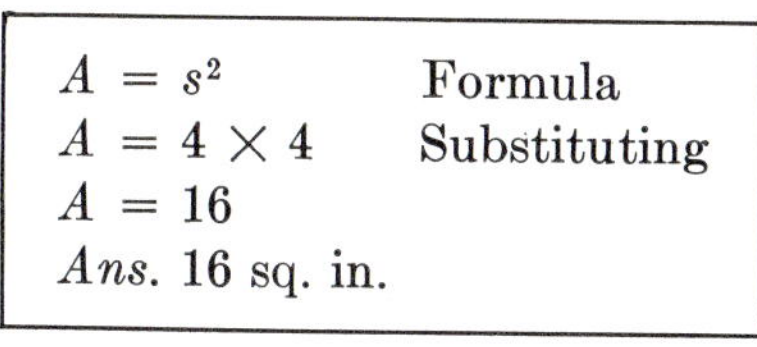

$A = s^2$	Formula
$A = 4 \times 4$	Substituting
$A = 16$	
Ans. 16 sq. in.	

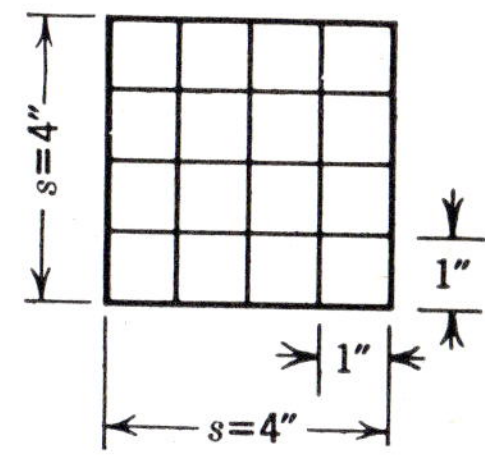

Note.—In the drawing, observe that the square inches in the area have been marked off. There are 4 rows with 4 square inch units in each row; hence, 4×4 sq. in. = 16 sq. in.

PROBLEMS

1. A wall of the shop shown in the diagram is 80 ft. long and 16 ft. high. How many square feet of space does the wall enclose?

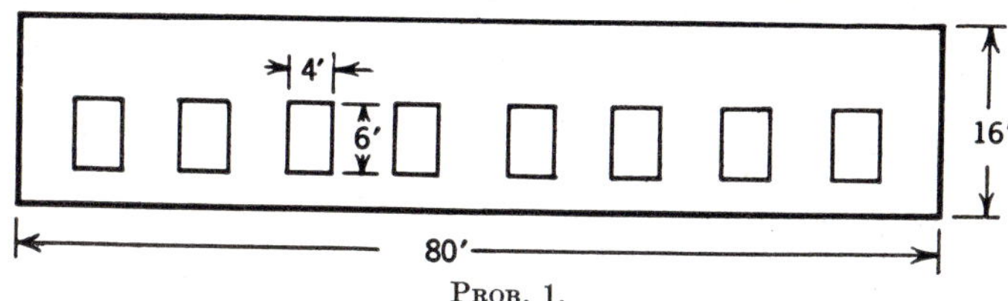

PROB. 1.

2. There are 8 windows 4 ft. wide by 6 ft. high in the wall of Prob. 1. How many square feet are occupied by windows? How many square feet of wall are there, not counting the areas of the windows?

3. What would it cost to plaster the wall in Prob. 2 at $1.90 per square yard?

4. A square column is $7\frac{3}{4}''$ on a side. Find the area of its cross section.

5. Figure the cost of a concrete walk 6 ft. wide and 75 ft. long at 28¢ per square foot.

6. The inside dimensions of a freight car are height, 8 ft., width, 8 ft., length, 40 ft. Find the number of square feet of inside surface (four walls, ceiling, and floor).

7. At a cost of 30 cents per square foot, determine the cost of paving a 6-ft. concrete walk around the outside of the building of which a diagram is shown.

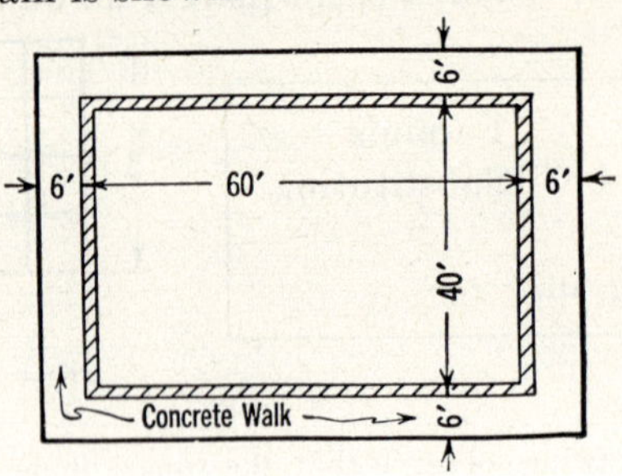

PROB. 7.

8. What is the area of a square $16\frac{1}{2}''$ on a side?

9. A rectangle contains 240 sq. in. It is 16 in. high. How long is the base?

10. The square vent pipe shown is 9 in. on a side. It connects to a rectangular vent, having the same area, in a 3-in. wall. How wide is the vent in the wall?

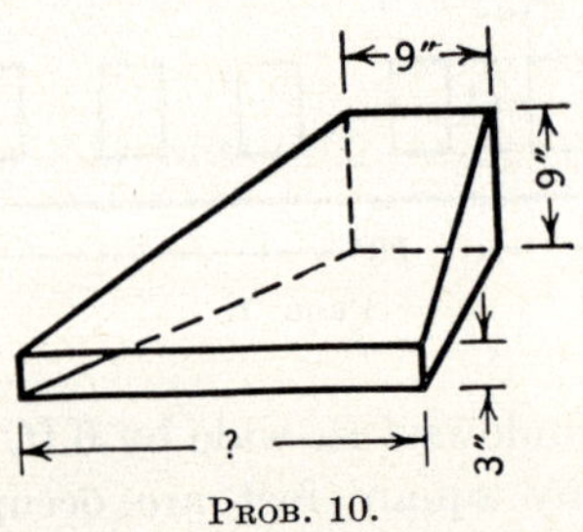

PROB. 10.

11. A room is 28′6″ long and 16′4″ wide. Find the area of the floor in square feet. In square inches. In square yards.

12. A baseball diamond is a square 90 ft. on a side. Figure the area in square feet. In square yards.

13. A rectangular duct 12″ wide has an area equivalent to the area of six square vents, placed side by side. Each square vent is 4″ on a side. How high is the rectangular duct? See the figure shown at the top of page 203.

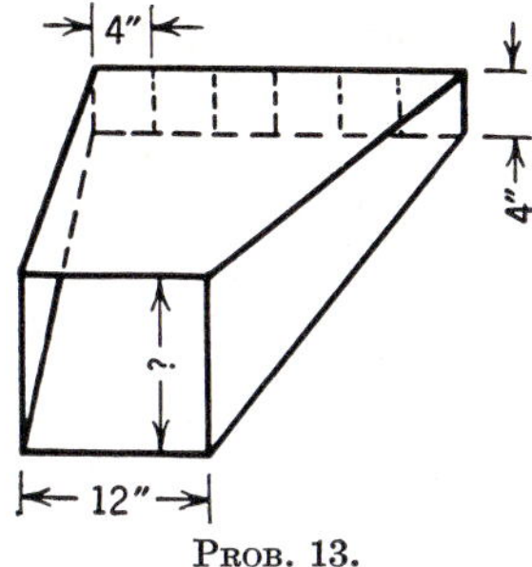

PROB. 13.

OTHER FORMULAS FOR THE SQUARE

Rule 2.—The side of a square is equal to the square root of the area.

Formula 2. $S = \sqrt{A}$.

Example 1.—Find the side of a square when the area is 81 sq. in.

$S = \sqrt{A}$	Formula
$S = \sqrt{81}$	Substituting
$S = 9$	Finding the square root
Ans. 9 in.	

?

81 Sq. In.

Rule 3.—The perimeter of a square is equal to 4 times the square root of the area.

Formula 3. $P = 4\sqrt{A}$, where P = perimeter and A = area.

Example 2.—Find the perimeter of a square when the area is 81 sq. in.

$P = 4\sqrt{A}$	Formula
$P = 4\sqrt{81}$	Substituting
$P = 4 \times 9$	
$P = 36$	
Ans. 36 in.	

81 Sq. In.

Note.—Remember that in a square all sides are equal. Perimeter means distance around a geometric figure.

Rule 4.—The diagonal of a square equals the square root of twice the area.

Let D = diagonal; then, we have the following:

Formula 4. $D = \sqrt{2A}$.

Example 3.—Find the diagonal of a square when the area is 81 sq. in.

$D = \sqrt{2A}$	Formula
$D = \sqrt{2 \times 81}$	Substituting
$D = \sqrt{162}$	Multiplying
$D = 12.73$	Finding the square root
Ans. 12.73 in.	

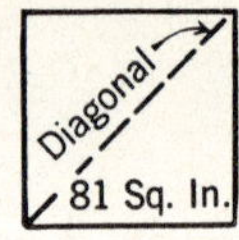

EXERCISES

Write the formula each time, and find the side, perimeter, and diagonal for each of the squares with the following areas. Where the answer is not a perfect square, carry the root to two decimal places. Square-root tables may be used.

1. 121 sq. in. **2.** 225 sq. in. **3.** 49 sq. ft.
4. 625 sq. ft. **5.** 2025 sq. ft. **6.** 3164 sq. in.
7. 8836 sq. ft. **8.** 939 sq. ft. **9.** 14,641 sq. ft.
10. 6889 sq. yd. **11.** 15,625 sq. ft. **12.** 22,201 sq. in.
13. 5.76 sq. in. **14.** 21.16 sq. ft. **15.** 82.81 sq. in.
16. 0.3136 sq. in. **17.** 0.0169 sq. in. **18.** 0.09 sq. ft.
19. 0.0256 sq. ft. **20.** 1600 sq. mi.

PROBLEMS

Draw a picture, state the formula or rule, and solve. Where the root is not a perfect square, carry the root out to two decimal places.

1. A square contains 425 sq. in. Find the length of the diagonal.

2. A square contains 9640 sq. in. Find the diagonal.

3. A square column has an area of 19 sq. in. Find the length of the side correct to two decimal places.

4. An acre of land contains 43,560 sq. ft. If the land is in the shape of a square, find the number of rods of fencing needed to surround it.

5. How long is the side of a square block that contains 5 acres? Find the answer in feet.

6. A vent in a wood partition has an area of 42 sq. in. Find the size of a square lead from this vent that will have the same area.

7. The hot-air duct leading from a furnace must contain 60 sq. in. If the duct is made in the shape of a square, how long are the sides?

8. Five ducts, each with an area of 42 sq. in., are to be fed by a large square duct having an area equal to the total area of the 5 ducts. What is the length of the side of the large square duct?

9. A square frame is 11 ft. on a side. Find the length of the braces that go from corner to corner.

10. The photograph shows workmen checking to see if the form for a building foundation is a perfect square. If the form is square, the diagonals will be the same length. What should the tape measure read if the building is 28′ on a side?

PROB. 10.

11. A rectangular room with a square floor contains 6561 cu. ft. The height is 9 ft. Find the side of the room. (The volume of a rectangular solid with a square base divided by its height equals the area of the base. Volume is measured in cubic units.)

12. A rectangular bin with a square base contains 5000 cu. ft. The height is 10 ft. Find the length of the side correct to the nearest 0.1 ft.

13. A rectangular tank holds 48 gal. It is 42 in. high. Find the length of the side of the square base (1 gal. equals 231 cu. in.).

14. A square gravel bunker holds 500 cu. yd. of crushed stone. If the bunker is 12 ft. deep, how long is it on a side?

15. A square tank contains 1,000,000 gal. The depth is 20 ft. Determine the length of the side of the tank (one cubic foot equals 7.5 gal.).

16. How much fencing is needed to enclose a square block containing 40,000 sq. ft.?

17. The side of a square is 12 in. Find the length of the side of a square containing four times as much area.

18. How long is the side of a square that has ten times as much area as a square 8″ on a side?

19. How far is it from one corner to the opposite corner of a square 12″ on a side?

20. Find the length of the side of a square that will have as much area as the sum of the areas of the following rectangles: 3″ × 14″, 5″ × 6″, and 8″ × 14″.

THE TRIANGLE

The **triangle** is a polygon having three sides.

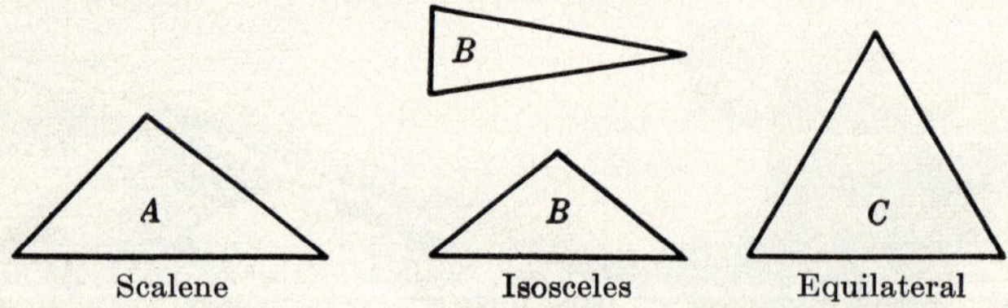

The types of triangle, classified with respect to their sides, are

1. **Scalene,** with no two sides equal (*A*).
2. **Isosceles,** with two sides equal (*B*).
3. **Equilateral,** with three sides equal (*C*).

The types of triangle, classified with respect to their angles, are

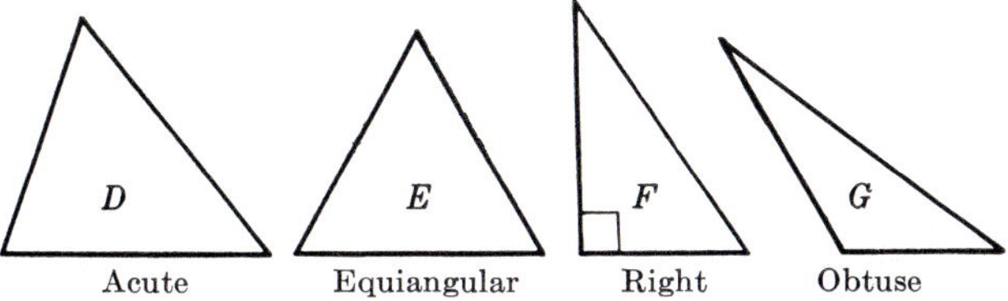

4. **Acute,** with all angles less than right angles (*D*).
5. **Equiangular,** with all angles equal, or 60° (*E*).
6. **Right,** with one right angle and two acute angles (*F*). The right angle is indicated by drawing a small square in the 90° vertex.
7. **Obtuse,** with one angle greater than a right angle (*G*).

AREA OF A TRIANGLE

Any triangle may be considered to be half of a rectangle or a parallelogram, as is shown below.

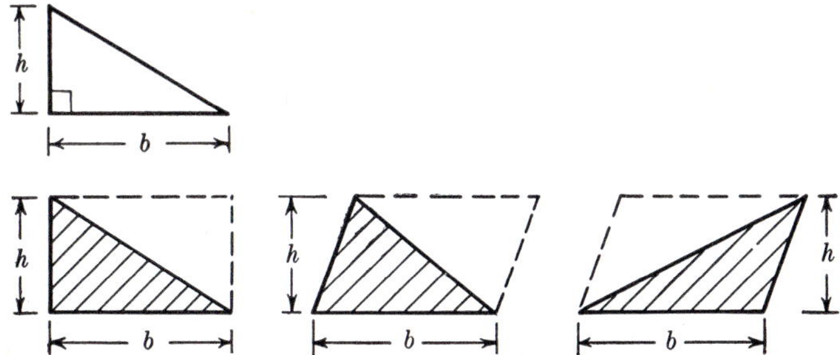

Since the area of a rectangle or a parallelogram is the product of the base and the height, the area of a triangle must be one-half the area of a rectangle or a parallelogram. Hence, the following may be stated:

Rule.—The area of a triangle equals one-half the product of the base and the height.

Formula. $A = \frac{1}{2} bh$, or $A = \frac{bh}{2}$.

Example 1.—Find the area of the triangle shown.

$A = \frac{1}{2}bh$	Formula
$A = \frac{1}{2} \times 8\frac{1}{2} \times 6$	Substituting
$A = 25\frac{1}{2}$	
Ans. $25\frac{1}{2}$ sq. in.	

6″

8½″

Example 2.—Determine the area of the triangle shown.

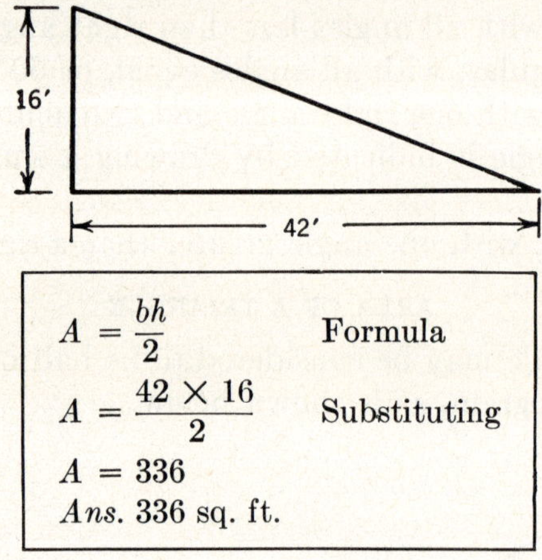

$A = \frac{bh}{2}$	Formula
$A = \frac{42 \times 16}{2}$	Substituting
$A = 336$	
Ans. 336 sq. ft.	

PROBLEMS

State the formula used and find the area of each of the following triangles:

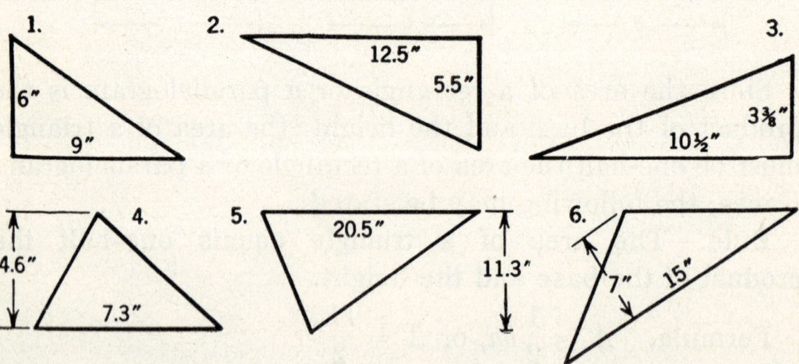

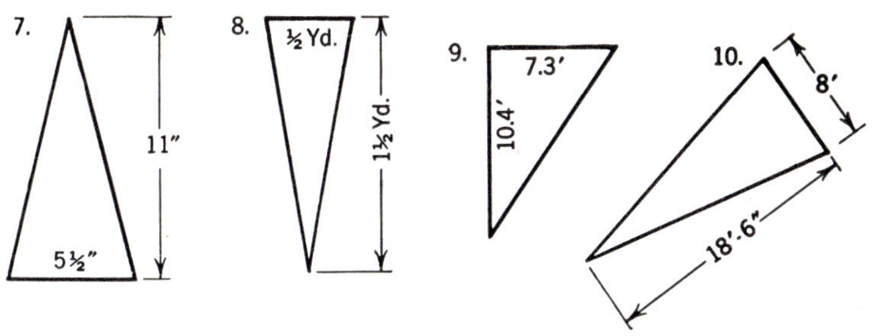

THE RIGHT TRIANGLE

If any two sides of a right triangle are known, the third side may be calculated.

A right triangle has one right angle. In the right triangle shown, the square drawn on the **hypotenuse, the side opposite the right angle,** is equal in area to the sum of the areas of the squares drawn on the other two sides. These sides of the triangle are called the base and altitude. The base is considered to be the side on which the triangle seems to stand. In the figure shown it is evident that

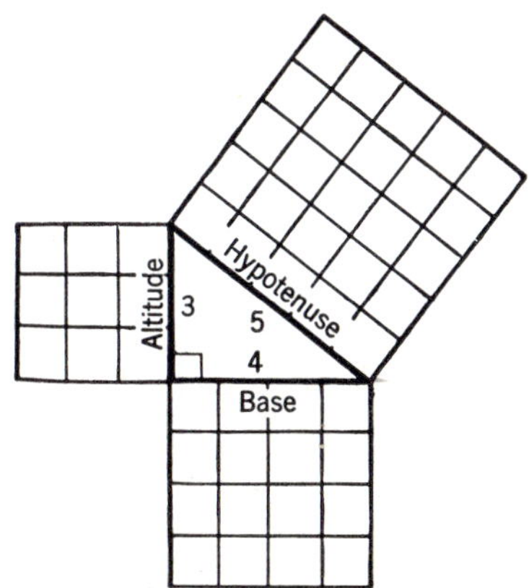

$(3 \text{ in.})^2 =$	9 sq. in.	Area of the square on the altitude
$(4 \text{ in.})^2 =$	16 sq. in.	Area of the square on the base
$(5 \text{ in.})^2 =$	25 sq. in.	Area of the square drawn on the hypotenuse

But
$$9 + 16 = 25$$
$$3^2 + 4^2 = 5^2$$

In words, this may be expressed as follows:

Rule.—The square on the hypotenuse equals the sum of the squares on the other two sides.

Note.—This rule is called the rule of Pythagoras.

c = hypotenuse, a = altitude, b = base

Formula. $c^2 = a^2 + b^2$.

The square or the square root of both sides of an equation may be taken without changing the equality.

Solving for c, a, and b gives

Formulas.
(1) $c = \sqrt{a^2 + b^2}$
(2) $a = \sqrt{c^2 - b^2}$
(3) $b = \sqrt{c^2 - a^2}$

Example 1.—In a right triangle the base is 32″ and the altitude is 24″. Find the hypotenuse.

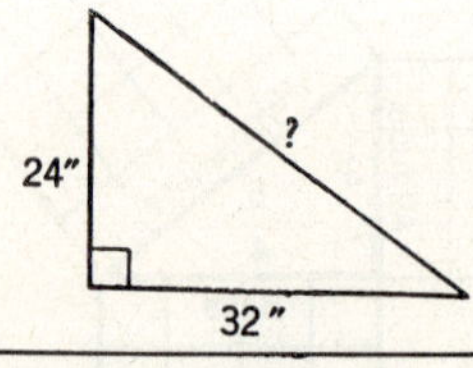

$c = \sqrt{a^2 + b^2}$	Formula
$c = \sqrt{32^2 + 24^2}$	Substituting
$c = \sqrt{1024 + 576}$	Squaring
$c = \sqrt{1600}$	Collecting
$c = 40$	Finding the square root
Ans. 40″	

Example 2.—In a right triangle, the hypotenuse is 66″. The base is 28″. Find the altitude to the nearest hundredth.

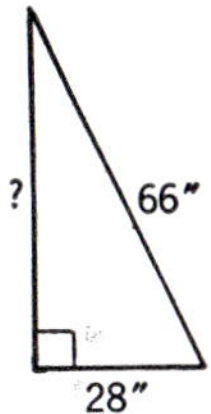

$a = \sqrt{c^2 - b^2}$	Formula
$a = \sqrt{66^2 - 28^2}$	Substituting
$a = \sqrt{4356 - 784}$	Squaring
$a = \sqrt{3572}$	Collecting
$a = 59.76$	Extracting the square root gives
Ans. 59.77−″	59.76 with a remainder greater than half of the last trial divisor. Hence,

the root is increased by 1 with a minus sign placed after it to indicate the change to the nearest two decimal places.

Use tables on pages 541–544 to get square roots.

PROBLEMS

In each of the triangles shown below and on page 212, find the length of the side not given to the nearest hundredth.

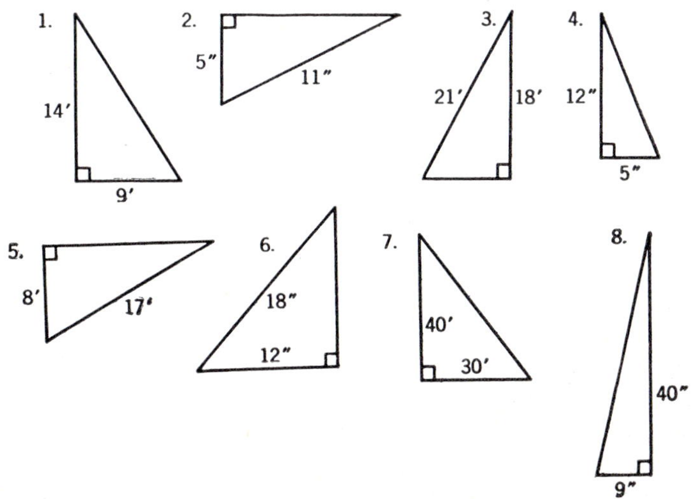

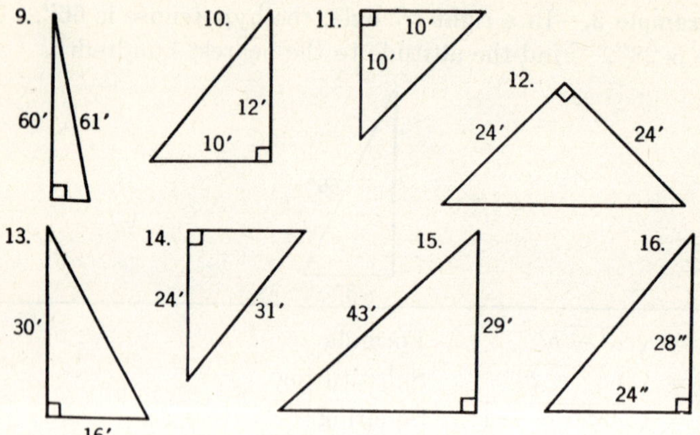

MISCELLANEOUS PROBLEMS

Draw a picture, label it, give the proper formula, and solve.

Note.—All the angles of a rectangle are right angles.

1. The base of a right triangle is 27″; the altitude is 36″. Find the hypotenuse.

2. The altitude of a right triangle is 15″; the base is 20″. Find the hypotenuse.

3. The hypotenuse of a right triangle is 29 in., and the altitude is 16 in. Find the base.

4. What is the base of a right triangle if the hypotenuse is 60 ft. and the altitude is 36 ft.?

5. One side of a right triangle is 50 ft. If the hypotenuse is 72 ft., what is the length of the other side?

6. If the hypotenuse of a right triangle is 746 ft. and one side is 524 ft., find the length of the other side.

7. To square the forms for a concrete foundation 62′ by 21′ a tape measure was stretched from corner to corner. What should the length of the diagonal read on the carpenter's tape measure? Find the answer to the nearest $\frac{1}{4}$″.

8. Find the diagonal of a rectangular building 120′ by 100′. Give the answer to the nearest $\frac{1}{4}$″.

9. How long will the braces be to reach from corner to corner of a rectangular frame 9 by 12 ft.?

10. How long are the diagonal braces for a square that is 10 ft. on a side?

11. If the hypotenuse of a right triangle is 60 ft. and the base is 50 ft., what is the altitude?

12. The base of a right triangle is 35 ft., and the hypotenuse is 37 ft. Find the altitude.

13. If the base of a right triangle is 84 ft. and the hypotenuse is 85 ft., find the altitude.

14. A rectangle has a diagonal 61 in. long. If one side is 60 in., how long is the other side?

15. A rectangle has a diagonal 41 ft. long. One side is 40 ft. Find the length of the other side.

16. How long is the diagonal of a rectangle when the base is 12 ft. and the height is 5 ft.?

17. How long is the diagonal of a room 50′ by 100′?

18. How long is the diagonal brace for a frame 16′ by 20′?

19. A window is 17′ from the ground. The base of a ladder is 5′ from the foot of the wall. If the ladder just reaches the window, how long is it?

20. How far from the wall should the base of a 16′ ladder be put just to reach the eaves of a house 13′ up?

21. How high up the wall will a 16′ ladder reach if its base is placed 6 ft. from the wall?

22. How much wire is needed to brace a telephone pole when it is stretched from a point 20′ above the ground to a point 20′ from the base, allowing 3′ on each end for fastening?

23. A radio aerial pole is 20′ high. Three braces run from the top and three from the center to points 40′ from the base. How much wire is needed for the six braces if a total of 12′ is added for fastening?

24. The length of a lake was wanted. A right angle was carefully laid out near the shore as shown. The sides are 540 ft. and 390 ft. How long is the lake?

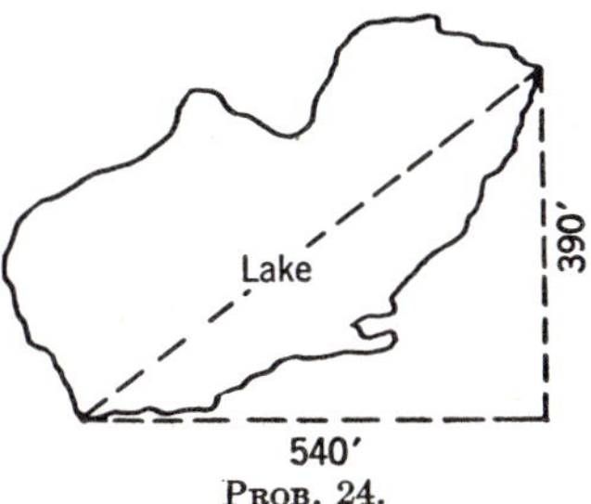

PROB. 24.

25. The gable of a house, as illustrated, shows a run of 6 ft. and rise of 6 ft. How long will the rafter be if 1 ft. extra is allowed for the overhang?

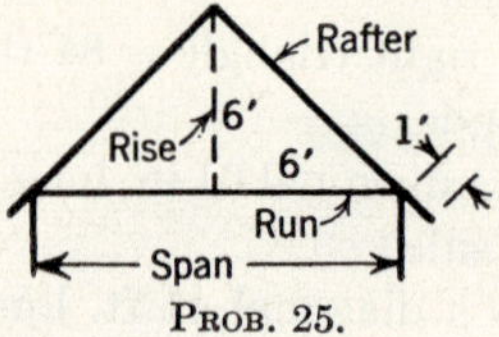

Prob. 25.

26. How long are the diagonal braces A and B on the bridge truss shown, if the truss is 6 ft. high?

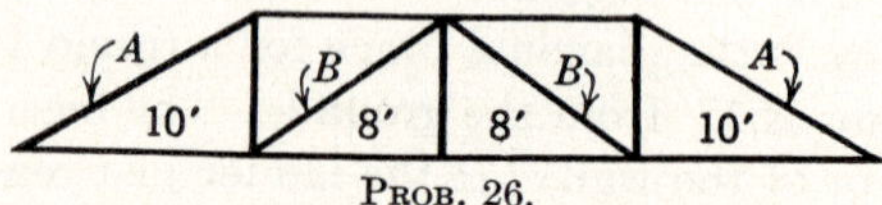

Prob. 26.

27. How long are the pieces A and B on the roof truss shown? The dotted lines divide the lines to which they are drawn into equal parts.

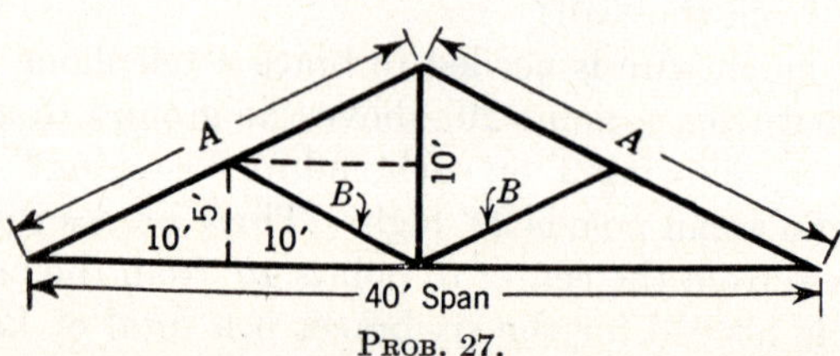

Prob. 27.

28. Determine the length of a rafter if the rise is 8′ and the run is 12′.

29. What is the rise of a rafter 20 ft. long if the span is 32 ft.? (The run is one-half the span.)

30. On the truss shown, find the lengths of the pieces A, B, C, D, and E.

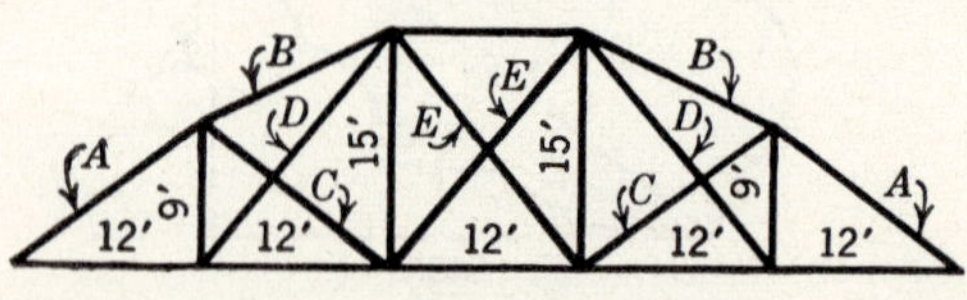

Prob. 30.

ANOTHER WAY TO FIND THE AREA OF A TRIANGLE

A formula for the area of a triangle, when the three sides are known, is as follows:

Formula. $A = \sqrt{s(s-a)(s-b)(s-c)}$
where $s = \frac{1}{2}(a+b+c)$, and a, b, and c are the lengths of the sides.

This is known as Hero's formula.

Example.—Find the area of the triangle shown.

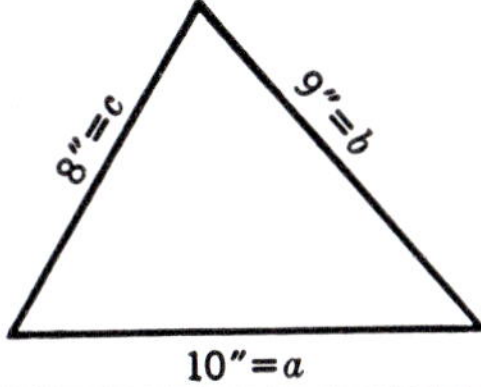

$s = \frac{1}{2}(10+9+8) = 13.5, \quad a = 10, \quad b = 9, \quad c = 8$	
$A = \sqrt{s(s-a)(s-b)(s-c)}$	Formula
$A = \sqrt{13.5(13.5-10)(13.5-9)(13.5-8)}$	Substituting
$A = \sqrt{13.5 \times 3.5 \times 4.5 \times 5.5}$	
$A = \sqrt{1169.4375}$	
$A = 34.2$	
Ans. 34.2 sq. in.	

PROBLEMS

In the following problems, divide the figures by drawing straight lines when it is necessary in order to obtain familiar figures. Name each figure, state the formula, and compute the total area for each of the figures below.

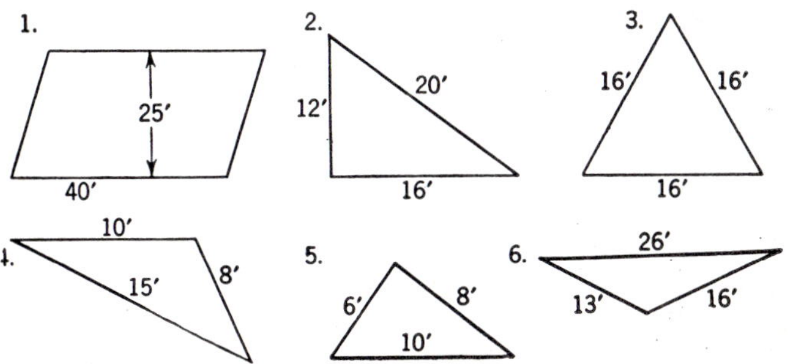

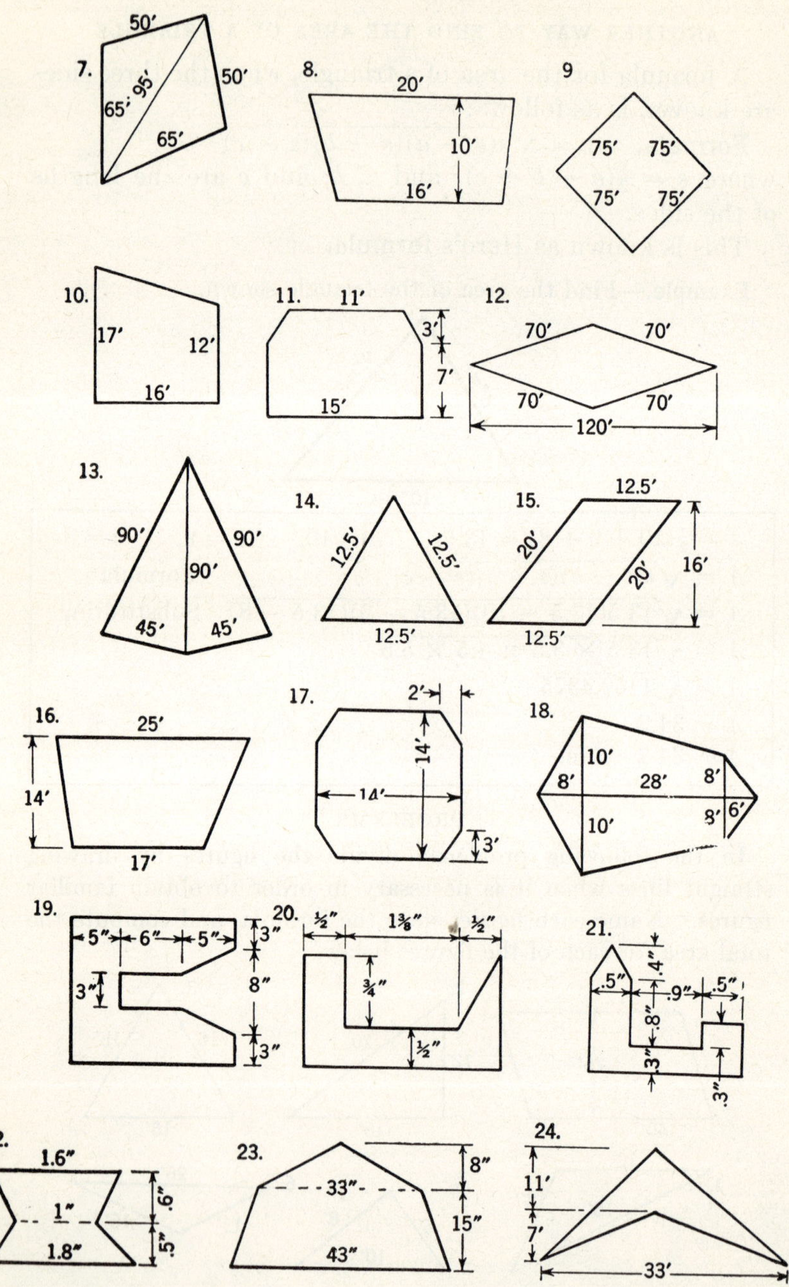
7.
50′
50′
65′
95′
65′
8.
20′
10′
16′
9
75′
75′
75′
75′
10.
17′
12′
16′
11.
11′
3′
7′
15′
12.
70′
70′
70′
70′
120′
13.
90′
90′
90′
45′
45′
14.
12.5′
12.5′
12.5′
15.
12.5′
20′
20′
16′
12.5′
16.
25′
14′
17′
17.
2′
14′
14′
3′
18.
10′
8′
28′
8′
10′
8′
6′
19.
5″
6″
5″
3″
3″
8″
3″
20.
½″
1⅜″
½″
¾″
½″
21.
.4″
.5″
.9″
.5″
.8″
.3″
.3″
22.
1.6″
1″
1.8″
.6″
.5″
23.
8″
33″
15″
43″
24.
11′
7′
33′

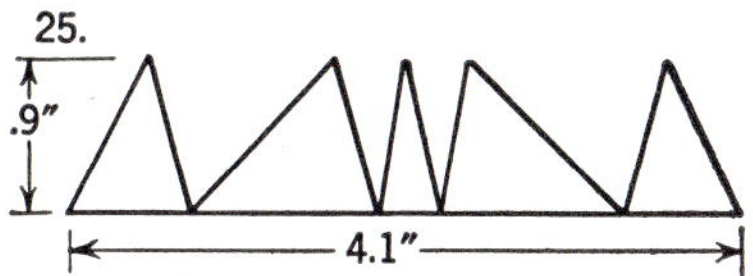

FINDING THE HEIGHT AND SIDE OF AN EQUILATERAL TRIANGLE

In an equilateral triangle, all the sides are equal.

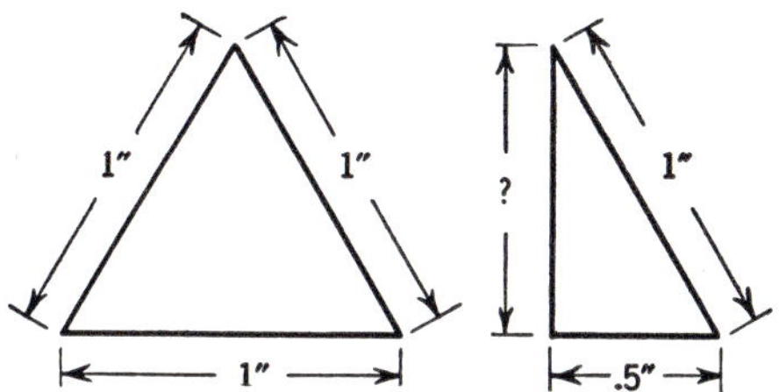

The height, or altitude, divides an equilateral triangle into two equal right triangles.

The equilateral triangle shown has sides 1 in. long. If the height of the triangle is drawn, the right triangle formed will have a base 0.5 in. long and a hypotenuse 1 in. long. The height h of this right triangle is equal to the height of the equilateral triangle. The formula for finding it is as follows:

Formula. $h = \sqrt{s^2 - \left(\frac{s}{2}\right)^2}.$

This formula is based on the rule of Pythagoras.

Substituting

$$h = \sqrt{1^2 - .5^2}$$
$$h = \sqrt{1. - .25}$$
$$h = \sqrt{.75}$$
$$h = .866$$

Ans. .866 in.

The height of any equilateral triangle 1 in. on a side is, therefore, 0.866 in.

An equilateral triangle 2 in. on a side has twice the height of the equilateral triangle 1 in. on a side.

Rule.—The height h of any equilateral triangle is .866 times the length of a side s.

Formula. $h = .866s$.

Solving the formula for the side gives $s = \frac{h}{.866}$, or, changing $\frac{1}{.866}$ to a decimal, $s = 1.155h$.

Rule.—The side s of any equilateral triangle equals 1.155 times the height h of the triangle.

Formula. $s = 1.155h$.

Example 1.—Find the height of an equilateral triangle 9″ on a side.

$h = .866s$	Formula
$h = .866 \times 9$	Substituting
$h = 7.794$	
Ans. 7.79″	

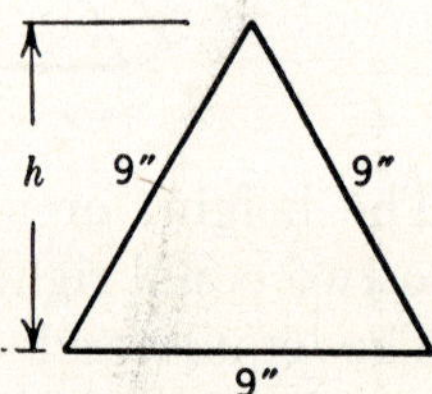

Example 2.—The height of an equilateral triangle is 15 in. Find the length of a side.

$s = 1.155h$	Formula
$s = 1.155 \times 15$	Substituting
$s = 17.325$	
Ans. 17.325 in.	

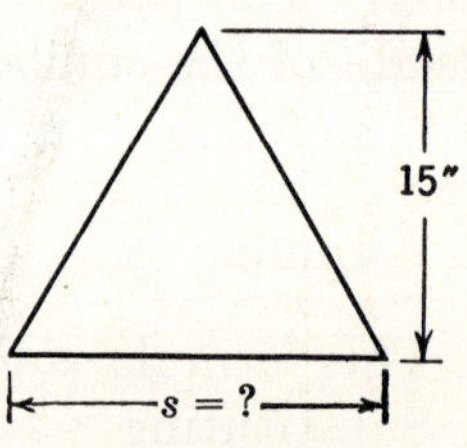

TO FIND THE AREA OF AN EQUILATERAL TRIANGLE

Rule.—To find the area of any equilateral triangle, multiply the length of a side s squared by .433.

Formula. $A = .433s^2$.

This formula is derived as follows:

$$A = \frac{bh}{2} \quad \text{Formula for the area of any triangle}$$

$$A = \frac{s \times .866s}{2}$$

Substituting $b = s$ and $h = .866s$ for any equilateral triangle

$$A = \frac{.866s^2}{2}, \qquad \text{or} \qquad A = .433s^2$$

Example.—Find the area of an equilateral triangle 9 in. on a side.

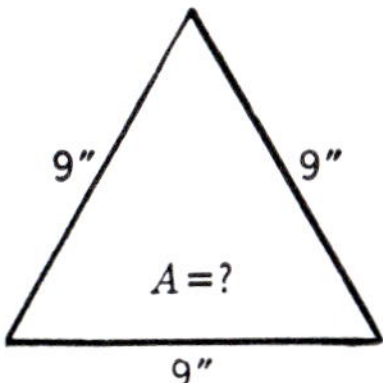

$A = .433s^2$	Formula
$A = .433 \times 9 \times 9$	Substituting
$A = 35.073$	
Ans. 35.07 sq. in., rounded off to two decimal places	

PROBLEMS

Use the formulas for equilateral triangles in solving each of the following problems. Round off the answers to two decimal places.

Number	Given	Find	Number	Given	Find
1	$s = 16''$	h	**6**	$s = 1.155''$	h
2	$s = 47'$	h	**7**	$s = 12''$	A
3	$h = 21''$	s	**8**	$s = 21''$	A
4	$h = 11''$	s	**9**	$h = 16''$	s
5	$h = 0.866''$	s	**10**	$s = 27'$	A

Illustrate, state the formula, and solve for each of the following:

11. What is the cost of making a shield in the shape of an equilateral triangle, 9″ on a side, of Monel metal at $1.00 per square foot?

12. Find the cost of making the shield in Prob. 11 if it is to be ½″ on a side.

13. What is the area of a highway road sign in the shape of an equilateral triangle 20″ on a side?

THE ISOSCELES TRIANGLE

An isosceles triangle has two equal sides.

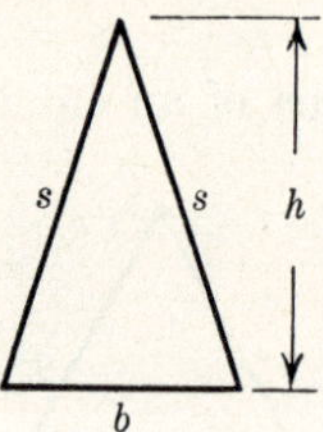

The height h of an isosceles triangle with base b and equal sides s can be found by using the following:

Formula. $h = \sqrt{s^2 - \left(\frac{b}{2}\right)^2}$.

The area A of an isosceles triangle with base b and equal sides s can be found by using the following:

Formula. $A = \frac{b}{2}\sqrt{s^2 - \left(\frac{b}{2}\right)^2}$.

Example.—Find the height and the area of the isosceles triangle shown.

$h = \sqrt{s^2 - \left(\frac{b}{2}\right)^2}$	Formula
$h = \sqrt{5^2 - (\frac{6}{2})^2}$	Substituting
$h = \sqrt{25 - 9}$	
$h = \sqrt{16}$	
$h = 4$, or 4 in.	
$A = \frac{1}{2}bh$	Formula
$A = \frac{1}{2} \times 6 \times 4$	Substituting
$A = 12$	
Ans. 12 sq. in.	

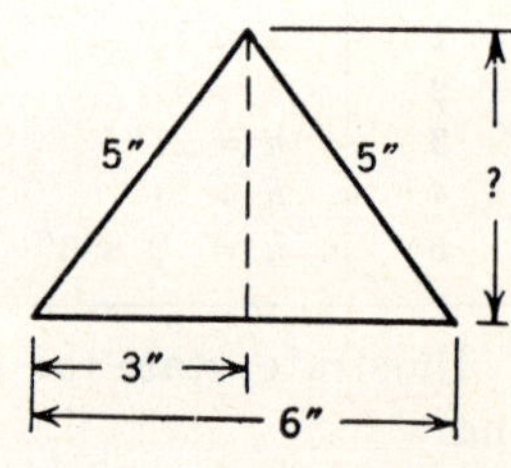

EXERCISES

Find the height and the area of each of the following isosceles triangles whose dimensions are listed on page 221.

Number	Equal sides	Base
1	16 in.	12 in.
2	30 in.	20 in.
3	12 in.	6 in.
4	15 in.	10 in.
5	10 ft.	15 ft.
6	72 ft.	60 ft.
7	100 ft.	50 ft.
8	50 ft.	90 ft.
9	72 ft.	140 ft.
10	48 yd.	18 yd.
11	32 yd.	24 yd.
12	18 rd.	28 rd.
13	10.2 in.	18.6 in.
14	11.1 in.	20 in.

THE CIRCLE

The circle is a very familiar figure. Geometrically, the circle is a closed curve, lying in a plane, that has all its points the same distance from a point inside the curve called its center. The line drawn from the center of the circle to the curved line is called the **radius.** The line drawn through the center of the circle, going completely across the circle, is called the **diameter.** The distance around the circle is called the **circumference.**

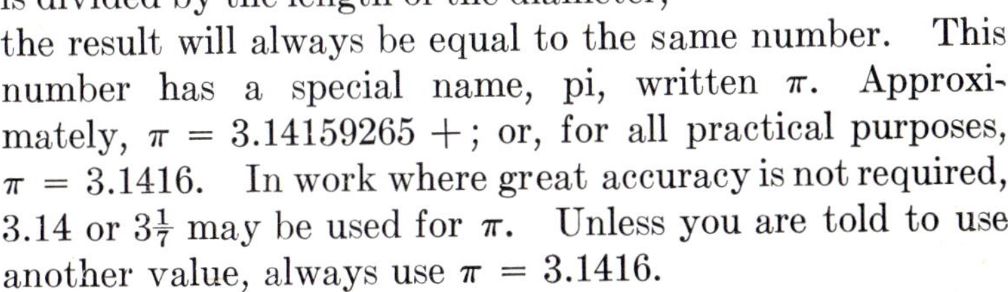

If, for any circle, the circumference is divided by the length of the diameter, the result will always be equal to the same number. This number has a special name, pi, written π. Approximately, $\pi = 3.14159265 +$; or, for all practical purposes, $\pi = 3.1416$. In work where great accuracy is not required, 3.14 or $3\frac{1}{7}$ may be used for π. Unless you are told to use another value, always use $\pi = 3.1416$.

Rule.—The diameter d of a circle is twice the radius r; the radius is equal to one-half the diameter.

Formulas. $d = 2r$ and $r = \frac{d}{2}$.

Since $\pi = \frac{c}{d}$, solving for c, we get a formula for the circumference in terms of the diameter, or $c = \pi d$. But $d = 2r$. Substituting $2r$ in the preceding formula,

$$c = \pi \times 2r,$$

and we get a formula for the circumference in terms of the radius.

Rule.—The circumference c **of a circle is equal to the product of** π **and the diameter** d **or to the product of** π **and twice the radius** r.

Formulas. $c = \pi d$, and $c = 2\pi r$.

Example 1.—The diameter of a wheel is 30″. Find its circumference.

$c = \pi d$	Formula
$c = 3.1416 \times 30$	Substituting
$c = 94.2480$	
Ans. 94.248″	

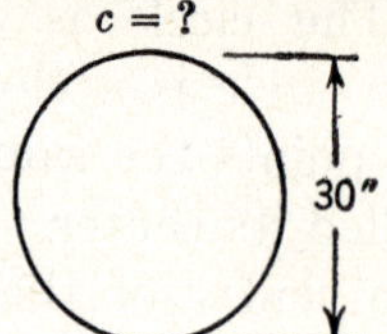

If we divide the circumference by π, we get the diameter.

Rule.—The circumference of a circle divided by π **is equal to the diameter.**

Formula. $d = \frac{c}{\pi}$.

Example 2.—The circumference of a wheel is 94.248″. Find its diameter.

$d = \frac{c}{\pi}$	Formula
$d = \frac{94.248}{3.1416}$	Substituting
$d = 30$	
Ans. 30″	

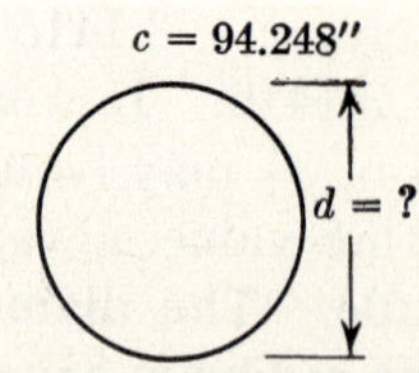

Example 3.—The radius of a circle is 15″. Find the circumference.

$c = 2\pi r$	Formula
$c = 2 \times 3.1416 \times 15$	Substituting
$c = 94.2480$	
Ans. 94.248″	

$c = ?$

15″

If we divide the circumference of a circle by 2π we get the radius.

Rule.—The circumference of a circle divided by 2π gives the radius.

Formula. $r = \dfrac{c}{2\pi}$.

Example 4.—The circumference of a circle is 94.248″. Find the radius.

$r = \dfrac{c}{2\pi}$	Formula
$r = \dfrac{94.248}{6.2832}$	Substituting
$r = 15$	
Ans. 15″	

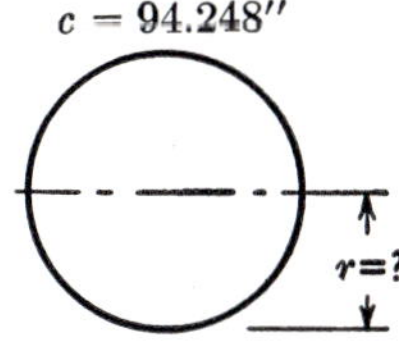

EXERCISES AND PROBLEMS

Select the right formula for each of the following, and solve. Round off the answers to two decimal places unless the problem states otherwise.

Number	Given	Find	Number	Given	Find
1	$c =$ 34″	r and d	**11**	$c =$ 44″	d and r
2	$d =$ 99″	c and r	**12**	$d =$ 21″	c and r
3	$c =$ 628″	r and d	**13**	$r =$ 48″	c and d
4	$c =$ 100″	r and d	**14**	$d =$ 762″	c and r
5	$r =$ 99″	c and d	**15**	$c =$ 96″	r and d
6	$c =$ 344″	r and d	**16**	$c =$ 100″	r and d
7	$c =$ 728″	r and d	**17**	$d =$ 77″	c and r
8	$r =$ 28″	c and d	**18**	$d =$ 32″	c and r
9	$r =$ 16″	c and d	**19**	$c =$ 314″	d and r
10	$c =$ 210″	r and d	**20**	$c =$ 2512″	r and d

21. The diameter of an automobile tire is 29″. Find its circumference.

22. The diameter of the pulley shown is 18″. Find its circumference.

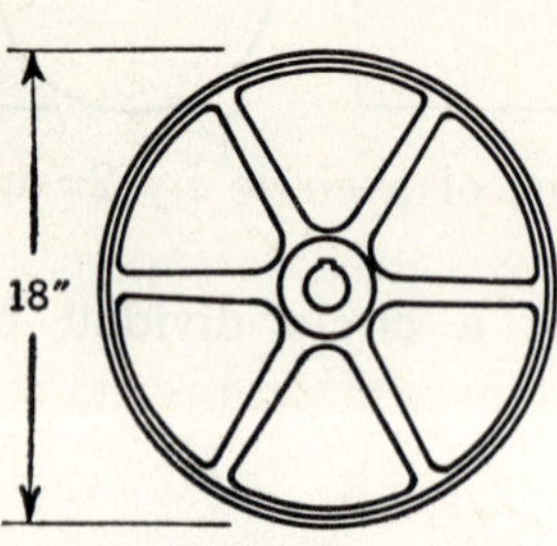

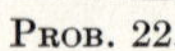
PROB. 22.

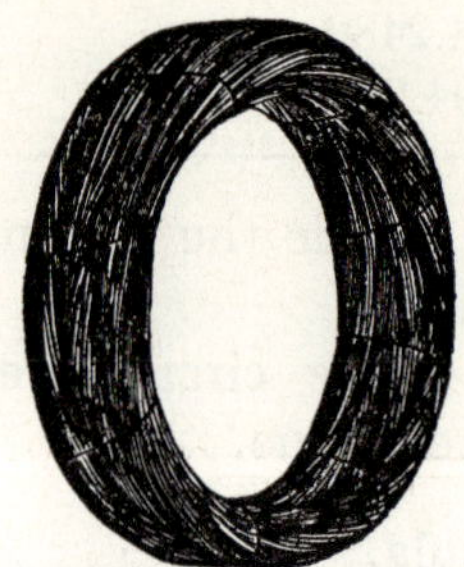
PROB. 27.

23. The circumference of a tank is 100 ft. Find its diameter and radius.

24. The radius of a flywheel is 66″. Find the number of feet in its circumference.

25. The circumference of a California redwood is 72.6 ft. Find its diameter.

26. The circumference of a quart can is 9″. What radius would a tinsmith use in laying out the end?

27. A coil of wire is shown. What is the length of 20 turns of the wire if the average diameter of the coil is 20″?

28. A customer asks for 200 ft. of aerial wire. How many turns will be needed if the average diameter of the coil is 14″? (Use $\frac{22}{7}$ for π.)

29. If the circumference of an automobile tire is 7.85 ft., what is its diameter expressed to the nearest whole inch?

30. An automobile tire is listed as 28″ × 6.25″. What is its outside circumference? What is its inside circumference?

31. How many turns would the tire in Prob. 30 make if the car traveled 1 mile? 8 miles?

32. How many turns would the tire in Prob. 29 make if it ran for 30,000 miles?

33. On the ring shown, the inner circle is 2 ft. in diameter. The outer circle is 4 ft. in diameter. How do the circumferences compare?

34. On the circular saw shown, the diameter is 22 in. If there are 25 teeth evenly spaced on the saw, what is the spacing between the points?

35. If the teeth of an inserted tooth circular saw are $4\frac{1}{4}''$ between points, what is the circumference of the saw if there are 56 teeth? What is the diameter of the saw?

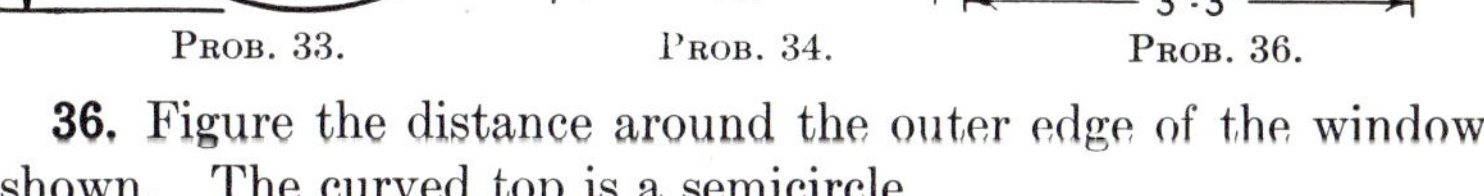

PROB. 33. PROB. 34. PROB. 36.

36. Figure the distance around the outer edge of the window shown. The curved top is a semicircle.

37. If the diameter of a locomotive drive wheel is 8 ft., what is its circumference?

38. How many turns will the wheel in Prob. 37 make when the locomotive travels 1 mile?

39. What is the length of a semicircle if the diameter of the circle is $6'6''$?

40. What is the length of a quarter circle if the diameter of the circle is 100 ft.?

AREAS OF CIRCLES

A piece of material, circular in shape, can be cut to form many narrow triangles, as shown. The altitude of each triangle thus formed would be the radius r. The sum of the bases would be the circumference c.

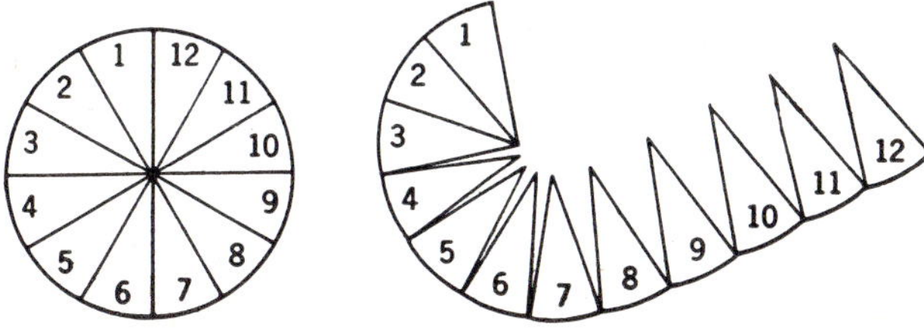

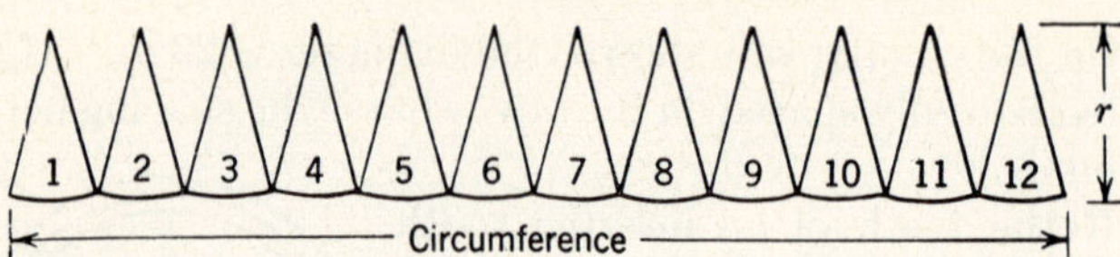

The area of each of the triangles is one-half the base multiplied by the height. Since r is the height of each triangle and c is the sum of the bases of all the little triangles, the entire area can be found as follows:

$A = \frac{1}{2}bh$	Formula for area of a triangle
$A = \frac{1}{2}rc$	Substituting r for h and c for b
$A = \frac{1}{2}r \times 2\pi r$	Substituting $2\pi r$ for c
$A = \pi r^2$	

Rule.—The area of a circle in terms of the radius is π times the radius squared.

Formula. $A = \pi r^2$.

Rule.—In terms of the diameter, the area of a circle is 0.7854 times the diameter squared.

Formula. $A = .7854d^2$.

This formula is derived from the preceding one as follows:

$A = \pi r^2$	Formula in terms of the radius
$A = 3.1416\left(\frac{d}{2}\right)^2$	Substituting $\frac{d}{2}$ for r and 3.1416 for π
$A = \frac{3.1416d^2}{4}$	
$A = .7854d^2$	

Note.—By measuring the area of a square with a planimeter, a special instrument used to measure areas, and then measuring the area of a circle inscribed within the square, the ratio of the areas is found to be .7854 to 1. This ratio serves as a check on the formula just given.

Example 1.—Find the area of a circle that has a radius of 12 in.

$A = \pi r^2$	Formula
$A = 3.1416 \times 12^2$	Substituting
$A = 3.1416 \times 144$	
$A = 452.39$	
Ans. 452.39 sq. in.	

Example 2.—Find the area of a circle that has a diameter of 15 in.

$A = .7854d^2$	Formula
$A = .7854 \times 15 \times 15$	Substituting
$A = 176.715$	
Ans. 176.72 sq. in.	

Rule.—In terms of the circumference, the area of a circle is the circumference squared divided by 4π.

Formula. $A = \frac{c^2}{4\pi}$, or $A = \frac{c^2}{12.57}$.

This formula was derived from the preceding one as follows:

$$A = \frac{\pi d^2}{4}$$

and substituting $d = \frac{c}{\pi}$

$$A = \frac{\pi}{4} \times \frac{c}{\pi} \times \frac{c}{\pi}$$

$$A = \frac{\not{\pi}}{4} \times \frac{c}{\not{\pi}} \times \frac{c}{\pi}$$

Canceling and collecting gives

$$A = \frac{c^2}{4\pi} = \frac{c^2}{4 \times 3.1416} = \frac{c^2}{12.5664} \text{ or } \frac{c^2}{12.57}$$

Example 3.—Find the area of a circle that has a circumference of 45 in.

$A = \frac{c^2}{12.57}$	Formula
$A = \frac{45 \times 45}{12.57}$	Substituting
$A = 161.09$	
Ans. 161.1 sq. in.	

EXERCISES AND PROBLEMS

Select the right formula for each of the following, and solve:

Number	Given	Find	Number	Given	Find
1	$d =$ 32 in.	A	**11**	$c =$ 628 in.	A
2	$d =$ 25 in.	A	**12**	$c =$ 942 in.	A
3	$r =$ 17 in.	A	**13**	$c =$ 1250 ft.	A
4	$r =$ 110 ft.	A	**14**	$d =$ 21 ft.	c and A
5	$d =$ 55 in.	A	**15**	$c =$ 21 ft.	A
6	$c =$ 100 ft.	A	**16**	$r =$ 45 in.	c and A
7	$c =$ 25 yd.	A	**17**	$d =$ 48 in.	c and A
8	$c =$ 314 in.	A	**18**	$c =$ 700 in.	d, r, and A
9	$r =$ 21 ft.	A	**19**	$c =$ 500 ft.	d, r, and A
10	$d =$ 47 in.	A	**20**	$r =$ 3 miles	A and c

21. The diameter of a circular shaft is 5″. Find the cross-sectional area.

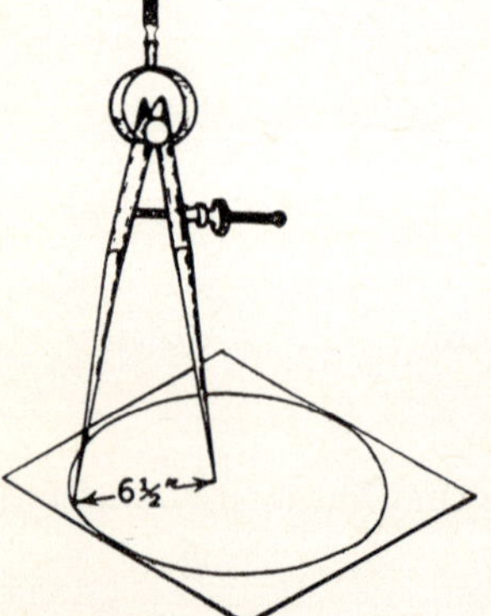

Prob. 23.

22. The radius of a circle is 4.9″. Find the area.

23. As shown in the illustration, a tinsmith set his compass at $6\frac{1}{2}''$ in laying out the bottom for a can. How many square inches did he use?

24. A round tank is 14 ft. in diameter. Compute the material needed to make the bottom, with no allowances for seams.

25. A concrete floor is to be put in a round silo 16 ft. in diameter at a cost of 25¢ per square foot. What will be the total cost?

26. A round mirror 28″ in diameter is to be silvered at a cost of 90¢ per square foot. Compute the total cost.

27. The diameter of a piston is $3\frac{3}{8}''$. Find the area of the head.

28. Find the cross-sectional area of a 1″ diameter rod.

29. How many square inches are there in the end of a $\frac{1}{2}''$ round rod?

30. How many square inches are there in the end of a $2''$ round rod?

31. Examine the accompanying figures, and compare the answers of Probs. 28, 29, and 30. Determine if one is any part of the others.

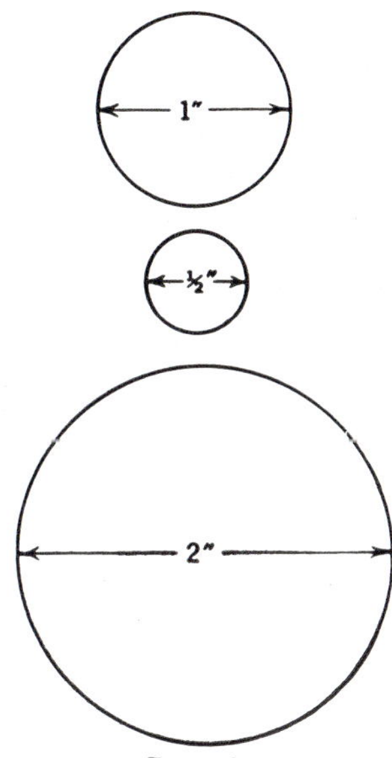

PROB. 31.

32. Give a rule by which you can figure the area of a circle when its diameter is twice as great as that of another known circle.

33. If the area of one circle is 9 times as great as the area of a smaller circle, how do their diameters compare?

34. Give a rule and write a formula by which you can figure the area of a circle when its diameter is 3 times as great as that of another known circle.

35. If the diameter of one circle is 4 times as great as the diameter of a smaller circle, how would the areas compare?

36. Give a rule and write a formula by which you can figure the area of a circle when the diameter is 4 times as great as that of another known circle.

OTHER FORMULAS FOR RADIUS

Since the area of a circle equals π multiplied by the radius squared, reversing the process will give the radius.

$$A = \pi r^2; \qquad r^2 = \frac{A}{\pi}; \qquad \text{hence,} \qquad r = \sqrt{\frac{A}{\pi}}$$

Rule.—The radius of a circle equals the square root of the area divided by π.

Formula. $r = \sqrt{\frac{A}{\pi}}$.

A shorter formula, although not so accurate, may be derived from this formula, as follows:

$$r = \sqrt{\frac{A}{\pi}} = \sqrt{A \times \frac{1}{3.14}} = \sqrt{A \times .318}$$

Formula. $r = \sqrt{.318A}$.

Example 1.—The area of a circle is 1256 sq. in. Find the radius. Find the diameter. (Use $\pi = 3.14$.)

Solution 1.

$r = \sqrt{\frac{A}{\pi}}$ Formula

$r = \sqrt{\frac{1256}{3.14}}$ Substituting

$r = \sqrt{400}$ $1256 \div 3.14 = 400$.

$r = 20$

Ans. 20 in. radius.

$d = 2r$, $2 \times 20 = 40$

Ans. 40 in. diameter.

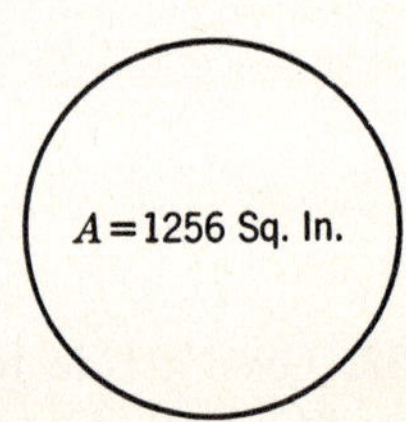

Solution 2.

$r = \sqrt{.318A}$ Formula

$r = \sqrt{.318 \times 1256}$ Substituting

$r = \sqrt{399.408}$

$r = 19.99$

Ans. 19.99 in. radius. This is approximately the same as the answer above and may be increased to 20 in.

EXERCISES

Use $\pi = 3.14$, and find the radius and diameter of circles having the following areas:

1. 500 sq. ft. **2.** 628 sq. ft. **3.** 2198 sq. ft.
4. 2512 sq. ft. **5.** 10 sq. ft. **6.** 150 sq. ft.
7. 240 sq. ft. **8.** 942 sq. ft. **9.** 2826 sq. ft.
10. 314. sq. ft.

COMBINED CIRCLES

It is often necessary to find the radius or diameter of a circle equal in area to the combined area of two or more circles.

Rule.—The radius R_c of a circle having the same area as several circles combined is equal to the square root of the several radii squared and added.

Formula. $R_c = \sqrt{r^2 + R^2 + \cdots}$.

Note.—This formula can be extended to take care of any number of circles. The three dots following the plus sign mean "and so on."

This formula was obtained as follows:

$\pi R_c^2 = \pi r^2 + \pi R^2$ R_c = radius of combined circle

$\frac{\not\pi R_c^2}{\not\pi} = \frac{\not\pi r^2 + \not\pi R^2}{\not\pi}$ Dividing each term by π

$R_c^2 = r^2 + R^2$

$R_c = \sqrt{r^2 + R^2}$ Extracting the square root

Example 1.—What are the radius and diameter of a circle whose area equals the combined areas of a 6″ and an 8″ circle?

$d = 6''$, $r = 3''$; $D = 8''$, $R = 4''$

$R_c = \sqrt{r^2 + R^2}$ Formula

$R_c = \sqrt{3^2 + 4^2}$ Substituting

$R_c = \sqrt{9 + 16}$

$R_c = \sqrt{25}$

$R_c = 5$

Ans. 5″, the radius of the combined circle

$D_c = 2R_c$, $5 \times 2 = 10$

Ans. 10″, the diameter of the combined circle

Similarly, it may be shown that the diameter of a combined circle is expressed in the following way:

Formula. $D_c = \sqrt{d^2 + D^2 + \cdots}$.

Example 2.—Find the diameter of a circle whose area equals the combined areas of a 6″ and an 8″ circle.

$D_c = \sqrt{d^2 + D^2}$	Formula
$D_c = \sqrt{6^2 + 8^2}$	Substituting
$D_c = \sqrt{36 + 64}$	
$D_c = \sqrt{100}$	
$D_c = 10$	
Ans. 10″	

Example 3.—Find the diameter of a circle having an area equal to the combined areas of three 4″ circles and three 6″ circles.

$D_c = \sqrt{d^2 + d^2 + d^2 + D^2 + D^2 + D^2}$	Formula. d and
$D_c = \sqrt{4^2 + 4^2 + 4^2 + 6^2 + 6^2 + 6^2}$	D are each used
$D_c = \sqrt{16 + 16 + 16 + 36 + 36 + 36}$	three times.
$D_c = \sqrt{156}$	Substituting
$D_c = 12.49$	

Ans. 12.49″. For practical purposes, this answer would be rounded off to 12.5″.

PROBLEMS

1. Find the radius of a circle whose area equals the combined area of a 5″ and a 9″ diameter circle.

2. What is the diameter of a rod that is to have the same area as four 1″ diameter rods?

3. Find the diameter of a combined circle equivalent to the areas of five circles each 6″ in diameter.

4. What is the diameter of a single water pipe used to replace three 6″ diameter pipes?

5. Find the radius and the area of a circle equal to the sum of two circles each 20″ in diameter.

6. What is the radius of a circle whose area equals the combined area of circles with radii of 3″, 4″, 5″, and 6″?

7. Find the diameter of a water main to have the same area as twelve 3″ diameter laterals.

8. What radius should you use to draw a circle whose area equals the combined area of five 12″ diameter circles?

9. What diameter gas line should you use to replace four 8″ diameter lines?

10. Find the diameter of a rod to have the same area as the combined area of six rods 0.25″ in diameter.

REVIEW PROBLEMS INVOLVING CIRCLES

1. The circumference of a circle is 100″. Find the area.

2. The area of a circle is 546 sq. in. Find its diameter and circumference.

3. The diameter of a circle is 8″. The diameter of another circle is 16″. How do the diameters compare? How do the areas compare?

4. The radius of a circle is 7″. The radius of another circle is 21″. How do the diameters compare? How do the circumferences compare?

5. If the radius of a circle is multiplied by any number, what is the ratio of increase for the diameters? For the circumferences? For the areas?

6. A main sewer line is to receive the sewage from four 12″ lines. Compute the diameter of the main.

7. A wall heater outlet is 10″ by 14″. Find the diameter of a round conductor pipe with the same area.

8. The floor outlet from a heater is 22″ × 22″. How large a round pipe will have the same area?

9. Find the cross-sectional area of a circular smokestack with an area equal to three 24″ diameter round boiler stacks.

10. Find the diameter of the main stack in Prob. 9.

11. Find the length of the side of a square that has the same area as an 18″ diameter circle.

12. Find the length of the side of a square chimney that has the same area as two 24″ round flues.

13. Find the side of a square chimney that has the same area as four 28″ diameter round stacks. (Use $\frac{22}{7}$ for π.)

14. Compute the diameter of a circle that has the same area as a rectangle 22″ by 32″.

15. A cast-iron column is 8 in. in diameter and 1 in. thick. Determine the cross-sectional area.

16. Figure the cost of beveling the edge of a 28-in.-diameter plate-glass mirror at 7¢ per linear foot.

17. Determine the diameter of a pipe that has the same cross-sectional area as two 3″ diameter pipes.

CIRCULAR-RING AREA

Rule.—The area of a circular ring equals the area of the outside circle minus the area of the inside circle.

Formulas.

$$A = \pi R^2 - \pi r^2 \qquad A = .7854(D^2 - d^2)$$
$$A = \pi(R^2 - r^2) \qquad A = .7854D^2 - .7854d^2$$
$$A = \pi(R + r)(R - r) \qquad A = .7854(D + d)(D - d)$$

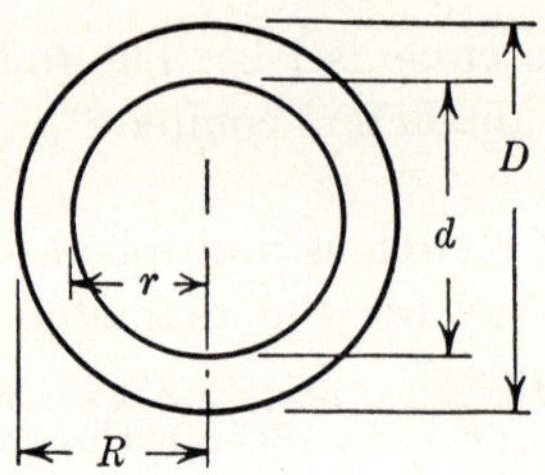

Example.—Outside diameter, $D = 6$ in.; inside diameter, $d = 5$ in. Find the area of the cross section.

$A = .7854(D + d)(D - d)$	Formula
$A = .7854(6 + 5)(6 - 5)$	Substituting
$A = .7854 \times 11 \times 1$	Collecting
$A = 8.6394$	
Ans. 8.64 sq. in. rounded off to two decimal places	

PROBLEMS

Round off the answers to two decimal places.

1. In a circular ring, the outside diameter is 8″ and the inside diameter is 6″. Find the area of the cross section.

2. In a circular ring, the outside diameter is 6″, and the inside diameter is $5\frac{1}{2}$″. Find the area of the cross section.

3. In a circular ring, the outside diameter is 10″ and the inside diameter is 7″. Find the area of the cross section.

4. A wood flume has a 16″ inside diameter. The wood is 1″ thick. Find the area of the wood in the end.

5. A hollow circular column is 2″ thick, and the outside diameter is 6″. Find the cross-sectional area.

6. A standard 1″ pipe is 1.315″ outside diameter and 1.049″ inside diameter. Find the area of the metal at the end.

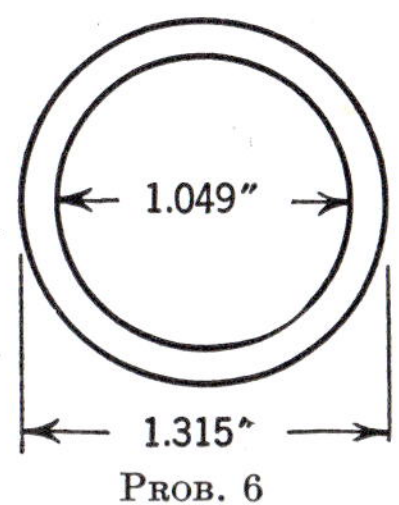

PROB. 6

7. A standard 6″ pipe has a diameter of 6.625″ outside and of 6.065″ inside. Find the area of the cross section.

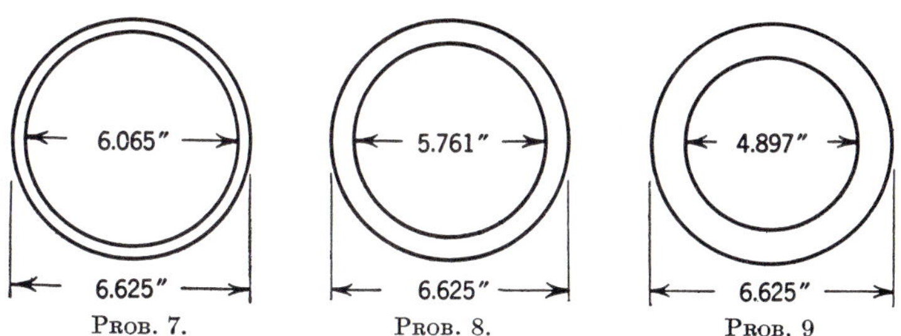

PROB. 7. PROB. 8. PROB. 9

8. An extra-strong 6″ pipe has an outside diameter of 6.625″ and an inside diameter of 5.761″. Find the area of the cross section.

9. On a double-extra-strong iron 6″ pipe the outside diameter is 6.625″ and the inside diameter is 4.897″. Find the area of the cross section.

10. What is the per cent of increase in material in the cross-sectional area of extra-strong pipe over standard pipe? (Examine the diagrams and the answers to Probs. 7 and 8.)

11. How much thicker is the material in an extra-strong 6″ pipe than in a standard 6″ pipe? (See Probs. 7 and 8.)

12. How much thicker is the wall of double-extra-strong 6″ pipe than the wall of standard pipe? (See Probs. 7 and 9.)

13. Double-extra-strong iron pipe, 6″ size, has what per cent increase in metal area as compared with the standard 6″ pipe? (See Probs. 7 and 9.)

MATERIAL NEEDED FOR CURVED PIECES

To find the length of a piece of metal needed for a curved bend, the center line of the piece should be used for measurement. The center line is commonly called the measuring line, the neutral line, or the neutral plane. On curved bends, the outside stretches and the inside compresses, but the neutral line remains the same. On work where the length must be accurate to a few thousandths of an inch, as in tool and die work, the neutral line is figured from one-third to two-fifths outward from the inside of the bend. Many technical formulas have been developed and should be used for this accurate work; but, for ordinary bending, one-half the thickness of the stock as the measuring line gives a sufficient degree of accuracy.

Rule 1.—For circular bends, find the circumference (or the part of the circumference used) on the neutral line.

Example 1.—Determine the length needed to make the circular ring shown. (Use $\pi = 3.14$.)

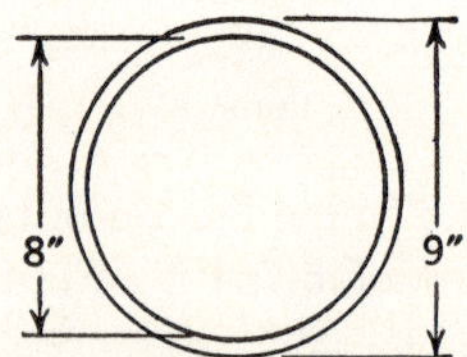

$d = 8.5$	Since 8″ is the inside diameter and 9″ is the outside diameter, $8\frac{1}{2}$″ is the diameter on the neutral line.
$c = \pi d$	Formula
$c = 3.14 \times 8.5$	Substituting
$c = 26.69$	
Ans. 26.69″	

Rule 2.—For curved bends, calculate the length of the part of the circle used, measuring along the neutral line, and add this length to the total length of the straight parts.

Example 2.—Calculate the material needed for the piece shown. Use $\pi = 3.14$.)

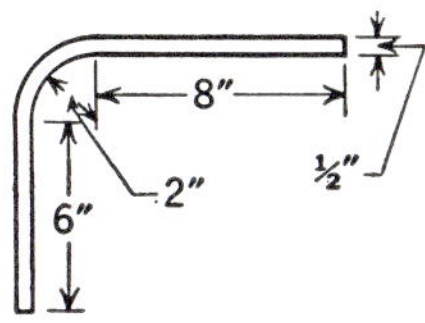

A 2″ radius for the inside circle on $\frac{1}{2}''$ stock makes the radius on the neutral line $2\frac{1}{4}''$.

$c = 2\pi r$ Formula

$c = 2 \times 3.14 \times 2\frac{1}{4}$ Substituting

$c = 14.13$ or $14.13''$

Since only one-fourth of the circle is used,

$$\tfrac{1}{4}c = \tfrac{1}{4} \times 14.13'' = 3.5325'',$$

or 3.53″, rounded off to two decimal places.

Adding the straight parts for the total length,

$$3.53'' + 6'' + 8'' = 17.53''$$

Ans. 17.53″

PROBLEMS

Calculate the length of stock needed for each of the pieces shown below. (Use 3.14 for π. Round off the answers to two decimal places.)

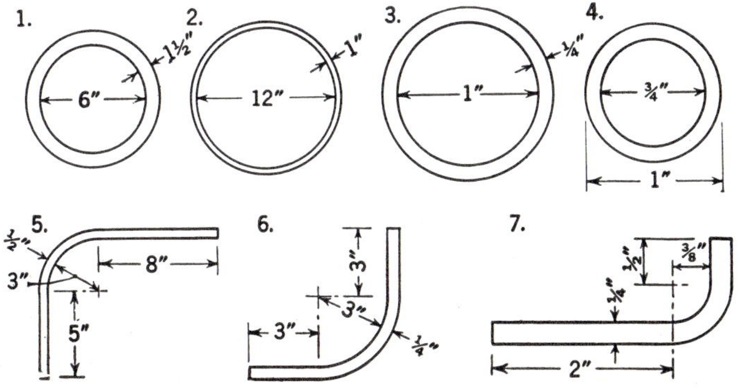

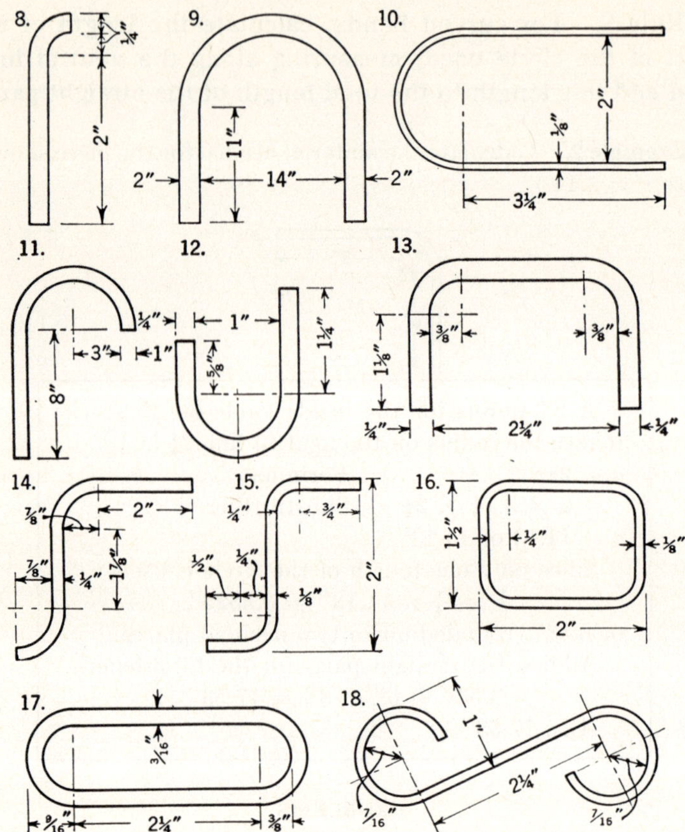

THE SURFACE AREA OF A CYLINDER

Rule.—The area of the curved surface S of a cylinder is equal to the product of the circumference and the height h of the cylinder.

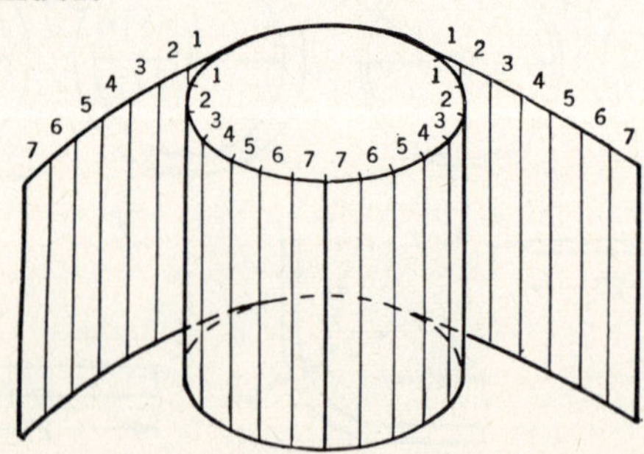

Since $S = ch$ and $c = \pi d$, or $c = 2\pi r$, the following formulas are obvious:

Formulas. $S = \pi dh$ or $S = 2\pi rh$.

Example 1.—Find the amount of metal needed to make the walls of a tank 12 ft. in diameter and 10 ft. high.

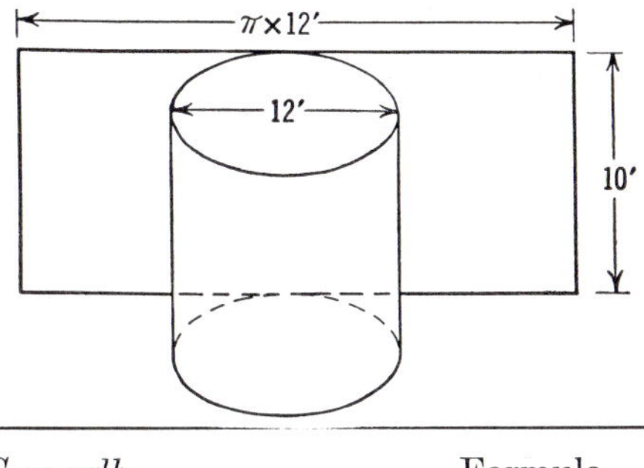

$S = \pi dh$	Formula
$S = 3.1416 \times 12 \times 10$	Substituting
$S = 376.992$	
Ans. 376.99 sq. ft.	

Example 2.—Find the number of square feet of material required for the entire surface of the tank in Example 1 including the two ends.

$A = 0.7854d^2$	Formulas for the area of one end
$A = 0.7854 \times 12 \times 12$	Substituting

$A = 113.0976$, or 113.1 sq. ft. for one end

The total surface area, T, equals the sum of the areas of the top, bottom, and curved surface.

$T = 113.1 + 113.1 + 376.99 = 603.19$

Ans. 603.19 sq. ft.

PROBLEMS

Give the formulas and solve for each of the following:

1. Find the area of the wall of a cylindrical tank 4 ft. in diameter and 6 ft. high.

2. Find the area of one end for the tank in Prob. 1.

3. Find the total area of the tank in Probs. 1 and 2.

4. What is the wall area of a cylinder 10 in. in diameter and 15 in. high?

5. Figure the wall area of a cylindrical can 8″ deep and 12″ in diameter.

6. Compute the total surface area of a cylindrical can 8″ deep and 12″ in diameter.

7. Figure the total area of a tank 4 ft. in diameter and 7 ft. deep.

8. How many square feet of material are needed for the curved surface of a tank 20 ft. in diameter and 8 ft. deep?

9. Find the total surface area of the tank in Prob. 8.

10. Determine the area of the wall surface of a cylinder 16″ in diameter and 6″ high.

11. Compute the area of the wall surface of a cylinder 6″ in diameter and 16″ high.

12. What is the wall area of a cylinder 9 ft. in diameter and 9 ft. deep?

13. Figure the number of square inches of surface in a 20-ft. length of 16-in.-diameter pipe.

14. Change the square inches in Prob. 13 to square feet.

15. Change the square feet in Prob. 14 to square yards. Round off to the nearest hundredth of a square unit.

16. Which has the greater surface, a round tank 20 ft. in diameter and 10 ft. deep, or a rectangular tank 20 ft. on a side and 10 ft. deep?

17. If two tanks are of the same depth but one has twice the diameter of the other, the larger has how many times the surface area of the smaller?

18. Two tanks have the same diameter, but one is twice as deep as the other. How do the areas of their surfaces compare?

19. Figure the curved surface area of a cylinder 1 ft. in diameter and 1 ft. deep.

20. Figure the area of each end of the cylinder in Prob. 19.

TANK PROBLEMS

Complete the following. Copy the tabulation. Do not write in this book.

Number	Height	Diameter	Wall area	Area of both ends	Total surface area
1	15′	16′			
2	12′	4′			
3	14′	9′			
4	25′	44′			
5	48′	56′			
6	100′	100′			
7	21′	44′			
8	16′	240′			
9	7′	7′			
10	8′	8′			
11	12′	12′			
12	10′	16′			
13	15′	24′			
14	25′	28′			
15	1′	1′			
16	21′	20′			
17	14′	10′			
18	16″	12″			
19	15″	22″			
20	48″	24″			

THE PYRAMID

A pyramid is a solid with triangular faces that meet in a common vertex and whose base is a polygon.

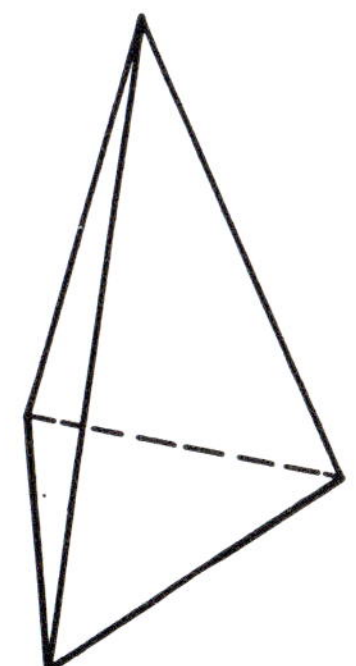

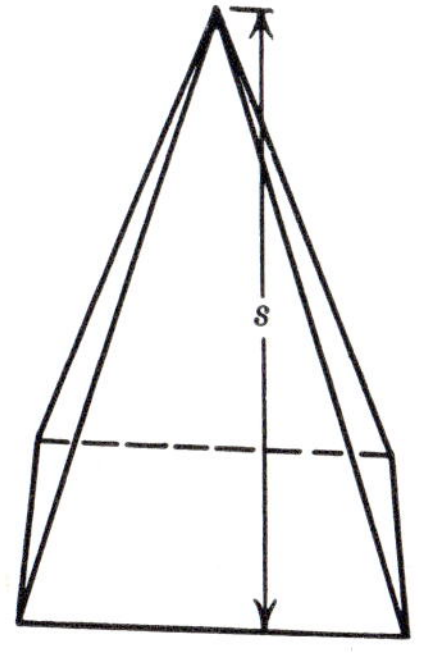

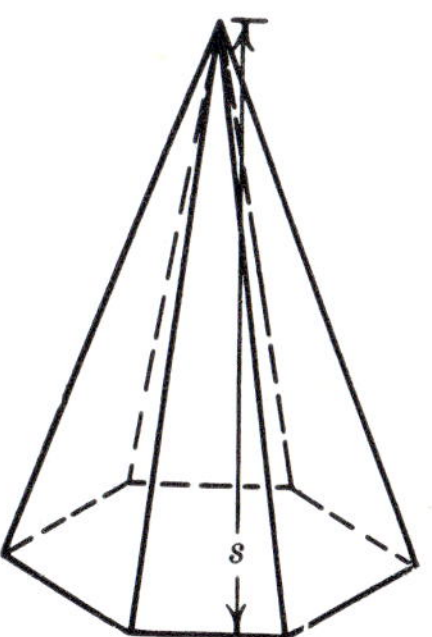

Let s = slant height, p = perimeter of base, L = lateral surface area.

Formulas. $L = \frac{ps}{2}$, $p = \frac{2L}{s}$, $s = \frac{2L}{p}$

Example.—Find the lateral surface area of a pyramid whose base is a square 12 in. on a side and whose slant height is 16 in.

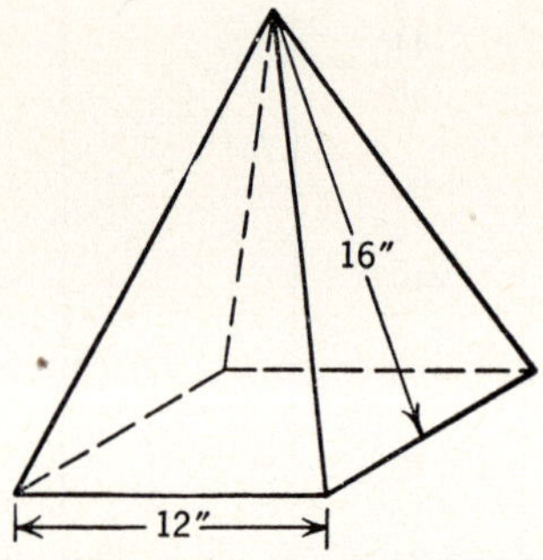

$L = \frac{ps}{2}$	Formula
$L = \frac{48 \times 16}{2}$	Substituting ($p = 4 \times 12 = 48$)
$L = 384$	
Ans. 384 sq. in.	

PROBLEMS

1. Find the lateral surface area of a pyramid with a 6-in. square base and with a slant height of 10 in.

2. Compute the lateral surface area of a pyramid with a square base 24 ft. on a side and with a slant height of 21 ft.

3. Calculate the lateral surface area of a pyramid with a hexagonal base 18 in. on a side and with a slant height of 25 in.

4. Find the lateral surface area of a church steeple in the shape of a pyramid whose base is an octagon 8 ft. on a side, and with a slant height of 36 ft.

5. What is the slant height of a pyramid with a square base 8 ft. on a side, if the lateral surface area is 640 sq. ft.?

6. Find the perimeter and the side of a hexagonal pyramid whose slant height is 14 ft. and whose lateral surface area is 700 sq. ft.

7. The side of one pyramid is 8 in., the side of another pyramid is 16 in., and both pyramids have 24-in. slant heights.

How do the lateral surface areas compare if both bases are squares?

8. List several practical applications of the pyramid.

THE CONE

The **cone** is a solid that tapers uniformly from a circular base to a point. The cone is sometimes called a pyramid with a circular base.

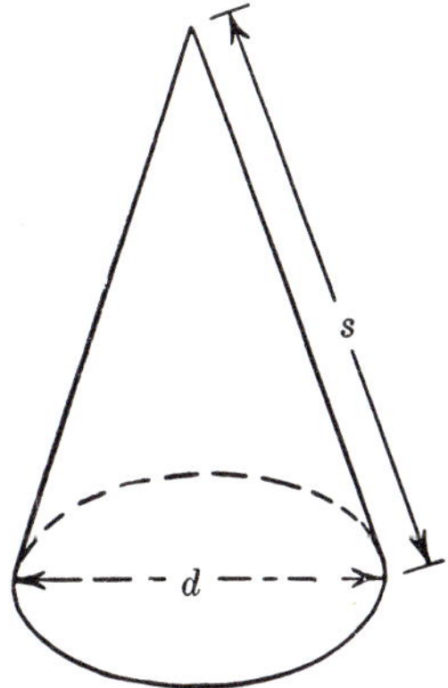

Let d = diameter of base, s = slant height, L = lateral surface area, r = radius of base.

Formulas. $L = \frac{\pi ds}{2}$ or $L = \pi rs$, $d = \frac{2L}{\pi s}$, $s = \frac{2L}{\pi d}$

Example.—Find the lateral surface area of a cone whose diameter is 18 in. and whose slant height is 33 in. (Use $\pi = 3.14$.)

$L = \frac{\pi ds}{2}$	Formula
$L = \frac{3.14 \times 18 \times 33}{2}$	Substituting
$L = 932.58$	
Ans. 932.58 sq. in.	

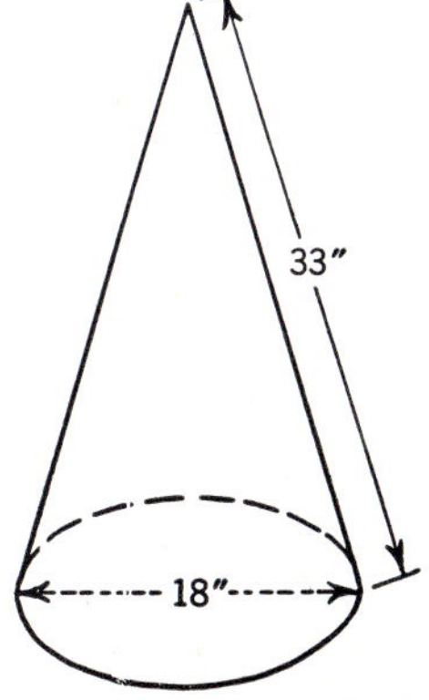

PROBLEMS

Use 3.14 for π.

1. Find the lateral surface area of a cone 24 ft. in diameter and with a slant height of 16 ft.

2. Find the lateral surface area of a cone whose base diameter is 77 ft. and whose slant height is 40 ft.

3. Find the lateral surface area of a cone with a base diameter of 12 ft. and a slant height of 15 ft.

4. What is the lateral surface area of a cone when the base diameter is 24 ft. and the slant height is 30 ft.?

5. How do the surface areas of Probs. 3 and 4 compare?

6. How many square yards of surface are there on a pile of sand in the shape of a cone with the base diameter 50 ft. and the slant height 30 ft.?

7. Find the number of square feet of metal needed to make a conical vessel 24 ft. in diameter and 15 ft. high.

8. What is the diameter of a cone if the slant height is 15 in. and the lateral surface area is 225 sq. in.?

9. What is the slant height of a cone if its diameter is 15 ft. and the lateral surface area is 2826 sq. ft.?

10. List several practical applications of the cone.

THE SPHERE

The sphere is a solid bounded by a surface whose every point is equally distant from a point within the solid called the center.

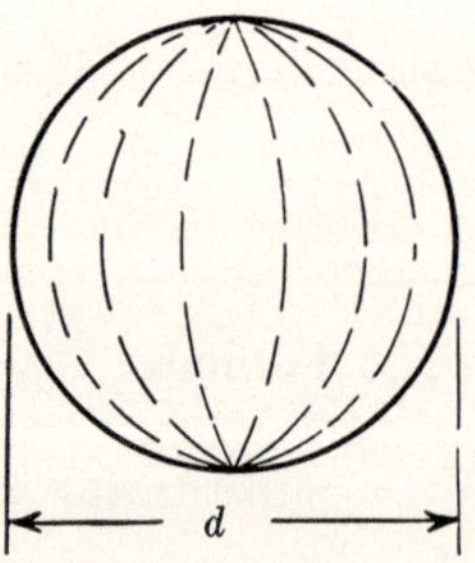

Let d = diameter, r = radius, S = surface area.

Formulas. $S = \pi d^2$ or $S = 4\pi r^2$

$$d = \sqrt{\frac{S}{\pi}}, \qquad d = \sqrt{0.318S}$$

Example.—Find the surface area of a sphere 16 in. in diameter. (Use $\pi = 3.14$.)

$S = \pi d^2$	Formula
$S = 3.14 \times 16 \times 16$	Substituting
$S = 803.84$	
Ans. 803.84 sq. in.	

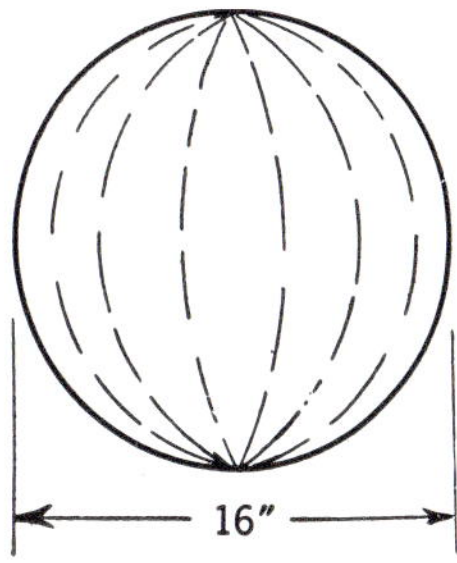

PROBLEMS

Use 3.14 for π.

Find the surface areas of spheres having the following diameters:

1. 18 in.
2. 12 ft.
3. 33 yd.
4. 15.5 ft.
5. 0.5 in.
6. 24 ft.

Find the surface areas of spheres having the following radii.

7. 9 in.
8. $\frac{1}{4}$ in.
9. 21 ft.
10. 12.5 ft.
11. 0.2 in.
12. 17 ft.

Find the diameters of spheres having the following surface areas:

13. 5000 sq. ft.
14. 2528 sq. ft.
15. 0.628 sq. in.
16. 31.4 sq. in.

SURFACE AREAS OF OTHER GEOMETRIC FIGURES

There are many other geometric figures used in construction work and designs.

For practical purposes, the shop man need not derive the rules and formulas he uses. In this section, only standard formulas are presented with their applications to practical problems.

A **circumscribed circle** is a circle that goes through each of the vertices of the polygon.

An **inscribed circle** is a circle drawn inside the polygon that just touches each side of the polygon at one point.

THE REGULAR HEXAGON

A regular hexagon is a polygon with six equal sides and six equal angles. The letters used in the formulas for hexagons are A for area, R for the radius of the circumscribed circle, r for the radius of the inscribed circle, and s for a side of the hexagon.

Formulas.

$$R = s, \qquad R = 1.155r$$
$$r = .866s, \qquad r = .866R$$
$$s = R, \qquad s = 1.155r$$
$$A = 2.598s^2, \qquad A = 2.598R^2, \qquad A = 3.464r^2$$

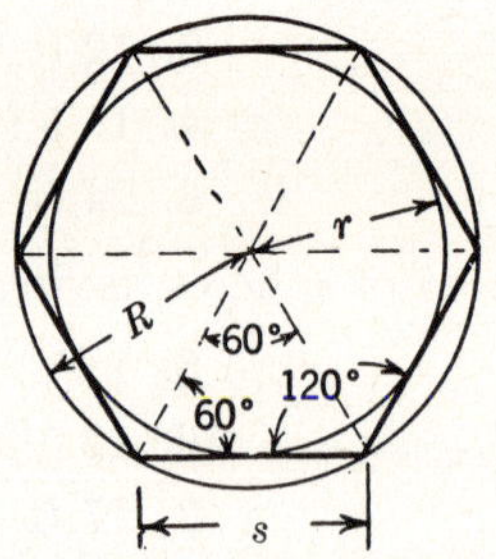

Example.—The side of a regular hexagon is 5 in. Find the area, the radius of the inscribed circle, and the radius of the circumscribed circle.

$A = 2.598s^2$	Formula
$A = 2.598 \times 5 \times 5$	Substituting
$A = 64.95$	
Ans. 64.95 sq. in.	
$r = .866s$	Formula
$r = .866 \times 5$	Substituting
$r = 4.33$	
Ans. 4.33 in.	
$R = s = 5$	
Ans. 5 in.	

EXERCISES AND PROBLEMS

In each of the following exercises, use the figure and formula given for the regular hexagon. State the formula used in each case, and solve.

1. The side of a regular hexagon is 9″. Find A, r, and R.

2. A regular hexagon is to be drawn about an inscribed circle of 11 in. radius. Find A, s, and R.

3. In a regular circumscribed hexagon, $r = 14''$. Find A, s, and R.

4. In a regular inscribed hexagon, $R = 15''$. Find A, s, and r.

5. The perimeter of a hexagon is 42″. Find s, A, r, and R.

6. The diameter of the inscribed circle of a regular hexagon is 20″. Find r, R, s, and A.

7. The diameter of the circumscribed circle of a regular hexagon is $\frac{1}{2}''$. Find r, R, s, and A.

8. Find the total number of degrees in the angles of a regular hexagon.

THE REGULAR OCTAGON

A regular octagon has eight equal sides and eight equal angles. The letters used in the formulas for octagons are A for area, R for the radius of the circumscribed circle, r for the radius of the inscribed circle, and s for a side of the octagon.

Formulas.

$$R = 1.307s, \qquad R = 1.082r$$
$$r = 1.207s, \qquad r = .924R$$
$$s = .765R, \qquad s = .828r$$
$$A = 4.828s^2, \qquad A = 2.827R^2, \qquad A = 3.314r^2$$

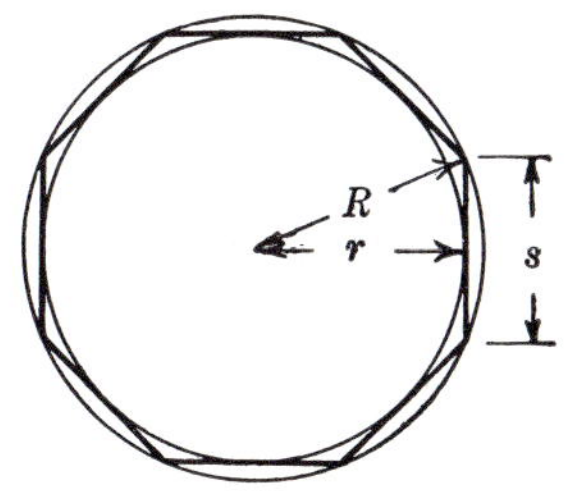

Example.—An octagon is inscribed in a circle 12 in. in diameter. Find

1. The radius of the circumscribed circle.
2. The radius of the inscribed circle.
3. A side of the octagon.
4. The area of the octagon.

(1) $R = \frac{D}{2}$	(2) $r = .924R$
$R = \frac{12}{2}$	$r = .924 \times 6$
$R = 6$	$r = 5.544$
Ans. 6 in.	*Ans.* 5.54 in.
(3) $s = .765R$	(4) $A = 2.827R^2$
$s = .765 \times 6$	$A = 2.827 \times 6 \times 6$
$s = 4.59$	$A = 101.772$
Ans. 4.59 in.	*Ans.* 101.77 sq. in.

PROBLEMS

In each of the following problems, use the formulas for the regular octagon:

1. An octagon is 5″ on a side. Find A, R, and r.

2. The radius of a circle inscribed in an octagon is 5″. Find A, R, and s.

3. The perimeter of an octagon is 32″. Find s, A, R, and r.

4. The diameter of a circle circumscribed about an octagon is 42″. Find R, r, A, and s.

5. On a regular octagon, R is 15″. Find A, s, and r.

6. On a regular octagon, r is 15″. Find A, s, and R.

7. The side of an octagonal piece of tool steel is $\frac{1}{4}$″. Find A, r, and R.

8. The distance across the flats of an octagonal steel tool is 1″. Find A, r, R, and s.

9. The diameter across the corners of an octagonal steel tool is 1″. Find A, R, r, and s.

10. Compute the total number of degrees in each angle of a regular octagon.

REGULAR POLYGONS

The area (A) of a regular polygon, the radius (r) of its inscribed circle and the radius (R) of its circumscribed circle, which has a lateral edge or side (s) may be found by the use of the formulas in the table below.

Name	Number of sides (s)	Area (A) of polygon	Radius of inscribed circle (r)	Radius of circumscribed circle (R)
Triangle (equilateral).....	3	$0.433s^2$	$0.289s$	$0.577s$
Tetragon (square)........	4	$1.000s^2$	$0.500s$	$0.707s$
Pentagon...............	5	$1.720s^2$	$0.688s$	$0.851s$
Hexagon................	6	$2.598s^2$	$0.866s$	$1.000s$
Heptagon...............	7	$3.634s^2$	$1.038s$	$1.152s$
Octagon................	8	$4.828s^2$	$1.207s$	$1.307s$
Nonagon................	9	$6.182s^2$	$1.374s$	$1.462s$
Decagon................	10	$7.694s^2$	$1.539s$	$1.618s$
Undecagon..............	11	$9.366s^2$	$1.703s$	$1.775s$
Dodecagon..............	12	$11.196s^2$	$1.866s$	$1.932s$

EXERCISES

In the following exercises on regular polygons, state the formula used in each case and solve.

1. Equilateral triangle, side 8 in. Get A, r, and R.

2. Tetragon, side $6\frac{3}{4}$ in. Get A, r, and R.

3. Pentagon, side 13 ft. Get A, r, and R.

4. Hexagon, side 9.67 in. Get A, r, and R.

5. Heptagon, side 3.5 ft. Get A, r, and R.

6. Octagon, side 2.22 in. Get A, r, and R.

7. Nonagon, side 0.625 in. Get A, r, and R.

8. Decagon, side 11 ft. Get A, r, and R.

9. Undecagon, side 14 ft. Get A, r, and R.

10. Dodecagon, side 0.33 in. Get A, r, and R.

11. List six practical uses for the hexagon.

12. List six practical uses for the octagon.

13. List as many practical uses for other polygons as you can.

14. The area of a hexagon is what per cent of the area of its circumscribed circle? Of its inscribed circle?

15. Write a rule for getting the perimeter of a polygon.

FRUSTUM OF A PYRAMID

The **frustum of a pyramid** is the remainder of a pyramid when the top is cut off by a plane parallel to the base.

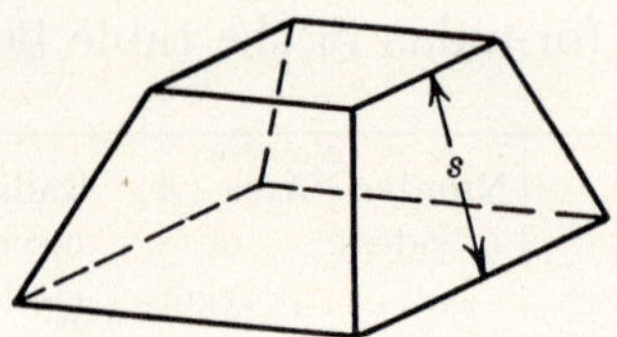

This figure is found in many practical problems. The lateral surface area of the frustum is equal to the sum of the areas of the several trapezoidal faces.

The letters used in the formulas for the frustum of a pyramid are P for the perimeter of the lower base, p for the perimeter of the upper base, s for the slant height, and L for the lateral surface area.

Formula. $L = \frac{s(P + p)}{2}$.

Example.—Find the surface area of the frustum of a square pyramid when the upper base is an 11-in. square, the lower base is a 15-in. square, and the slant height is 12 in.

$L = \frac{s(P + p)}{2}$	Formula
$L = 12 \times \frac{60 + 44}{2}$	Substituting
$L = 12 \times 52$	
$L = 624$	
Ans. 624 sq. in.	

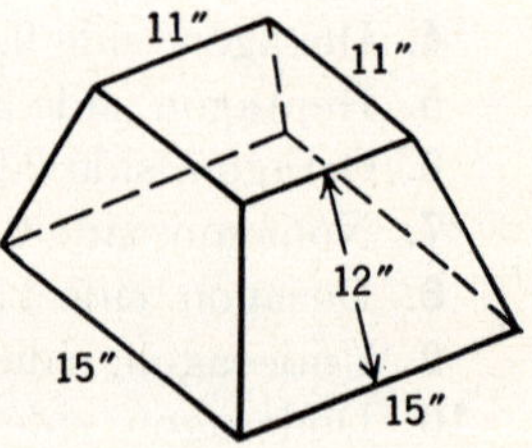

PROBLEMS

1. Find the lateral surface area of the frustum of a pyramid when the slant height is 8 in., the upper base is a 12-in. square, and the lower base is an 18-in. square.

2. What is the lateral surface area of the frustum of the square pyramid shown when the upper base has 6-in. sides, the lower base has 9-in. sides, and the lateral height is 16 in.?

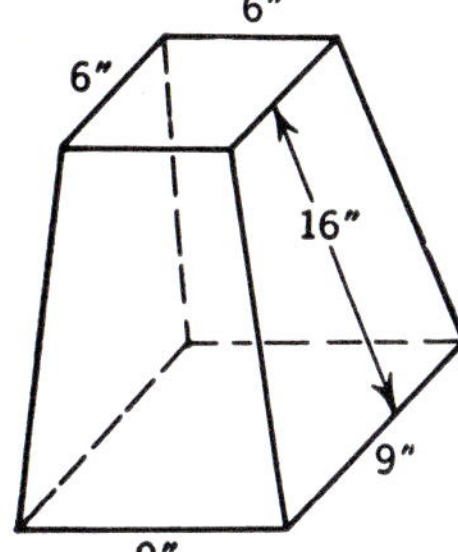

PROB. 2.

3. Find the number of square inches of material needed to make a frustum similar to that in Prob. 2 if the upper base has 4-in. sides, the lower base has 7-in. sides, and the slant height is 9 in.

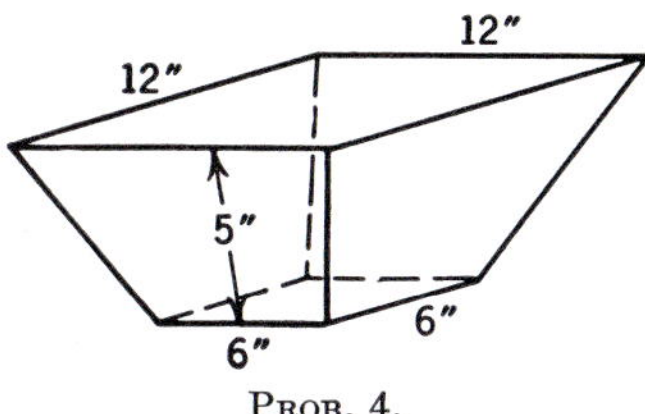

PROB. 4.

4. Find the number of square inches of sheet metal needed to make the transition piece shown. The ends are squares.

5. Find the number of square inches of sheet metal needed to make the transition piece shown. The ends are rectangular.

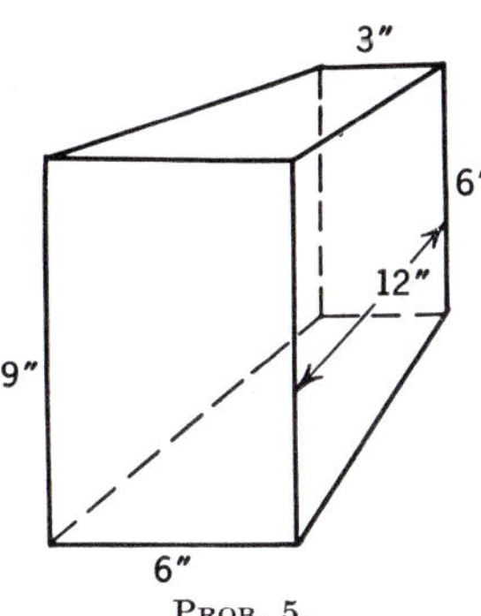

PROB. 5.

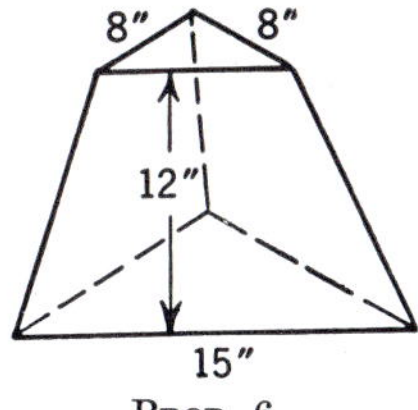

PROB. 6.

6. The base of a pyramid is an equilateral triangle 15 in. on a side. It is cut off parallel to the base to form another equilateral triangle 8 in. on a side. The slant height of the frustum formed is 12 in. Find the lateral surface area of the frustum (see the accompanying figure).

7. Find the lateral surface area of a frustum of a pyramid with slant height of 8 in. when the upper base is a hexagon 9 in. on a side and the lower base is a hexagon 12 in. on a side.

8. What is the lateral surface area of the frustum of a pyramid when the slant height is 15 ft., the upper base is an octagon 12 ft. on a side, and the lower base is an octagon 20 ft. on a side?

9. List several practical applications of the frustum of a pyramid.

FRUSTUM OF A CONE

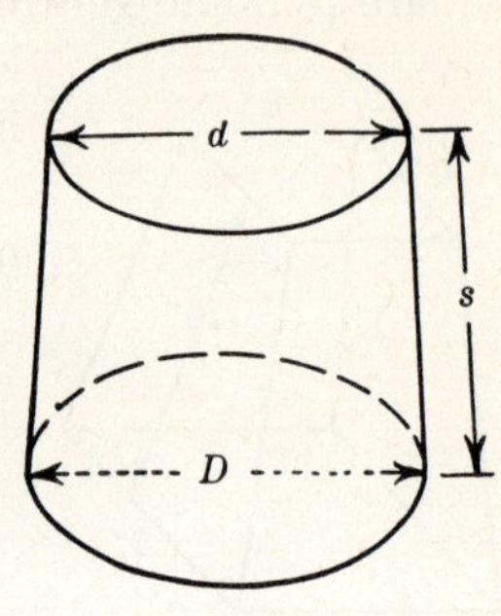

The **frustum of a cone** is the remainder of a cone when the top has been cut off by a plane parallel to the base. The letters used in the formulas for the frustum of a cone are s for slant height, L for the lateral surface area, D for the diameter of the large end, and d for the diameter of the small end.

Formulas. $L = \frac{\pi s(D + d)}{2}$, $L = 1.57s(D + d)$, or

$$L = \pi s(R + r).$$

Example.—Find the lateral surface area of the frustum of a cone with a slant height of 8 in.; the diameter of the small end is 10 in., and the diameter of the large end is 12 in. (Use $\pi = 3.14$.)

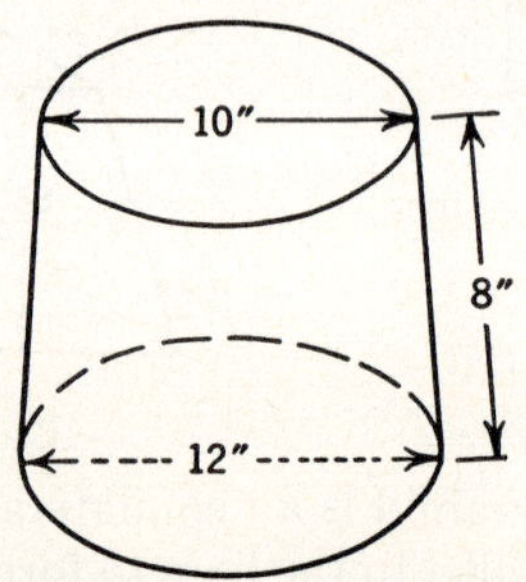

$L = \frac{\pi s(D + d)}{2}$	Formula
$L = \frac{3.14 \times 8(12 + 10)}{2}$	Substituting
$L = \frac{3.14 \times 8 \times 22}{2}$	
$L = 276.32$	
Ans. 276.32 sq. in.	

PROBLEMS

Use 3.14 for π.

1. What is the lateral surface area of the frustum of a cone when its slant height is 15 ft. and the end diameters are 14 and 9 ft., respectively?

2. A bucket has a 10-in. top diameter. The bottom diameter is 8.5 in. The slant height is 9 in. Find the lateral surface area.

3. A funnel is 6 in. in diameter at the top and 1 in. in diameter at the neck. The slant height is 8 in. Find the lateral surface area.

4. The spout for the funnel in Prob. 3 is 1 in. in diameter at the top and $\frac{3}{8}$ in. in diameter at the bottom. The spout is 4 in. long. Find its lateral surface area.

What is the lateral surface area of each of the following conic frustums?

Number	Slant height	Diameter of small end	Diameter of large end
5	12 in.	15.5 in.	21 in.
6	20 ft.	16 ft.	22 ft.
7	15 ft.	10 ft.	16 ft.
8	1.4 in.	2.2 in.	3.6 in.
9	$\frac{1}{2}$ in.	$\frac{3}{8}$ in.	$\frac{3}{4}$ in.

10. List several applications of the conic frustum.

THE CIRCULAR SECTOR

A **circular sector** is a figure bounded by two equal radii and a part of the circumference of a circle.

Such a section of a circumference is called an **arc.**

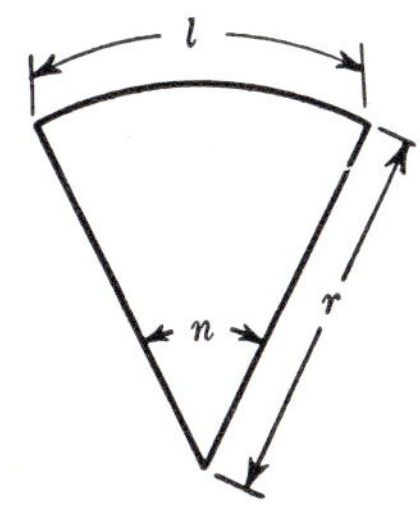

The letters used in the formulas for the circular sector are A for area, l for the length of the arc, r for the radius, n for the number of degrees in the central angle.

Formulas.

$$A = \frac{\pi n r^2}{360} \quad \text{or} \quad A = .0087nr^2 \qquad r = \frac{2A}{l}$$

$$l = \frac{\pi n r}{180} \quad \text{or} \quad l = .01745rn \qquad l = \frac{2A}{r}$$

$$n = \frac{57.3l}{r} \qquad r = \frac{57.3l}{n} \qquad n = \frac{A}{.0087r^2}$$

Example 1.—In a circular sector the radius of the circle is 2″ and the central angle is 45°. Find A and l.

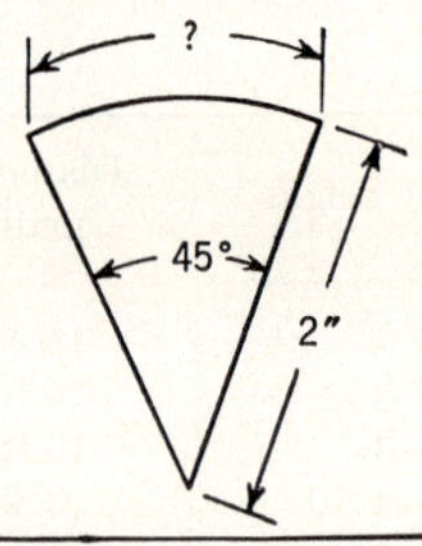

$A = .0087\ nr^2$	Formula
$A = .0087 \times 2 \times 2 \times 45$	Substituting
$A = 1.566$	
Ans. 1.57 sq. in.	
$l = .01745rn$	Formula
$l = .01745 \times 2 \times 45$	Substituting
$l = 1.5705$	
Ans. 1.57″	

Example 2.—In a circular sector the length of the arc is 5″ and the central angle is 28°. Find the radius and the area.

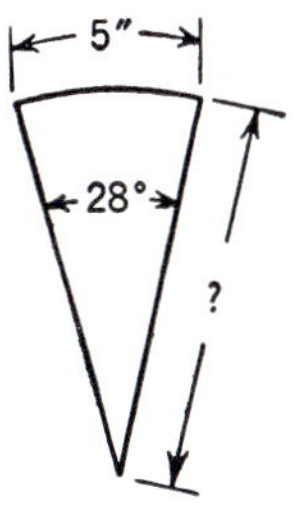

$r = \frac{57.3l}{n}$	Formula
$r = \frac{57.3 \times 5}{28}$	Substituting
$r = 10.232$	
Ans. 10.23″	
$A = .0087nr^2$	Formula
$A = .0087 \times 28 \times 10.23 \times 10.23$	Substituting
$A = 25.493+$	
Ans. 25.49 sq. in.	

PROBLEMS

In each of these problems the figure involved is a circular sector. Choose the correct formula, and solve. Use 3.14 for π.

Number	Given	Find
1	$r = 3''$ and $n = 38°$	A and l
2	$r = 5''$ and $n = 16°$	A and l
3	$r = 6''$ and $l = 8''$	A and n
4	$l = 7''$ and $n = 33°$	r and A
5	$l = 24''$ and $n = 65°$	r and A
6	$l = 1''$ and $r = 1''$	A and n
7	$A = 16$ sq. in. and $r = 4$ in.	l and n
8	$A = 120$ sq. in. and $l = 10$ in.	r and n
9	$n = 56°$ and $d = 10''$	l and A
10	$n = 21°$ and $l = 6\frac{1}{2}'$	r and A
11	$r = \frac{1}{2}$ in. and $A = 0.1$ sq. in.	l and n
12	$r = 68'$ and $n = 44°$	l and A

THE ELLIPSE

On the ellipse the distance from X to X' is called the major axis, long axis, major diameter, or long diameter.

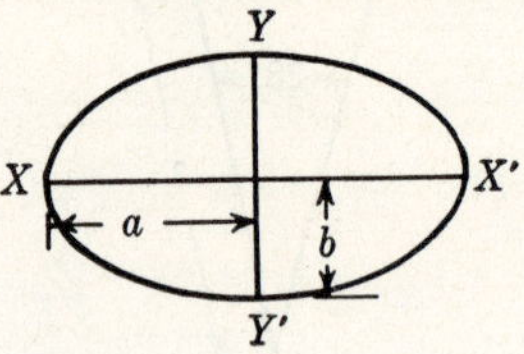

The distance from Y to Y' is called the minor axis, short axis, minor diameter, or short diameter. The letters used in the formulas for the ellipse are a for one-half the long axis, b for one-half the short axis, A for the area, P for the perimeter.

Formulas.

$$A = \pi ab$$
$$P = \pi \sqrt{2(a^2 + b^2)} \text{ (approximately)}$$

Example.—On the ellipse shown, find the area and the perimeter. (Use $\pi = 3.14$.)

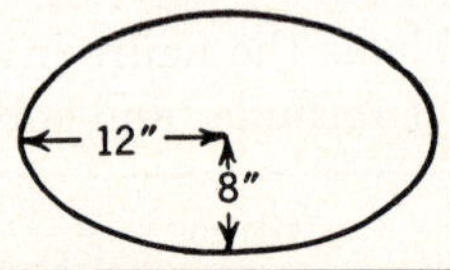

$A = \pi ab$	Formula
$A = 3.14 \times 12 \times 8$	Substituting
$A = 301.44$	
Ans. 301.44 sq. in.	
$P = \pi \sqrt{2(a^2 + b^2)}$	Formula
$P = 3.14 \times \sqrt{2(12^2 + 8^2)}$	Substituting
$P = 3.14 \times \sqrt{2(144 + 64)}$	
$P = 3.14 \times \sqrt{416}$	
$P = 3.14 \times 20.4$	
$P = 64.056$	
Ans. 64.06″ rounded off to two decimals	

PROBLEMS

Use 3.14 for π.

1. Find the area and the perimeter of an ellipse that has a long axis of 14 in. and a short axis of 10 in.

2. What is the area of an ellipse whose major axis is 15 in. and whose minor axis is 11 in.?

3. Find the perimeter of an ellipse that has a long axis of 72 ft. and a short axis of 50 ft.

4. What is the area of an ellipse when a is 37 ft. and b is 18 ft.?

5. Compute the area of an ellipse whose long diameter is 1 in. and whose short diameter is 0.5 in.

6. Find the perimeter of an ellipse that has a major axis of 3 ft. and a minor axis of 1 ft.

7. Find the area and the perimeter of an ellipse when a is 0.24 in. and b is 0.18 in.

8. Compute the long diameter of an ellipse with an area of 628 sq. in. when the short diameter is 10 in. (Solve the area formula for a.)

9. Find the long radius a of an ellipse that has an area of 100 sq. in. and a short radius b of 4 in.

10. What is the short diameter of an ellipse if the area is 1000 sq. ft. and the long diameter is 55 ft.?

11. Determine the major axis of an ellipse that contains 1 acre and has a short diameter of 90 ft.

12. List several practical applications of the ellipse.

CHAPTER 10

VOLUME

VOLUME OF A RECTANGULAR SOLID

In measuring volume, or capacity, three dimensions must be considered, namely, length, width, and height. Volume is measured by the number of cubic units contained. A cube is a solid that has six equal square sides. The edges are perpendicular to the base of the cube. A cubic foot is equal to a cube one foot long, one foot wide, and one foot high.

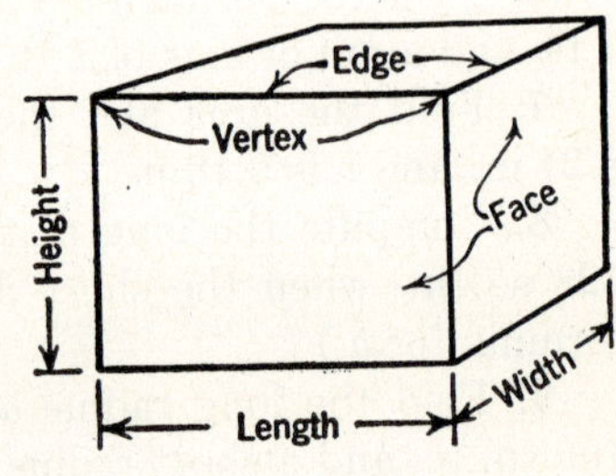

The most familiar solid is the **rectangular parallelepiped,** usually referred to as a **rectangular solid.** It has six

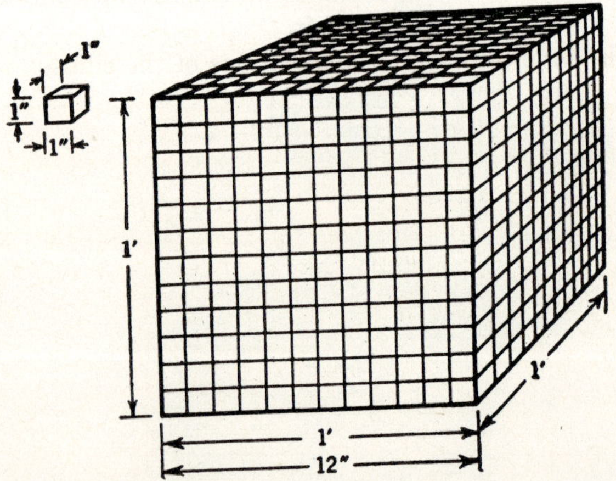

rectangular faces which meet to form right angles. To find the volume of the rectangular solid shown, we must see how many cubic inches are contained in the cubic foot. The base of the cubic foot is 12 in. long and 12 in. wide. Therefore, we could cover the base with a layer of 144 one-inch cubes, each cube being 1 in. long and 1 in. wide. Each of these little cubes is 1 in. high. Hence, it would take 12 layers with 144 cubes to the layer to fill up the cubic foot. There would be 1728 cu. in. in the cubic foot. In other words, to find the volume, first we find the area of the base (length times width) and multiply it by the height. This may be expressed as follows:

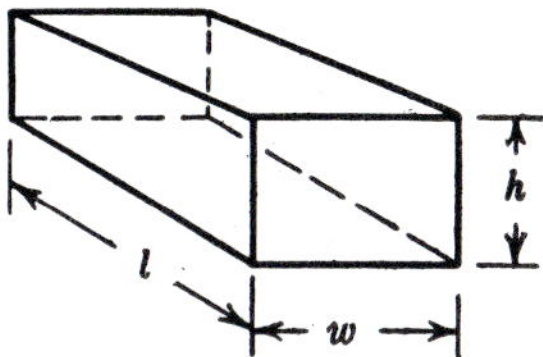

Rule.—The volume of a rectangular solid equals the product of its length, width, and height.

The letters used in the formula for the volume of a rectangular solid are V for volume, l for length, w for width, and h for height.

Formula. $V = lwh.$

Example 1.—Find the volume of a rectangular solid 20 ft. wide, 28 ft. long, and 16 ft. high.

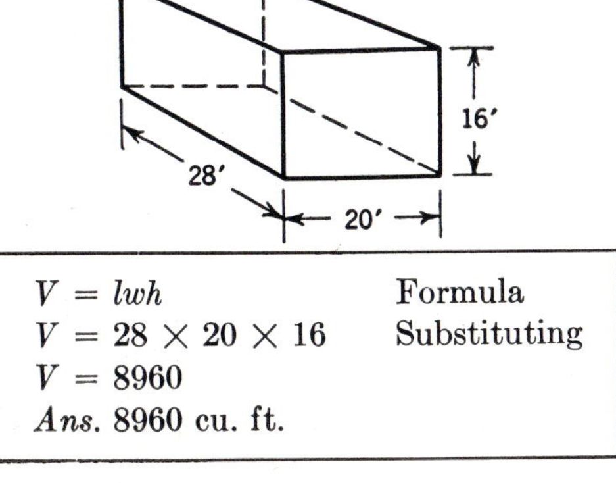

$V = lwh$	Formula
$V = 28 \times 20 \times 16$	Substituting
$V = 8960$	
Ans. 8960 cu. ft.	

The formula $V = lwh$, solved for h, is as follows:

Formula. $h = \frac{V}{lw}$.

Example 2.—Find the height of a rectangular tank 8 ft. wide and 12 ft. long that contains 1440 cu. ft.

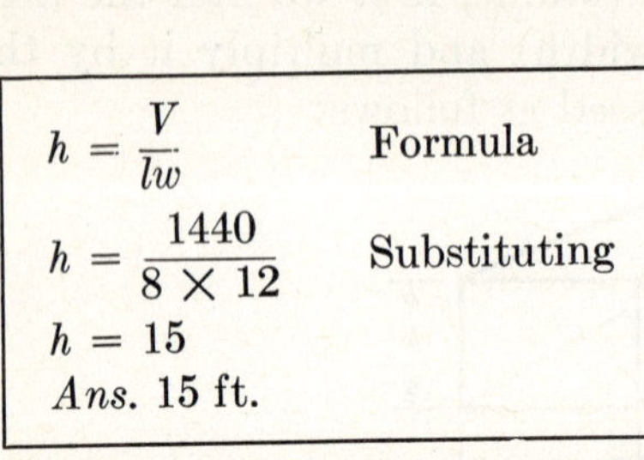

$h = \frac{V}{lw}$ Formula

$h = \frac{1440}{8 \times 12}$ Substituting

$h = 15$

Ans. 15 ft.

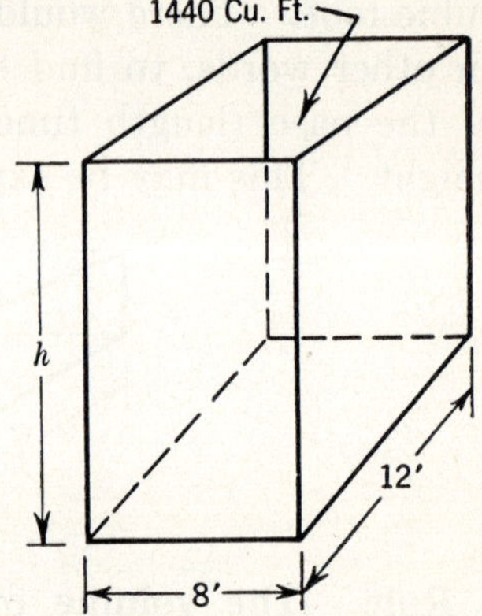

VOLUME OF A CUBE

A cube is a rectangular solid in which all the dimensions are equal. Since the length, the width, and the height are all equal, one letter s is used instead of l, w, and h in the formula for the volume.

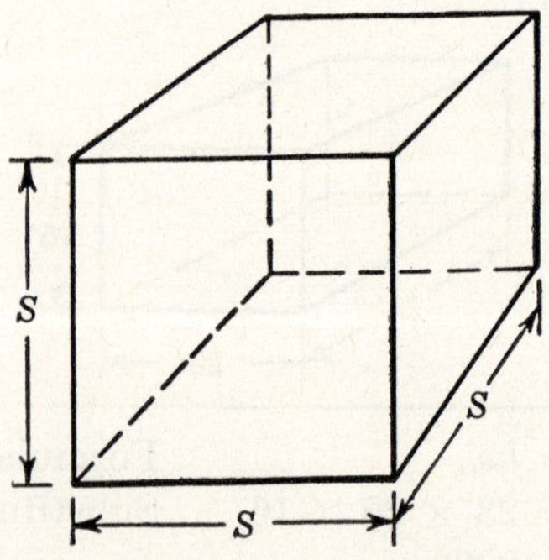

Formula. $V = s \times s \times s$, or $V = s^3$.

Example.—Find the volume of a cube 3 in. on a side.

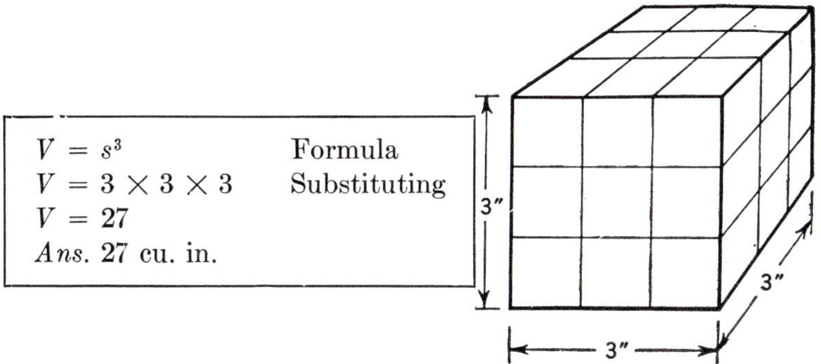

PROBLEMS

Draw the figure involved in each problem, choose the correct formula, and solve.

1. A truck bed is 7 ft. wide, 4 ft. deep, and 12 ft. long. Find the volume in cubic feet. In cubic yards.

2. Find the volume of a basement 7 ft. deep, 18 ft. wide, and 38 ft. long. How many cubic yards does it contain?

3. A box is 15″ long, 10″ wide, and 7″ deep. Find its volume in cubic inches.

4. A trench is 3 ft. wide, 5 ft. deep, and 1 mile long. How many cubic feet are there in the trench? How many cubic yards?

5. Determine the number of cubic feet in a paved platform 77 ft. wide, 96 ft. long, and 6 in. thick.

6. A wagon bed is 3 ft. wide, 10 ft. long, and 26 in. deep. How many cubic inches are there? How many cubic feet?

7. A room is 28′6″ wide, 45′ long, and 12′ high. How many cubic feet are there in the room? How many cubic yards?

8. At 95¢ per cubic yard, find the total cost of a cellar 15′ deep, 12′ wide, and 30′ long.

9. A box is $27\frac{1}{2}$″ wide, 36″ long, and $12\frac{1}{2}$″ high. Find its volume in cubic inches. In cubic feet.

10. How many tons of coal are there in a car $7\frac{1}{2}$ ft. wide and 30 ft. long when it is filled to a depth of 6 ft.? Coal weighs 50 lb. per cu. ft.

11. To what depth will the car in Prob. 10 be filled with cast-iron scrap that averages 400 lb. per cu. ft. when it holds the same weight as in Prob. 10? Carry your work to three decimal places.

12. At 60¢ per cubic foot, determine the cost of paving a walk 4′ wide, 75′ long, and 4″ thick.

13. How long is a trench 2 ft. wide and 4 ft. deep if it contains 100 cu. yd.?

14. To what depth must a rectangular pit 6 ft. 6 in. × 8 ft. be dug to contain 364 cu. ft.?

15. Find the weight of an iron casting 3 in. wide, 5 in. thick, and 19 in. long if a cubic inch of cast iron weighs 0.26 lb.

16. A box $6\frac{1}{2}$ in. long, $5\frac{1}{2}$ in. wide, and $3\frac{1}{2}$ in. deep is filled with brass shavings weighing 0.3 lb. per cu. in. What is the weight of the shavings in the box?

17. Steel weighs 485 lb. per cu. ft. Find the weight of a piece of steel 3 in. wide, 9 in. thick, and $14\frac{1}{2}$ in. long.

18. Find the number of cubic feet in a cellar 24 ft. wide, 32 ft. long, and 7 ft. deep.

19. If the dirt from the cellar in Prob. 18 is spread evenly over all of a lot 50 by 150 ft. except the part occupied by the cellar, how deep will the fill be?

VOLUME OF A RIGHT PRISM

A **prism** is a solid whose bases, or ends, are polygons that are equal and parallel to each other and whose sides are parallelograms. In a right prism the edges are perpendicular to the bases. Since the sides of a prism are all parallelograms, the prism is named by its base.

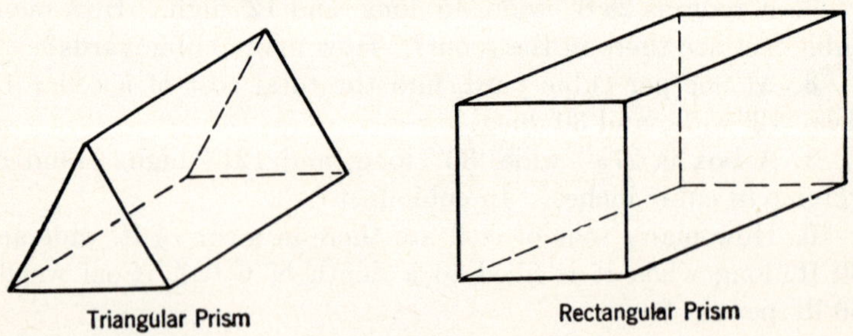

Triangular Prism   Rectangular Prism

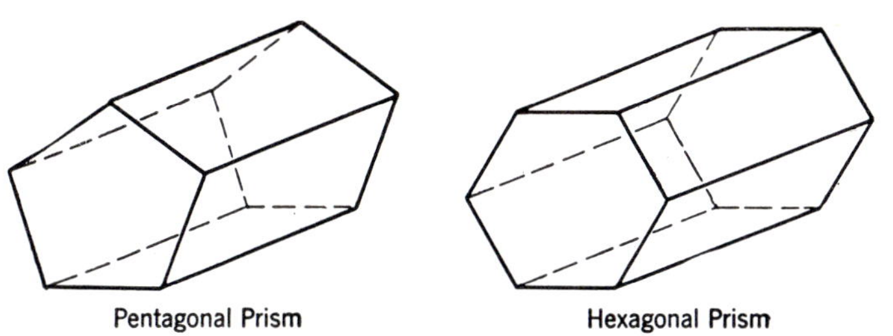

In finding the volume of a rectangular solid, you will remember that you find the area of the base and multiply it by the height; that is, $(l \times w) \times h$. The rectangular solid is a special right prism. The formula could be restated thus: Volume V equals the area of base B times height h, or $V = Bh$. This formula can be used for any right prism. The area of the base B is found by means of one of the area formulas. The height h of any right prism is the perpendicular distance between the bases.

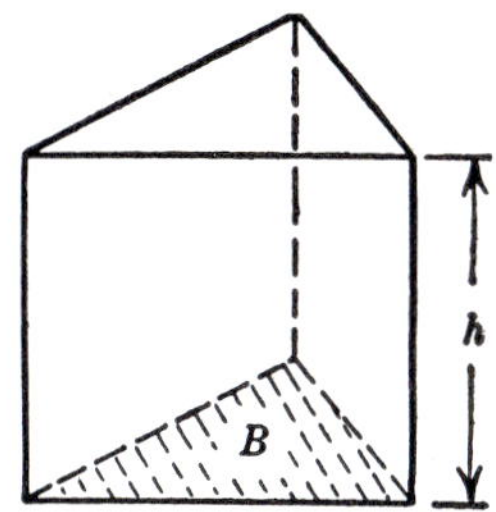

Rule.—The volume of a right prism is the product of the area of the base and the height of the prism.

Formula. $V = Bh$.

Example 1.—The area of the base of a pentagonal prism is 88 sq. in. The height is 12 in. Find the volume of the prism.

$V = Bh$	Formula
$V = 88 \times 12$	Substituting
$V = 1056$	
Ans. 1056 cu. in.	

Example 2.—Find the volume of the triangular prism shown.

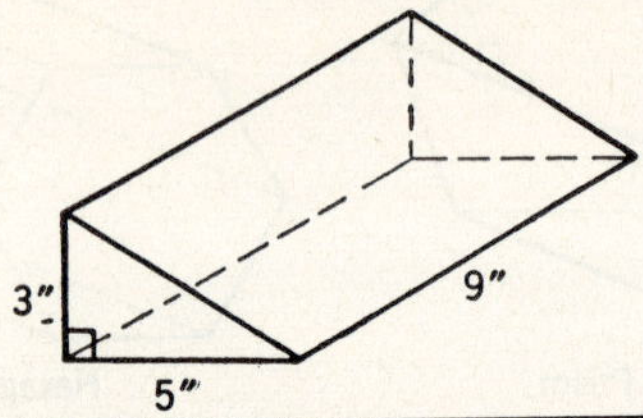

$V = Bh$	Formula
$V = \frac{3 \times 5}{2} \times 9$	Substituting (the area of the base B is the area of the triangle, $\frac{3 \times 5}{2}$)
$V = 67.5$	
Ans. 67.5 cu. in.	

Example 3.—Find the volume of the regular hexagonal prism shown which is 3 in. on a side and 11 in. long.

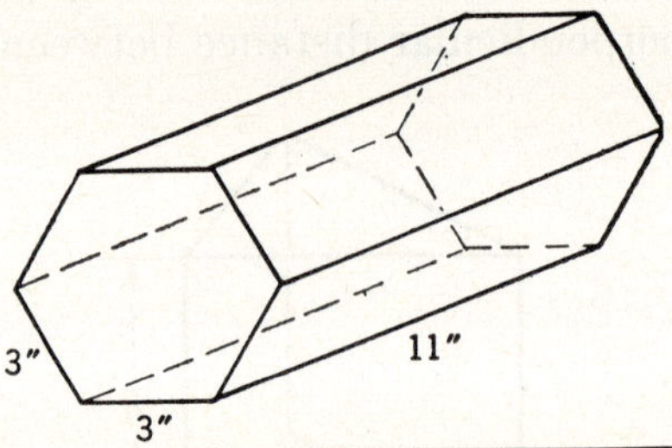

$V = Bh$	Formula
$V = 2.598s^2h$	
$V = 2.598 \times 3 \times 3 \times 11$	Substituting (the area of the base is the area of a hexagon, $A = 2.598s^2$).
$V = 257.202$	
Ans. 257.2 cu. in.	

PROBLEMS

For each of the right prisms shown on page 265, find (*a*) the volume, and (*b*) the total surface area.

The total surface area equals the area of the ends plus the area of all the sides. Use the formula $s = ph$, where p stands for the perimeter of the base, h for the height of the prism, s for the total surface area of the sides.

In the figures numbered 7 to 12, each of the bases is a regular polygon.

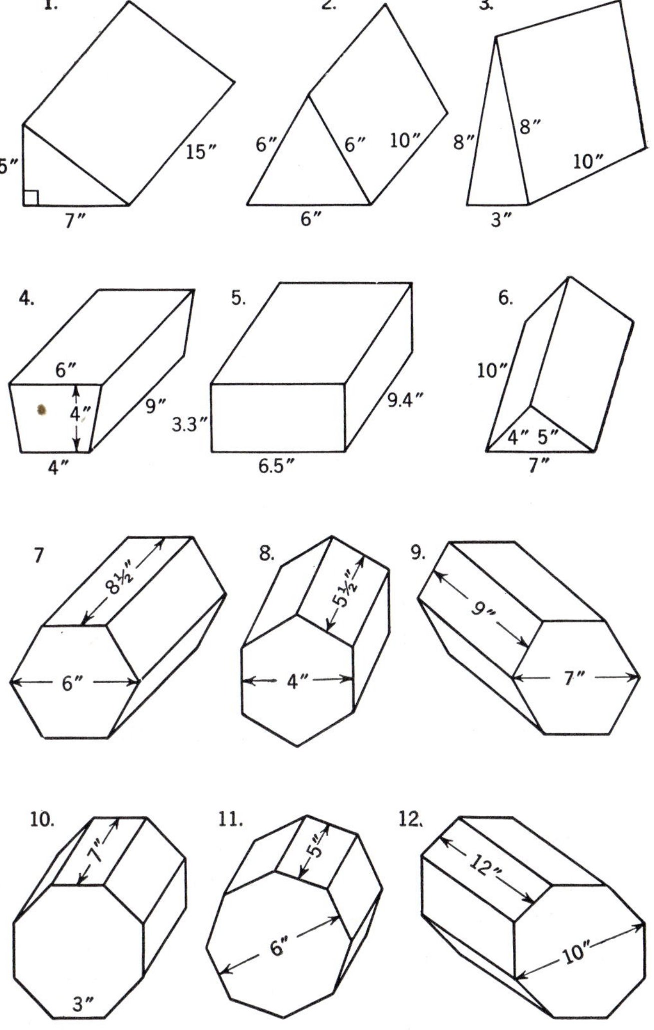

13. Find the weight of the solid shown if it is a solid piece of cast iron that weighs 0.26 lb. per cu. in.

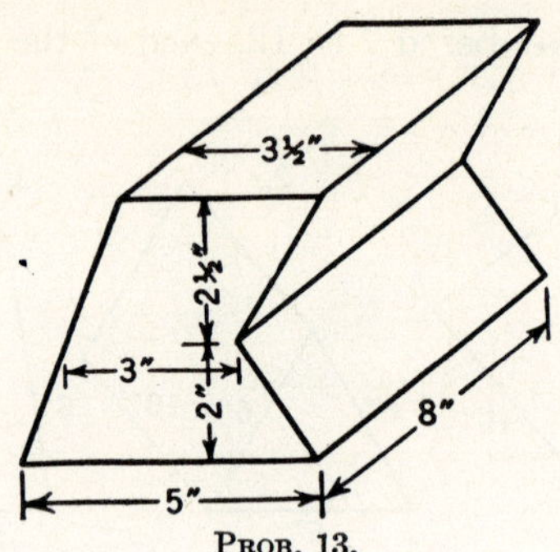

PROB. 13.

14. Find the weight of the V block shown if it is a solid piece of steel weighing 0.2833 lb. per cu. in.

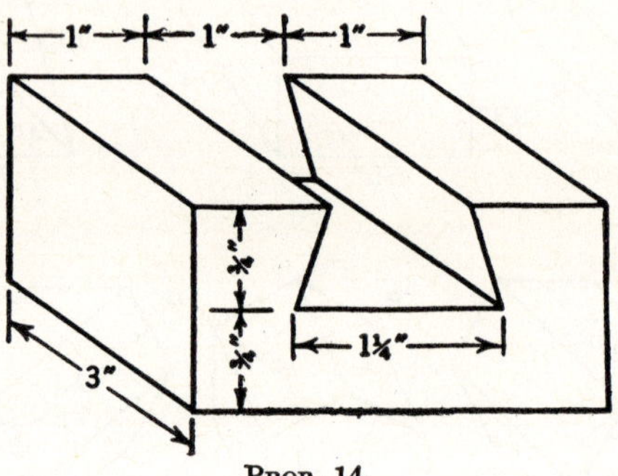

PROB. 14.

CAPACITY OF A CYLINDRICAL TANK

A cylindrical tank is a cylinder, or a right prism with a circular base. Hence, the following may be stated:

Rule.—The capacity, or volume, of a cylindrical tank is found by multiplying the area of the circular base by the height or length of the tank.

The letters used in the formulas are V for the volume, B for the area of the base, d for the diameter of the base, r for the radius of the base, and h for the height of the cylinder.

Since $V = Bh$ (see page 263), B in the case of a cylindrical tank equals the area of a circle, πr^2; then, $B = \pi r^2$, or $B = .7854d^2$.

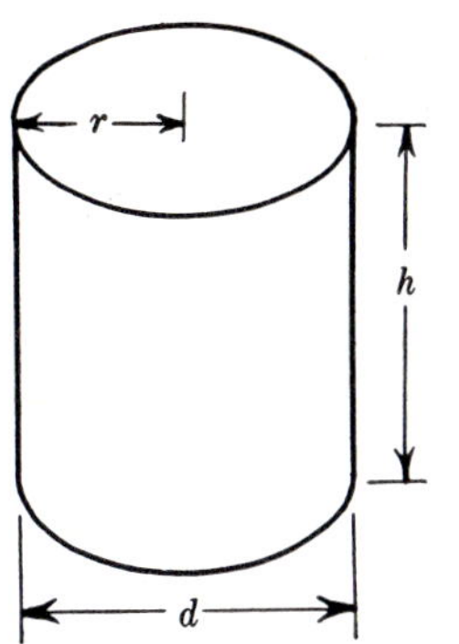

Substituting in the formula $V = Bh$,
Formulas. $V = 3.1416r^2h$ or $V = .7854d^2h$

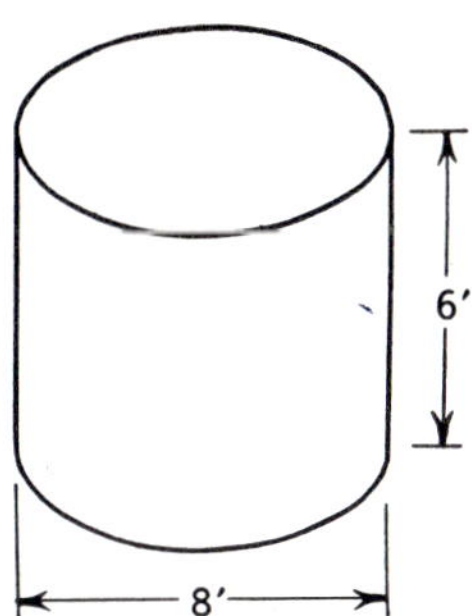

Example 1.—Find the capacity of a cylindrical tank 8 ft. in diameter and 6 ft. high.

$V = .7854d^2h$	Formula
$V = .7854 \times 8 \times 8 \times 6$	Substituting
$V = 301.5936$	
Ans. 301.6 cu. ft. rounded off to one decimal place	

Example 2.—Find the volume in cubic inches and the capacity in gallons of an oil drum 18 in. in diameter and 3 ft. high.

$V = .7854d^2h$	Formula
$V = .7854 \times 18 \times 18 \times 36$	Substituting
$V = 9160.9056$, or 9160.9 cu. in.	
$\frac{9160.9}{231} = 39.65$	Dividing the total number of cubic inches by the number of cubic inches in 1 gal.
Ans. 39.7 gal.	

Example 3.—Find the capacity in gallons of a cylindrical tank car 9 ft. in diameter and 38 ft. long.

$V = .7854d^2h$	Formula
$V = .7854 \times 9 \times 9 \times 38$	Substituting
$V = 2417.46$, or 2417.46 cu. ft.	
$2417.46 \times 7.5 = 18{,}130.95$	Multiplying the total number of cubic feet by the number of gallons in 1 cu. ft.
Ans. 18,131 gal.	

Example 4.—Find the capacity in barrels of an oil tank 200 ft. in diameter and 30 ft. high.

$V = .7854d^2h$	Formula
$V = .7854 \times 200 \times 200 \times 30$	Substituting
$V = 942{,}480$, or 942,480 cu. ft.	
$\frac{942{,}480}{5.6} = 168{,}300$	Dividing by the number of cubic feet in 1 bbl.
Ans. 168,300 bbl.	

PROBLEMS

Round off the answers to one decimal place.

1. Find the cubic-inch capacity of an oil drum 32″ in diameter and 48″ high.

2. Find the number of cubic feet of gas in a standpipe 36′ in diameter and 60′ high.

3. How many cubic feet of water are there in a round cistern 12′ in diameter and 6′ deep? How many gallons will it hold?

4. A railroad water tank is 16′ in diameter and 12′ high. How many gallons of water will it hold?

5. If the water in Prob. 4 weighs 62.4 lb. per cu. ft., how much will a tankful weigh?

6. The area of the base of a cylinder is 100 sq. ft. and the height is 100 in. Find the volume. How many gallons does it contain when full?

7. A fire extinguisher is 6″ in diameter and 22″ high. How much fluid will it hold? Express the answer in gallons.

8. If a tank is 64 ft. high and 20 ft. in diameter, how many cubic feet does it contain?

9. How many gallons of water would the tank of Prob. 8 hold?

10. How many pounds of water would there be in the tank of Prob. 8 if it were full?

11. An artesian well is driven into the ground 1200 ft. The pipe is 4 in. in diameter. How many gallons of water are needed to fill the well?

12. One of the world's deepest oil wells is 15,680 ft. deep and has an average diameter of 9 in. How many cubic feet of material have been removed from the well?

13. An 18″ oil line runs from Bakersfield to the Los Angeles harbor, a distance of 143 miles. Find the number of gallons of oil needed to fill the oil line. (*Suggestion.* 18″ = 1.5′.)

14. A 12-in. oil line runs from Oklahoma to Chicago, a distance of 685 miles. How many gallons of oil will the pipe hold?

15. How many barrels of oil are there in Prob. 14?

16. How many barrels of oil are there in a tank 12 ft. in diameter and 24 ft. long?

17. The Los Angeles Aqueduct is 287 miles long and 8 ft. in diameter. How many gallons does it contain?

18. How many barrels of oil are there in a storage tank 60 ft. in diameter and 25 ft. high?

19. Compute the number of barrels of oil in a tank 32 ft. in diameter and 20 ft. long.

20. On a tank farm, there are 36 tanks, or reservoirs, 500 ft. in diameter and 16 ft. deep. Find the capacity in barrels of each tank. Find the capacity in barrels of the entire tank farm.

FINDING THE HEIGHT OR DIAMETER OF A TANK

It is often necessary to find the height or diameter of a tank when its capacity is known. This is done by solving the formula for the value wanted.

The formula $V = .7854d^2h$, solved for h, is as follows:

Formula. $h = \frac{V}{.7854d^2}.$

Example 1.—Find the height of a cylindrical tank that is 16″ in diameter and holds 100 gal.

$h = \dfrac{V}{.7854d^2}$	Formula
$h = \dfrac{100 \times 231}{.7854 \times 16 \times 16}$	Substituting (since d is given in inches, change gallons to cubic inches by multiplying 100 by 231)
$h = 114.89$, or 114.89″	
$h = 9'6.89''$	Changing inches to feet and inches
Ans. 9′6.89″	

Example 2.—Find the height of a cylindrical tank 3 ft. in diameter and holding 300 gal.

$h = \dfrac{V}{.7854d^2}$	Formula
$h = \dfrac{300}{7.5 \times 0.7854 \times 3 \times 3}$	Substituting (since d is given in feet, $V = 300$ gal. or $300 \div 7.5$ for cubic feet)
$h = 5.65$	
Ans. 5.65 ft.	

The formula, $V = .7854d^2h$, solved for d, is as follows:

Formula. $d = \sqrt{\dfrac{V}{.7854h}}$

Example 3.—Find the diameter of a tank that holds 50 gal. when the height is 40 in.

$d = \sqrt{\dfrac{V}{.7854h}}$	Formula
$d = \sqrt{\dfrac{50 \times 231}{.7854 \times 40}}$	Substituting (V changed to cubic inches is 50×231)
$d = \sqrt{367.64}$	
$d = 19.18$, or 19.18″	
$d = 1'7.18''$	Changing the inches to feet and inches
Ans. 1′7.18″	

Example 4.—If a tank holds 6000 gal. and is 20 ft. high, find its diameter.

$$d = \sqrt{\frac{V}{.7854h}}$$ Formula

$$d = \sqrt{\frac{6000}{7.5 \times .7854 \times 20}}$$ Substituting (in cubic feet, $V = 6000 \div 7.5$)

$$d = \sqrt{50.92}$$

$$d = 7.14$$

Ans. 7.14 ft.

PROBLEMS

Find the unknown for each of the following. Remember to change correctly to cubic feet or cubic inches. Substitute in the formula, and solve.

Number	Capacity	Diameter	Height
1	1,000 gal.	16 ft.	?
2	54 gal.	?	30 in.
3	75 gal.	?	3 ft.
4	20 gal.	20 in.	?
5	6,666 gal.	?	11 ft.
6	21.5 gal.	?	33 in.
7	1 qt.	?	6 in.
8	25 gal.	?	1 ft.
9	12,600 gal.	10 ft.	?
10	1 gal.	?	6 in.
11	2 gal.	?	2 ft.
12	545 gal.	3 ft.	?
13	1,000 gal.	?	6 ft.
14	14 qt.	?	9 in.
15	1 qt.	3 in.	?
16	375 gal.	?	15 ft.
17	31,416 gal.	?	77 in.
18	65,000 gal.	30 ft.	?
19	200 gal.	?	2 ft. 1 in.
20	500 gal.	?	5 ft.

THE PYRAMID

The letters used in the formulas for the volume of the pyramid are V for volume, B for the area of the base, and h for height.

Formula. $V = \frac{Bh}{3}$.

Example.—What is the volume of a pyramid that has a square base 4 in. on a side? The height of the pyramid is 12 in.

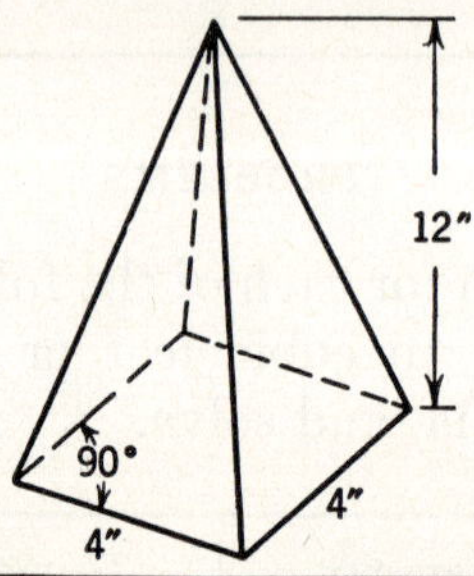

$V = \frac{Bh}{3}$	Formula
$V = \frac{4 \times 4 \times 12}{3}$	Substituting ($B = s^2$)
$V = 64$	
Ans. 64 cu. in.	

PROBLEMS

Find the volume of each of the pyramids shown. The base of each pyramid is a regular polygon.

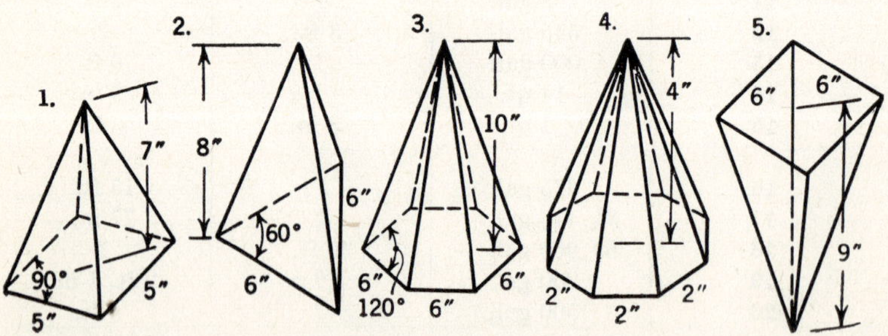

THE CONE

The cone may be thought of as a pyramid with a circular base. Hence, $B = \pi r^2$; and, substituting in the general formula for the volume of a pyramid, we obtain the formulas given here. The letters used in the formulas are V for volume, d for diameter, r for radius, and h for height.

Formulas.

$$V = \frac{\pi r^2 h}{3}$$

$$V = .2618d^2h$$

Example.—Find the volume of a cone when the base diameter is 12 in. and the height is 16 in.

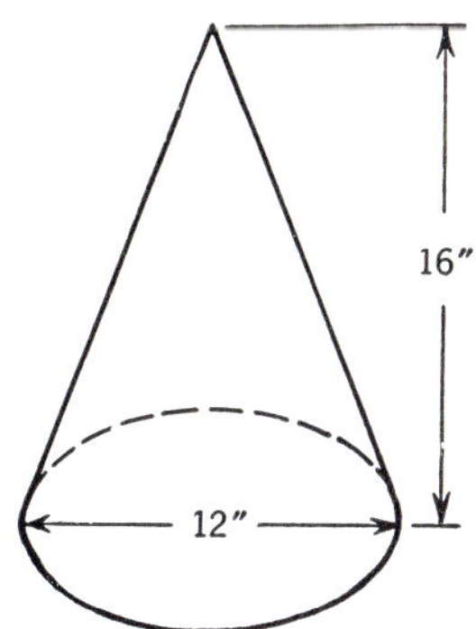

$V = .2618d^2h$	Formula
$V = .2618 \times 12 \times 12 \times 16$	Substituting
$V = 603.1872$	
Ans. 603.2 cu. in.	

PROBLEMS

Find the volume of each of the cones on page 274. Round off the answers to one decimal place.

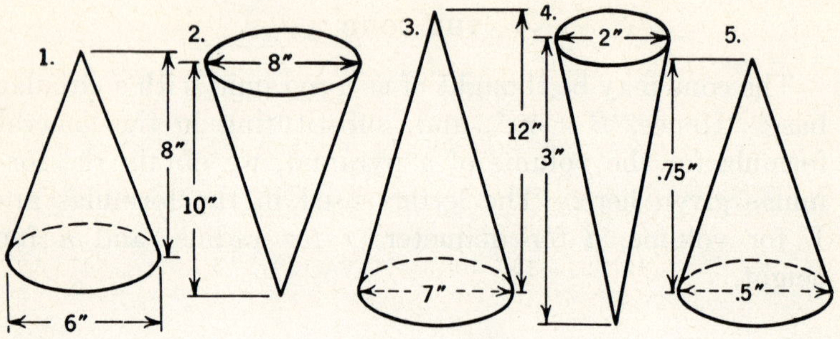

THE SPHERE

The letters used in the formulas for the volume of the sphere are V for volume, d for diameter, and r for radius.

Formulas.

$$V = \frac{\pi d^3}{6} \quad \text{or} \quad V = .5236d^3$$

$$V = \frac{4\pi r^3}{3} \quad \text{or} \quad V = 4.1888r^3$$

Example.—Find the volume of a sphere 16 in. in diameter.

$V = .5236d^3$	Formula
$V = .5236 \times 16 \times 16 \times 16$	Substituting
$V = 2144.6656$	
Ans. 2144.7 cu. in.	

16"

PROBLEMS

For each of the spheres below, find (*a*) the volume and (*b*) the weight if they are made of cast iron weighing 0.26 lb. per cu. in. Round off to one decimal place.

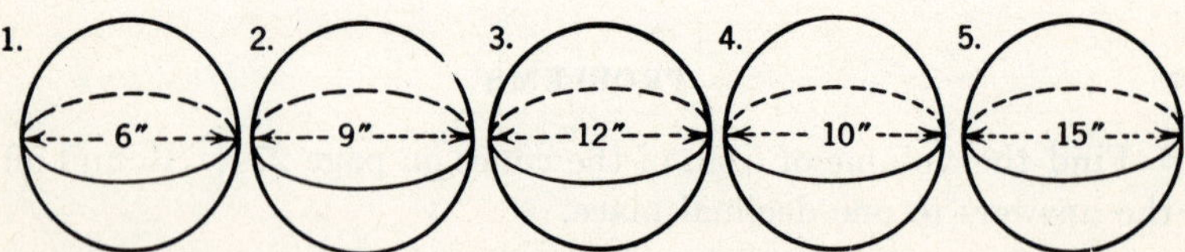

FRUSTUM OF A PYRAMID OR A CONE

The letters used in the formula for the volume of the frustum of a pyramid or a cone are V for the volume, h for the height, B for the area of the large base, and b for the area of the small base. For the frustum of a pyramid, S is the side of the large base and s is the side of the small base. For the frustum of a cone, R is the radius of the large base and r is the radius of the small base.

Formula. $V = \frac{h}{3}(B + b + \sqrt{Bb})$.

Example 1.—Find the volume of the frustum of the pyramid shown.

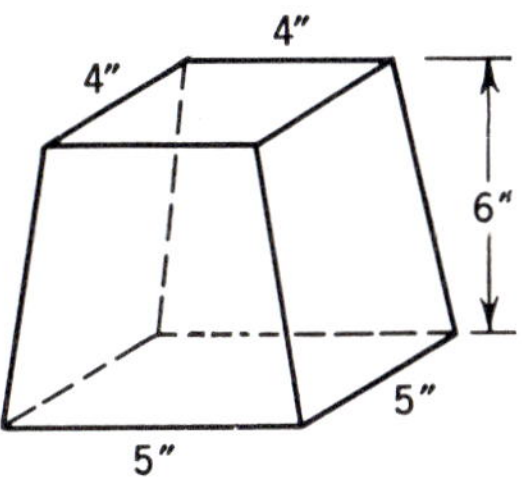

$V = \frac{h}{3}(B + b + \sqrt{Bb})$	Formula
$V = \frac{6}{3}(25 + 16 + \sqrt{25 \times 16})$	Substituting ($B = S^2$;
$V = 2(41 + \sqrt{400})$	$b = s^2$)
$V = 2(41 + 20)$	
$V = 2 \times 61$	
$V = 122$	
Ans. 122 cu. in.	

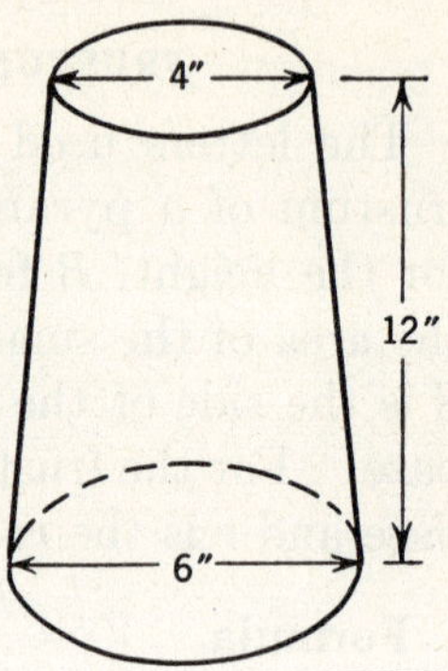

Example 2.—Find the volume of the frustum of a cone when the small base is 4 in. in diameter and the large base is 6 in. in diameter. The height of the cone is 12 in. ($\pi = 3.14$.)

$$V = \frac{h}{3}(B + b + \sqrt{Bb}) \quad \text{Formula}$$

$$V = \tfrac{12}{3}[3.14 \times 3 \times 3 + 3.14 \times 2 \times 2 + \sqrt{(3.14 \times 3 \times 3) \times (3.14 \times 2 \times 2)}]$$

Substituting ($B = \pi R^2$; $b = \pi r^2$)

$$V = \tfrac{12}{3}(28.26 + 12.56 + \sqrt{28.26 \times 12.56})$$

$$V = 4(40.82 + \sqrt{354.9456})$$

$$V = 4(40.82 + 18.84)$$

$$V = 4 \times 59.66$$

$$V = 238.64$$

Ans. 238.6 cu. in.

PROBLEMS

Find the volume of each of the frustums of pyramids shown. The bases of each are regular polygons.

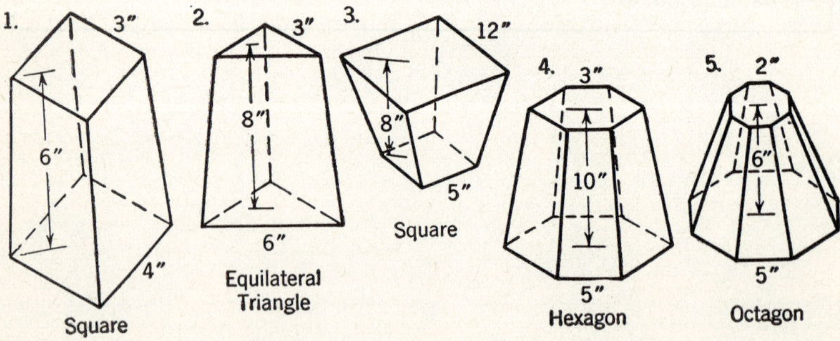

Find the volume of each of the conic frustums shown below.

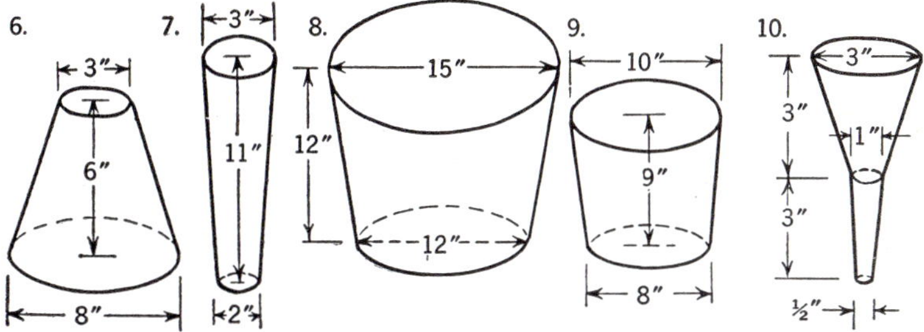

REVIEW PROBLEMS

Find the weight of each of the solids shown below if it is made of cast iron weighing 0.26 lb. per cu. in.

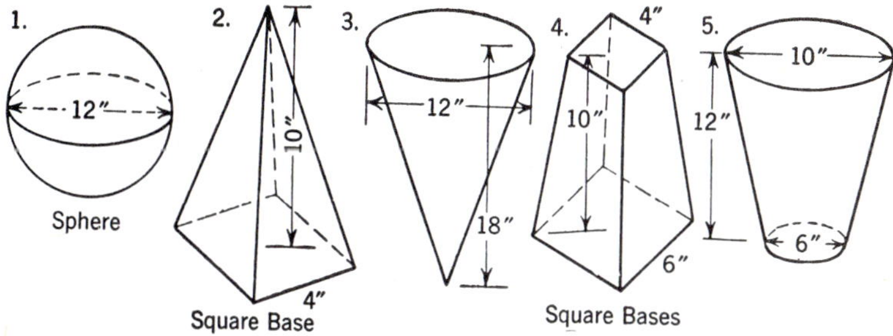

FINDING THE WEIGHT OF IRON SHEETS

There are some useful short-cut methods for determining the weight of iron and steel sheets and bars. Since wrought iron weighs 480 lb. per cu. ft., a piece 1 in. thick and 1 ft. square weighs one-twelfth of 480 lb., or 40 lb.

Rule.—To find the weight of sheet iron per square foot, multiply the thickness in inches by 40.

Example.—Find the weight of a sheet of iron 24 × 96 in. and 0.05 in. thick.

0.05 × 40 = 2, or 2 lb.	Weight of the sheet per square foot
24 in. = 2 ft., 96 in. = 8 ft.	
2 × 8 = 16, or 16 sq. ft.	Area of the sheet
2 × 16 = 32	
Ans. 32 lb.	

FINDING THE WEIGHT OF MILD-STEEL SHEETS

Mild steel is slightly heavier than wrought iron (sheet iron); hence, the following may be stated:

Rule.—To find the weight of a sheet of mild steel per square foot, multiply the thickness in inches by 40.8.

Example.—Find the weight of a sheet of mild steel 24 by 96 in. if it is 0.05 in. thick.

$0.05 \times 40.8 = 2.04$, or 2.04 lb. per sq. ft.
24×96 in. $= 2 \times 8$ ft.
$2 \times 8 = 16$, or 16 sq. ft. Area of the sheet
$2.04 \times 16 = 32.64$
Ans. 32.64 lb.

FINDING THE WEIGHT OF SQUARE MILD-STEEL BARS

Rule.—To find the weight per linear foot for square mild-steel bars, multiply the side squared by 3.4.

Example.—Find the weight of a 2-in. square bar 12 ft. long.

$2 \times 2 \times 3.4 = 13.6$, or 13.6 lb. per lin. ft.
$13.6 \times 12 = 163.2$
Ans. 163.2 lb.

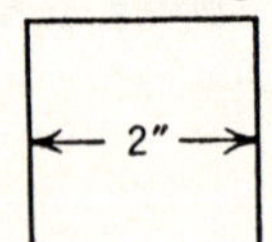

FINDING THE WEIGHT OF ROUND MILD-STEEL BARS

Rule.—To find the weight per linear foot of round mild-steel bars, multiply the diameter squared by 2.68.

Example.—Find the weight of a round mild-steel bar 2 in. in diameter and 12 ft. long.

$2 \times 2 \times 2.68 = 10.72$, or 10.72 lb. per lin. ft.
$10.72 \times 12 = 128.64$
Ans. 128.64 lb.

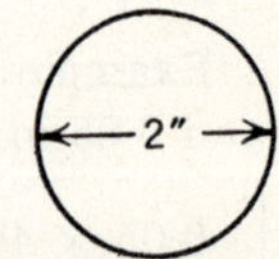

FINDING THE WEIGHT OF HEXAGONAL MILD-STEEL BARS

Rule.—To find the weight per linear foot of hexagonal mild-steel bars, multiply the distance across the flats squared by 2.945.

Example.—Find the weight of a 2-in. hexagonal bar 12 ft. long.

$2 \times 2 \times 2.945 = 11.780$, or 11.78 lb. per lin. ft.
$11.78 \times 12 = 141.36$
Ans. 141.36 lb.

FINDING THE WEIGHT OF OCTAGONAL MILD-STEEL BARS

Rule.—To find the weight per linear foot for octagonal mild-steel bars, multiply the distance across the flats squared by 2.82.

Example.—Find the weight of a 2-in. octagonal mild-steel bar 12 ft. long.

$2 \times 2 \times 2.82 = 11.28$, or 11.28 lb. per lin. ft.
$11.28 \times 12 = 135.36$
Ans. 135.36 lb.

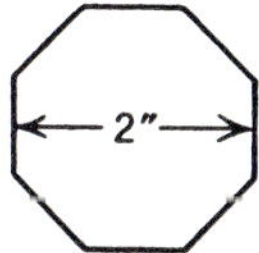

PROBLEMS

1. Find the weight of a steel plate $30'' \times 96''$ if it is $\frac{1}{4}''$ thick.

2. Determine the weight of a steel plate $42'' \times 120''$ and 0.0375″ thick.

3. What is the weight of a piece of sheet steel

$$\tfrac{1}{2} \text{ in.} \times 2 \text{ ft.} \times 5 \text{ ft.?}$$

4. Get the weight of a wrought-iron plate $\frac{3}{16}'' \times 30'' \times 96''$.

5. What is the weight of a $\frac{1}{2}$ in. round mild-steel bar 8 ft. long?

6. Compute the weight of a 3-in. square mild-steel bar 8 ft. long.

7. Find the weight of 50 hexagonal mild-steel bars each 12 ft. long and $\frac{3}{4}$ in. across the flats.

8. What will 1 mile of $\frac{1}{4}$-in. round mild-steel wire weigh?

9. What is the total weight of 24 octagonal mild-steel bars each 20′ long and $\frac{3}{4}''$ across the flats?

10. What is the weight of 6 $\frac{1}{2}$-in. round mild-steel bars, each 30 ft. long?

11. What is the weight of a steel shaft $3\frac{3}{4}$ in. in diameter and 16 ft. long?

12. How much will a $\frac{3}{8}$-in. octagonal bar of steel 12 ft. long weigh?

13. What is the weight of a sheet of wrought iron

$$0.0625'' \times 30'' \times 96''?$$

14. Get the weight of a $\frac{1}{8}$-in. mild-steel sheet for the walls of a round tank 6 ft. in diameter and 4 ft. high. ($\pi = 3.14$.)

15. Find the weight of the two ends for the tank in Prob. 14.

16. Compute the weight of a coil of mild-steel rod $\frac{1}{4}$ in. in diameter if there are 30 turns each 2 ft. in diameter.

17. Get the weight of 1000 ft. of $\frac{1}{16}$-in. round mild-steel wire.

18. Find the weight per foot of 1″ round mild steel, 1″ square mild steel, 1″ octagonal mild steel, 1″ hexagonal mild steel.

19. How do the weights in Prob. 18 compare?

20. How do the end areas of the bars in Prob. 18 compare?

CAPACITY OF A CYLINDRICAL TANK ON ITS SIDE

Cylindrical tanks, such as tank trucks and trailers, underground storage tanks at gasoline service stations, and railroad tank cars, are usually placed horizontally.

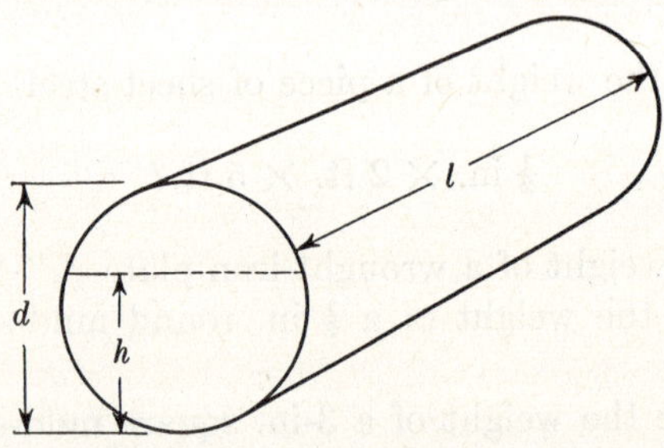

When tanks so placed are only partly filled, it is often necessary to determine the contents in cubic units or gallons. The formula below is a very close approximation. The letters used in the formula are d for diameter of the tank, l for length of the tank, h for the height to which the tank is filled, and C for capacity.

Formula. $C = lh\sqrt{.017d + 1.7dh - h^2}$ (approximately)

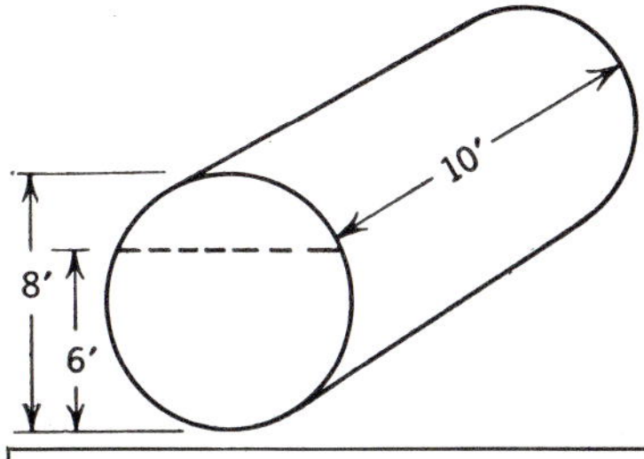

Example 1.—How many cubic feet and how many gallons are there in a round tank on its side if the diameter is 8 ft., the length 10 ft., and the depth to which it is filled 6 ft.?

$C = lh\sqrt{.017d + 1.7dh - h^2}$	Formula
$C = 10 \times 6\sqrt{.017 \times 8 + (1.7 \times 8 \times 6) - (6 \times 6)}$	
$C = 60\sqrt{.136 + 81.6 - 36}$	Substituting
$C = 60\sqrt{45.736}$	
$C = 60 \times 6.76$	
$C = 405.60$, or 405.6 cu. ft.	
Number of gallons = 405.6 × 7.5	Multiplying the number of cubic feet by the number of gallons in a cubic foot
Number of gallons = 3042	
Ans. 3042 gal. approximately	

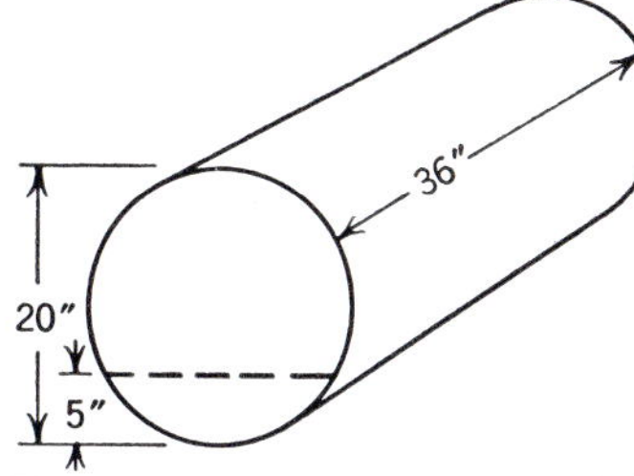

Example 2.—Find the number of cubic inches and the number of gallons in a round tank on its side when the diameter is 20 in., the length 36 in., and the depth filled 5 in.

$C = lh\sqrt{.017d + 1.7dh - h^2}$	Formula
$C = 36 \times 5\sqrt{.017 \times 20 + (1.7 \times 20 \times 5) - (5 \times 5)}$	
$C = 180\sqrt{.34 + 170 - 25}$	Substituting
$C = 180\sqrt{145.34}$	
$C = 180 \times 12.05$	
$C = 2169.0$, or 2169.0 cu. in.	
2169 ÷ 231 = 9.4	Dividing by the number of cubic inches in a gallon
Ans. 9.4 gal.	

PROBLEMS

1. Find the number of cubic inches in a round tank on its side when the diameter is 16″, the length is 20″, and the liquid is 10″ deep.

2. How many gallons are there in Prob. 1?

3. How many cubic feet are there in a tank car 7 ft. in diameter, 34 ft. long, and filled to a depth of 5 ft.? How many gallons is this?

4. A service-station operator, measuring his gasoline in an underground tank, finds the liquid 4 ft. deep in a 6-ft. diameter tank that is 10 ft. long. How many gallons has he?

5. In another tank, the same size as that of Prob. 4, the depth of fluid is only 14 in. How many gallons does it contain?

6. An oil drum, 20″ in diameter and 32″ deep, is lying on its side. If the depth of the liquid is 12″, how many gallons does the drum contain?

7. By the same formula, express in gallons the amount of space above the liquid in Prob. 6.

8. Determine the total capacity of the same drum by the formula $\pi r^2 l \div 231$. Compare this answer with the sum of the two answers of Probs. 6 and 7.

9. The actual capacity of the drum is given by your answer to Prob. 8. The total of Probs. 6 and 7 gives the approximate capacity. What is the per cent of error of the approximate answer?

Find the volume and the number of gallons contained in each of the following round tanks lying on its side:

Number	Diameter	Length	Depth filled
10	8 ft.	10 ft.	5 ft.
11	8 ft.	10 ft.	6 ft.
12	10 ft.	12 ft.	8 ft.
13	6 ft.	8 ft.	3 ft.
14	36 in.	30 in.	24 in.
15	66 in.	60 in.	54 in.
16	55 in.	90 in.	50 in.
17	48 in.	70 in.	40 in.

CHAPTER 11

CONSTRUCTING GEOMETRIC FIGURES

INTRODUCTION

When geometric figures are drawn with a compass and straightedge (or unmarked ruler) they are said to be **constructed.** Geometric construction has many practical applications in shopwork and in the building trades. Plato, the ancient Greek philosopher and mathematician, is responsible for limiting the tools used in geometric constructions to straightedge and compasses.

All geometric figures are made up of **line segments.** Technically a line has an indefinite length. A line segment is a line of definite length. It is common practice to speak of a line segment simply as a line, for example, a $3\frac{1}{2}$-in. line, a 5-ft. line.

Lines are straight, curved, or a combination of straight and curved lines. When a $3\frac{1}{2}$-in. line is considered, it is assumed to be a straight line segment $3\frac{1}{2}$ in. long.

All the surfaces studied thus far are enclosed by lines. A moving line generates a surface. All solid figures are enclosed by surfaces.

The geometric constructions we shall study deal with plane figures. The logical proofs of geometric constructions are included in any high-school plane geometry course. Only the actual geometric constructions will be given here.

BISECTION

To bisect means to divide into two equal parts.

Construction 1. To bisect the line segment *AB*.

With *A* and *B* as centers and with any radius greater than half the line *AB*, draw arcs that intersect at points *C* and *D*. Draw a straight line joining points *C* and *D*.

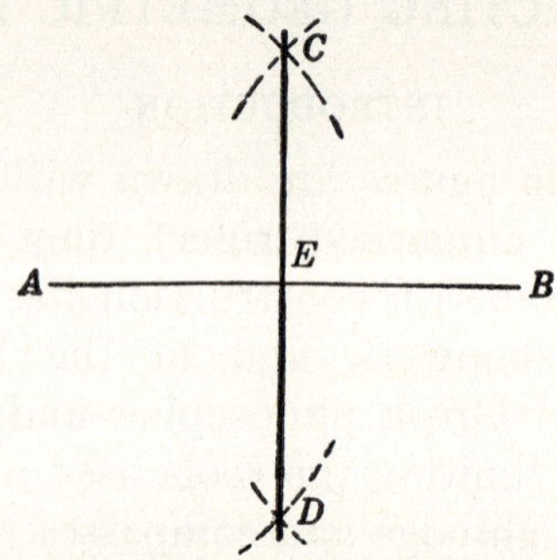

This line *CD* bisects the line *AB* at point *E*. *CD* is also perpendicular to *AB*.

Note.—Do not change the radius when drawing the arcs from *A* and *B* as centers.

Construction 2. To bisect the arc *AB*.

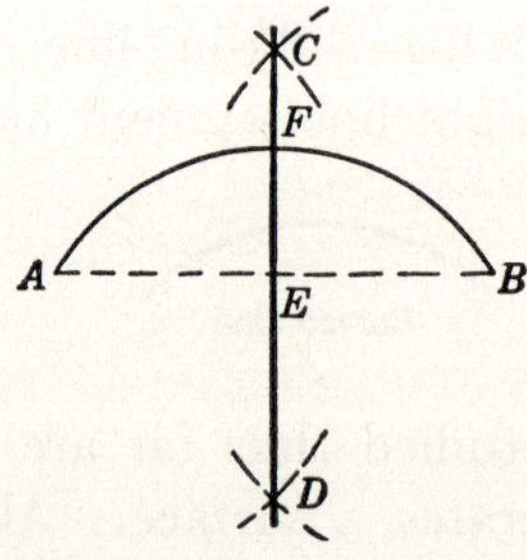

Draw a straight line from *A* to *B* making chord *AB*. Bisect the chord *AB* as in Construction 1. *CD* bisects chord *AB* at *E* and arc *AB* at *F*.

Construction 3. To bisect the angle *ABC* (on page 285).

With the vertex *B* as a center and any convenient radius, draw arc *DE*. With points *D* and *E* as centers and with the same radius greater than half the distance from D to E, draw arcs that intersect at point F. The line joining the vertex B and point F bisects the angle ABC.

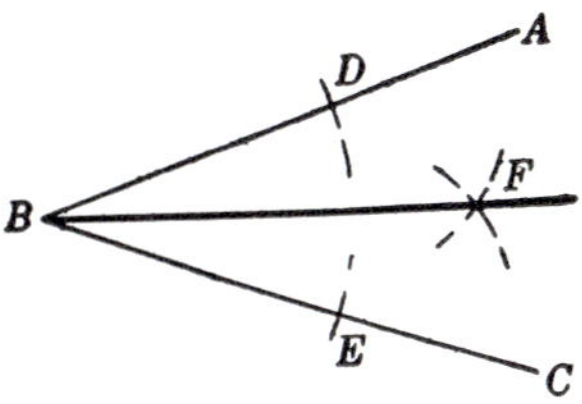

Note.—Use the same radius for the arcs from D and E as centers.

PROBLEMS

1. What similarity is there in the constructions for bisecting a line, an arc, a chord, and an angle?

2. Draw three horizontal lines of different lengths, and construct the perpendicular bisector of each.

3. Draw three vertical lines of different lengths, and construct the perpendicular bisector of each.

4. In Construction 3, how large are angles ABF and FBC (*a*) if angle ABC is 38°? (*b*) If angle ABC is 66°? (*c*) If angle ABC is 135°?

5. Check your work in Probs. 2 and 3 with your ruler.

6. Show how you would divide a line into four equal parts by constructing perpendicular bisectors.

7. Divide a line 5″ long into 8 equal parts by constructing perpendicular bisectors.

8. With your compass, draw three arcs each having a different radius and length. Bisect each arc.

9. Draw three different arcs by following the edge of a coin, a can, and a bottle. Bisect each arc.

10. Draw at least three different acute angles (angles less than 90°), and bisect each.

11. Draw at least three different obtuse angles (angles greater than 90° and less than 180°), and bisect each.

12. Check Probs. 10 and 11 by drawing chords and measuring their lengths with your ruler.

13. How should you divide an angle into four equal parts?

14. Draw an angle, and bisect it. Extend the sides of the angle through the vertex, and bisect the new angle formed. What kind of line do the two bisectors form?

PERPENDICULARS

Perpendicular lines are two lines that meet to form right angles, or angles of 90°. It is often necessary to construct a perpendicular to a given line.

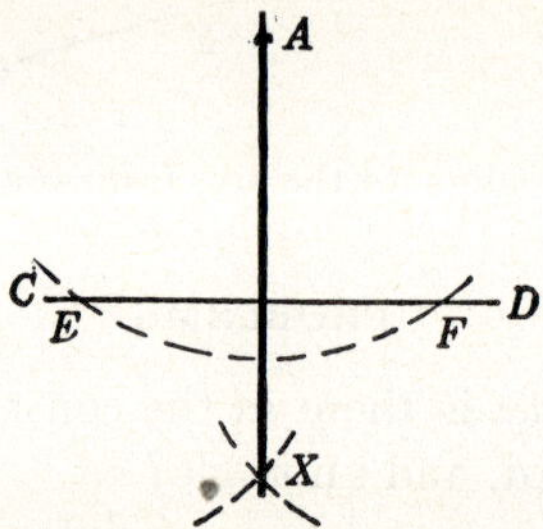

Construction 4. To construct a perpendicular from a point not on a given line to the line.

From point A, not on the line CD as a center, and with a radius greater than the distance from point A to the line CD, draw an arc cutting CD at points E and F. With the same radius greater than half the length of line EF and with points E and F as centers, draw arcs intersecting at point X. Line AX is the required perpendicular.

Construction 5. To construct a perpendicular to a line at a point on the line.

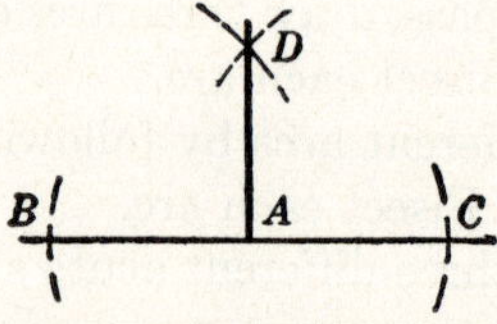

With point A as a center and with any convenient radius, lay off AB and AC of equal length. With points B and C as centers and with a radius greater than AB, draw arcs intersecting at point D. Line AD is the required perpendicular.

Construction 6. To draw a perpendicular to a line at the end of the line.

First Method.—From any point A, not in the given line XY, and with AY as a radius, draw the arc XYZ. From point X, draw line XA. Extend XA until it cuts the arc

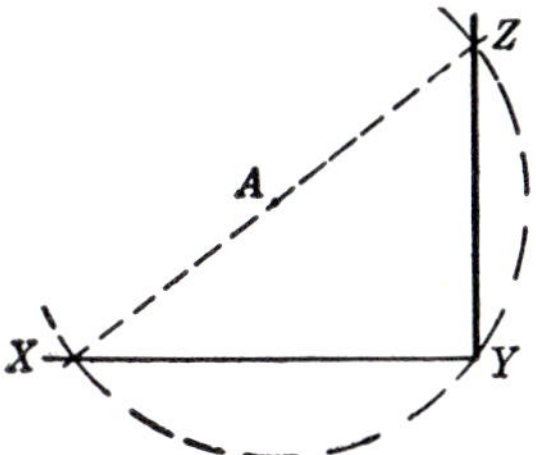

at point Z. Draw line ZY. Line ZY is the required perpendicular at the end of the line.

Note.—By extending the line through Y and applying Construction 5, ZY may be constructed perpendicular to XY at Y.

Second Method.—From point P at end of line, lay off PB equal to 4 units of any scale. From P and B as centers, lay off PA equal to 3 units and BA equal to 5 units. AP is the required perpendicular (see page 209).

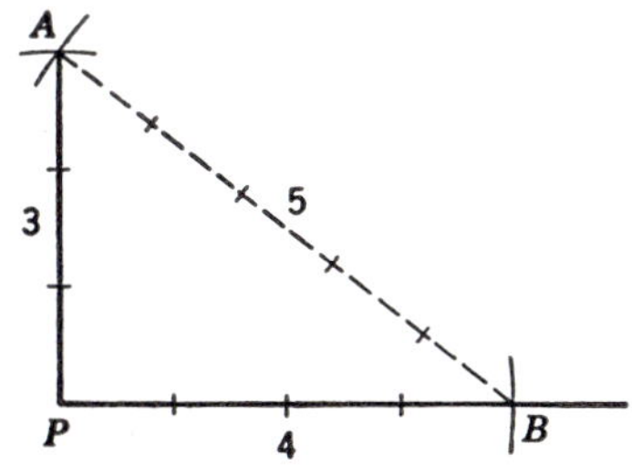

PROBLEMS

1. In what respect is Construction 4 similar to bisecting a line as in Construction 1?

2. In what respect is Construction 5 similar to bisecting a line?

3. For at least three horizontal lines of different lengths, construct perpendiculars from points not on the lines.

4. For at least two vertical lines of different lengths, construct perpendiculars from points not on the lines.

5. Draw at least four lines of different lengths, some horizontal, some vertical, and some at an angle. Mark a point on each line, and construct a perpendicular at this point.

6. Draw at least four lines of different lengths, some horizontal, some vertical, and some at an angle. Construct a perpendicular at one end of each.

7. Ask a carpenter to tell you two methods he uses to determine if a wall is plumb (perpendicular).

8. See if you can find out how a brickmason tests his work to see if it is perpendicular.

9. How can you tell that a flagpole is perpendicular to the ground?

10. What kinds of lines in nature are perpendicular to the earth's surface?

11. List several perpendicular lines in your classroom, and tell to what each line is perpendicular.

12. See if you can construct a perpendicular from a point to a line using your ruler only.

DUPLICATION

It is often necessary to construct angles, triangles, and other geometric figures equal to given angles, triangles, etc. This is called duplicating.

Construction 7. To duplicate an angle.

Construct an angle on line YZ equal to angle ABC.

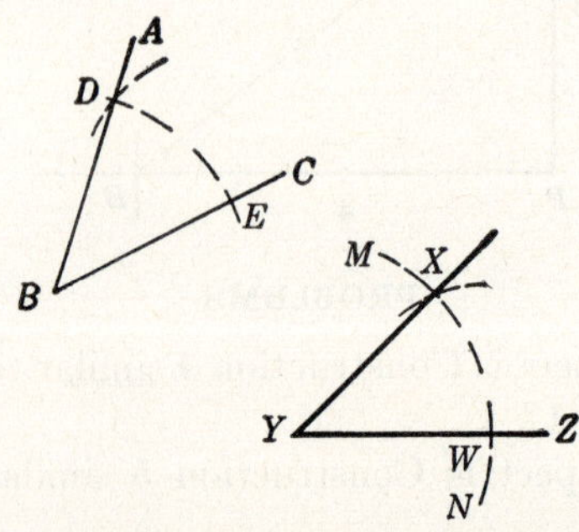

With B as a center and any convenient radius, draw an arc cutting the sides of the given angle at D and E.

With this same radius and Y as a center, draw arc MN cutting YZ at W.

With the distance DE as radius and W as center, draw an arc cutting arc MN at X. Draw XY. Angle XYZ is equal to angle ABC.

Construction 8. To duplicate triangle ABC.

First Method.—Draw any straight line. Choose a point Y on this line. With Y as center and BC as radius, lay off YZ equal to BC. With Y as center and BA as radius, draw an arc. With Z as center and CA as radius, draw an arc intersecting the other arc at X. Draw XY and XZ. Triangle XYZ is equal to triangle ABC. This method uses the three sides of the triangle.

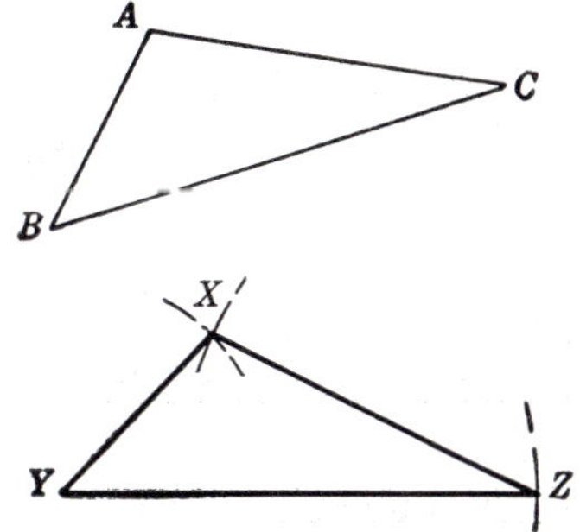

Second Method.—Construct line YZ equal to line BC. At point Y, construct an angle equal to the angle at B in triangle ABC. At Z, construct an angle equal to the

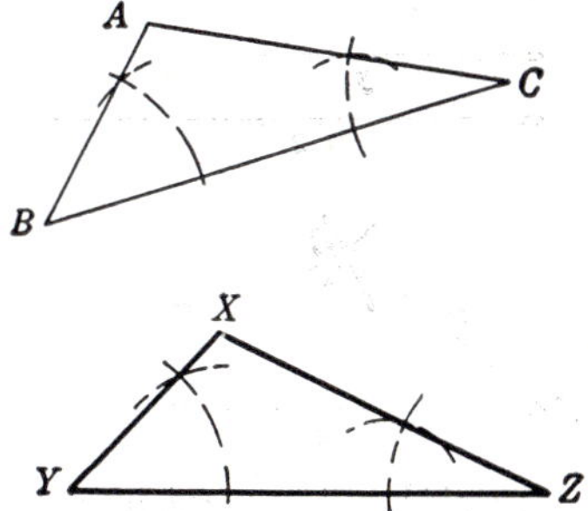

angle at C in triangle ABC. Extend the sides of these angles to meet at X. Triangle XYZ is equal to triangle

ABC. This method uses two angles and the included side of the triangle.

Third Method.—Construct line YZ equal to side BC. At point Y, construct an angle equal to the angle at B in triangle ABC. With AB as radius and Y as center, mark off YX equal to AB. Connect X and Z. Triangle XYZ is equal to triangle ABC. This method uses two sides and the included angle of the triangle.

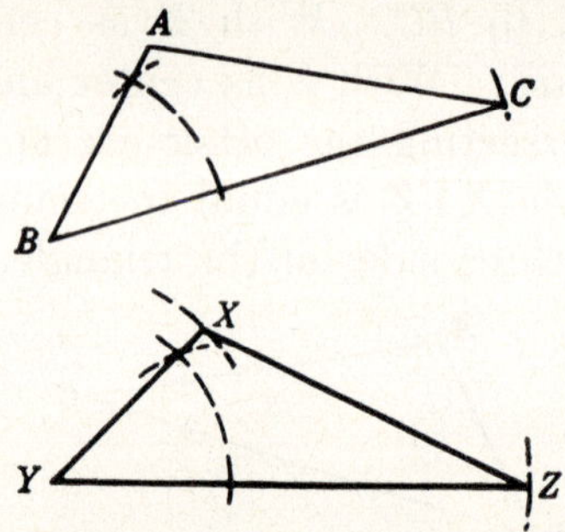

When figures are exactly the same, as are these triangles, they are said to be congruent.

Construction 9. To construct a triangle when the length of the sides is known.

First Method.—When the sides are unequal.

On any line, lay off AB equal to line segment z. With A as center and a radius equal to line segment x, draw an arc. With B as center and a radius equal to line segment

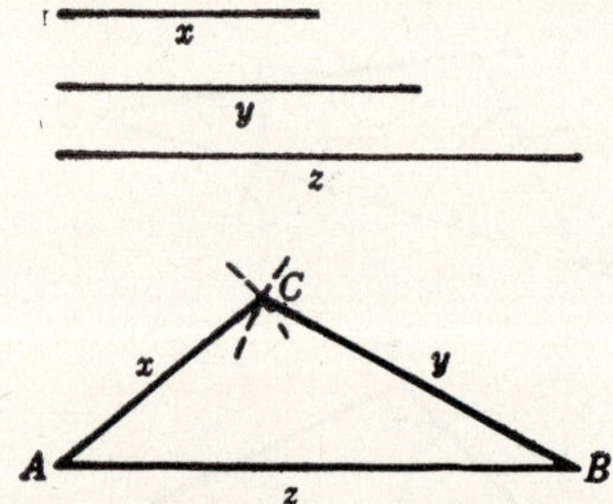

y, draw another arc cutting the first arc at point C. Draw lines AC and CB. Triangle ABC is the required triangle having its sides equal to lines x, y, and z.

Second Method.—When the sides are equal.

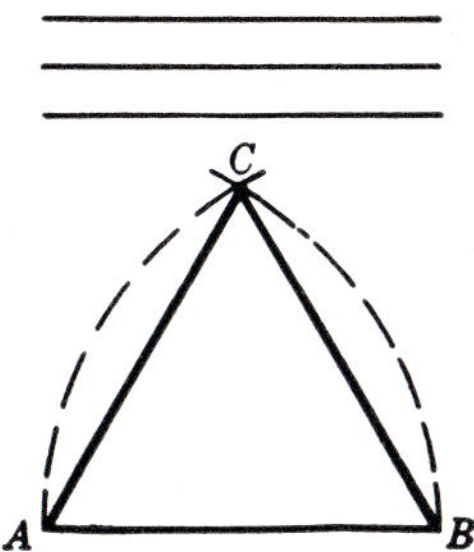

On a line, lay off length AB equal to one of the sides. With points A and B as centers and with length AB as the radius, draw arcs intersecting at C. Connect C with A and B. Triangle ABC is the required triangle. It is an equilateral triangle.

PROBLEMS

1. Draw an acute angle, and duplicate it.

2. Draw an angle equal to the angle formed at the corner of your work sheet.

3. Draw an obtuse angle. Construct an angle of the same size.

4. Draw an acute angle. Construct an angle adjacent to it and of the same size. (Adjacent angles have the same vertex and a common side between them.)

5. Draw a scalene triangle, and duplicate it.

6. Draw an isosceles triangle. Construct a triangle of the same size.

7. Construct a triangle with sides 1″, $1\frac{1}{2}$″, and 2″ long.

8. Construct a triangle with all sides 1″ long. Construct a triangle with all sides $1\frac{1}{2}$″ long. What kind of triangles are these?

9. Construct a right triangle when the sides including the right angle are $\frac{3}{4}$″ and 1″. Measure the length of the hypotenuse.

10. Draw two acute angles of different sizes. Duplicate these angles at the ends of a line 2″ long so as to form a triangle.

11. Draw a triangle. Duplicate it by using one of its angles and the two sides adjacent to the chosen angle.

12. Draw a triangle. Duplicate it by using one side and the angles at its ends.

PARALLEL LINES

Parallel lines are lines in the same plane that will not meet no matter how far they are extended.

Construction 10. To construct a line parallel to a given line and a required distance away.

Construct a parallel line $\frac{3}{4}$ in. from AB.

Choose any two points E and F on AB. At E and F, construct perpendiculars to AB. With E and F as centers, and a radius of $\frac{3}{4}$ in., draw arcs cutting the perpendiculars at C and D, respectively. Join C and D. Line CD is parallel to AB and $\frac{3}{4}$ in. away from it.

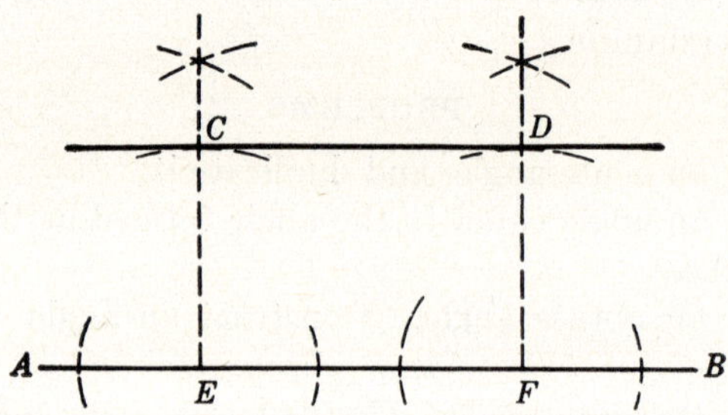

Remember that distances are always measured on perpendiculars.

Construction 11. To construct a line parallel to a given line AB through a given point P.

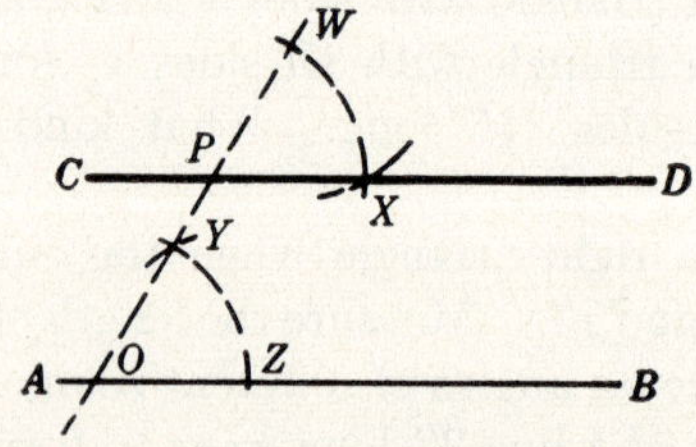

Through point P, draw line OW at any convenient angle to AB. At P, construct an angle WPX equal to angle

WOB. Draw line *PX*. Line *PX*, or the extension of line *PX*, is the required parallel line to *AB* through point *P*.

PROBLEMS

1. Construct a line *CD* parallel to and $1\frac{1}{4}''$ away from line *AB*.

2. On your drawing of Prob. 1, construct a line *EF* parallel and $\frac{3}{4}''$ away from line *CD*.

3. Show that line *EF* and line *AB* of Probs. 1 and 2 are parallel.

4. Construct a line parallel to a given line *AB* through a point *P* not on *AB*.

5. Draw an acute angle. Construct a line parallel to each side of the angle and $\frac{1}{2}''$ away on the outside of each side.

6. In Prob. 5, extend the lines constructed parallel to the sides of the given angle until they intersect. How does the angle thus formed compare with the original angle?

7. List five examples of parallel lines (*a*) in your classroom and (*b*) in shop equipment.

8. See if you can learn how a draftsman draws horizontal parallel lines. Why are the lines thus drawn parallel?

9. See if you can learn how a carpenter draws parallel lines.

10. Show why two vertical lines are parallel.

11. Draw parallel lines through any two points *A* and *B*.

Construction 12. To divide a given line *AB* into any number of equal parts: for example, into five equal parts.

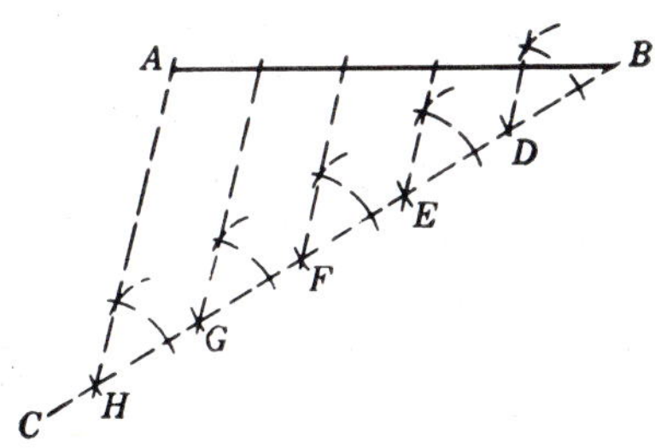

Draw line *BC* of indefinite length, forming the acute angle *ABC*. With a compass, mark off five equal divisions of any convenient length along *BC*. Connect the last point H with A. Draw lines through points G, F, E, and D, parallel to line AH. These lines divide AB into five equal parts.

PROBLEMS

1. Divide a line $3\frac{1}{4}''$ long into 4 equal parts, by the method explained in Construction 12.

2. Divide a line $3\frac{1}{4}''$ long into 4 parts by first bisecting it and then by bisecting each half.

3. Which method do you prefer and why, that of Prob. 1 or that of Prob. 2?

4. Divide the paper you are working on into 5 columns of equal width.

5. Explain or show how you would divide a board $5\frac{1}{2}''$ wide into 4 equal parts.

6. Show how you should divide a wire $5''$ long to make it into an equilateral triangle.

Construction 13. To divide the area of (*a*) a triangle, or (*b*) a trapezoid into any number of equal parts: for example, into three equal parts.

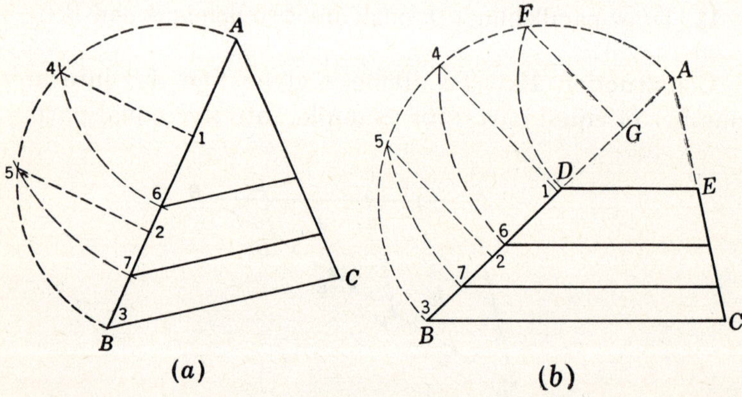

(*a*) Draw a semicircle on side AB as a diameter. Divide side AB into three equal parts, at 1 and 2. Through points 1 and 2 draw perpendicular lines to AB to intersect the semicircle at points 4 and 5. Through points 4 and 5 draw circular arcs with center at A. Through points 6 and 7, where the circular arcs intersect the side AB, draw lines parallel to side BC. This divides the area of the triangle into three equal parts.

(b) To divide the trapezoid $BCDE$ into three equal parts, extend the sides of the trapezoid to form triangle ABC. With AB as a diameter, draw a semicircle. With A as a center and AD as a radius, strike an arc to intersect the semicircle at F. From F draw a perpendicular to AB to intersect at G. Divide GB into three equal parts, and complete the construction as shown and described for the triangle (a). Note the only difference: divide GB, not AB, into the number of equal parts desired.

PROBLEMS

1. Draw a right-angle triangle with sides 3″, 4″, and 5″. Divide it into three equal areas with lines parallel to the 3″ side.

2. Draw another right-angle triangle with sides 3″, 4″, and 5″. Divide it into three equal areas with lines parallel to the 4″ side.

3. Draw another right-angle triangle with sides 3″, 4″, and 5″. Divide it into equal areas with lines parallel to the hypotenuse.

4. Draw an equilateral triangle 3″ on a side, and divide it into four equal areas.

5. Draw an isosceles triangle with base 3″ and equal sides 4″. Divide it into five equal areas with lines parallel to the base.

6. Divide the tract of land shown below into five lots of equal areas with sides parallel to the 250′ side. (Draw side AB 2.5″ and side BC 6″.)

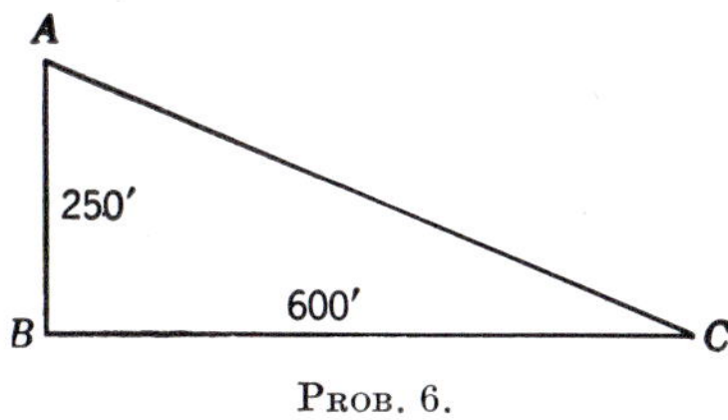

PROB. 6.

7. Draw a trapezoid with top 2″, bottom 3″, and height 2″. Divide it into four equal trapezoids with lines parallel to the bottom.

8. Divide the plot shown below into two equal lots. Draw it with sides 5″, 3″, and 3″. Measure the elements of each part, and figure the areas to check the accuracy of your work.

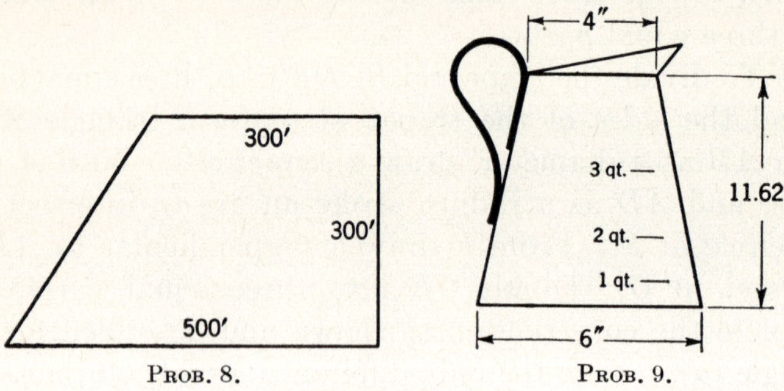

PROB. 8. PROB. 9.

9. The tapered can shown holds 1 gal. Make a full sized drawing and show where to mark the wall for 1, 2, and 3 qt.

CIRCLES

It is often necessary to find the center of a circle or the center of an arc.

Construction 14. To construct a circle through three given points A, B, and C, not in a straight line.

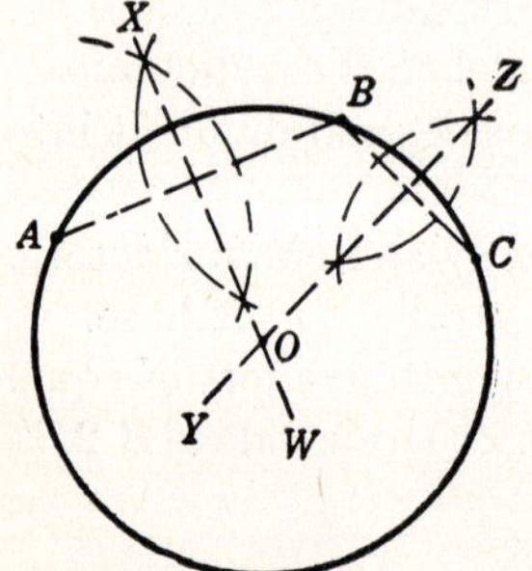

Draw lines AB and BC. Construct the perpendicular bisector WX of line AB and the perpendicular bisector YZ of line BC. The intersection of WX and YZ at O is the center of the required circle. With O as the center and OA as the radius, construct the required circle.

Construction 15. To find the center of the circle on which a given arc lies.

Draw two chords AB and CD. Construct perpendicular bisectors to both chords. The perpendicular bisectors intersect at O, the center of the circle on which the given arc lies.

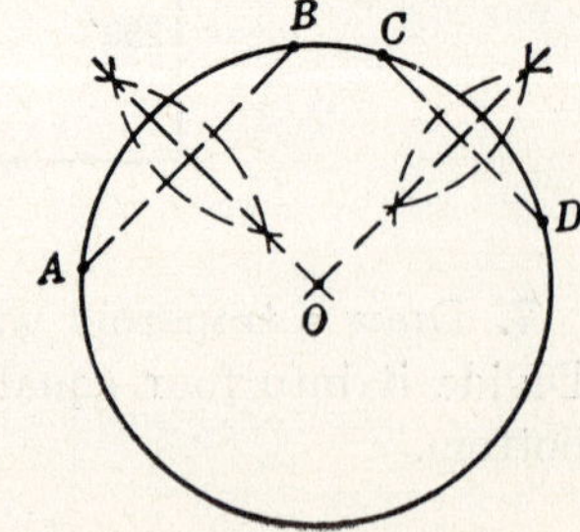

PROBLEMS

1. Draw a line AB 1″ long and a line BC $\frac{3}{4}$″ long forming an obtuse angle. Construct a circle through points A, B, and C.

2. Draw an acute angle ABC, and construct a circle that passes through points A, B, and C.

3. Draw a circle around a round coin. Find the center of the circle.

4. Show how you would find the center of the circle at the end of a 2″ diameter steel bar.

5. Draw an arc part way around a circular object, and then find the center of the circle. Complete the circle with your compass. Place the round object over the constructed circle to see how they coincide.

6. In what respects are Constructions 14 and 15 similar?

7. With your compass, draw an arc that is about three-fourths of a circle. Find the center of the circle by construction. Check its position with the center mark made by your compass.

8. Draw an arc halfway around a large coin. Draw intersecting chords from each end of the arc that cut segments about two-thirds of the arc. Find the center of the circle by finding the intersection of the perpendicular bisectors of these chords.

9. Draw a scalene triangle ABC. Construct a circle passing through all the vertices.

10. Draw a right triangle ABC. Construct a circle passing through all the vertices.

11. In Prob. 10, what side of the right triangle coincides with the diameter of the circle?

12. Draw a semicircle, and draw several triangles in it like the triangle of Prob. 11. What kind of triangles are these?

TANGENTS

A tangent to a circle is a line that touches the circle at only one point. Tangent circles touch at only one point.

Construction 16. To construct a tangent to a circle at a given point A on the circle.

Draw radius OA, and extend it through A to X. At A, construct BC perpendicular to XO. BC is tangent to the circle at point A.

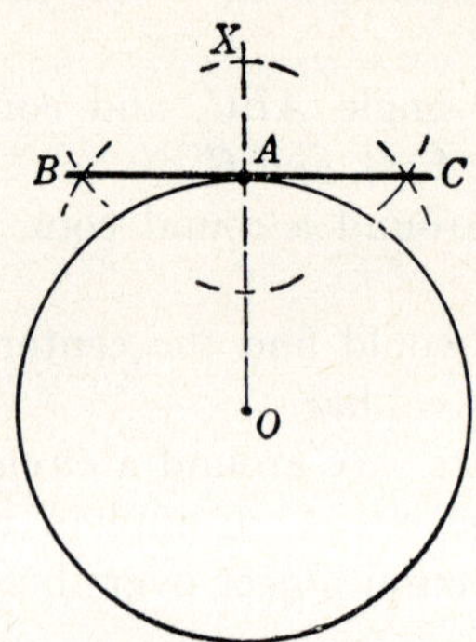

Construction 17. To construct tangents to a circle from a point A outside the circle.

Draw line AO connecting point A with the center O of the circle. Bisect AO at point X. With X as a center and a radius equal to XO, draw arcs, cutting the circle

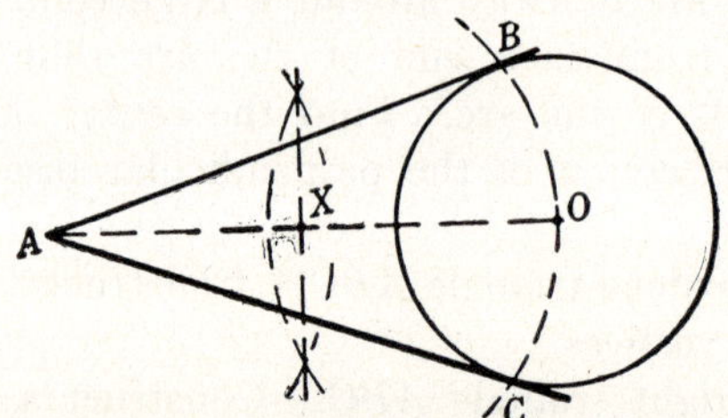

at points C and B. Draw AB and AC. Both AB and AC are tangents to the circle from point A.

Construction 18. To construct tangent circles.

Circles Tangent Externally.—Extend the radius AB of one circle the length of the radius BC of the other circle.

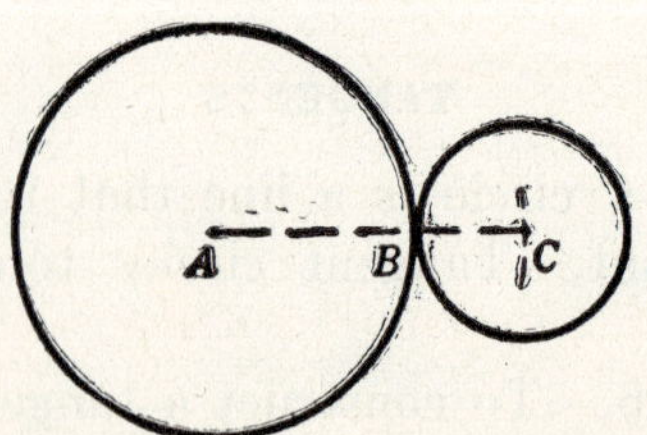

With C as center and BC as radius, draw circle C. It is tangent to circle A.

Circles Tangent Internally.—On the radius AB of the larger circle, with B as center, mark off BC equal to the radius of the smaller circle. With C as center and BC as radius, construct the inscribed circle. The circles are tangent at point B.

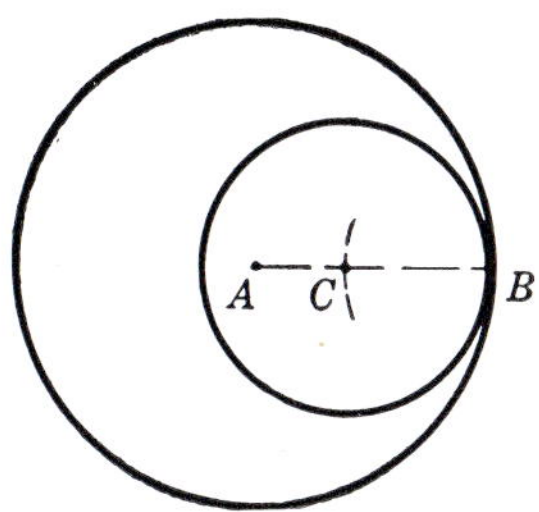

PROBLEMS

1. Draw a 2″ diameter circle. Construct a tangent to the circle at a point on the circle.

2. Why cannot two tangents be drawn through the same point on a circle?

3. Draw a circle with a $\frac{3}{4}$″ radius. Construct tangents to the circle from a point 2″ outside the circle.

4. Draw a 2″ diameter circle, and construct a tangent to it so that the tangent will be approximately a vertical line.

5. Draw a line and see if you can construct a circle tangent to it.

6. Construct a $\frac{3}{4}$″ diameter circle tangent externally to a $1\frac{1}{2}$″ diameter circle.

7. Draw circles of the same size as in Prob. 6, tangent internally.

8. Construct three circles of different diameters, all tangent at one point.

9. See whether you can find some lines tangent to circles in the shop.

10. See whether you can find some circles that are tangent in the shop.

POLYGONS

Many geometric shapes are readily constructed.

Construction 19. To inscribe a regular hexagon in a circle.

The radius of the circle is equal to AB. Starting with any point on the circle, as B, apply the radius as a chord

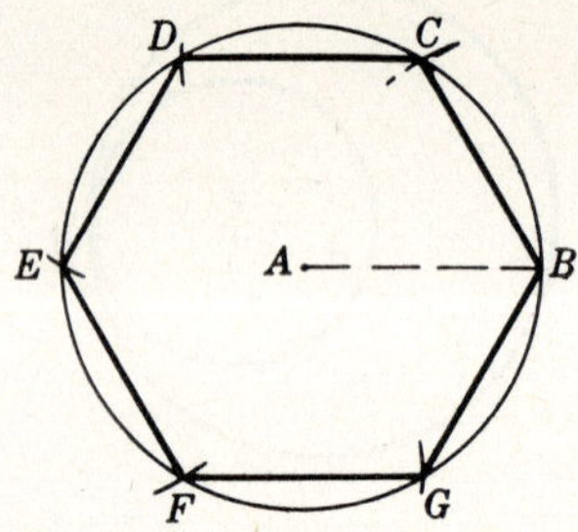

successively around the circle. Connect the successive points with straight lines to form the required hexagon $BCDEFG$.

Construction 20. To inscribe a regular octagon in a circle.

Draw a circle with diameter AB. Construct diameter CD perpendicular to diameter AB at O, the center of the

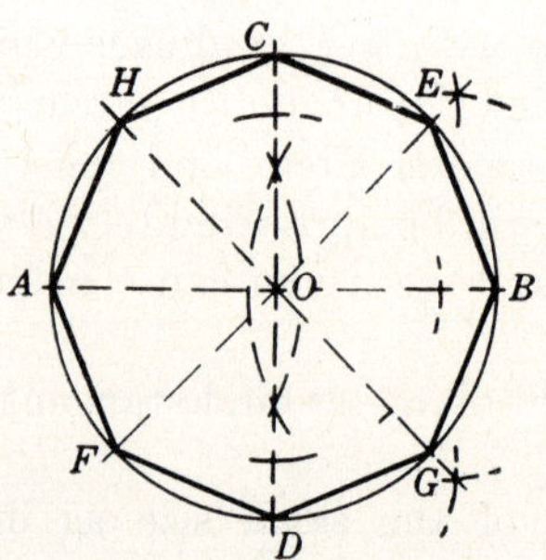

circle. Bisect angles COB and BOD. Extend these bisectors to the opposite side of the circle. Connect the successive points A, H, C, E, B, G, D, and F with straight lines to get the required octagon.

Construction 21. To inscribe a regular octagon in a square $WXYZ$ (page 301).

Draw the diagonals *WY* and *XZ* intersecting at *O*. With a radius equal to *WO* and with *W* as a center, draw arc *CH*. With the same radius *WO*, draw similar arcs with *X*, *Y*, and *Z* as centers. Connect the successive points

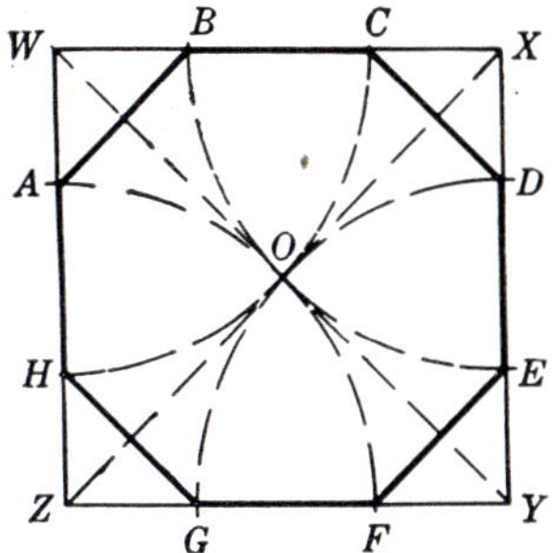

A, *B*, *C*, *D*, *E*, *F*, *G*, and *H* with straight lines to form the required octagon.

Construction 22. To construct any regular polygon on a side *AB*, for example, a regular heptagon (7 sides).

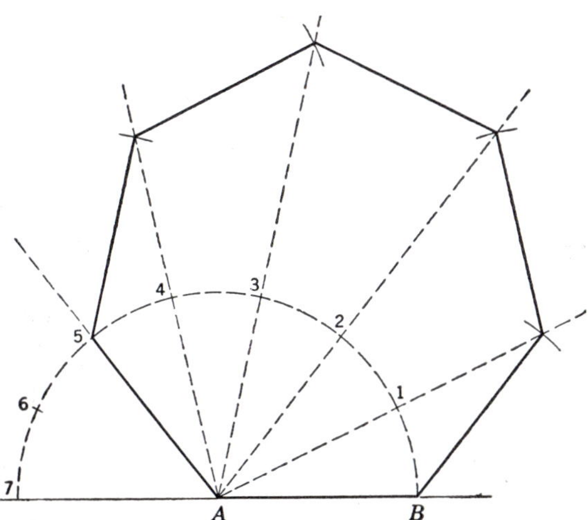

With *A* as a center and *AB* as a radius, draw a semicircle and divide it into seven equal parts, of which $(n - 2)$, counting from *B*, are to be used. In this case 5 parts are to be used. Draw ray lines through points 1, 2, 3, 4, and 5, and complete the heptagon by using the side AB stepped around the ray lines.

PROBLEMS

1. Inscribe a hexagon in a circle when the radius of the circle is $1\frac{1}{4}''$.

2. Construct a heptagon with sides $\frac{3}{4}''$ long.

3. Construct a hexagon with the distance across the corners equal to $2''$.

4. Examine the hexagon in Construction 19, and show how you would construct an equilateral triangle in a circle.

5. Construct a regular polygon with 9 sides.

6. Construct an octagon in a circle with a $2''$ diameter.

7. The distance across the corners of an octagon is $3''$. Construct the octagon.

8. Examine carefully Construction 20, and see what points you would connect to form a square in a circle.

9. Construct a regular pentagon.

10. Construct a square $2''$ on a side. Inscribe an octagon in this square.

11. Show how you would inscribe a figure with 16 sides in a circle.

12. Show how you would find the center of an octagon.

MISCELLANEOUS PROBLEMS

1. What is wrong with the construction shown? This was an attempt to draw tangents to the circle from point A.

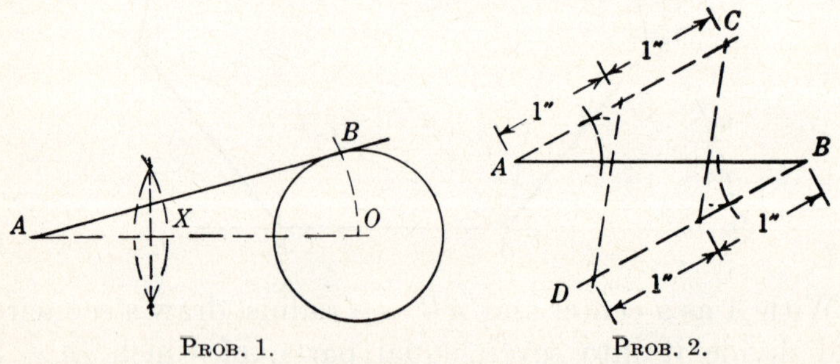

PROB. 1. PROB. 2.

2. Can you duplicate and explain the construction above which divides the line AB into 3 equal parts?

3. Using the same method as that used in Prob. 2, divide a line 3″ long into 5 equal parts.

4. See whether you can divide a semicircle into sectors of 15° each, as shown.

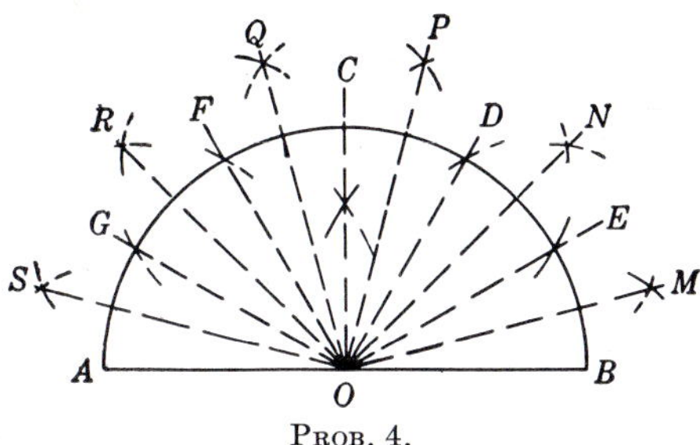

Prob. 4.

5. Explain why the carpenter's square placed on the board, as shown, divides the width into 10 equal parts.

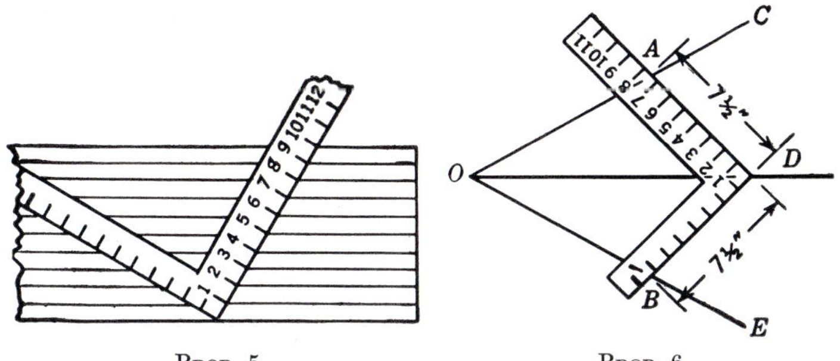

Prob. 5. Prob. 6.

6. Explain why the carpenter's square, used as shown, bisects the angle *COE*.

7. See whether you can explain and construct the octagon, as shown, with the side *AB* equal to 1 in.

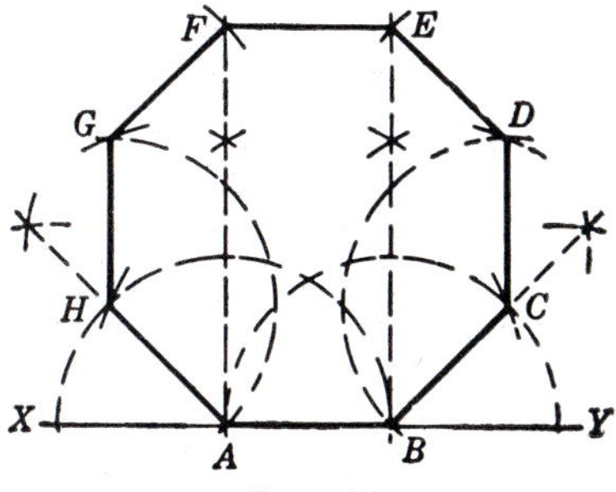

Prob. 7.

See whether you can construct each of the figures below with your compass and ruler and explain your method.

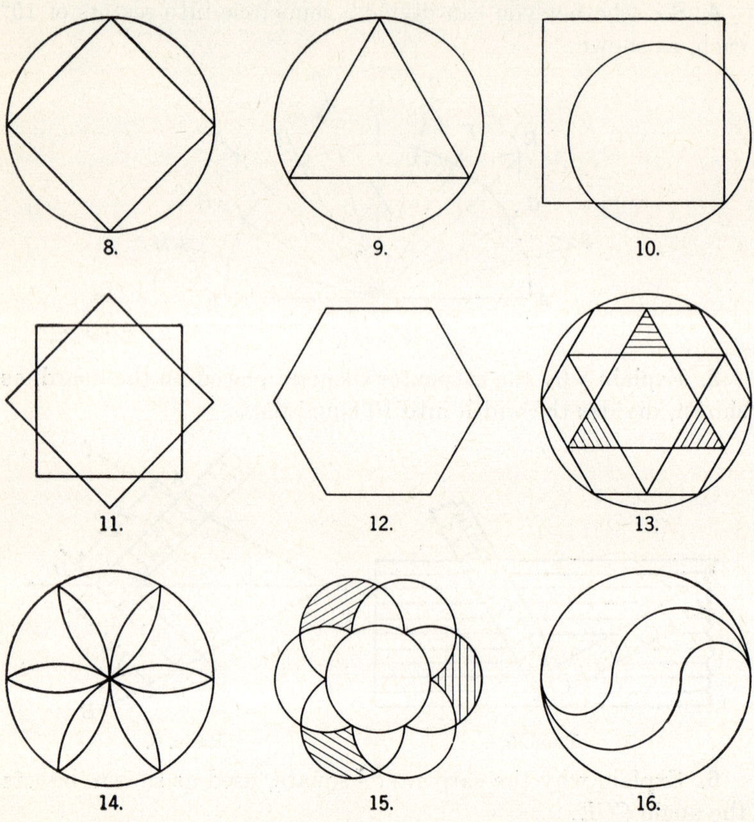

CHAPTER 12

DIRECT MEASUREMENT

MEASURING INSTRUMENTS USED BY MECHANICS

Skill in measuring must be learned by every mechanic. The practical use of the measuring instruments discussed here must be acquired by careful shop application. Properly reading the instruments used in measuring is a much neglected skill.

The following figure shows part of a 6-in. quick-reading **rule.** One edge is graduated in thirty-seconds and the

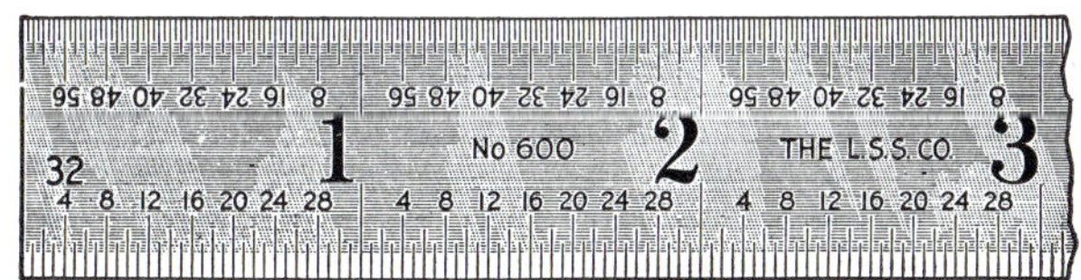

other in sixty-fourths of an inch. On the back of this rule, graduations are in eighths and sixteenths. Other rules are made in lengths from $\frac{1}{4}$ in. to 72 in. A rule is often incorrectly called a scale.

All rules read similarly except that some are graduated with more divisions per inch than others. To read a rule,

First, determine the size of the divisions on it by one of the following methods:

a. Referring to the small figure near the end, close to the division lines.
b. Counting the divisions in 1 in.
c. Comparison with another rule on which divisions are given.
d. Judgment acquired after considerable practice.

Second, count the divisions of the size indicated from zero or the last whole inch, and add to the whole inches. Reduce your answer if possible.

Example 1.—On the drawing of the rule shown, how far is it from O to A?

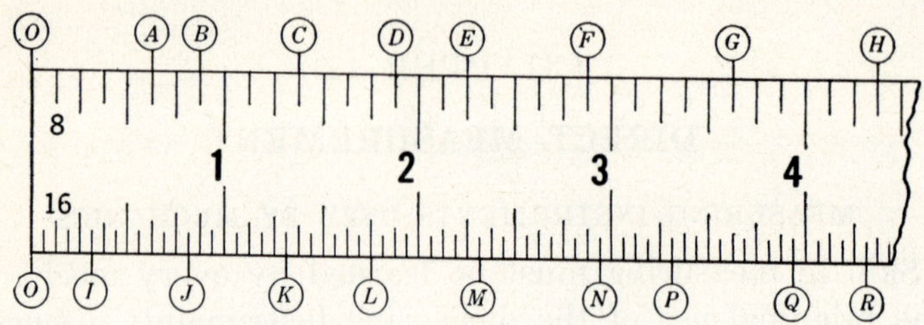

First. The small 8 indicates that the divisions are $\frac{1}{8}''$.
Second. From O at the end to A, there are 5 spaces; hence, $\frac{5}{8}''$ is the measurement.

Example 2.—On the same drawing, how far is it from K to N?

First. The small 16 indicates that the divisions are $\frac{1}{16}''$.
Second. (*a*) By counting, there are 26 spaces from K to N.

$$\frac{26}{16}'' = 1\frac{10}{16}'' = 1\frac{5}{8}''$$

(*b*) K is at $1\frac{5}{16}$; N is at $2\frac{15}{16}$.

$$2\frac{15}{16}'' - 1\frac{5}{16}'' = 1\frac{10}{16}'' = 1\frac{5}{8}''$$

(*c*) From K to $2''$ is $\frac{11}{16}''$; from $2''$ to N is $\frac{15}{16}''$.

$$\frac{11}{16}'' + \frac{15}{16}'' = \frac{26}{16}'' = 1\frac{10}{16}'' = 1\frac{5}{8}''.$$

Note.—Method *a*, *b*, or *c* may be used.

PROBLEMS

On the rule shown above, graduated in eighths and sixteenths of an inch, read the dimensions indicated.

1. O to B. **2.** O to D. **3.** O to E. **4.** O to C.
5. O to H. **6.** O to G. **7.** O to F. **8.** A to C.
9. G to C. **10.** F to H. **11.** O to J. **12.** O to M.
13. O to P. **14.** O to Q. **15.** O to R. **16.** J to M.
17. L to M. **18.** L to Q. **19.** M to R. **20.** J to L.
21. D to H. **22.** A to E. **23.** H to B. **24.** K to P.
25. N to R.

On the rule shown below, graduated in thirty-seconds and sixty-fourths of an inch, read the dimensions indicated.

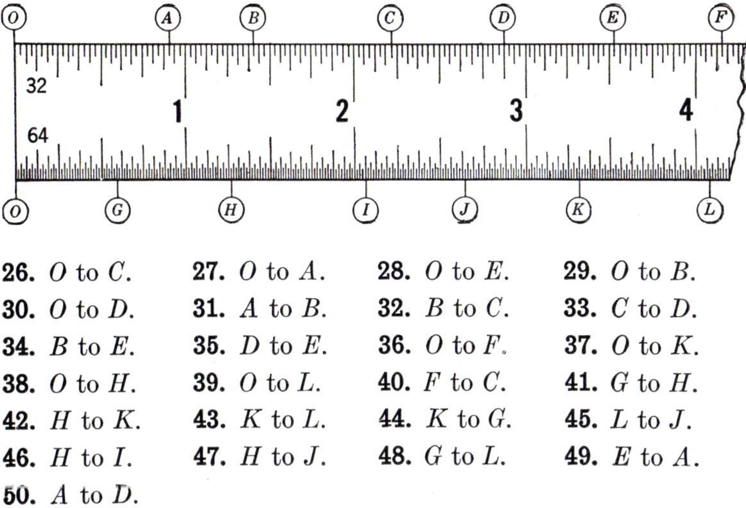

26. *O* to *C*. **27.** *O* to *A*. **28.** *O* to *E*. **29.** *O* to *B*.
30. *O* to *D*. **31.** *A* to *B*. **32.** *B* to *C*. **33.** *C* to *D*.
34. *B* to *E*. **35.** *D* to *E*. **36.** *O* to *F*. **37.** *O* to *K*.
38. *O* to *H*. **39.** *O* to *L*. **40.** *F* to *C*. **41.** *G* to *H*.
42. *H* to *K*. **43.** *K* to *L*. **44.** *K* to *G*. **45.** *L* to *J*.
46. *H* to *I*. **47.** *H* to *J*. **48.** *G* to *L*. **49.** *E* to *A*.
50. *A* to *D*.

The **tape,** as shown, is a convenient rolled-up rule. Tapes are usually graduated in eighths or sixteenths of an

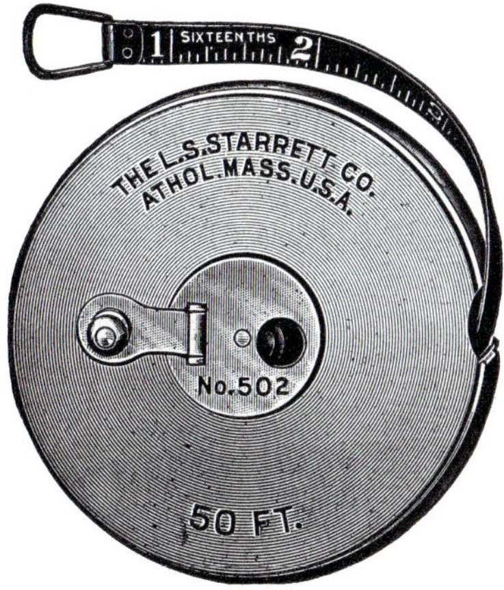

inch. Surveyor's tapes are graduated in tenths of an inch. Tapes come in lengths of 3 ft. long to 100 ft.

The **combination square** combines a rule with several sliding heads, including a square and miter head, a centering head, and a protractor head for measuring angles.

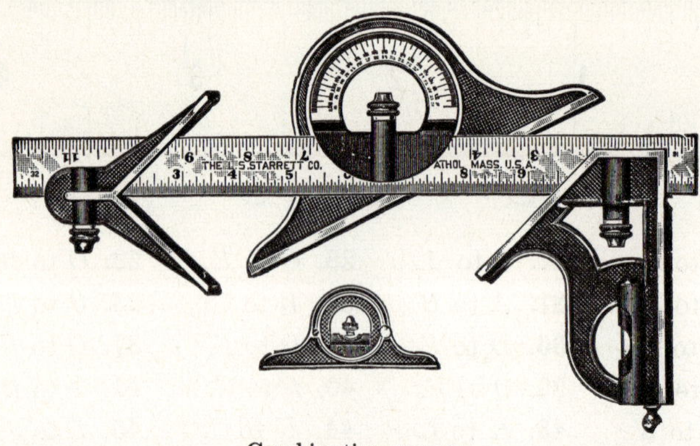

Combination square.

Outside calipers, as shown, are used to measure outside diameters, etc. **Inside calipers,** as shown, are used to measure inside diameters and distances.

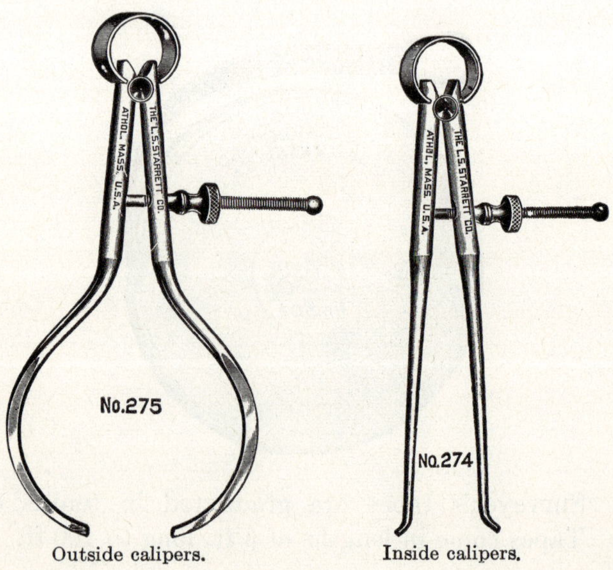

Outside calipers.

Inside calipers.

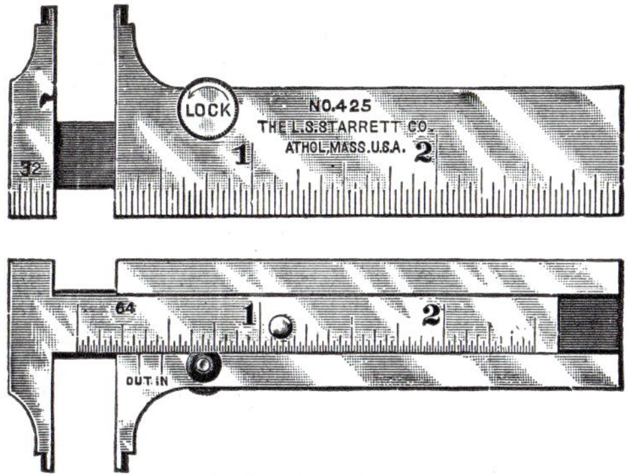

The **caliper rule** shown above serves for both inside and outside measurements.

The Micrometer Caliper

The **micrometer caliper,** the **vernier micrometer,** the **vernier caliper,** and the **vernier protractor** are used when great accuracy is required. These are precision instruments and require the most careful handling.

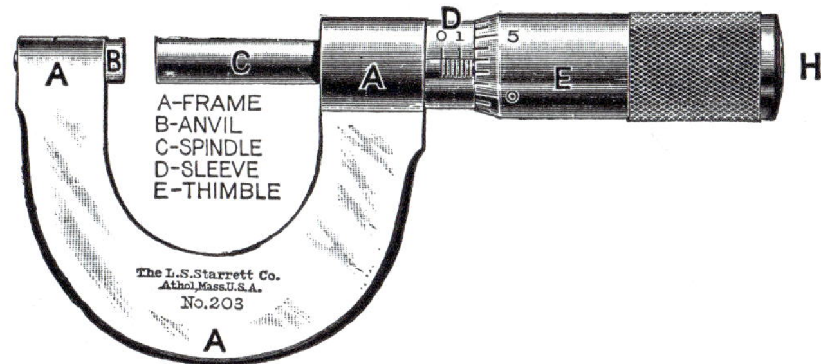

In the micrometer caliper shown above on the hidden end of the spindle C is a screw which is solid at H with the thimble E. This screw has 40 threads per inch, and it turns in the sleeve D.

Since there are 40 threads per inch, one turn of the thimble E moves the spindle C $\frac{1}{40}$ in. One-fortieth $(1 \div 40)$ is twenty-five thousandths in. (0.025″). The sleeve D is marked with 40 lines per inch; hence, the distance between the marks is also 0.025 in. Since four of these spaces make 0.100 in., every fourth mark is graduated in tenths of an inch, or hundred-thousandths. The thimble's beveled edge is divided into 25 equal parts, each equal to

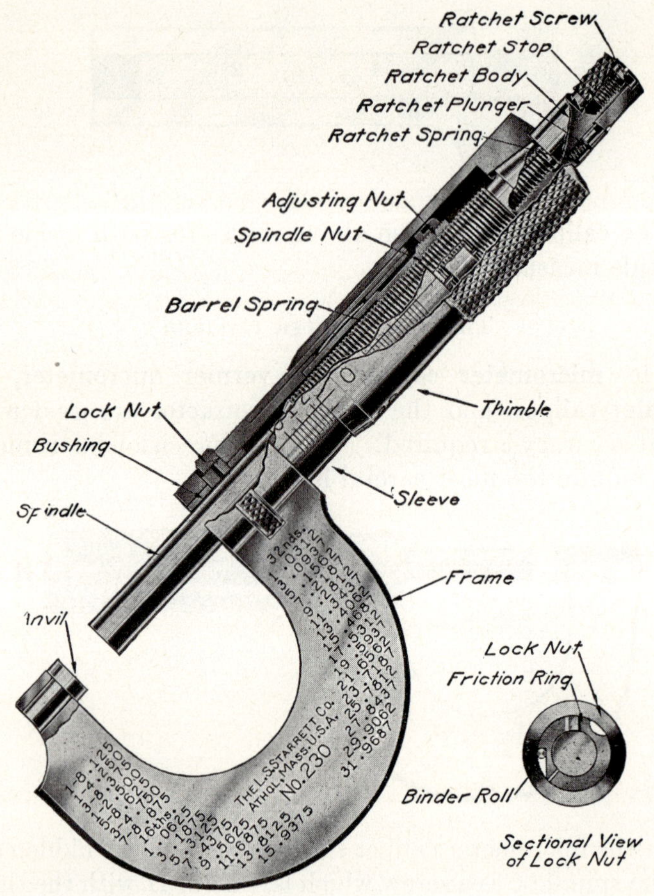

0.001 in. Rotating the thimble one division is 0.001 in.; two divisions equals 0.002 in.; etc. One complete turn of the thimble is 0.025 in., and the first mark on the sleeve *D* is in view. Two complete turns of the thimble reveal two marks on the sleeve; $2 \times 0.025 = 0.050$ in., etc.

Rule.—To read the micrometer, multiply the vertical divisions visible on the sleeve by .025, and add to this the number of divisions on the bevel of the thimble from 0 to the line that coincides with the horizontal index line on the sleeve.

Example 1.—Read the setting shown in the figure on page 309.

7 × .025 = .175″	There are 7 divisions visible on the sleeve
3 = .003″	The third line on the thimble coincides with the index line; hence, add .003
.178″	Total

Example 2.—Read a micrometer that is set so that 11 on the thimble coincides with the third division past the graduated 7 on the sleeve.

7 × .100 = .700″	7 numbered graduations
3 × .025 = .075″	3 unnumbered graduations
11 = .011″	Coinciding thimble line
.786″	Total

Example 3.—Find the reading if the thimble end is between 0.775 and 0.800 and line 5 on the thimble coincides.

.775″	Spindle reading
.005″	Thimble reading
.780″	Total

Note.—All these calculations can be made mentally after a little practice.

PROBLEMS

Read the settings shown.

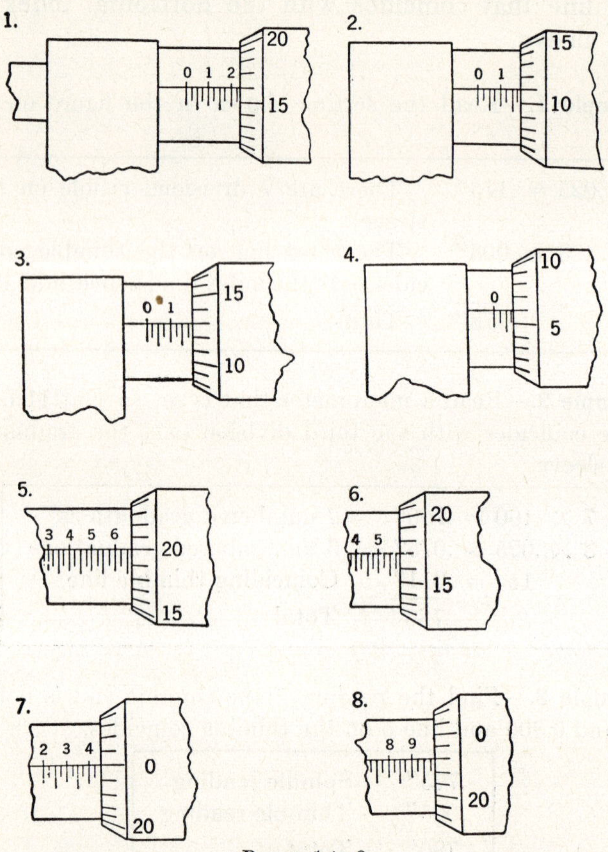

PROBS. 1 to 8.

Illustrate like Probs. 1 to 8 the following settings:

9. 0.540″ **10.** 0.366″ **11.** 0.684″ **12.** 0.333″

13. $\frac{3}{8}$″ **14.** $\frac{7}{16}$″ **15.** $\frac{21}{32}$″ **16.** $\frac{45}{64}$″

Read the following settings:

17. Spindle between 0.0 and 0.025; coinciding line of thimble is 12.

18. Spindle between 0.500 and 0.525; coinciding line of thimble is 7.

19. Spindle between 0.275 and 0.3; coinciding line of thimble is 22.

20. Spindle is between 0.75 and 0.775; coinciding thimble line is 16.

21. Spindle is between 0.375 and 0.4; coinciding thimble line is 12.

22. Spindle is between 0.15 and 0.175; coinciding thimble line is 15.

23. Spindle is between 0.65 and 0.675; coinciding thimble line is 11.

24. Spindle is between 0.975 and 1.000; coinciding thimble line is 24.

Set the following readings on a micrometer:

25. 0.016″	**26.** 0.026″	**27.** 0.035″	**28.** 0.053″
29. 0.078″	**30.** 0.095″	**31.** 0.136″	**32.** 0.263″
33. 0.380″	**34.** 0.546″	**35.** 0.777″	**36.** 0.898″
37. $\frac{3}{8}''$	**38.** $\frac{7}{16}''$	**39.** $\frac{25}{32}''$	**40.** $\frac{41}{64}''$

THE VERNIER MICROMETER

For more accurate work than is possible with the micrometer, the **vernier micrometer** is used. This instrument can be read to ten-thousandths of an inch.

On the ten-thousandths micrometer, there is another scale, called a **vernier.**

On the ten-thousandths vernier micrometer, there are 10 divisions around the sleeve just above the regular

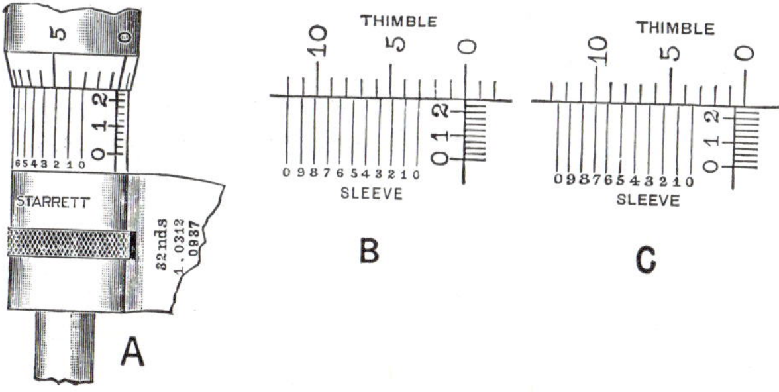

graduations on the sleeve, as shown at A in the accompanying figure. These 10 divisions occupy the same space as 9 divisions on thimble. The difference between the width of one of the 10 spaces on the sleeve and one of the 9 spaces on the thimble is one-tenth of a thimble space, or $0.1 \times 0.001 = 0.0001$, as shown at B.

Rule.—To read a ten-thousandths micrometer: (1) Read the thousandths, as on the ordinary micrometer. (2) Observe the line on the vernier that coincides exactly with any line on the thimble. The coinciding number on the vernier is the number of ten-thousandths to be added to the ordinary reading.

Example 1.—Read the vernier micrometer setting shown at C, in the preceding figure.

.200″	2 numbered graduations, 2 × .100
.050″	2 unnumbered graduations, 2 × .025
.0	Thimble graduations just above 0
.0007″	Line 7 on the vernier coincides
.2507″	Total

Example 2.—Read the setting shown on the vernier micrometer below.

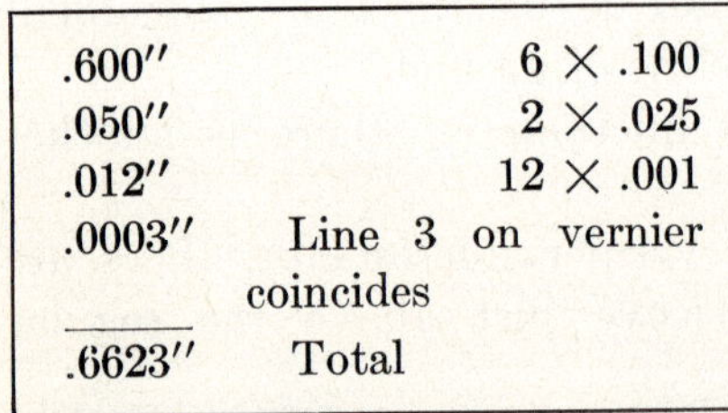

.600″	6 × .100
.050″	2 × .025
.012″	12 × .001
.0003″	Line 3 on vernier coincides
.6623″	Total

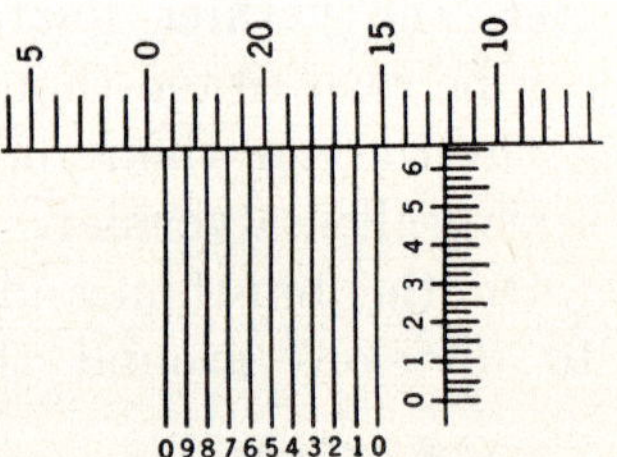

Example 3.—Read a vernier micrometer when the thimble is between 0.375 and 0.400, index line is between 15 and 16 on the thimble, and the seventh line of the vernier coincides.

.375″	Spindle reading
.015″	Thimble reading
.0007″	Vernier reading
.3907″	Total

PROBLEMS

Read the vernier micrometer settings shown.

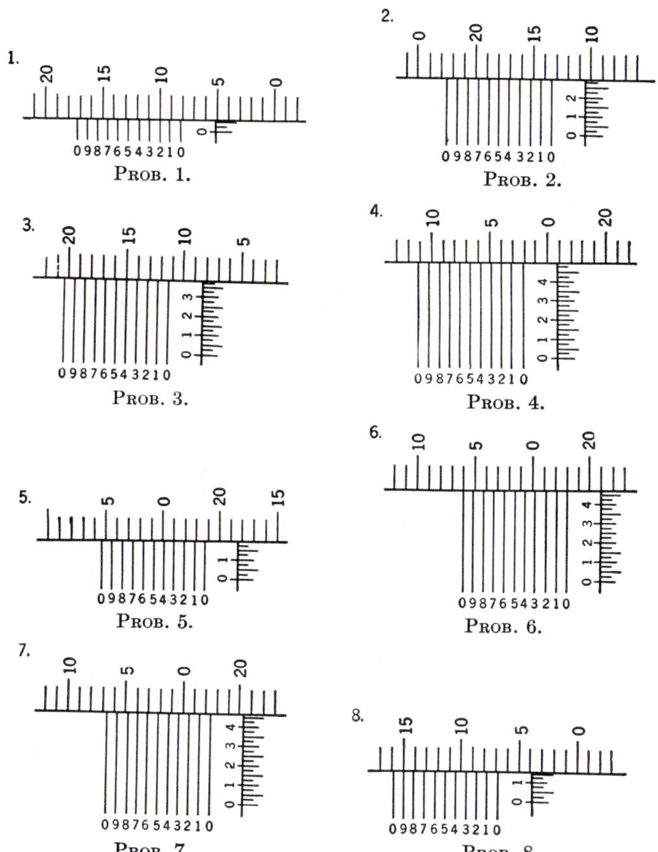

PROB. 1. PROB. 2. PROB. 3. PROB. 4. PROB. 5. PROB. 6. PROB. 7. PROB. 8.

Read the following vernier micrometer settings:

Number	Thimble end is between	Index line is between	Coinciding vernier line is	Number	Thimble end is between	Index line is between	Coinciding vernier line is
9	0.250 and 0.275	12 and 13	5	**17**	0.275 and 0.300	7 and 8	8
10	0.0 and 0.025	24 and 25	9	**18**	0.025 and 0.050	12 and 13	5
11	0.7 and 0.725	8 and 9	6	**19**	0.8 and 0.825	10 and 11	7
12	0.225 and 0.250	15 and 16	3	**20**	0.75 and 0.775	5 and 6	8
13	0.650 and 0.675	7 and 8	4	**21**	0.625 and 0.650	20 and 21	9
14	0.775 and 0.8	17 and 18	2	**22**	0.675 and 0.7	16 and 17	1
15	0.900 and 0.925	0 and 1	0	**23**	0.825 and 0.850	15 and 16	6
16	0.950 and 0.975	0 and 1	1	**24**	0.975 and 1.0	24 and 25	9

Set the following readings on a vernier micrometer.

25. 0.1234″ **26.** 0.7878″ **27.** 0.5432″ **28.** 0.0008″

29. 0.5006″ **30.** 0.4003″ **31.** 0.0158″ **32.** 0.0999″

33. 0.6757″ **34.** 0.8959″ **35.** 0.9111″ **36.** 0.9999″

37. $\frac{1}{16}$″ **38.** $\frac{9}{32}$″ **39.** $\frac{15}{16}$″ **40.** $\frac{61}{64}$″

THE VERNIER CALIPER

The **vernier caliper** is similar to the vernier micrometer except that it reads only to thousandths of an inch. An

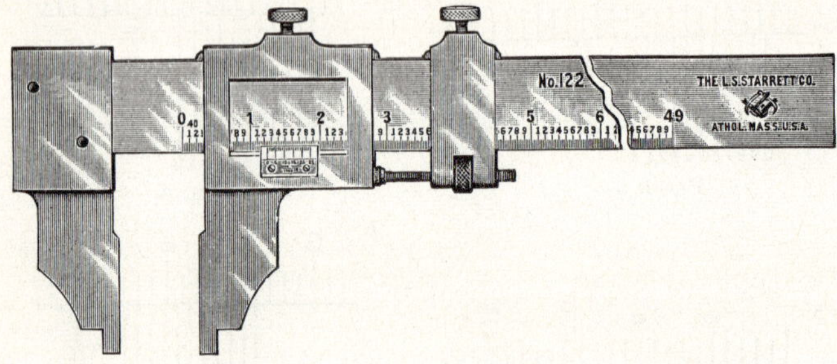

enlarged drawing of a vernier caliper is shown below. The bar of this instrument is graduated in fortieths (0.025)

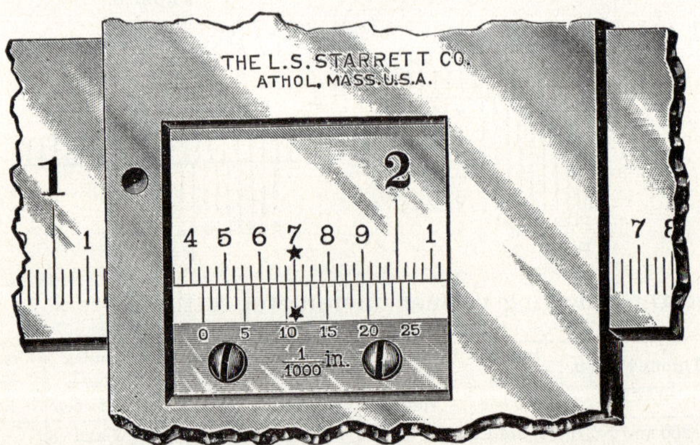

of an inch. Every fourth division, representing one-tenth, or 0.100, of an inch is numbered. On the sliding vernier plate is a space divided into 25 equal parts, numbered 0, 5, 10, 15, 20, 25. These 25 divisions on the vernier occupy the same space as 24 divisions do on the bar. The difference is, therefore, $\frac{1}{25}$ of $\frac{1}{40}$, or $\frac{1}{1000}$ in. (0.001).

Rule.—To read a vernier caliper: (1) Read the whole inches, the tenths (or .100), and the fortieths (or .025). These dimensions are all on the bar and read from the left to right. (2) Find a line on the sliding vernier that coincides with a line on the bar. The line on the vernier represents thousandths. (3) Add all the readings.

Example 1.—Read the setting shown in the preceding figure.

1.000″	Inch mark at the left is 1.
.400″	Numbered tenths are 4 spaces.
.025″	Unnumbered thousandth is 1 space.
.011″	Eleventh vernier line coincides with a bar line.
1.436″	Total

Example 2.—Read a vernier caliper when 2″ show; the index is between 0.375 and .400, and the coinciding vernier line is 21.

2.000″	Numbered inches
0.375″	Numbered thousandths inches
0.021″	Vernier thousandths
2.396″	Total

Example 3.—Set the vernier caliper for 2.478″.

> Set the vernier frame to the right of 2″.
> Set the index slightly to the right of .475″.
> Set the slide so that the vernier line 3 coincides with a bar line.

PROBLEMS

Read the vernier caliper settings shown.

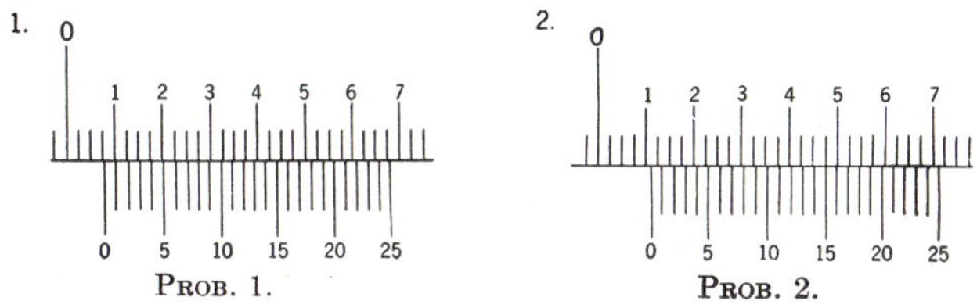

PROB. 1. PROB. 2.

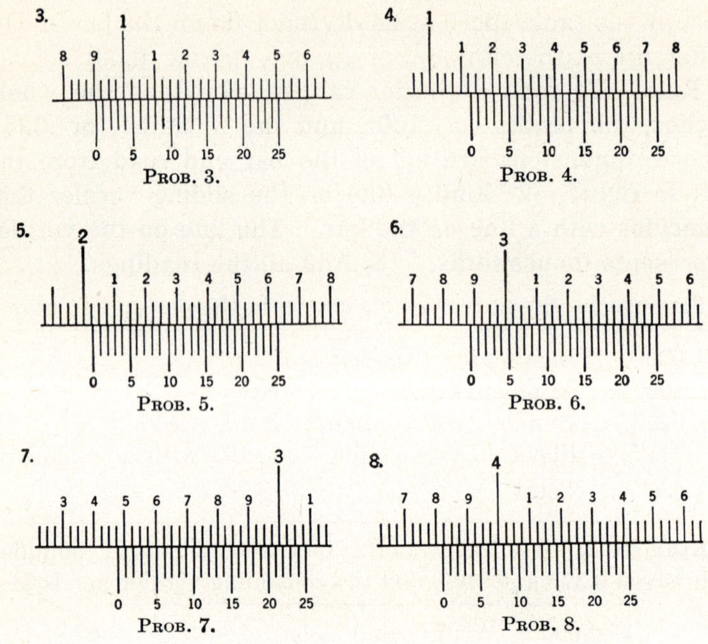

Prob. 3. Prob. 4. Prob. 5. Prob. 6. Prob. 7. Prob. 8.

Read the following vernier caliper settings:

Number	Inches	Index line is between	Coinciding line on vernier	Number	Inches	Index line is between	Coinciding line on vernier
9	1	0.0 and 0.025	5	**17**	3	0.150 and 0.175	11
10	3	0.125 and 0.15	10	**18**	0	0.025 and 0.05	18
11	2	0.0 and 0.025	15	**19**	1	0.075 and 0.1	5
12	2	0.1 and 0.125	20	**20**	3	0.1 and 0.125	8
13	0	0.5 and 0.525	25	**21**	2	0.2 and 0.225	12
14	3	0.575 and 0.6	24	**22**	1	0.175 and 0.2	1
15	2	0.725 and 0.75	17	**23**	2	0.875 and 0.9	22
16	0	0.975 and 1.00	24	**24**	0	0.975 and 1.00	0

Set the following readings on a vernier caliper:

25. 0.156″ **26.** 2.130″ **27.** 2.544″ **28.** 3.080″

29. 2.642″ **30.** 1.456″ **31.** 1.599″ **32.** 2.005″

33. 3.024″ **34.** 1.137″ **35.** 0.742″ **36.** 1.201″

37. $1\frac{1}{8}''$ **38.** $3\frac{7}{16}''$ **39.** $2\frac{5}{32}''$ **40.** $\frac{41}{64}''$

THE VERNIER PROTRACTOR

Whole angles can be laid off with the protractor head placed on the blade of the combination square shown on

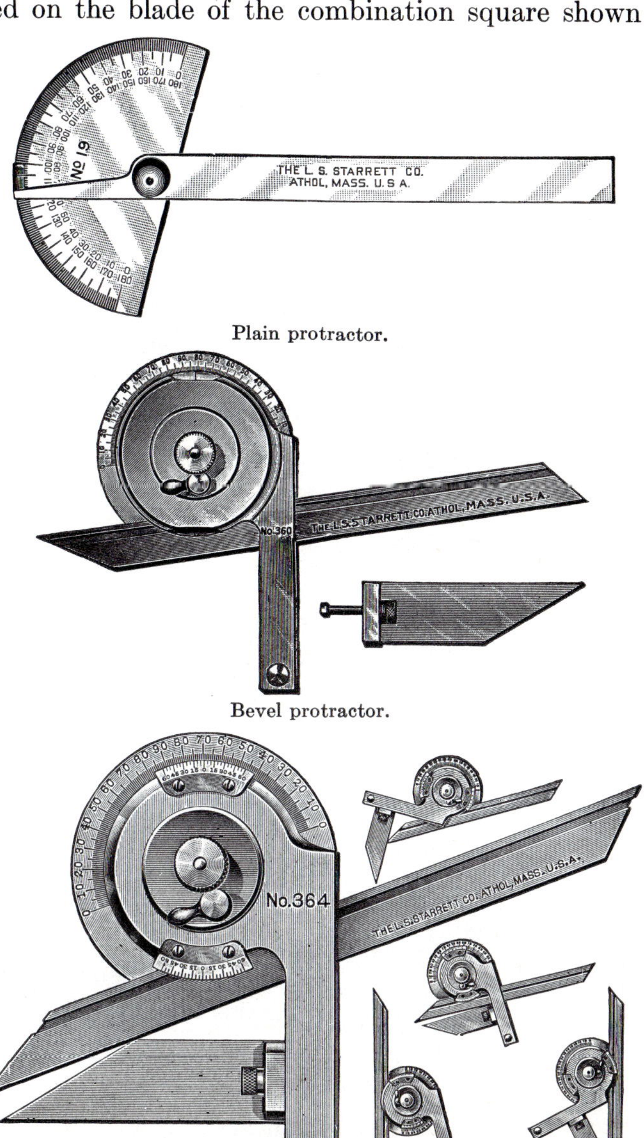

Plain protractor.

Bevel protractor.

Vernier protractor.

page 308, the plain protractor, the bevel prctractor, or the vernier protractor.

The **vernier protractor** can be read to $\frac{1}{12}$ degree, or 5 minutes. On the vernier protractor the disk, or main scale, is graduated in degrees, from 0 to 90° each way. The vernier plate is graduated so that 12 divisions on the vernier occupy the same space as 23 divisions on the disk. The difference between the width of 1 of the 12 spaces on the venier and 2 of the 23 spaces on the disk is $\frac{1}{12}°$ or 5′.

Rule.—To read a vernier protractor: (1) Read the number of whole degrees on the disk between the 0 on the disk and the 0 on the vernier. (2) Reading in the same direction, above the 0 on the vernier, locate the vernier line that coincides with a line on the disk. The coinciding vernier line will be the minutes.

Example 1.—Read the vernier protractor setting shown.

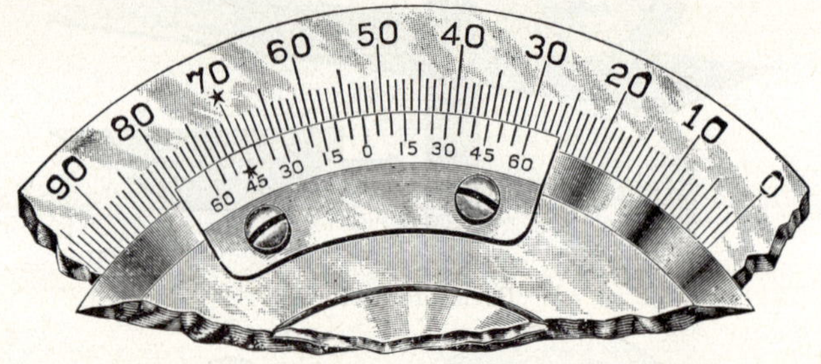

Vernier protractor.

52°	Whole degrees on the disk between 0 on the disk and 0 on the vernier
45′	Vernier line 45 coincides with a line on the disk
52°45′	Total

Example 2.—The 0 line on the vernier is between 40 and 41° on the disk. The coinciding vernier line is numbered 35. Read the setting.

40°	Whole degrees on the disk
35′	Minutes on the vernier
40°35′	Total

PROBLEMS

Read the vernier protractor settings shown.

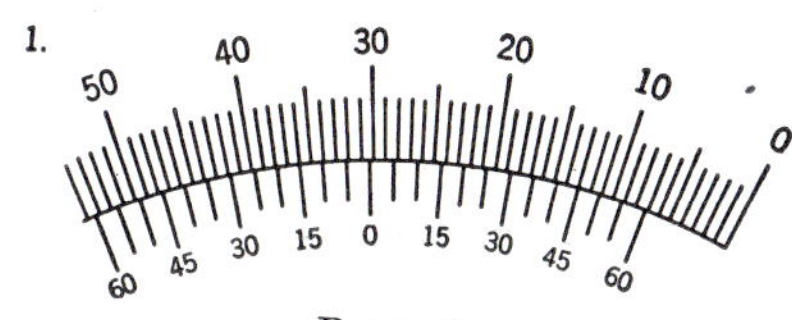

PROB. 1.

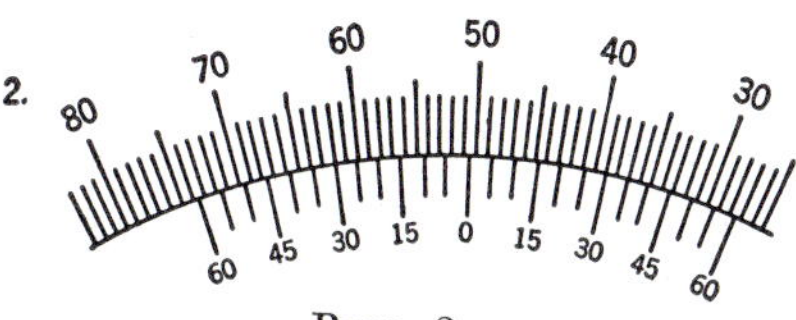

PROB. 2.

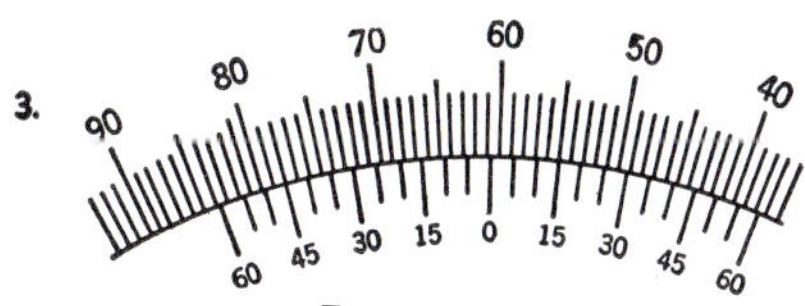

PROB. 3.

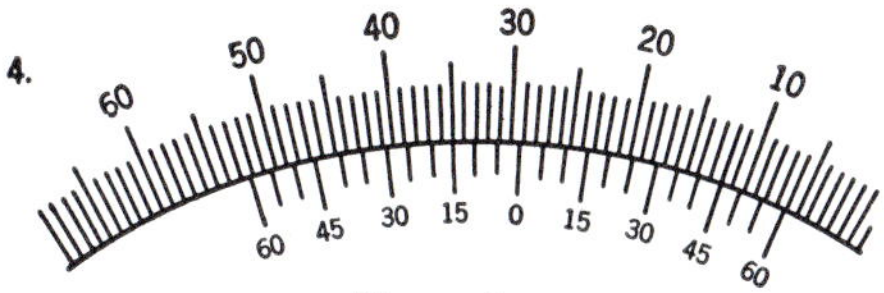

PROB. 4.

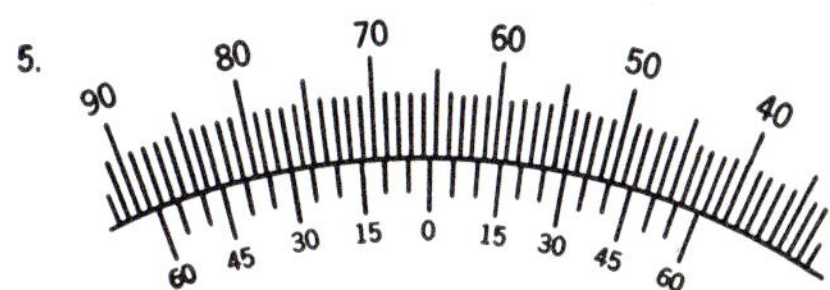

PROB. 5.

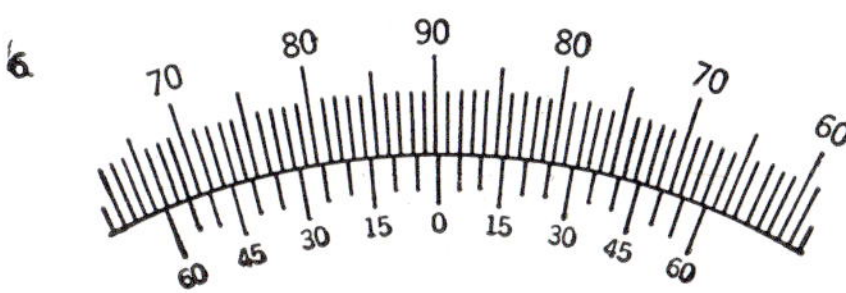

PROB. 6.

Read the following vernier protractor settings:

Number	Vernier 0 is between	Coinciding Vernier line	Number	Vernier 0 is between	Coinciding Vernier line
7	15 and 16	20	**15**	10 and 11	25
8	40 and 41	35	**16**	75 and 76	40
9	55 and 56	45	**17**	61 and 62	30
10	79 and 80	25	**18**	58 and 59	55
11	85 and 86	15	**19**	43 and 44	5
12	88 and 89	10	**20**	32 and 33	15
13	0 and 1	55	**21**	21 and 22	35
14	17 and 18	50	**22**	19 and 20	20

Set the following readings on a vernier protractor:

23. 16°45′ **24.** 27°30′ **25.** 49°55′ **26.** 59°5′

27. 88°50′ **28.** 40°40′ **29.** 50°50′ **30.** 0°30′

31. 60°55′ **32.** 72°40′ **33.** 64°20′ **34.** 35°50′

35. 27°30 **36.** 41°5′ **37.** 16°15′ **38.** 45°45′

METRIC MEASUREMENTS

The **metric system** of measurement was originated by the French about 1800 and is used exclusively in many foreign countries. It was legalized in the United States as early as 1866 and is used in most scientific work, but it has never been adopted by American industry except when manufacturers make articles for foreign sale or when they purchase foreign products.

Because the mechanic will occasionally work the metric measurements, he should become familiar with the most common of these and should be able to convert them into American equivalents.

In the metric system the fundamental unit of length is the **meter.** This is said to be exactly one ten-millionth of the distance from the equator to either pole. One meter is 39.37 in.

It is easy to understand the metric system. There are only three units of measure. Each of these uses the same prefixes to designate larger and smaller values.

The **meter** is the unit of length. The square meter is the unit of area.

The **gram** is the unit of mass and weight. A gram is the weight of one cubic centimeter of water at its maximum density.

The **liter** is the unit of volume. A liter is slightly greater than one cubic decimeter (1.000027). It is the volume of one kilogram of water at its maximum density. The cubic meter is also a unit of volume.

The most common prefixes and their values are

milli	0.001 of a unit	deka	10 times a unit
centi	0.01 of a unit	hecto	100 times a unit
deci	0.1 of a unit	kilo	1,000 times a unit
		myria	10,000 times a unit

Unit: 1.0

Thus,

A millimeter is 0.001 meter.

A kilometer is 1000 meters.

A centigram is 0.01 gram.

A hectoliter is 100 liters.

Etc.

The **metric tables** in the Appendix (pages 525 to 527) give the divisions, multiples and abbreviations of the most important metric units and other tables of common American equivalents.

From a careful study of the prefixes and Table III, it is evident that any unit is ten times the next smaller unit.

Rule 1.—To change metric units to the next smaller units, move the decimal point one place to the right.

Examples.

Length: 25 m. = 250 dm., = 2500 cm. = 25,000 mm.
Mass: 12 g. = 120 dg. = 1200 cg. = 12,000 mg.
Volume: 20 l. = 200 dl. = 2000 cl. = 20,000 ml.

Rule 2.—To change metric units to the next larger units, move the decimal point one place to the left.

Examples.

Length: 25 m. = 2.5 dkm. = 0.25 hm. = .025 km., etc.
Mass: 40 g. = 4 dkg. = 0.4 hg. = 0.04 kg., etc.
Volume: 125 l. = 12.5 dkl. = 1.25 hl. = .125 kl., etc.

EXERCISES AND PROBLEMS

Refer to Tables III and IV, on pages 525 to 527.

1. Express 5.5 m. as centimeters; as millimeters.

2. Change 65.65 km. to meters; to dekameters; to decimeters.

3. How many grams in 5 kg.? 5 dg.? 5 dkg.?

4. Change 0.56 sq. hm. to square meters; to square millimeters.

5. How many liters in 66.6 dl.? 8 ml.?

6. Express 540 cu. m. as cubic decimeters; as cubic millimeters.

7. 0.006 ml. is how many centiliters?

8. Express 16.55 kg. as hectograms; as decigrams.

9. Change 840 cu. m. to cubic millimeters.

10. Change 8328 mm. to centimeters; to kilometers.

11. Change 568 cu. m. to cubic centimeters.

12. 65.6 dg. plus 3 g. plus 56.6 mg. (change to milligrams).

13. Add 0.4 km., 66 dm., 55.3 m. and 7 dkm. (change to decimeters).

14. Add 514 sq. m. 0.62 sq. dm. and 0.25 sq. cm.

15. Subtract 546 mm. from 8 m.

16. From 3.546 kg. subtract 88.3 dg.

17. Subtract 67.8 dl. from 5.3 hl.

18. Multiply 51.3 m. by 8, and give result in hectometers.

19. How many milligrams are 56.6 cg. multiplied by 0.8?

20. Multiply 16.3 sq. cm. by 66 dm., and give result in cubic meters.

21. Multiply 18.4 m. by 120 dm., and give result in square centimeters.

22. Divide 0.04 cu. cm. by 0.5, and give quotient in cubic millimeters.

23. Divide 4.666 km. by 549 m. Carry result to three decimals.

24. Divide 734 kg. by 5.3, and express result to three decimals in grams.

25. How many pieces 75 mm. long can be sheared from 1.5 m.?

METRIC AND AMERICAN CONVERSIONS

To change metric units into American units or to change American units into metric units, it will be necessary to make a careful study of the *conversion tables* in the Appendix on pages 526 and 527.

Rule 1.—To change metric values into American values, select the American equivalent for one metric unit and multiply this by the number of metric units.

Example 1.—Change 75 mm. to inches.

Solution 1.
1 mm. = 0.03937 in. Table IV(*a*), page 526
75 × 0.03937 = 2.95275
Ans. 2.9528− in.
Solution 2. By direct reading in Table IV(*c*), page 526
Opposite the 7 in column 1 and below the 5 in column 7 read 2.9528
Ans. 2.9528 in.

Example 2.—Change 50 liters to gallons.

Solution 1.
1 liter = 0.2642 gal. Table IV(*f*), page 527
50 × .2642 = 13.21
Ans. 13.21 gal.
Solution 2.
50 liters = 10 × 5 liters
5 liters = 1.321 gal. Table IV(*f*), page 527
50 liters = 13.21 gal. Moving decimal point one place to the right.

Rule 2.—To change American values into metric values, select the metric equivalent for one American unit and multiply this by the number of American units.

Example 1.—Change 15 gal. to liters.

1 gal. = 3.785 liters	Table IV(*f*), page 527
3.785 × 15 = 56.775	Multiplying by the number of gallons
Ans. 56.775 liters	

Example 2.—Express 6.5 sq. in. as square millimeters.

1 sq. in. = 6.452 sq. cm.	Table IV(*d*), page 527
6.452 sq. cm. = 645.2 sq. mm.	Changing square centimeters to square millimeters by multiplying by 100
645.2 × 6.5 = 4193.8	Multiplying by the number of square inches
Ans. 4193.8 sq. mm.	

EXERCISES AND PROBLEMS

Use the conversion tables in the Appendix on pages 526 and 527.

1. Change 27 mm. to inches.

2. Convert 6.8 in. into millimeters.

3. Express 0.88 in. as millimeters.

4. How many miles are 75 km.?

5. 1000 yd. are how many meters?

6. Change 68 sq. yd. to square meters.

7. How many square miles are there in 1000 sq. km.?

8. Change 145 cu. cm. to cubic inches.

9. What are 500 cu. in. expressed as cubic centimeters?

10. Change 48 liquid ounces to cubic centimeters.

11. How many gallons are 88 liters?

12. How many hectoliters are there in 58 bu.?

13. A car of coal weighs 34.6 short tons. Express this as metric tons.

14. Change 5 ft. 6 in. to millimeters; to meters.

15. One yard is how many decimeters? How many millimeters?

16. Three pints are how many liters? How many deciliters?

17. Change 4 cu. m. to cubic feet; to cubic inches.

18. Change 545 sq. ft. to square meters; to square decimeters.

19. Change 8 yd. 2 ft. 8 in. to centimeters.

20. Change 5 miles 700 yd. to kilometers.

21. Express 2250 kg. as pounds; as ounces.

22. How many hectoliters in 5 bu.?

23. Express 15 hectares as acres; as square feet.

24. Change 50 lb. to kilograms. How many grams is this?

25. Change 1 bbl. to kiloliters.

26. Change 16 grams to ounces; to grains.

27. What is the metric equivalent for 1000 gal. of gasoline?

28. What is the American equivalent for the French 75 gun (75 mm.)?

29. 50,000 lb. has what metric equivalent?

30. An oil drum holds 54.5 gal. This is how many liters?

31. Change 2240 lb. to its metric equivalent.

32. Express 555 mm. in feet, inches, and sixteenths of an inch.

33. Change 4 ft. $5\frac{1}{8}$ in. to millimeters.

34. Express 27 ft. $3\frac{1}{2}$ in. as meters, decimeters, centimeters, and millimeters.

35. Change 5 m. 27 dm. 7 cm. 6 mm. to feet, inches, and thousandths of an inch.

36. A pipe is 16 in. in diameter. Find its area in square centimeters.

CHAPTER 13

LUMBER MEASURE

ORDERING LUMBER

Special usages in respect to volume measure are found in the lumber industry and carpenter trade. The volume of lumber is measured in **board feet.** A board foot of lumber is a piece of lumber 1′ long, 1′ wide, and 1″ thick. Lumber less than 1″ thick is usually considered as being 1″ thick.

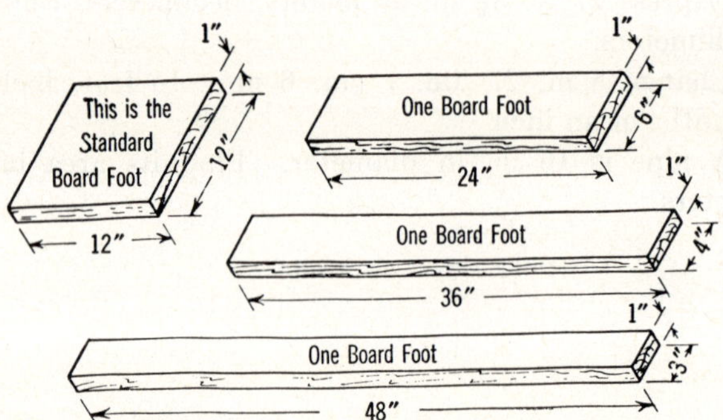

In ordering lumber, it is customary to state

1. Number of pieces.
2. Thickness in inches.
3. Width in inches.
4. Length in feet (in school shop the length may be given in inches).
5. Grade or quality of lumber.
6. Kind of wood.
7. Finish of each piece.

For example, "16 pc. 1″ × 8″ × 16′ No. 1, O.P., S4S," means "16 pieces, 1″ thick, 8″ wide, 16′ long, No. 1 quality, Oregon pine, surfaced four sides."

Of this information only the first four items 16 pcs. 1″ × 8″ × 16′, are needed to figure the number of board feet, or **footage,** as it is called in the trade.

The following table gives the fraction of board foot for each foot length of various-sized stock.

Size of stock, inches	Board feet for each linear foot	Size of stock, inches	Board feet for each linear foot
1 × 2	$\frac{1}{6}$	2 × 2	$\frac{1}{3}$
1 × 3	$\frac{1}{4}$	2 × 3	$\frac{1}{2}$
1 × 4	$\frac{1}{3}$	2 × 4	$\frac{2}{3}$
1 × 6	$\frac{1}{2}$	2 × 6	1
1 × 8	$\frac{2}{3}$	2 × 8	$1\frac{1}{3}$
1 × 10	$\frac{5}{6}$	2 × 10	$1\frac{2}{3}$
1 × 12	1	2 × 12	2

Standard thicknesses are 1″, $1\frac{1}{4}$″, $1\frac{1}{2}$″, 2″, 3″, 4″, 6″, etc., increased by even inches. Less than 1″ thick is considered 1″ thick, except in special cases, as thin stock which is often made by resawing 1″ stock.

Standard widths are given in an even number of inches except that 3″ and 5″ widths are given on 1″ finish stock only, and hardwoods are given in random widths, that is, in exactly the width that each piece measures.

Standard lengths are given in an even number of feet except for 3′, 5′, 7′, 9′, and 11′ lengths, which can be cut from twice their lengths, and hardwoods which are random lengths.

Finish sizes are computed according to the original rough sizes. Owing to shrinkage and more particularly to surfacing, stocks always measure less than the nominal sizes stated; for example, 1″ is about $\frac{13}{16}$″.

Plywood and **veneer** are sold by the square foot. The thickness, number of plies (or layers), the quality (or kind) of surface, and the size of the piece determine the price per square foot.

The basic rule for finding the number of board feet is as follows:

Rule.—The number of board feet equals the product of the number of pieces multiplied by the thickness in inches, multiplied by the width in feet, multiplied by the length in feet. (If the thickness is less than 1″, it is considered as equal to 1″.)

Let t'' = thickness in inches, w' = width in feet, w'' = width in inches, l' = length in feet. pc. = pieces, bd. ft. = board feet

Formulas. (1) $\text{Bd. ft.} = \text{pc.} \times t'' \times w' \times l'$

$$\text{(2) Bd. ft.} = \frac{\text{pc.} \times t'' \times w'' \times l'}{12}$$

Example 1.—Find the number of board feet in 16 pieces of $1'' \times 1' \times 14'$

$\text{Bd. ft.} = \text{pc.} \times t'' \times w' \times l'$	Formula (1)
$= 16 \times 1 \times 1 \times 14 = 224$	Substituting
Ans. 224 bd. ft.	

Example 2.—Find the number of board feet in 16 pieces of lumber $1'' \times 12'' \times 14'$.

$\text{Bd. ft.} = \dfrac{\text{pc.} \times t'' \times w'' \times l'}{12}$	Formula (2)
$= \dfrac{16 \times 1 \times 12 \times 14}{12} = 224$	Substituting
Ans. 224 bd. ft.	

Though it is not standard commercial practice, it is commom practice in school shopwork for the student to figure the footage for some project when lengths are given in inches. In this case, the formula for school practice is as follows:

Formula. (3) $\text{Bd. ft.} = \dfrac{\text{pc.} \times t'' \times w'' \times l''}{12 \times 12}.$

Example.—Find the number of board feet in 4 pieces of lumber $1'' \times 5'' \times 30''$.

$$\text{Bd. ft.} = \frac{\text{pc.} \times t'' \times w'' \times l''}{12 \times 12} \qquad \text{Formula (3)}$$

$$= \frac{4 \times 1 \times 5 \times 30}{12 \times 12} = 4\tfrac{1}{6} \qquad \text{Substituting}$$

Ans. $4\frac{1}{6}$ bd. ft.

With ordinary lumber, it is common practice to take the nearest whole number of board feet when an answer does not come out in an even whole number of board feet. Some firms, however, always take the next greater whole number of board feet for any fraction of a foot, even when it is less than $\frac{1}{2}$ bd. ft.

With the more expensive hardwoods, as those used in the cabinet shop, it is common practice to charge for parts of a board foot as figured.

EXERCISES AND PROBLEMS

Find the number of board feet in each of the following. Consider any fraction of a board foot as a whole board foot.

1. 45 pc. $2'' \times 8'' \times 16'$.
2. 96 pc. $2'' \times 10'' \times 14'$.
3. 144 pc. $2'' \times 4'' \times 10'$.
4. 120 pc. $1'' \times 8'' \times 16'$.
5. 88 pc. $1'' \times 6'' \times 16'$.
6. 260 pc. $1'' \times 3'' \times 9'$.
7. 67 pc. $1'' \times 8'' \times 20'$.
8. 55 pc. $1'' \times 14'' \times 14'$.
9. 33 pc. $2'' \times 8'' \times 20'$
10. 6 pc. $1'' \times 14'' \times 14'$.
11. 16 pc. $2'' \times 4'' \times 12'$.
12. 45 pc. $1'' \times 8'' \times 18'$.
13. 100 pc. $2'' \times 10'' \times 10'$.
14. 66 pc. $2'' \times 8'' \times 14'$.
15. 85 pc. $1'' \times 8'' \times 12'$.
16. 40 pc. $1'' \times 4'' \times 10'$.
17. 72 pc. $2'' \times 3'' \times 14'$.
18. 33 pc. $4'' \times 4'' \times 16'$.
19. 45 pc. $1'' \times 8'' \times 18'$
20. 42 pc. $2'' \times 10'' \times 24'$.

21. How many board feet of lumber are there in 56 posts $4'' \times 4'' \times 7'$?

22. A bridge used 24 pc. of lumber $8'' \times 12'' \times 26'$. Figure the number of board feet used.

23. A warehouse floor required

68 pc. of lumber, 2″ × 10″ × 18′,
148 pc. of lumber, 2″ × 12″ × 16′.

Find the total number of board feet.

24. How many board feet of lumber are there in a fence that contains 44 posts 4″ × 4″ × 7′, 88 pc. of lumber 2″ × 4″ × 10′, and 440 pc. of lumber 1″ × 12″ × 4′?

25. A contractor ordered the following for framing the garage shown below:

36 studs, 2″ × 4″ × 8′.
4 sills, 2″ × 4″ × 20′.
4 plates, 2″ × 4″ × 20′.
24 rafters, 2″ × 4″ × 14′.
60 pc. 1″ × 6″ × 22′ for sheathing.

PROB. 25.

Find the number of board feet required for each item and the total.

26. A truck was loaded with 440 pc. of lumber 2″ × 4″ × 20′. Find the number of board feet of lumber in the load.

27. A boxcar was loaded with 1344 pc. of lumber

2″ × 4″ × 16′.

Find the number of board feet in the carload.

EXERCISES AND PROBLEMS

In the first 10 exercises in this lesson, only even widths, as 2″, 4″, 6″, will be considered. Any odd widths will be counted as the next even-inch width.

The same principle applies to odd-foot lengths. Any odd-foot length will be figured as the next even-foot length above it.

The following thicknesses are standard: 1″, $1\frac{1}{4}$″, $1\frac{1}{2}$″, 2″, 3″, 4″, and thicker, increased by even inches. Any other thickness must be taken as the next thickness.

Before you substitute in the formula, convert the given sizes into the proper larger dimensions, then figure as usual.

1. 4 pc. $\frac{3}{4}'' \times 6'' \times 16'$.

2. 10 pc. $1\frac{1}{8}'' \times 8'' \times 14'$.

3. 8 pc. $\frac{3}{4}'' \times 7'' \times 12'$.

4. 15 pc. $1'' \times 7\frac{5}{8}'' \times 16'$.

5. 12 pc. $2'' \times 4'' \times 15'$.

6. 24 pc. $1\frac{1}{4}'' \times 3\frac{3}{4}'' \times 12'$.

7. 12 pc. $\frac{5}{8}'' \times 3\frac{5}{8}'' \times 9'$.

8. 22 pc. $\frac{3}{4}'' \times 3\frac{1}{2}'' \times 24'$.

9. 16 pc. $1\frac{1}{4}'' \times 4'' \times 17'$.

10. 14 pc. $1\frac{1}{8}'' \times 4\frac{1}{2}'' \times 15'$.

11. Find the number of board feet in the following bill if all inch dimensions are to be net. Net means exact finished size. (Take each answer to the nearest whole board foot.)

6 pc. $\frac{5}{8}'' \times 3\frac{3}{8}'' \times 4'$.
12 pc. $\frac{3}{8}'' \times 7\frac{5}{8}'' \times 12'$.
19 pc. $1\frac{1}{8}'' \times 7\frac{1}{2}'' \times 10'$.

12. Find the number of board feet in the following bill. (Take each answer to the nearest whole board foot.)

6 pc. 1″ net $\times$ 5″ $\times$ 14′.
10 pc. $1\frac{1}{4}$″ net $\times$ 4″ $\times$ 12′.
11 pc. $1\frac{1}{2}$″ net $\times$ $5\frac{1}{2}$″ net $\times$ 10′.

13. In the following bill, thickness, width, and length are net. (Find the answer to the nearest whole board foot.)

15 pc. $\frac{3}{4}'' \times 9\frac{1}{4}'' \times 11'$.
21 pc. $1\frac{5}{8}'' \times 6\frac{1}{4}'' \times 9'$.
12 pc. $\frac{3}{8}'' \times 7\frac{1}{2}'' \times 6'8''$.

In the following exercises involving expensive hardwoods, all dimensions are to be considered standard. None of the previous rules in this lesson applies. Use the exact figures given, and get the answer to the nearest hundredth of a board foot.

In this case, use the formula

$$\text{Bd. ft.} = \frac{t'' \times w'' \times 1''}{12 \times 12}$$

14. 4 pc. $\frac{3}{4}'' \times 8'' \times 36''$.
15. 3 pc. $2'' \times 3'' \times 18''$.
16. 10 pc. $\frac{7}{8}'' \times 5'' \times 32''$.
17. 6 pc. $1\frac{1}{2}'' \times 4'' \times 6'$.
18. 4 pc. $2\frac{1}{2}'' \times 2\frac{1}{2}'' \times 42''$.
19. 3 pc. $\frac{7}{8}'' \times 3'' \times 17''$.
20. 10 pc. $\frac{13}{16}'' \times 9'' \times 56''$.
21. 4 pc. $1\frac{1}{2}'' \times 2\frac{1}{2}'' \times 30''$.

LUMBER COST

Except in small amounts, as in school shopwork, lumber is always quoted at the price per 1000 bd. ft., as $144.00 per M means $144.00 per 1000 bd. ft. (M is the Roman numeral for 1000.) In lumber measure, "per foot" means per board foot.

It is evident that $144.00 per M is the same as $14.40 per hundred, and $0.144 per foot, or 14.4 cents per foot.

To find the total cost of lumber, figure the number of board feet; then multiply by the unit price selected.

Example.—Determine the cost of 14 pc. $2'' \times 6'' \times 16'$ at $144.00 per M.

$$\text{Bd. ft.} = \frac{\text{pc.} \times t'' \times w'' \times 1'}{12} \qquad \text{Formula}$$

$$= \frac{14 \times 2 \times 6 \times 16}{12} = 224$$

$$= 224 \text{ bd. ft.}$$

Solution 1.
224 bd. ft. = .224 thousand board feet
.224 × $144 = $32.256, or $32.26

Solution 2.
224 bd. ft. = 2.24 hundred board feet
2.24 × $14.40 = $32.256, or $32.26

Solution 3.
224 bd. ft. = 224 board feet
224 × $0.144 = $32.256, or $32.26

Ans. $32.26

EXERCISES AND PROBLEMS

In the following exercises, find the cost. Consider any fraction of a board foot as a whole board foot.

1. 20 pc. 1″ × 10″ × 10′ @ $156.00 per M.
2. 16 pc. 2″ × 8″ × 16′ @ $166.00 per M.
3. 41 pc. 3″ × 8″ × 14′ @ $12.60 per hundred.
4. 24 pc. 2″ × 4″ × 20′ @ $14.80 per hundred.
5. 32 pc. 4″ × 6″ × 12′ @ $0.15 per foot.
6. 200 pc. 2″ × 18″ × 20′ @ $0.27 per foot.
7. 44 pc. 2″ × 3″ × 16′ @ $14.25 per hundred.
8. 16 pc. 1″ × 5″ × 14′ @ $168.00 per M.
9. 22 pc. 1″ × 7″ × 12′ @ $0.096 per foot.
10. 24 pc. 3″ × 3″ × 16′ @ $0.09½ per foot.
11. 77 pc. 3″ × 6″ × 14′ @ $145.00 per M.
12. 14 pc. 1″ × 10″ × 12′ @ $164.00 per M.
13. 6 pc. 1″ × 8″ × 16′ @ $0.16 per foot.
14. 20 pc. 2″ × 4″ × 8′ @ $116.00 per M.
15. 12 pc. 1″ × 12″ × 12′ @ $12.20 per hundred.
16. 10 pc. 10″ × 10″ × 10′ @ $14.60 per hundred.
17. 25 pc. 2″ × 6″ × 20′ @ $187.50 per M.
18. 33 pc. 2″ × 8″ × 16′ @ $170.00 per M.
19. 100 pc. ½″ × 3″ × 8′ @ $14.80 per hundred.
20. 48 pc. 6″ × 6″ × 16′ @ $85.00 per M.
21. 6 pc. 12″ × 12″ × 12′ @ $159.00 per M.
22. 24 pc. 10″ × 10″ × 16′ @ $0.185 per foot.
23. 50 pc. 2″ × 2″ × 18′ @ $0.13 per foot.
24. 45 pc. 2″ × 8″ × 14′ @ $0.166 per foot.

On page 336 is a typical bill of lumber as made out for the yardman to load for delivery. The terms "clear," "No. 1 common," "No. 2 common," "D select," etc., refer to a grade or quality of lumber. Abbreviations for mountain pine, yellow pine, Oregon pine, redwood, Douglas fir, cypress, sugar pine are easy to recognize. These all refer to the kind of wood. Finishes are rough: S1S1E (surfaced one side one edge); S4S (surfaced four sides); S1S2E (surfaced one side and two edges). V.G. means vertical grain. T & G means "tongued and grooved." Flg. means flooring.

Hayward Lumber & Investment Co.
GEN. OFFICE & LOS ANGELES YARD
410 SAN FERNANDO ROAD
P.O. Box 155-LOS ANGELES, CALIF.

LUMBER
SASH & DOORS
HARDWARE
PAINT
PIPE, FITTINGS
FENCING
BOLTS
CORRUGATED IRON
GLASS
MILL-WORK

______________ 19 __

Name ______________
Address ______________
Instructions ______________

Lin. Feet	PIECES	SIZE	Length	DESCRIPTION	Feet B. M.	PRICE		EXTENSION	
	32	2″×4″	8′	No. 1 O.P. S4S		$144	00		
	20	2″×4″	14′	No. 2 O.P. S1S1E		128	00		
	22	2″×6″	12′	No. 2 M.P. S2E		116	00		
	14	1″×12″	12′	Clear W.P. S4S.		95	00		
	320	1″×6″	16′	No. 3 M.P. rough		88	00		
	45	1″×8″	12′	No. 2 RW. S4S		182	00		
	21	1″×4″	14′	No. 3 V.G. D.F. T&G. Flg.		204	00		
	120	2″×4″	18′	No. 2 O.P. S1S1E		110	00		
	36	1″×10″	10′	D Sel. M.P.		244	00		
	21	2″×6″	14′	No. 3 RW. rough		174	00		
	66	1″×10″	16′	Clear O.P.		246	00		
	55	2″×8″	18′	No. 1 M.P. rough.		88	00		
	84	½″×4″	8′	Clear M.P. bits.		88	00		
	42	1″×5″	7′	Clear O.P. Casing		220	00		

We agree to furnish the items listed above for the sum of $__________ F. O. B. __________
provided this estimate is accepted within__________ from this date.
Date__________ 19 ____ By__________

Figure the number of board feet for each item above and the cost. Find the total bill. Find the net bill, if 5% discount is allowed for cash. Do not write in this book.

FLOORING AND CEILING LUMBER

The lumber used for flooring and ceiling usually has a **tongued and grooved,** or matched, joint as shown. The board footage for such lumber is figured from the rough lumber sizes. So-called 1″ × 4″ flooring is reduced to a $3\frac{1}{4}''$ net width to make the tongue and groove.

Since nearly one-fourth of the width is used in matching, it is customary to add 25 per cent of the area figured when

1″ × 4″ flooring is used. Lumber 2″ wide makes less than $1\frac{1}{2}$″ flooring, and about one-third of the area is usually added. One-inch flooring is $\frac{13}{16}$″ to $\frac{7}{8}$″ thick. For cutting

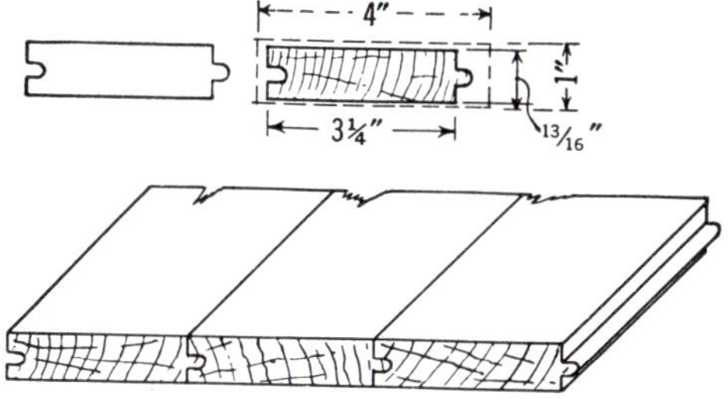

and waste, an additional allowance of 2 to 3 per cent is often made.

Example.—Find the number of board feet of 1″ × 4″ flooring needed and the cost at $210.00 per M for a room 12 ft. wide and 18 ft. long, allowing 25% for matching and 3% for waste.

12 × 18 = 216, or 216 sq. ft. in the area	Finding the number of square feet in the room
.25 + .03 = .28	Allowance for matching and waste
.28 × 216 = 60.48, or 61 sq. ft. 216 sq. ft. + 61 sq. ft. = 277 sq. ft.	Total number of square feet required
277 × 1 = 277, or 277 bd. ft.	Since the flooring is 1″ thick, multiplying 277 × 1 gives the number of board feet required
277 bd. ft. = .277 M bd. ft. .277 × 210 = 58.170 *Ans.* $58.17	The cost

PROBLEMS

1. Allowing 25% for matching and waste, find the number of board feet and the cost of 1″ × 4″ flooring needed for a room 16′ by 22′. The price of the lumber is $255.00 per M bd. ft.

2. A shop floor is 50′ by 60′. It is to be covered with 1″ × 6″ matched flooring, which is priced at $224.00 per M. Find the number of board feet and the price if 20% is allowed for matching and waste.

3. The ceiling of a porch is 8′ by 48′. It is to be "ceiled" with $\frac{5}{8}$″ matched material, listed at $148.00 per M. Allowing 25% for waste, what is the cost?

4. The shop plan shown is to be floored with $\frac{7}{8}$″ × 2″ matched maple flooring, priced at $378.00 per M. Allowing one-third the surface area for matching and waste, what are the footage and cost?

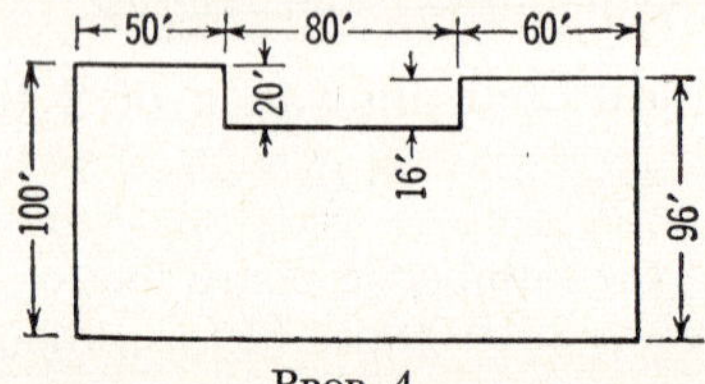

Prob. 4.

5. In the house plan shown, all the rooms except the kitchen, porch, and bath are floored with $\frac{13}{16}$″ × 2″ oak, listed at $245.00 per M. Allow one-third for matching the oak.

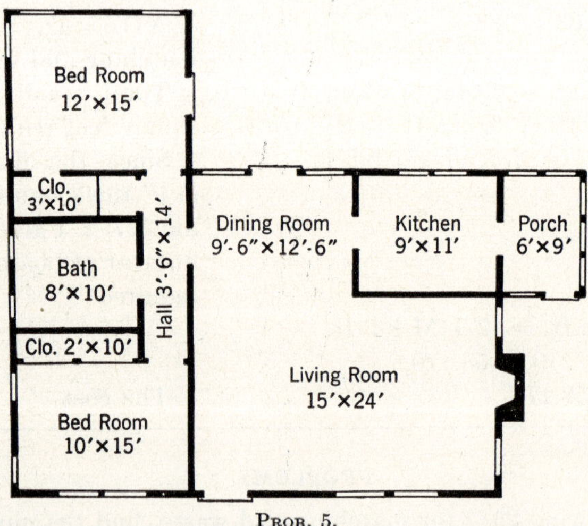

Prob. 5.

The bath, kitchen, and porch are floored with 1″ × 4″ V.G. Oregon pine, listed at $196.00 per M. Allow 20% for matching the pine. Find the number of board feet of each kind of lumber required and the total cost.

SIDING LUMBER

The finished lumber used to cover the exterior walls of a building is designated as **siding.** Several varieties

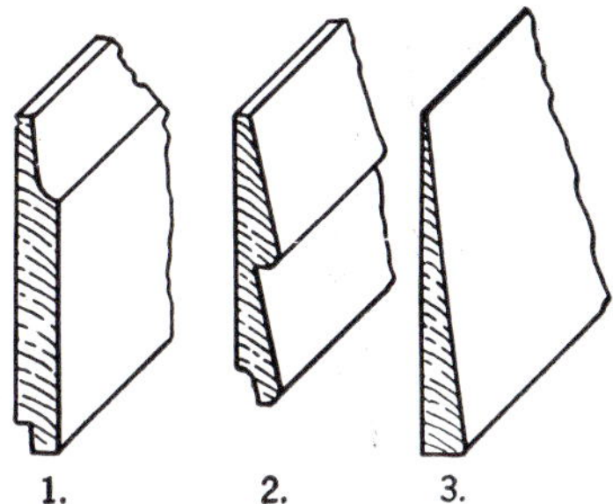

of matched siding are shown. Figure 1 is called channel siding or channel rustic. It is made in widths of 6″ to

House construction using 1″ × 10″ channel rustic.

12″. Figure 2 is called novelty siding. Figure 3 is called clapboard siding. There are many other varieties.

Because the siding is matched, an allowance of 10 to 40 per cent must be added to the number of square feet to be covered.

It is common practice to deduct only one-half the area of the openings. This deduction is made because of the waste in cutting around the openings.

Example.—Find the number of board feet of 1″ × 12″ channel rustic needed to cover a wall 12 ft. high and 42 ft. long, if there are 60 sq. ft. of openings and 20% is added for waste and matching. Figure the cost at $298.00 per M.

12 × 42 = 504, or 504 sq. ft.	Finding the wall area
½ of 60 sq. ft. = 30 sq. ft.	Deduction for openings
504 sq. ft. − 30 sq. ft. = 474 sq. ft.	Area to be covered
474 sq. ft. × .20 = 94.8, or 95 sq. ft.	Allowance for matching
474 sq. ft. + 95 sq. ft. = 569 sq. ft.	Total number of square feet needed
569 sq. ft. = 569 bd. ft.	Total number of board feet needed
569 bd. ft. = .569 M bd. ft.	
.569 × 298 = $169.56	Cost of lumber

PROBLEMS

Deduct only one-half the area of openings.

1. Find the number of board feet of 1″ × 6″ novelty rustic needed to cover a building 10 ft. high that has a perimeter of 172 ft. Allow 25% for matching and 120 sq. ft. for openings.

2. Compute the cost of the siding in Prob. 1 at $166.00 per M.

3. How much siding is needed to cover the exterior of a rectangular building 32 ft. wide and 45 ft. long if the walls are 9 ft. high? There are 100 sq. ft. of openings. Allow 20% for matching and waste.

4. Find the cost of the siding in Prob. 3 at $172.00 per M.

5. A wall is 10 ft. high and 80 ft. long. There are four windows, each 4 ft. by 6 ft. Figure the amount of siding needed for the wall if 20% is allowed for matching and waste.

6. The perimeter of a building is 182′. The wall to be sided is 8′6″ high. There are 10 windows, each 4′ by $4\frac{1}{2}$′, and 3 doors, each 3′ by 7′, in the walls. Find the amount of siding needed if 25% is allowed for matching and waste.

7. A building is 36 ft. wide and 46 ft. long. It is to be covered 9 ft. high with 1″ × 10″ clapboard siding. There are 60 sq. ft. of openings, and 10% is to be added for matching and waste. How many board feet are needed?

8. Find the cost of the siding in Prob. 7 at $187.50 per M.

9. A residence is to be sided with 1″ × 8″ channel rustic. The house is 36′6″ wide and 44′6″ long and will be covered to a height of 9′6″. There are eight windows, each 3′ by $4\frac{1}{2}$′; four windows, each 2′ by 3′; 1 door, $3\frac{1}{2}$′ by 7′; and two doors, each $2\frac{1}{2}$′ by 7′. Allowing 15% for matching and waste, calculate the number of board feet needed.

10. Find the cost of the rustic in Prob. 9 at $214.00 per M.

SHINGLES

Shingles are used for roof covering and wall siding. Wood shingles are usually figured as being 4 in. wide and 16 or 18 in. long. Thickness is listed as "five to 2 in.," "five to $2\frac{1}{4}$ in.," etc., meaning that the butt, or thick, end requires five shingles to make the thickness stated.

Wood shingles are estimated by the thousand and are put up in bundles of 250 shingles per bundle. The shingles vary from 4 to 12 in. in width, and a bundle will cover

1000 in. of width. Wide shingles are usually split so as to be less than 6 in. wide.

Asphalt shingles are 9 in. wide and $12\frac{1}{2}$ in. long and are packed in bundles to cover 25 sq. ft. when exposed 4 in. to the weather.

Asphalt "4-in-1" strip shingles, shown below, and hexagonal strip shingles are packed in bundles to cover 50 sq. ft. when exposed 4 in. to the weather.

Asphalt "4-in-1" strip shingles. *(Courtesy of Johns-Manville Company.)*

Asbestos shingles vary in size from 8 by 16 in. to strips 10 by 36 in.

Asbestos siding shingles are 12 in. high and 24 in. wide and are packed 19 to the bundle. Three bundles cover one square (an area of 100 sq. ft.). Johns-Manville asbestos siding shingles are shown on the opposite page.

Asphalt roll roofing is put up in rolls 32 in. and 36 in. wide and long enough to cover 100 sq. ft. of surface, allowing 1 in. on each side for lapping. Lightweight rolls weigh 35 lb. and the heaviest weigh up to 90 lb. per roll. The heaviest grades are also crated in flat sheets 32 by 80 in., with 6 sheets to cover 100 sq. ft. Crates contain 24 or 30 sheets, depending on the weight.

Asbestos siding shingles. *(Courtesy of Johns-Manville-Company.)*

Wood shingles are **exposed** to the weather 4, $4\frac{1}{2}$, or 5 in., depending on the pitch, or slope, of the roof. Steeper roofs have greater exposure.

The figure on page 341 shows how wood shingles are laid on a roof.

The **square**, (100 sq. ft.) is used in estimating the shingles needed for a job. The table below gives the number of shingles needed according to the exposure.

Amount of shingles exposed, inches	Square feet covered by 1000 shingles	Number of shingles needed for one square (100 sq. ft.)
5	138	720
$4\frac{1}{2}$	125	800
4	111	900

In estimating the quantity of shingles needed, any portion of a bundle is considered as a whole bundle.

Example 1.—Find the number of bundles of wood shingles needed and the cost at $15.80 per M to cover a roof 28 by 42 ft. with the shingles exposed $4\frac{1}{2}$ in.

28 × 42 = 1176, or 1176 sq. ft.	Find the area.
1176 sq. ft. = 11.76 squares	100 sq. ft. = 1 square.
11.76 × 800 = 9408, the number of shingles needed	
9408 ÷ 1000 = 9.408 M, or 9.5 M	Number of shingles to buy.
9500 ÷ 250 = 38, or 38 bundles	9500 shingles divided by the number of shingles per bundle gives the number of bundles to buy.
9.5 × $15.80 = $150.10	Number of M shingles times cost per M equals total cost.

Example 2.—Find the number of bundles of 4-in-1 asphalt strip shingles and the cost at $2.45 per bundle to cover a roof 36 by 54 ft. if a bundle covers 50 sq. ft. when exposed 4 in. to the weather.

36 × 54 = 1944, or 1944 sq. ft.	Area of the roof.
1944 ÷ 50 = 38.8, or 39 bundles	Number of bundles of shingles required.
39 × $2.45 = $95.55, total cost	

PROBLEMS

Remember that any part of a bundle is figured as a whole bundle.

1. A roof is 32 by 48 ft. It is to be covered with wood shingles, exposed 5 in. to the weather. Five-to-$2\frac{1}{4}$-in. shingles costing $16.25 per M are to be used. Find the number of bundles needed and the cost.

2. How many bundles of 4-in-1 asphalt strip shingles will be needed to cover the roof in Prob. 1 if a bundle covers 50 sq. ft. when exposed 4 in. to the weather?

3. Compute the cost of the shingles in Prob. 2 at $2.95 per bundle.

4. A roof having 4620 sq. ft. is to be shingled with 9- by 12½-in. asphalt shingles. If each bundle covers 25 sq. ft., calculate the number of bundles needed.

5. Find the cost of the shingles in Prob. 4 at $2.79 per bundle.

6. Find the number of bundles of 4-in-1 strip shingles needed to cover the roof of Prob. 4 if a bundle covers 50 sq. ft.

7. What is the cost of the shingles in Prob. 6 at $3.55 per bundle?

8. Find the cost of covering the roof in Prob. 4 with wood shingles exposed 4 in. to the weather if the price is $15.80 per M.

9. On the house shown, determine the cost of the shingles with

PROB. 8.

a. Wood shingles exposed 5 in. to the weather at $16.75 per M.

b. 4-in-1 asphalt strip shingles at $2.95 per bundle.

c. 9- by 12½-in. asphalt shingles at $1.85 per bundle.

10. What is the difference between the highest and the lowest cost in Prob. 9?

11. What per cent is saved if the lowest-cost shingles are used?

CHAPTER 14

PROTECTIVE COATINGS

To protect outside surfaces of buildings from corrosion and decay and for decorative effects, both inside and outside, a good quality of paint should be applied. The right paint for the purpose should be selected and applied according to the manufacturers' specifications.

The rate at which a definite quantity of paint will cover a certain area is called the **spreading rate.** The area depends on the consistency of the paint and the roughness and porosity of the surface.

The following table, gives the spreading rate for the most common products.

Spreading Rate of Paints*

Kind of paint	Surface condition	1-gal. coverage, square feet		
		1 coat	2 coats	3 coats
Oil paints	Wood, smooth	600	325	225
	Wood, rough	350	200	150
	Metal	700	350	250
	Soft brick, concrete	300	175	125
	Cement stucco	150	75	
	Cement floors	250	150	
	Plaster, smooth	300	175	125
	Plaster, rough	250	150	
Varnish	Interior wood and linoleum	450	250	175
Shingle paint		125	75	
Asphalt roof paint	Rough surface	150	100	
	Smooth surface	250		
Calcimine, 5 lb.	Plaster	400		

* *U.S. Department of Agriculture Bulletin* 1452.

It is common practice to deduct only one-half the area of openings. Windows, doors, and frames are figured separately.

Paint is put up in quantities varying from quarts to gallons. Special decorative paints and enamels may also be purchased in $\frac{1}{8}$-, $\frac{1}{4}$-, $\frac{1}{2}$-pint and pint sizes. If an estimate shows that $2\frac{1}{3}$ qt. are needed, for example, and the paint is available in quarts, 3 qt. must be bought. If the paint can be purchased only in gallons, it is necessary to buy 1 gal.

The quantity of paint needed is estimated as follows:

Rule. To find the quantity of paint needed to cover a given surface, divide the total area by the spreading rate.

Example 1.—Determine the amount of paint and the total cost at $6.25 per gallon to apply two coats to a concrete wall 12 ft. high and 146 ft. long. There are four windows, each 3 by 5 ft., and two doors, each 3 by 7 ft.

$146 \times 12 = 1752$	Number of square feet of area
$4 \times 3 \times 5 = 60$	Number of square feet of window area
$2 \times 3 \times 7 = 42$	Number of square feet of door area
$60 + 42 = 102$	Number of square feet of openings
$102 \div 2 = 51$	Number of square feet to be deducted for openings
$1752 - 51 = 1701$	Number of square feet to be covered

175 sq. ft. per gal. is the spreading rate

$1701 \div 175 = 9.7$, or 10 gal. needed for 2 coats

$\$6.25 \times 10 = \$62.50$, total cost of the paint

Example 2.—Determine the amount of paint and the cost at $5.40 per gallon to apply two coats of oil paint to the two sides

$50 + 150 + 150 = 350$	Number of linear feet of wall
$350 \times 16 = 5600$	Number of square feet of area
$24 \times 5 \times 6 = 720$	Number of square feet of openings
$720 \div 2 = 360$	Number of square feet to be deducted for openings
$5600 - 360 = 5240$	Number of square feet to be covered

200 sq. ft. per gallon is the spreading rate

$5240 \div 200 = 26.2$, or 27 gal. needed for 2 coats

$\$5.40 \times 27 = \$145.80$, total cost of the paint

and rear of a rough wood building 16 ft. high, 50 ft. wide, and 150 ft. long. There are 24 openings, each 5 by 6 ft.

Example 3.—Find the cost of paint for two coats on smooth plaster in a room 12 ft. by 15 ft. and $8\frac{1}{2}$ ft. high. There are three windows, each 3 by 5 ft., and two doors, each 3 by 7 ft. The paint is sold in gallons at $5.20 and in quarts at $1.69.

$12 + 12 + 15 + 15 = 54$	Number of feet in perimeter
$54 \times 8.5 = 459$	Number of square feet of wall area
$15 \times 12 = 180$	Number of square feet of ceiling area
$459 + 180 = 639$	Total number of square feet of area
$3 \times 3 \times 5 = 45$	Number of square feet of windows
$2 \times 3 \times 7 = 42$	Number of square feet of doors
$45 + 42 = 87$	Total number of square feet of openings
$87 \div 2 = 43.5$, or 44 sq. ft.	To be deducted for openings
$639 - 44 = 595$	Number of square feet to be covered
$595 \div 175 = 3.4$	Number of gallons needed
$.4 \times 4 = 1.6$, or 2 qt.	Number of quarts in 0.4 gal.

Hence, 3 gal. and 2 qt. are needed.

3 × $5.20 = $15.60

2 × $1.69 = $ 3.38

= $18.98, total cost of the paint

PROBLEMS

See the table, page 346, for spreading rate.

1. How many gallons of paint are needed and what is the cost at $5.15 per gallon for two coats of paint on both sides of a rough board fence 6 ft. high and 466 ft. long?

2. How many gallons of varnish are needed, and what is the cost at $4.25 per gallon, to cover a gymnasium floor 66 ft. wide and 96 ft. long. Three coats are to be applied.

3. A sign painter is to spread two coats of paint in preparation for advertising signs on the two sides and the rear end of a brick building 21 ft. high, 50 ft. wide, and 92 ft. long. If no allowance is made for openings, how many gallons are needed? What is the cost at $3.88 per gallon?

4. The cement floor of a storage garage 48 by 148 ft. is to receive one coat of quick-drying floor paint. How many gallons are needed, and what is the cost at $6.65 per gallon?

5. Asphalt paint is to be applied to a rough flat roof 60 ft. wide and 154 ft. long. How much paint is needed for one coat? What is the cost at $2.25 per gallon?

6. How many gallons of shingle paint are needed for one coat on a shed roof with a slant height of 19 ft. and a length of 42 ft.? Find the cost at $3.52 per gallon.

7. Find the cost for two coats of oil paint for the smooth-plastered walls and ceiling of a room 17 ft. by 25 ft. and 9 ft. high. Paint at $5.68 per gallon and $1.71 per quart is to be used. There are three windows, each 4 by 5 ft., and two doors, each 3 by 7 ft.

8. The interior of a 5-story office building is 46 ft. wide and 97 ft. long. If no allowance is made for partitions, how many gallons of linoleum varnish are needed for one coat to be applied on each of the floors? Figure the cost at $4.66 per gallon.

9. What will be the cost of one coat of calcimine at 95¢ for a 5-lb. package to be applied to the interior of a storeroom 32 ft. wide, 68 ft. long, and 14 ft. high? No allowance is to be made for openings.

10. A concrete retaining wall is 6 ft. high and 596 ft. long. Find the amount of paint needed for a 2-coat job. Estimate the cost at $3.88 per gallon.

11. A sheet-metal warehouse with a flat, rough asphalt roof is 44 ft. wide, 100 ft. long, and 24 ft. high. It is to receive two coats of paint all over on the outside. The doors are metal and will be painted. There are five windows, each 4 by 4 ft., on each side. Oil paint costing $4.65 per gallon is to be used for the walls, and asphalt paint costing $2.15 per gallon will be applied to the roof. Find the amount of each kind of paint needed and the total cost.

CHAPTER 15

PULLEY SPEEDS AND DIAMETERS

RULES AND FORMULAS

The mechanic is often required to solve problems dealing with the transmission of power by means of belts over pulleys, chain drives, rope drives, and gears in mesh with each other.

There are definite mathematical relationships governing the transmission of power by these means. Rules and formulas have been derived that explain how unknown quantities can be determined.

When the speed of one of two pulleys connected by a belt is known, the speed of the other pulley can be determined by the following rule:

Rule 1.—Diameters of pulleys are inversely proportional to their r.p.m.

D = diameter of large pulley.
d = diameter of small pulley.
R = r.p.m. of large pulley.
r = r.p.m. of small pulley.

$D:d = r:R$ (the inverse proportion as stated by Rule 1).

A formula found by multiplying the means and the extremes of the proportion above is as follows:

Formula. $DR = dr$.

A rule expressing this formula is as follows:

Rule 2.—When two pulleys are connected by a belt, the diameter of the large pulley multiplied by its r.p.m. equals the diameter of the small pulley multiplied by its r.p.m.

Formula, $DR = dr$, solved for each term gives

Formulas.

$$D = \frac{dr}{R}, \qquad R = \frac{dr}{D}, \qquad d = \frac{DR}{r}, \qquad r = \frac{DR}{d}$$

Example 1.—A 12″ diameter pulley turning 100 r.p.m. is driving a 6″ diameter pulley. Get the r.p.m. of the 6″ driven pulley.

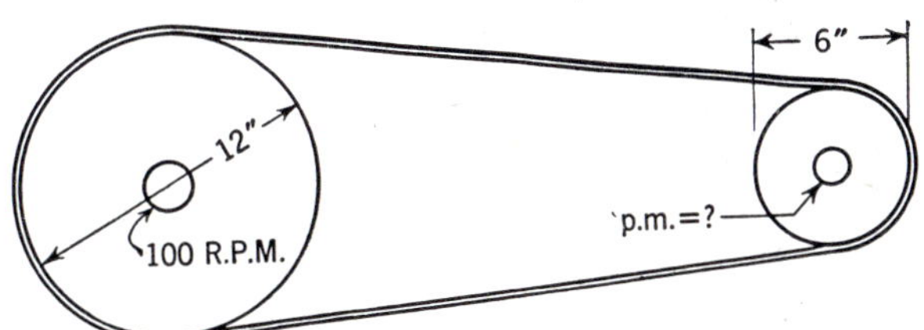

$r = \frac{DR}{d}$	Formula for r.p.m. of small pulley
$r = \frac{12 \times 100}{6}$	Substituting
$r = 200$ r.p.m. of small pulley	

It is evident in Example 1 that the following are the solutions for the other values:

$D = \frac{dr}{R}$,	$R = \frac{dr}{D}$,	$d = \frac{DR}{r}$
$D = \frac{6 \times 200}{100}$,	$R = \frac{6 \times 200}{12}$,	$d = \frac{12 \times 100}{200}$
$D = 12$ in. diameter	$R = 100$ R.P.M.,	$d = 6$ in. diameter

The solution of these examples by the inverse proportion $D:d = r:R$, as given in Rule 1, is as follows:

D unknown	R unknown	r unknown	d unknown
$D:6 = 200:100$	$12:6 = 200:R$	$12:6 = r:100$	$12:d = 200:100$
$100D = 1200$	$12R = 1200$	$6r = 1200$	$200d = 1200$
$D = 12''$	$R = 100$	$r = 200$	$d = 6''$

PROBLEMS

Get r.p.m. to nearest whole number.
Get diameter to two decimals.

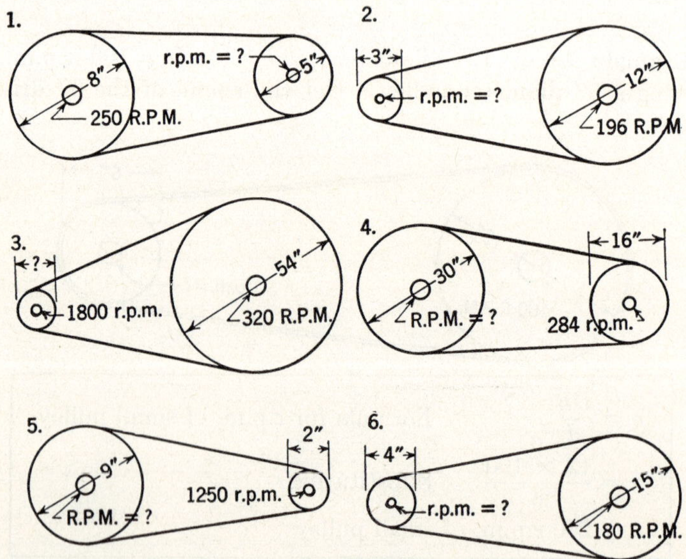

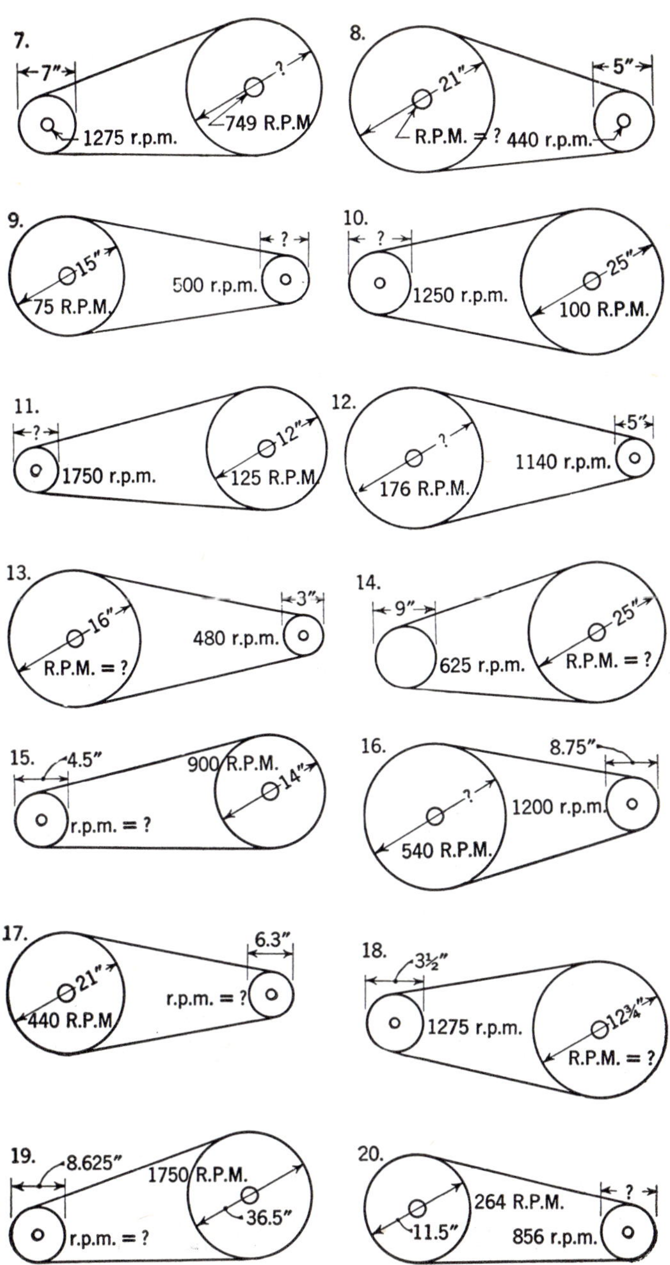
7.
7″
?
1275 r.p.m.
749 R.P.M
8.
21″
5″
R.P.M. = ?
440 r.p.m.
9.
15″
?
500 r.p.m.
75 R.P.M.
10.
?
25″
1250 r.p.m.
100 R.P.M.
11.
?
12″
1750 r.p.m.
125 R.P.M.
12.
?
5″
1140 r.p.m.
176 R.P.M.
13.
16″
3″
480 r.p.m.
R.P.M. = ?
14.
9″
25″
625 r.p.m.
R.P.M. = ?
15.
4.5″
900 R.P.M.
14″
r.p.m. = ?
16.
8.75″
?
1200 r.p.m.
540 R.P.M.
17.
6.3″
21″
r.p.m. = ?
440 R.P.M.
18.
3½″
1275 r.p.m.
12¾″
R.P.M. = ?
19.
8.625″
1750 R.P.M.
36.5″
r.p.m. = ?
20.
?
264 R.P.M.
11.5″
856 r.p.m.

PROBLEMS

Draw a figure to illustrate each of these problems, give the formula, and solve.

1. A 24″ pulley at 224 r.p.m. will drive a 9″ pulley at what r.p.m.?

2. The motor on a table saw runs at 1250 r.p.m. It has a 6″ driver pulley on it. At how many r.p.m. does the saw revolve if it carries a 5″ pulley solid to the saw arbor?

3. A line shaft runs at 196 r.p.m. It is driven by a motor that runs at 1200 r.p.m. The motor pulley is 4″ in diameter. Get the diameter of the line-shaft driven pulley.

4. A driving pulley has a diameter of 15″, and its speed is 240 r.p.m. Find the speed of a 9″ driven pulley.

5. A circular saw with a 5″ pulley must run at 1250 r.p.m. The countershaft, which acts as the driver, runs at 200 r.p.m. How large is countershaft driver pulley?

6. The nibbler shown has a 12″ pulley that must run at 450 r.p.m. The motor is running at 1800 r.p.m. What diameter of driver pulley is needed on the motor?

Prob. 6.

7. A jackshaft running at 252 r.p.m. has a 4″ driver pulley connected with a 22″ pulley on a grindstone. What is the r.p.m. of the grind stone? If the belt slippage is 3%, what is the r.p.m.?

8. An engine runs at 92 r.p.m. and has a pulley of 4′ diameter. It drives a 30″ pulley on a jackshaft at what r.p.m.?

9. A motor runs at 1800 r.p.m. What diameter of pulley will be needed to run a countershaft with a 24″ pulley at 210 r.p.m.? Find the answer to the nearest $\frac{1}{4}$″. (Carry out to three decimals and change to fourths, as instructed on page 72.)

10. The 4″ motor pulley on a table saw revolves at 3750 r.p.m. Find the speed of the saw if its pulley is 7″ diameter.

11. Find the diameter to the nearest $\frac{1}{8}$″ of a pulley on a motor that revolves at 1750 r.p.m. to drive a 12″ diameter pulley at 350 r.p.m.

12. The step-cone pulleys on the spindle of the drill press shown are 4″, $5\frac{1}{2}$″, 7″, and $8\frac{1}{2}$″. The motor-cone pulleys are 2″, $3\frac{1}{2}$″, 5″, and $6\frac{1}{2}$″. Find the different speeds of the spindle if the motor revolves at 1750 r.p.m.

PROB. 12.

13. On a step-cone pulley of a lathe the diameters are 6″, 7″, and 8″, respectively. Arranged in reverse order a similar cone pulley on the countershaft revolves at 225 r.p.m. Find the speed of the lathe spindle for each position of the belt, beginning with the smallest driver.

Prob. 13.

GEAR SPEEDS AND SIZES

Gears in mesh are used in close work, instead of pulleys and belts. Gears have an advantage in that there can

be no slippage. Like pulleys, the diameters of gears can be used in figuring speeds, but the work is simplified by using the number of teeth around the periphery.

Rule 1.—**The r.p.m. multiplied by the number of teeth on the large gear equals the r.p.m. multiplied by the number of teeth on the small gear.**

R = r.p.m. of large gear.
r = r.p.m. of small gear.
T = teeth on large gear.
t = teeth on small gear.

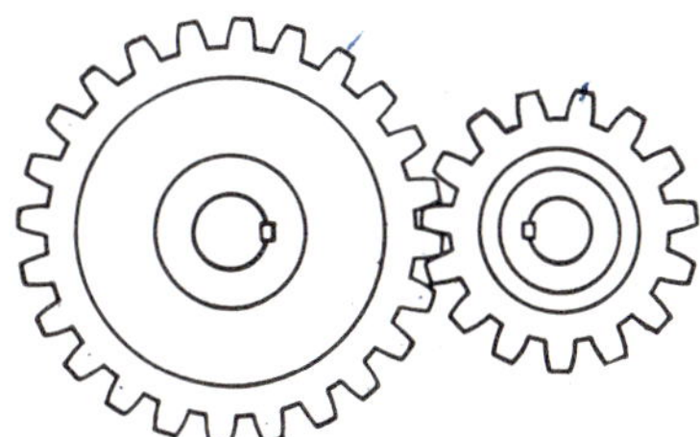

Formula. $R \times T = r \times t$, or $RT = rt$.
The formula solved for each term is as follows:

Formulas. $R = \frac{rt}{T}$, $T = \frac{rt}{R}$, $r = \frac{RT}{t}$, $t = \frac{RT}{r}$

Example.—A driving gear with 75 teeth running at 32 r.p.m. will give what r.p.m. to a driven gear with 24 teeth?

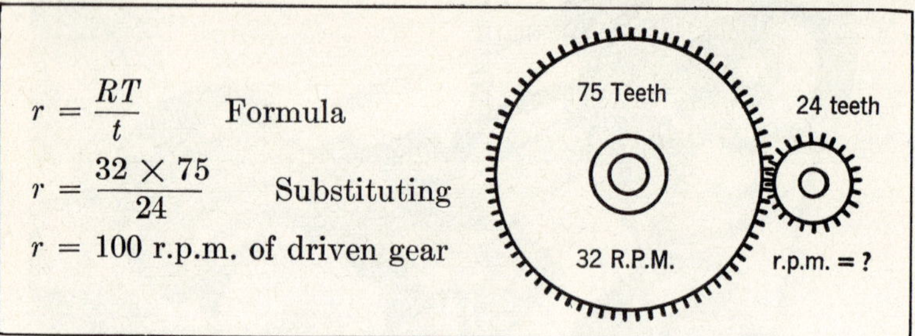

The following are the solutions for the other values:

$R = \frac{rt}{T}$,	$T = \frac{rt}{R}$,	$t = \frac{RT}{r}$
$R = \frac{100 \times 24}{75}$,	$T = \frac{100 \times 24}{32}$,	$t = \frac{32 \times 75}{100}$
$R = 32$,	$T = 75$,	$t = 24$

By inverse proportion, the following may be stated:

Rule 2.—The speeds of gears are inversely proportional to the number of their teeth.

$T:t = r:R$ (the inverse proportion as stated by Rule 2).

T unknown	t unknown	r unknown	R unknown
$T:24 = 100:32$	$75:t = 100:32$	$75:24 = r:32$	$75:24 = 100:R$
$32T = 2400$	$100t = 2400$	$24r = 2400$	$75R = 2400$
$T = 75$	$t = 24$	$r = 100$	$R = 32$

PROBLEMS

Solve the following problems, either by formula or by inverse proportion. Give each result to the nearest whole number.

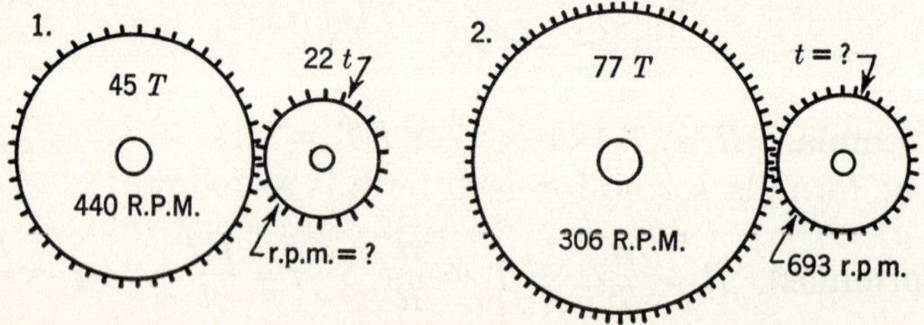

Probs. 1 and 2.

Conventional gears are shown without teeth, as in the following problems:

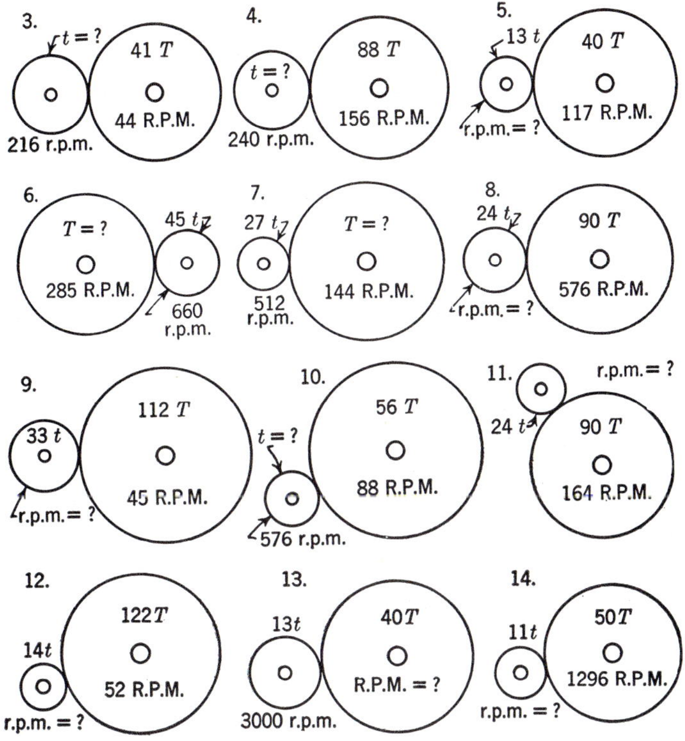

IDLERS, OR INTERMEDIATE GEARS

A slight variation in gear problems is met in the case of an idler, or intermediate gear. Such a gear is used when it is necessary to transmit power between two points that are too far apart to use only two gears of conventional size. As may be seen in the example below, the idle gear does not affect the speed of the driven gear. In fact, the teeth on the idle gear need not be known.

$$\frac{200 \times 20}{80} = 50 \text{ r.p.m. of idler,}$$

$$\frac{50 \times 80}{40} = 100 \text{ r.p.m. of driven}$$

Omitting the idler,

$$\frac{200 \times 20}{40} = 100 \text{ r.p.m. of driven}$$

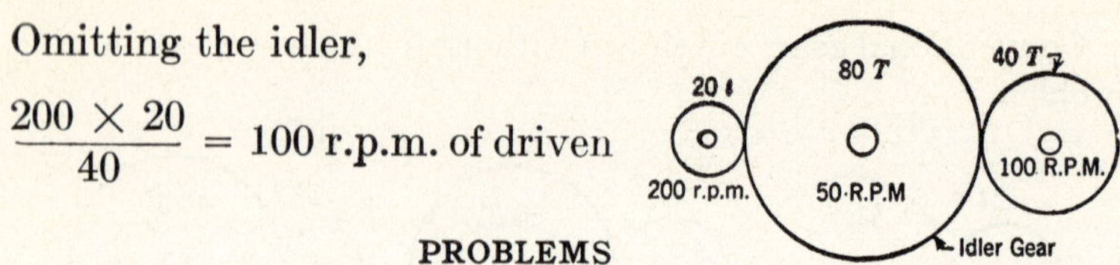

PROBLEMS

Solve the following gear problems, which include idler gears.

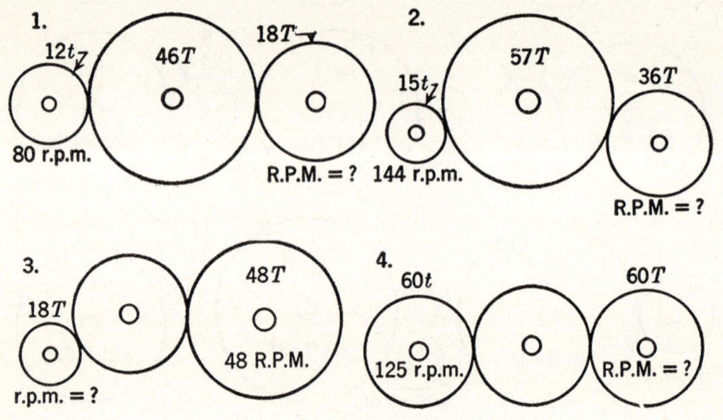

PROBLEMS

Illustrate, give the formula, and solve.

1. A gear having 96 teeth meshes with one having 27 teeth. Get the speed of the small one if the large one makes 108 r.p.m.

2. A driving gear with 16 teeth meshes with a driven gear with 75 teeth. The driving gear makes 60 r.p.m. Get the r.p.m. of the driven gear.

3. The ring gear on the axle of an automobile shown has 46 teeth; the pinion gear on the drive shaft has 13 teeth. If the drive shaft revolves at 300 r.p.m., get the r.p.m. of the axle.

PROB. 3.

4. How many teeth has a driven gear if it revolves at 108 r.p.m. and meshes with a driver gear with 16 teeth revolving at 81 r.p.m.?

5. On a pump the large gear has 40 teeth and turns at 30 r.p.m. Get the teeth on the driving gear if it turns at 100 r.p.m.

6. At how many r.p.m. would a 65-tooth gear revolve if meshed with a 22-tooth gear at 210 r.p.m.?

7. Figure the r.p.m. of a small gear with 12 teeth if it meshes with a large gear with 100 teeth at 90 r.p.m.

8. A motor running at 1260 r.p.m. has a gear with 21 teeth. It drives a drill press on which is a 105-tooth gear. Find the speed of the drill-press gear.

9. On a lawn mower the pinion gear attached to the cutter blades has 15 teeth. It meshes with an internal gear of the driving wheels that has 87 teeth. If the drive wheel revolves at 40 r.p.m., how fast do the cutter blades revolve?

10. On a hack saw the driving gear with 14 teeth is keyed to the axle of a motor that revolves at 1200 r.p.m. How many teeth are on the gear attached to the saw crank if the saw makes 56 strokes per minute?

11. A concrete mixer has an engine that revolves at 300 r.p.m. How fast does the drum revolve if it has a 108-tooth gear in mesh with a 12-tooth gear on the engine?

12. Determine the teeth needed on a gear fixed to a shaft to reduce the r.p.m. of a motor from 1250 to 250 if the motor gear has 16 teeth.

13. Chain drive is computed the same as gears in mesh. The gears are spoken of as sprockets. The illustration shows a chain drive from the crankshaft to the camshaft of an automobile. The crankshaft sprocket has 20 teeth and the camshaft sprocket has 40 teeth. Get the speed of the cam shaft when the crankshaft is running at (*a*) 2550 r.p.m.; (*b*) 4428 r.p.m.

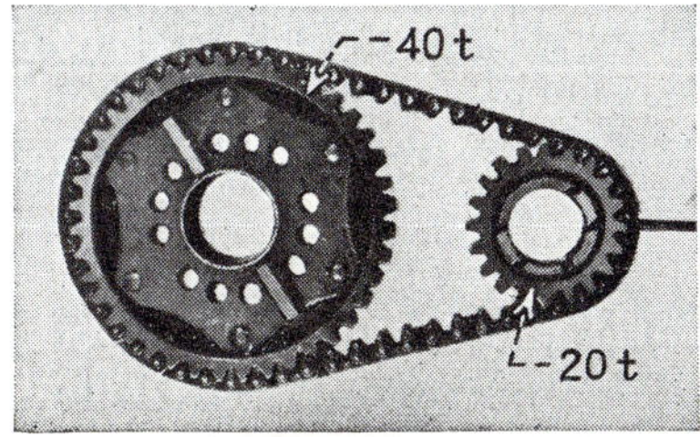

PROB. 13.

14. A gear with 56 teeth running at 700 r.p.m. meshes with an 88-tooth gear. Find the r.p.m. of the 88-tooth gear.

15. How many teeth are on a driven gear turning at 550 r.p.m. if the driver has 75 teeth and turns at 2200 r.p.m.?

16. Figure the r.p.m. of a large gear with 88 teeth if it meshes with a 21-tooth gear turning at 1760 r.p.m.?

17. The ring gear (large) on the axle of an automobile has 56 teeth. Find its r.p.m. if the pinion gear (small) on the drive shaft has 13 teeth and turns at 1000 r.p.m.

18. The sprockets on the jackshaft of a certain truck have 13 teeth. The rear-wheel sprockets have 42 teeth. Determine the r.p.m. of the rear wheels when the jackshaft revolves (*a*) at 600 r.p.m.; (*b*) at 925 r.p.m.

19. On the racing bicycle shown the pedal sprocket has 26 teeth and the rear-wheel sprocket has 10 teeth. What is the r.p.m. of the rear wheel if pedals revolve at (*a*) 20 r.p.m.? (*b*) 33 r.p.m.?

(*Courtesy of Arnold Schwinn & Company.*)
PROB. 19.

20. An electric motor running at 1750 r.p.m. has a sprocket with 15 teeth connected by chain to the sprocket with 75 teeth on a band saw. Get the r.p.m. of the band saw.

PULLEY TRAINS

When a series of pulleys are connected by belting with the power coming from one of the pulleys, it is called a pulley train. The train may serve as a means for great change of speed, or it may be used simply to transmit power at varying speeds to different shafts and machines.

The following rule and formulas may be used to determine the factors in pulley trains:

Rule.—The r.p.m. of the first small pulley multiplied by the diameters of all the small pulleys equals the r.p.m. of the last large pulley multiplied by the diameters of all the large pulleys.

r = r.p.m. of first small pulley.
d = diameter of first small pulley.
d_1 = diameter of second small pulley.
d_2 = diameter of third small pulley.

If there are other small pulleys, designate them as d_3, d_4, etc.

R = r.p.m. of last large pulley.
D = diameter of first large pulley.
D_1 = diameter of second large pulley.
D_2 = diameter of third large pulley.

If there are other large pulleys, designate them as D_3, D_4, etc.

Formula. $rdd_1d_2 = RDD_1D_2$. This basic formula may be solved for each term as follows:

Formulas.

$$r = \frac{RDD_1D_2}{dd_1d_2}, \qquad d = \frac{RDD_1D_2}{rd_1d_2},$$

$$d_1 = \frac{RDD_1D_2}{rdd_2}, \qquad d_2 = \frac{RDD_1D_2}{rdd_1},$$

$$R = \frac{rdd_1d_2}{DD_1D_2}, \qquad D = \frac{rdd_1d_2}{RD_1D_2},$$

$$D_1 = \frac{rdd_1d_2}{RDD_2}, \qquad D_2 = \frac{rdd_1d_2}{RDD_1}$$

Example 1.—In the pulley train shown above, the r.p.m. of the 24-inch pulley at the right is unknown. Solve for it.

$$R = \frac{rdd_1d_2}{DD_1D_2} \qquad \text{The formula for } R$$

$$R = \frac{1800 \times 6 \times 8 \times 10}{15 \times 20 \times 24} \qquad \text{Substituting}$$

$R = 120$, the r.p.m. of the 24″ pulley

Likewise, the r.p.m. of any pulley or its diameter may be solved for.

Example 2.—In the pulley train shown, assume that D_1 is unknown, and solve for it.

$$D_1 = \frac{rdd_1d_2}{RDD_2} \qquad \text{The formula for } D_1$$

$$D_1 = \frac{1800 \times 6 \times 8 \times 10}{120 \times 15 \times 24} \qquad \text{Substituting}$$

$D_1 = 20''$, the diameter

PROBLEMS

1. Find the r.p.m. of the last driven pulley at the right.

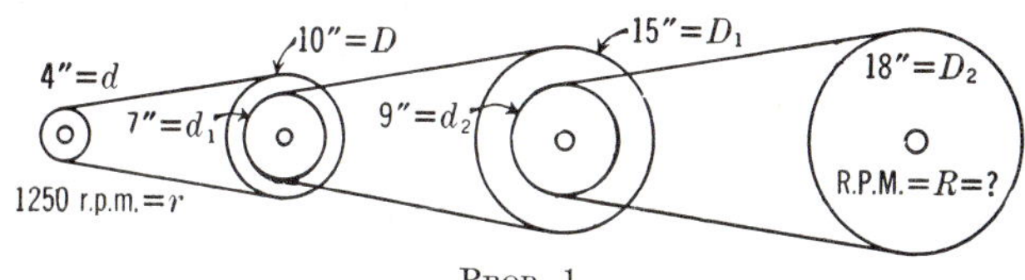

PROB. 1.

2. Suppose the right-hand pulley is the driver at 210 r.p.m. Get the r.p.m. of the 4″ pulley at the left. See figure above.

3. Find the r.p.m. of the last driven pulley (12″) at the right.

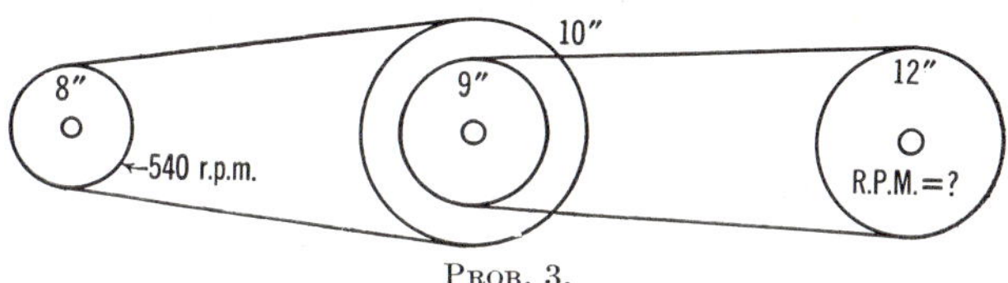

PROB. 3.

4. In the figure for Prob. 3, find the r.p.m. of the 8″ pulley if the 12″ pulley is running at 166 r.p.m.

5. In the figure below, find the r.p.m. of the lathe, if the motor is running 1750 r.p.m.

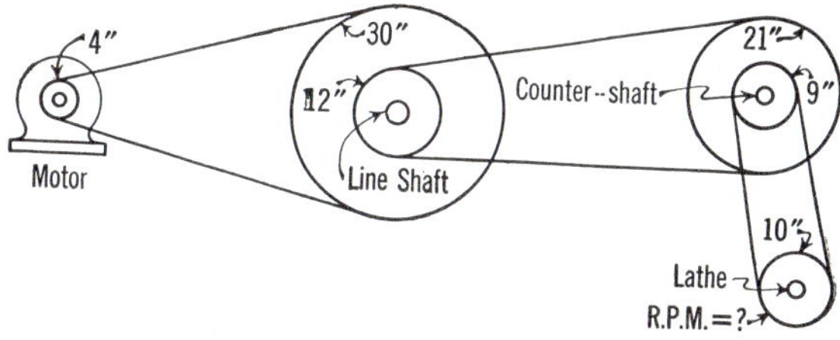

PROB. 5.

6. In the figure for Prob. 5, suppose it is necessary to run the lathe at 196 r.p.m. What would the r.p.m. of the motor be changed to?

7. Suppose the lathe needed to run at 160 r.p.m. and a new pulley is to be put on the motor of Prob. 5, what is its diameter?

8. In Prob. 5, suppose the motor runs at 1250 r.p.m. Find the size of driver pulley to replace the 9″ pulley on the counter-shaft so as to drive the lathe at 200 r.p.m.

9. Find the speed of the last driven shaft at the right in the figure below.

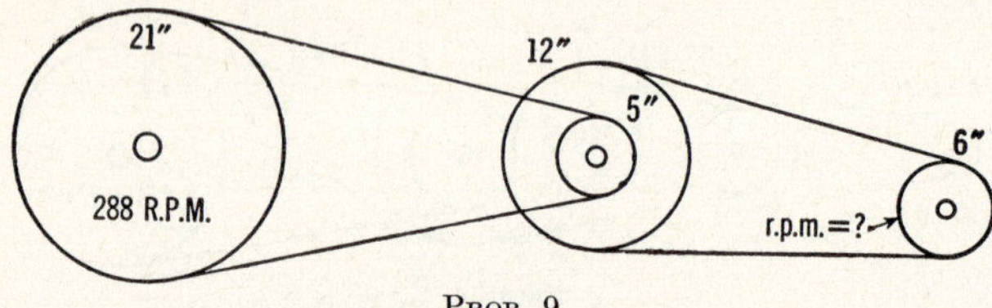

Prob. 9.

10. In Prob. 9, suppose the right-hand pulley is running at 3000 r.p.m. Find the speed of the last driven pulley at the left.

11. The figure below shows a plan view of a pulley train. Get the speed of the last driven pulley (the 7″) in the upper right-hand corner.

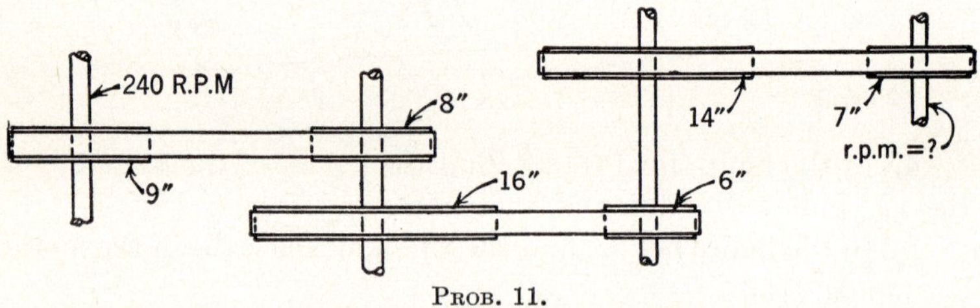

Prob. 11.

12. If the upper right-hand pulley (the 7″) is the original driver running at 750 r.p.m., at what r.p.m. will the shaft and pulley at the extreme left be running? See figure above.

13. In Prob. 11, put on a new driven pulley to replace the 7″ pulley on the last driven shaft so that its r.p.m. will be 1000.

GEAR TRAINS

Gears are also compounded into gear trains, to change both the speed and the gear ratio when considerable change is needed. Rules and formulas applying to gear trains are similar to those for pulley trains.

Rule.—The r.p.m. of the first small gear multiplied by the teeth on all the small gears equals the r.p.m. of the last large gear multiplied by the teeth on all the large gears.

r = r.p.m. of first small gear.
t = number of teeth on first small gear.
t_1 = number of teeth on second small gear.
t_2 = number of teeth on third small gear.

If there are other small gears, designate their number of teeth as t_3, t_4, etc.

R = r.p.m. of last large gear.
T = number of teeth on first large gear.
T_1 = number of teeth on second large gear.
T_2 = number of teeth on third large gear.

If there are other large gears, designate their number of teeth as T_3, T_4, etc.

Formula. $rtt_1t_2 = RTT_1T_2$.

This basic formula may be solved for each term, thus:

Formulas. $R = \dfrac{rtt_1t_2}{TT_1T_2}$, $\quad T = \dfrac{rtt_1t_2}{RT_1T_2}$, . . .

$r = \dfrac{RTT_1T_2}{tt_1t_2}$, $\quad t = \dfrac{RTT_1T_2}{rt_1t_2}$, . . .

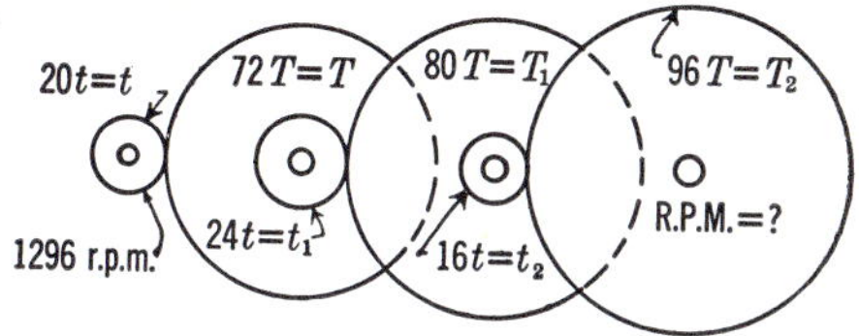

Example 1.—In the gear train shown above, assume the r.p.m. of the 96-tooth gear unknown and solve for it.

$R = \dfrac{rtt_1t_2}{TT_1T_2}$	Formula
$R = \dfrac{1296 \times 20 \times 24 \times 16}{72 \times 80 \times 96}$	Substituting
$R = 18$, the r.p.m. of large gear	

Likewise, any speed or number of teeth in the train may be solved for.

Example 2.—In the gear train above, suppose the number of teeth t on the left-hand gear is unknown, and solve for it.

$t = \dfrac{rt_1t_2}{RTT_1T_2}$	Formula
$t = \dfrac{18 \times 72 \times 80 \times 96}{1296 \times 24 \times 16}$	Substituting
$t = 20$ teeth	

PROBLEMS

1. Find the r.p.m. of the 108(T_2) gear, that is, the last driven gear at the right.

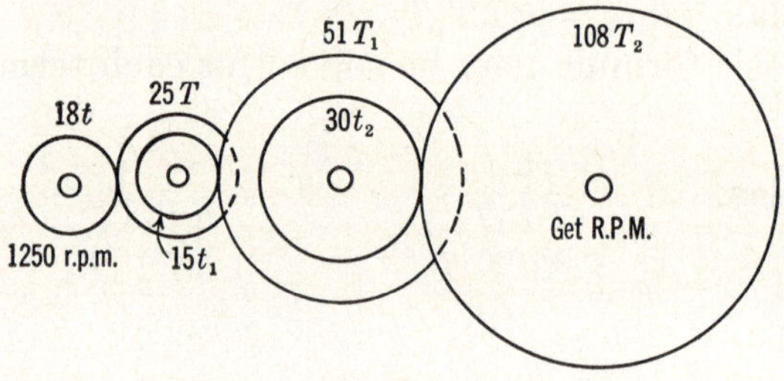

Prob. 1.

2. If the right-hand gear is the driver running 30 r.p.m. in the above problem, what will be the r.p.m. of the last driven or left-hand gear, that is, the 18-tooth gear?

3. In the figure below, find the r.p.m. of the 60-tooth gear.

$$t = 8, \quad t_1 = 10, \quad t_2 = 12$$
$$T = 32, \quad T_1 = 50, \quad T_2 = 60$$

T_1 T_2 T t_1 t_2 t 1200 r.p.m. R.P.M.=?

Prob. 3.

4. In the change gears illustrated at the top of page 369, find the r.p.m. of the screw gear.

PROB. 4.—Lathe change gears. (*Courtesy of South Bend Lathe Works.*)

5. In the gear train shown below, find the speed of the 50-tooth gear.

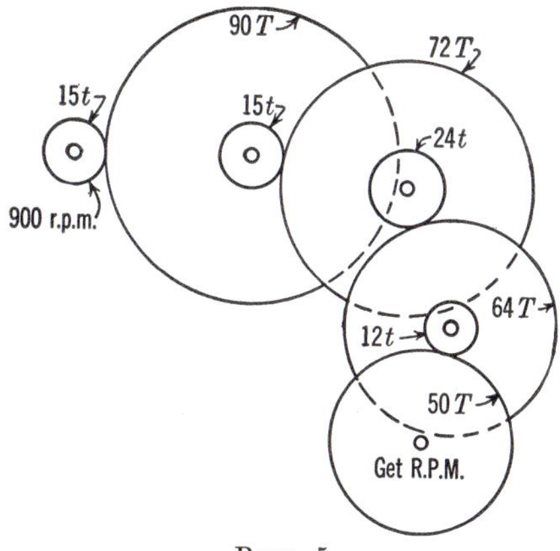

PROB. 5.

6. In Prob. 4, suppose the screw gear runs at 252 r.p.m. Find the speed of the stud gear.

7. In Prob. 5, if the 50-tooth gear is the driver at 10 r.p.m., find the r.p.m. of the 15-tooth gear at the left.

8. In Prob. 4, with what should you replace the 26-tooth gear if the screw gear must run at 60 r.p.m.?

9. In Prob. 4, suppose the stud gear has 100 teeth, what would be the speed of the screw gear?

10. In Prob. 5, suppose the 72-tooth gear is changed to a 96-tooth gear, what would be the speed of the 50-tooth gear?

11. In Prob. 5, a gear of how many teeth would replace the 90-tooth gear so that the 50-tooth gear runs at 5 r.p.m.?

12. In the figure below a plan view of a gear train is shown. Get the speed of the 60-tooth gear.

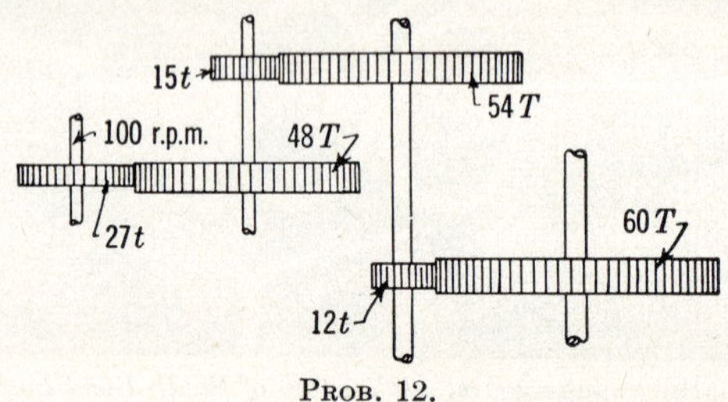

Prob. 12.

13. In the figure of Prob. 12 above, suppose the 60-tooth gear revolves at 25 r.p.m. Find the speed of the 27-tooth gear.

14. In Prob. 12, change the 15-tooth gear to a 21-tooth gear, and figure the speed of the 60-tooth gear.

15. In Prob. 12, change the 54-tooth gear so that the 27-tooth gear revolves at 300 r.p.m. and the 60-tooth gear revolves 15 r.p.m.

THE LATHE

The back-geared screw-cutting engine lathe affords many problems in the transmission of power by the use of pulleys and gears.

The drawing shows a lathe driven from an electric motor by belts to the line shaft, to the counter-shaft, and finally to the lathe cone pulley.

Four changes of r.p.m. of the work are possible simply by shifting the belts on the cone pulleys. This is called direct cone drive. A locking pin secures the spindle and work to the cone pulleys.

Four slower speeds are possible by the use of the back gears. When the back gears are engaged, the locking pin is disengaged. This permits the r.p.m. of the cone pulley to be greatly reduced.

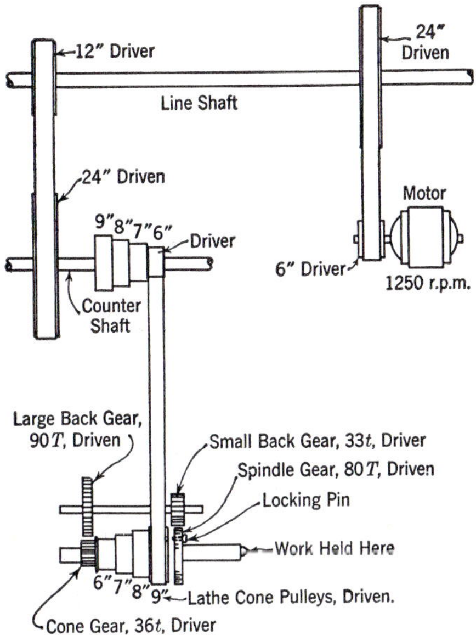

As in all pulley-train problems, the following applies:

Formula. $R = \dfrac{rdd_1d_2}{DD_1D_2}.$

Example 1.—On the lathe shown, find the r.p.m. of the spindle with the belt on the largest of the lathe cone pulleys.

$R = \dfrac{rdd_1d_2}{DD_1D_2}$	Formula for R
$R = \dfrac{1250 \times 6 \times 12 \times 6}{24 \times 24 \times 9}$	Substituting
$R = 104$, r.p.m. of spindle, or work	

When the locking pin is withdrawn or out, and the back gears are engaged, the speed of the work is further reduced because of the reduction due to the back gears. The formula is as follows.

Formula. $R = \dfrac{rdd_1d_2}{DD_1D_2} \times \dfrac{tt_1}{TT_1}.$

Example 2.—On the lathe shown, find the r.p.m. of the spindle with the belt on the large lathe cone pulley with the back gears engaged.

$R = \dfrac{rdd_1d_2tt_1}{DD_1D_2TT_1}$	Formula
$R = \dfrac{1250 \times 6 \times 12 \times 6 \times 36 \times 33}{24 \times 24 \times 9 \times 90 \times 80}$	Substituting
$R = 17.1$, or 17 r.p.m.	

PROBLEMS

If necessary for clearness, illustrate; then solve.

1. On the lathe shown, find the r.p.m. of the spindle when in direct drive but with belt shifted to each of the other three pulleys. (Three answers.)

2. On the same lathe, find the r.p.m. of the spindle when the belt is on each of the three other pulleys of the cones, with the back gears engaged. (Three answers.)

3. Starting with a 14″ driver on the line shaft, which revolves at 240 r.p.m., the driven pulley on the countershaft is 21″. The cone pulleys are 5″, 6″, 7″, and 8″ on the countershaft and the same, in reverse order, on the lathe cone. The large back gear has 88 teeth, and the small back gear has 21 teeth. The cone gear has 18 teeth, and the spindle gear has 56 teeth. Find the four speeds in direct drive and the four speeds with the back gears in mesh.

PROBLEMS

In the following problems, find the eight possible speeds for each lathe. Note that Probs. 2, 3, and 6 start with the countershaft r.p.m.

	No. 1	No. 2	No. 3	No. 4	No. 5	No. 6
Motor r.p.m.	1750			1800	1250	
Motor pulley (driver)	4″			5″	6″	
Line shaft pulley (driven)	25″			18″	20.5″	
Line shaft pulley (driver)	12″			8″	8.5″	
Countershaft pulley (driven) . . .	36″			15″	18″	
Countershaft r.p.m.		152	200			180
Countershaft cone pulleys (drivers).	5″, 6″, 7″, 8″	6″, 7″, 8″, 9″	7″, 9″, 11″, 13″	9″, 10″, 11″, 12″	6″, $6\frac{1}{2}$″, 7″, $7\frac{1}{2}$″	7″, 8″, 9″, 10″
Lathe cone pulleys (driven).	8″, 7″, 6″, 5″	9″, 8″, 7″, 6″	13″, 11′, 9″, 7″	12″, 11″, 10″, 9″	$7\frac{1}{2}$″, 7″, $6\frac{1}{2}$″, 6″	10″, 9″, 8″, 7″
Lathe cone gear (driver)	35 teeth	19 teeth	22 teeth	33 teeth	18 teeth	15 teeth
Spindle gear (driven)	90 teeth	84 teeth	77 teeth	77 teeth	84 teeth	75 teeth
Large back gear (driven)	88 teeth	76 teeth	96 teeth	100 teeth	72 teeth	88 teeth
Small back gear (driver)	25 teeth	21 teeth	30 teeth	20 teeth	19 teeth	20 teeth

AUTOMOBILE TRANSMISSIONS

While many makes of automobiles are changing to some type of automatic transmission drive, a great majority of cars in operation have selective transmissions. By means of a gearshift lever it is possible to engage the different gears and thus to change the r.p.m. of the drive shaft. This process is called **shifting gears.**

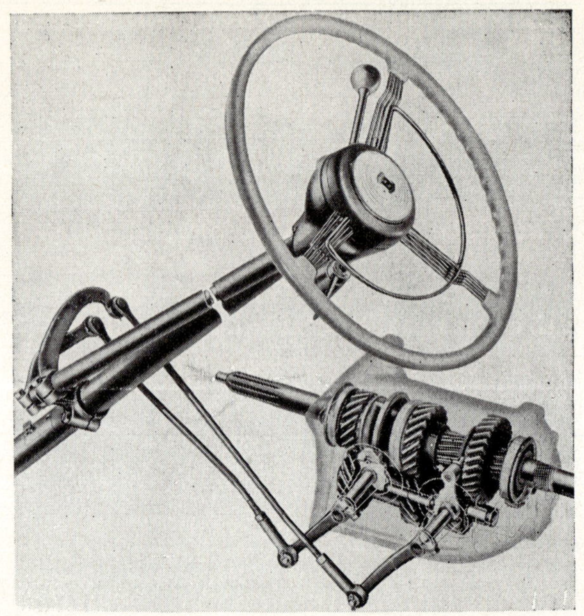

Transmission assembly. (*Courtesy of Cadillac Motor Company.*)

The following figures show a selective type of transmission as used in an automobile. It has three forward speeds and one reverse speed.

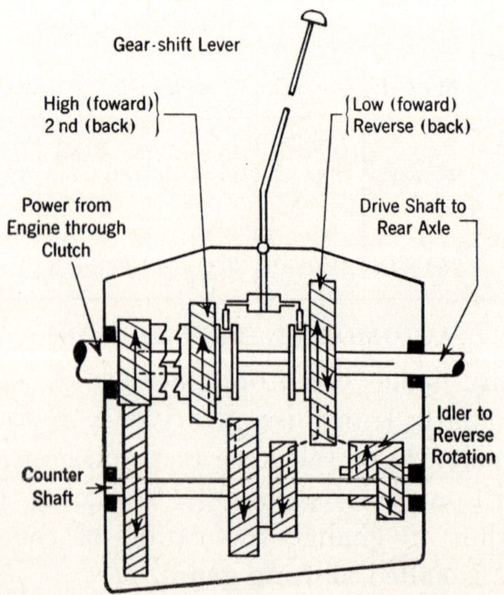

In this gearbox, the gear solid to the clutch drive shaft is always in mesh with the gear on the countershaft; hence, the countershaft always revolves when the clutch is engaged.

High Gear.—When the sliding shaft is moved forward, dogs on its end engage similar dogs on the end of the clutch shaft, as shown below. When thus engaged, the drive

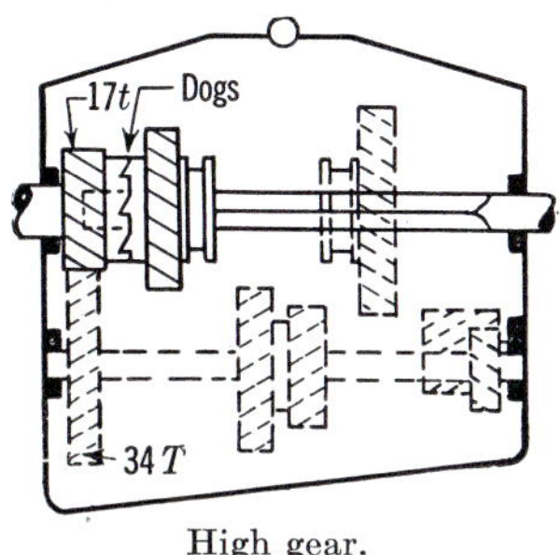

High gear.

shaft revolves at the same r.p.m. as the engine. This is high gear.

Intermediate Gear.—With the dogs disengaged and the 48-tooth gear on the sliding shaft slid back and meshed

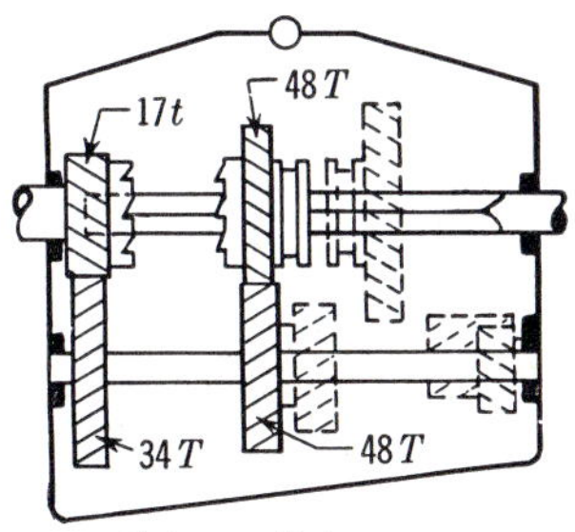

Intermediate gear.

with the 48-tooth gear on the countershaft the car is in intermediate, or second, gear.

Low Gear.—With the dogs disengaged and the 30-tooth gear slid forward to mesh with the 18-tooth gear, the car is in low, or first, gear.

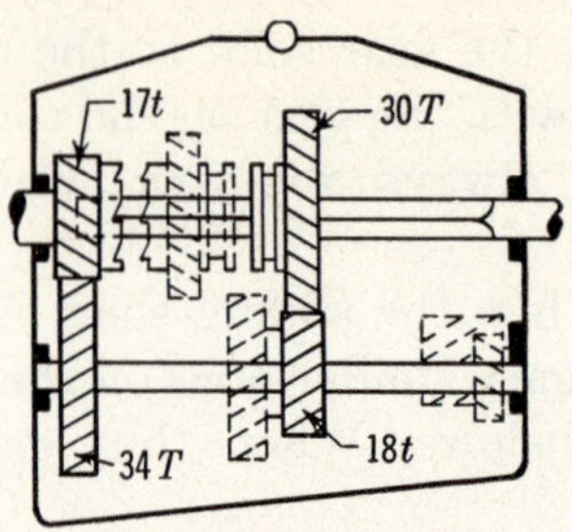

Low gear.

Reverse Gear.—With the dogs disengaged and the 30-tooth gear on the sliding shaft slid backward to engage the idler gear behind the 12-tooth gear with which it

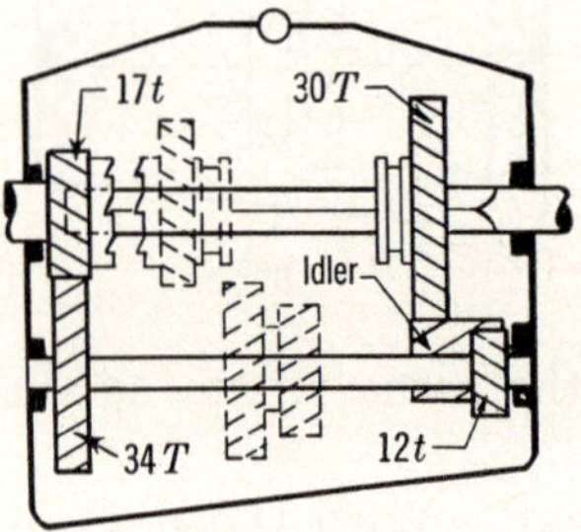

Reverse gear.

meshes, the car is in reverse gear. The teeth on the idler need not be considered. Idler gears simply serve to change the direction of rotation of the shaft.

Example 1.—Find the r.p.m. of the drive shaft with the different gears engaged, when the engine shown in the preceding illustrations runs at 1000 r.p.m.

High With the dogs engaged, the drive is direct. The drive shaft will revolve at 1000 r.p.m.

$$\text{Low} = \frac{1000 \times 17 \times 18}{34 \times 30} = 300 \text{ r.p.m.}$$

$$\text{Intermediate} = \frac{1000 \times 17 \times 48}{34 \times 48} = 500 \text{ r.p.m.}$$

$$\text{Reverse} = \frac{1000 \times 17 \times 12}{34 \times 30} = 200 \text{ r.p.m.}$$

Idler is not considered; it simply reverses the direction of rotation.

PROBLEMS

Find the low speed, intermediate speed, and reverse speed of the drive shaft for each transmission shown below, with all engine speeds at 3000 r.p.m.

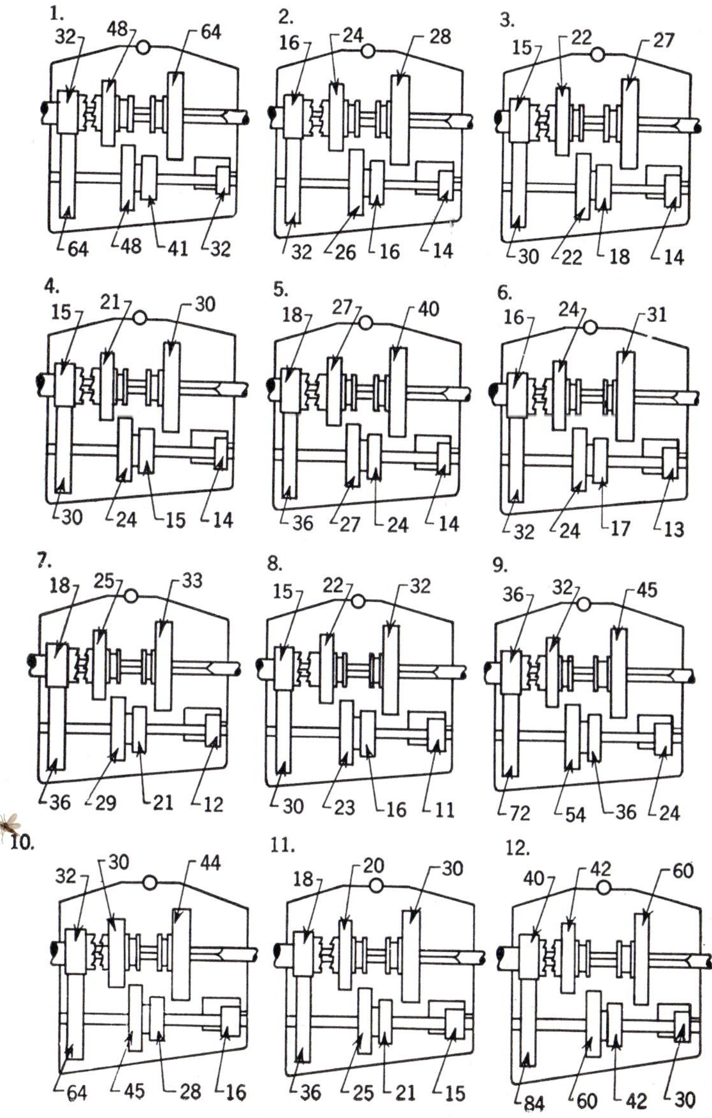

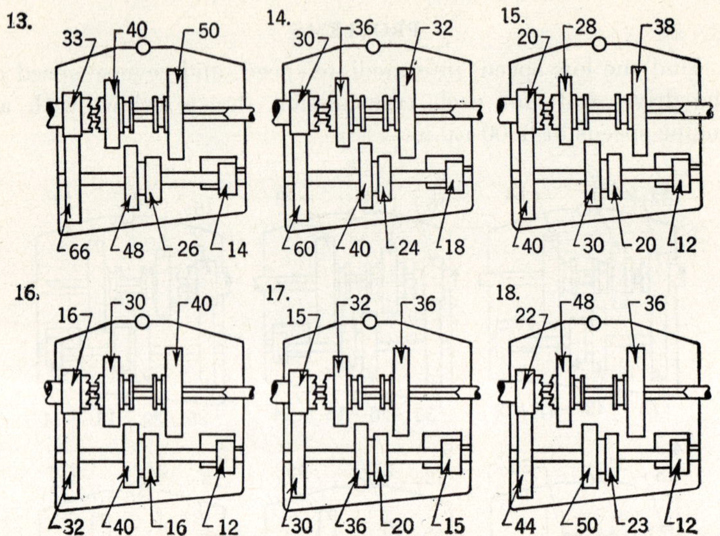
13.
33
40
50
66
48
26
14
14.
30
36
32
60
40
24
18
15.
20
28
38
40
30
20
12
16.
16
30
40
32
40
16
12
17.
15
32
36
30
36
20
15
18.
22
48
36
44
50
23
12

CHAPTER 16

VELOCITY, OR SURFACE SPEED, IN FEET PER MINUTE

RULES AND FORMULAS

In addition to learning to figure the r.p.m. of pulleys and gears, their diameters, and teeth, it is often necessary to determine the **velocity** in **f.p.m.** (feet per minute) of a

point on the surface of pulleys or the speed of the belt running on the pulleys. The speed of grinding wheels and of cutting tools on the lathe, mill, shaper, planer, drill press, etc., is also expressed in f.p.m. Surface speed is often spoken of as velocity in f.p.m.

A rule for getting f.p.m. when r.p.m. and diameter are known is as follows:

Rule 1.—F.p.m. on the surface of a revolving part is found by multiplying the circumference in feet by the r.p.m.

V = velocity, f.p.m.
D = diameter, feet.
πD = circumference.
R = r.p.m.
d = diameter, inches. 3.14 may be used for π.

Formula 1. $V = \pi DR.$

Example 1.—Determine the surface velocity in f.p.m. of a wheel 3 ft. in diameter at 500 r.p.m.

$V = \pi DR$	Formula
$V = 3.14 \times 3 \times 500$	Substituting
$V = 4710$ f.p.m. surface speed	

The size of most cutting tools and equipment is usually given in inches. When the diameter is expressed in inches, it is necessary to divide the values by 12. The formula is as follows:

Formula 2. $V = \frac{\pi dR}{12}.$

Example 2.—Find the surface velocity in f.p.m. of a grinding wheel of 6″ diameter at 1500 r.p.m.

$V = \frac{\pi dR}{12}$	Formula
$V = \frac{3.14 \times 6 \times 1500}{12}$	Substituting
$V = 2355$ f.p.m. surface speed	

Example 3.—Find the r.p.m. of a $\frac{3}{4}$″ drill that should have a surface speed of 80 f.p.m.

$R = \frac{12V}{\pi d}$	Formula solved for r.p.m.
$R = \frac{12 \times 80}{3.14 \times .75}$, or $\frac{12 \times 80}{3.14 \times \frac{3}{4}}$	Substituting
$R = 407.6+$ or 408 r.p.m.	

PROBLEMS

1. Solve Formulas 1 and 2 for each symbol.

Solve each of the following problems:

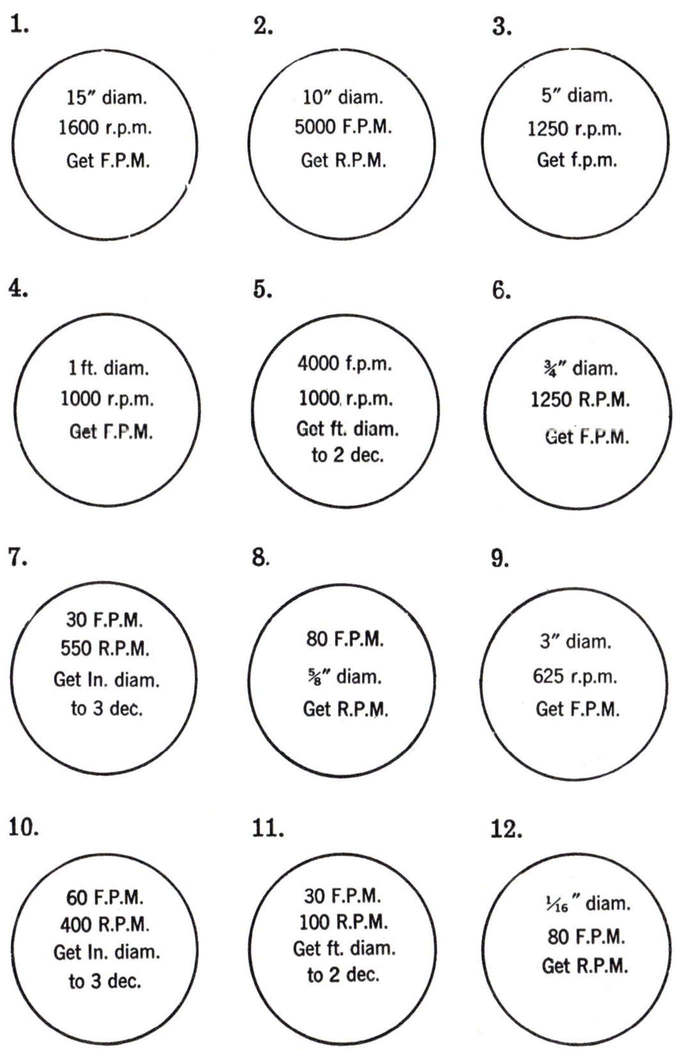

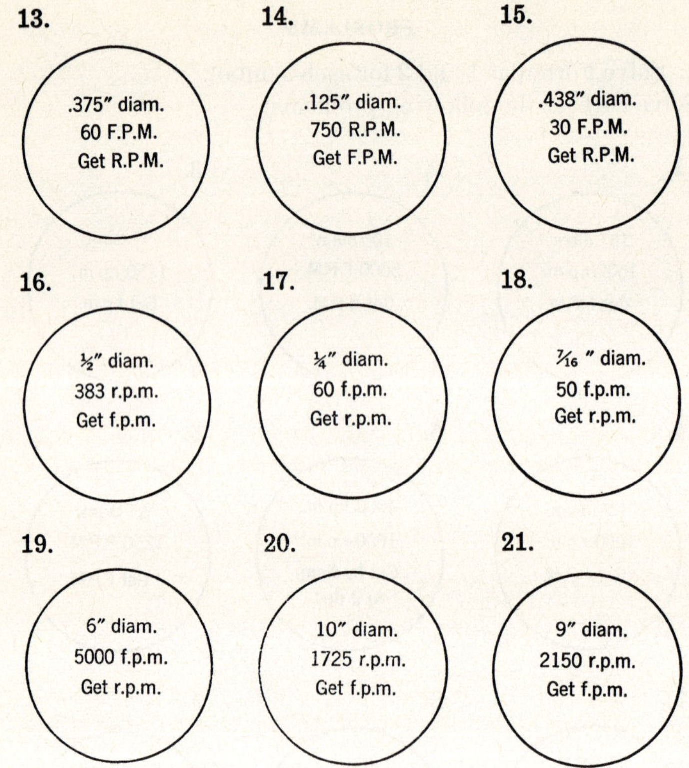
13.
.375″ diam.
60 F.P.M.
Get R.P.M.
14.
.125″ diam.
750 R.P.M.
Get F.P.M.
15.
.438″ diam.
30 F.P.M.
Get R.P.M.
16.
½″ diam.
383 r.p.m.
Get f.p.m.
17.
¼″ diam.
60 f.p.m.
Get r.p.m.
18.
7/16 ″ diam.
50 f.p.m.
Get r.p.m.
19.
6″ diam.
5000 f.p.m.
Get r.p.m.
20.
10″ diam.
1725 r.p.m.
Get f.p.m.
21.
9″ diam.
2150 r.p.m.
Get f.p.m.

PROBLEMS

Solve as indicated by (?). Get the diameter to two decimals.

Number	Diameter	R.p.m.	F.p.m.
1	15 in.	1200	?
2	5 ft.	40	?
3	?	240	4400
4	$\frac{1}{2}$ in.	?	60
5	3 in.	1750	?
6	$\frac{3}{4}$ in.	300	?
7	4 ft.	?	40
8	$\frac{1}{4}$ in.	?	80
9	?	10,000	50
10	$2\frac{1}{2}$ in.	150	?
11	?	100	50
12	$3\frac{1}{2}$ ft.	?	140
13	10 in.	1440	?
14	12 in.	?	2512
15	?	1728	942
16	7 in.	?	45
17	3 ft.	88	?
18	?	3600	40
19	$\frac{5}{8}$ in.	?	80
20	5 in.	70	?
21	?	5000	50
22	8 in.	36	?

PROBLEMS

1. A grinding wheel of 10″ diameter is attached to a motor running at 1750 r.p.m. Find surface f.p.m. of the wheel.

2. The 10″ wheel has a surface speed of 5000 f.p.m. Find its r.p.m.

3. The surface of a 12-ft. cast-iron flywheel should not run faster than 1 mile per min. Find its r.p.m.

4. A cast-iron flywheel of 36″ diameter is to revolve at 600 r.p.m. Find its surface speed in f.p.m.

5. A twist drill of $\frac{3}{8}$″ diameter is to travel at 40 f.p.m. Find its r.p.m.

6. The r.p.m. of a twist drill is 1700 if the diameter is $\frac{1}{2}''$. Find its surface speed.

7. A No. 80 drill is 0.013″ in diameter. If 50 f.p.m. is required, find the r.p.m.

8. Brass should be cut at about 80 f.p.m. How fast should you turn a 3″ diameter bar in a lathe?

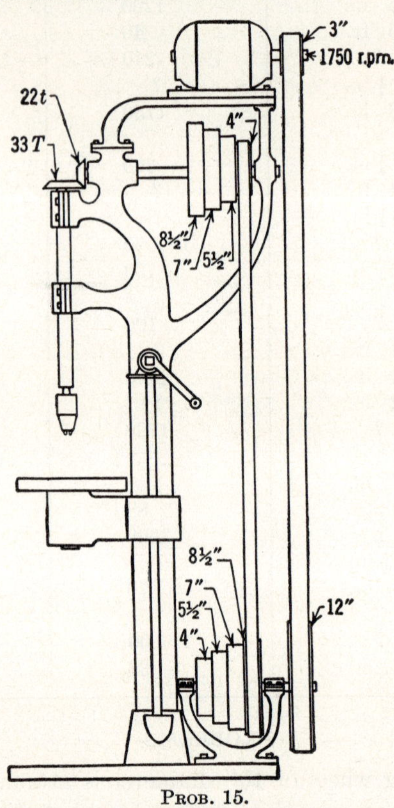

Prob. 15.

9. Find the velocity of a $\frac{1}{2}''$ piece of steel in a lathe if the r.p.m. is 540.

10. Find the surface speed of a 12″ grinding wheel mounted on motor revolving at 1250 r.p.m.

11. Find the surface speed of a $\frac{3}{4}''$ drill if it revolves at 360 r.p.m.

12. A motor at 1500 r.p.m. with a 3″ pulley drives a line shaft with a 21″ pulley; on the line shaft is a 14″ pulley driving a 5″

pulley on a grinder arbor. Find the r.p.m. of the grinder. Find the f.p.m. of an 8″ wheel on the arbor.

13. On the above machinery a 12″ wheel is to run at 5000 f.p.m. What is its r.p.m.? What should you change the motor speed to, to give this?

14. If a 1″ drill should have a periphery speed of 40 f.p.m. when cutting steel, what r.p.m. would you use?

15. On the drill press shown on page 410, compute the r.p.m. of the spindle for each position of the belt on the cone pulleys.

16. With the drill press of Prob. 15, which speed should you use to run the drill in Prob. 14?

17. With the drill press of Prob. 15, at which speed should you run the drill in Prob. 5?

PROBLEMS

On the following, solve as indicated:

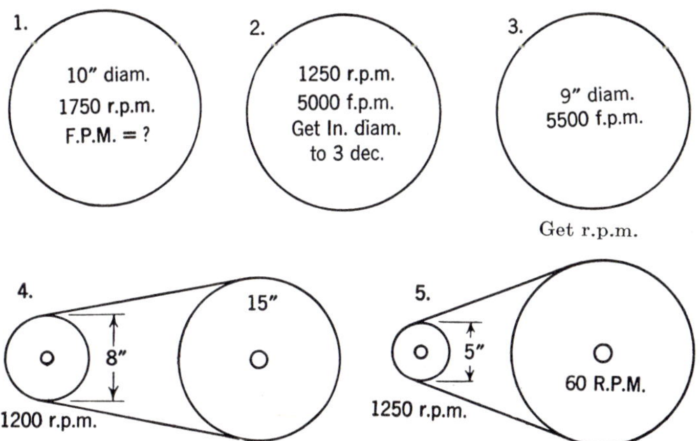

a. Find the f.p.m. of the 8″ pulley.
b. Find the r.p.m. of the 15″ pulley.
c. Find the f.p.m. of the 15″ pulley.
d. How do the f.p.m. of the 8″ and 15″ pulleys compare.

a. Find the f.p.m. of the 5″ pulley.
b. Find the diameter of the driven pulley.
c. Find the f.p.m. of the driven pulley.
d. How do the f.p.m. of the two pulleys compare?

6.

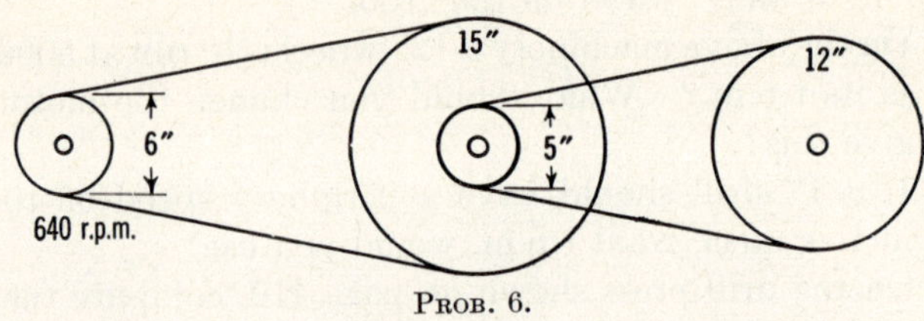

Prob. 6.

a. Find the r.p.m. of each shaft.

b. Find the f.p.m. of each pulley.

7.

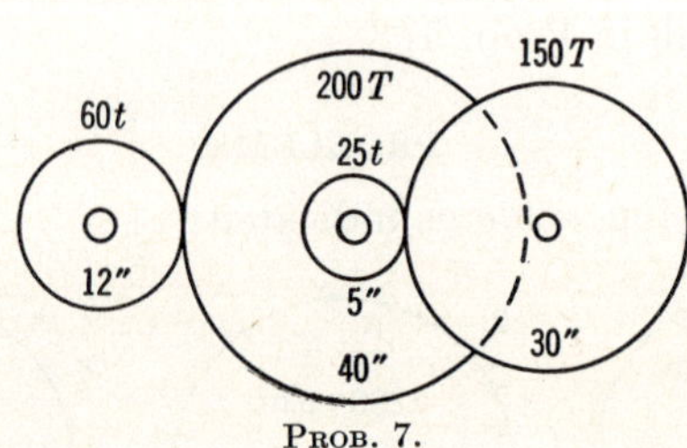

Prob. 7.

a. Find the r.p.m. of each shaft if the 60-tooth gear turns 600 r.p.m.

b. Find the f.p.m. on each gear.

8.

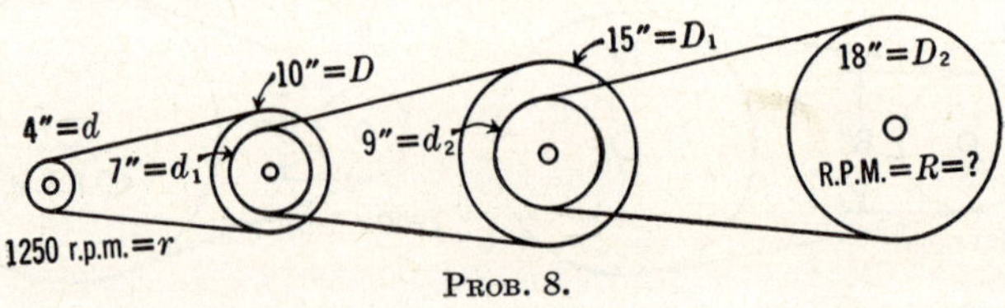

Prob. 8.

a. Find the r.p.m. of each shaft.

b. Find the f.p.m. of each pulley.

CUTTING SPEED

"Cutting speed," in f.p.m., for such tools as twist drills, lathe tools, and milling cutters, is another term used that means the same as velocity in f.p.m., as considered on page 380. The same rules and formulas may be used in computing cutting speed, but the symbol C is usually substituted for V.

Rule.—The cutting speed C of revolving tools or work in f.p.m. is found by multiplying the largest circumference in feet by the r.p.m.

C = cutting speed, f.p.m.
D = diameter, feet.
d = diameter, inches.
R = r.p.m.

Formulas. $C = \pi DR$, or $C = \dfrac{\pi dR}{12}$.

Example 1.—Determine the cutting speed in f.p.m. for a 3″ diameter cast-iron piston in a lathe, as shown if r.p.m. is 225.

(Courtesy of South Bend Lathe Works.)

$C = \dfrac{\pi dR}{12}$	Formula
$C = \dfrac{3.14 \times 3 \times 225}{12}$	Substituting
$C = 176.625$, or 177 f.p.m. cutting speed	

Example 2.—Find the r.p.m. of 2″ annealed tool steel in a lathe if the tool is high-speed steel, which has an allowable cutting speed of 70 f.p.m.

$$R = \frac{12C}{\pi d} \qquad \text{Formula solved for } R$$

$$R = \frac{12 \times 70}{3.14 \times 2} \qquad \text{Substituting}$$

$$R = 133.7, \text{ or } 134 \text{ r.p.m.}$$

Trade practice has determined the best cutting speeds to be used on each machine for the different materials and with the two common types of tool, those made of high-speed steel and of carbon steel.

The table below shows the proper f.p.m. on the various machines with each type of tool on the most common materials.

Cutting-speed Chart for Various Machines and Materials

Material to be cut	Machine used	High-speed steel tools, f.p.m.	Carbon steel tools, f.p.m.
Annealed tool steel	Lathe	50 to 70	25 to 35
	Drill press	50 to 60	20 to 30
	Miller	50 to 60	20 to 30
Cold-finished machine steel	Lathe	100 to 150	50 to 70
	Drill press	100 to 120	50 to 60
	Miller	100 to 125	50 to 70
Cast iron (C.I.)	Lathe	75 to 175	40 to 80
	Drill press	100 to 170	40 to 80
	Miller	100 to 150	60 to 80
Brass and bronze	Lathe	150 to 300	70 to 150
	Drill press	200 to 300	100 to 150
	Miller	150 to 250	80 to 125
Aluminum	Lathe	200 to 300	100 to 150
	Drill press	200 to 300	100 to 150
	Miller	200 to 350	100 to 175

These speeds correspond to those used in general practice, and the fastest may be used in figuring problems. Use the f.p.m. for high-speed steel unless the problem states otherwise.

PROBLEMS

1. Solve the formula $C = \frac{\pi d R}{12}$ for d and R.

2. If cold-rolled steel is to be cut at 80 f.p.m., at what r.p.m. should a 3-in. bar revolve in a lathe?

3. A bar of 2″ annealed tool steel is to be turned in a lathe at 45 f.p.m. Find the proper r.p.m.

(Courtesy of South Bend Lathe Works.)

Prob. 3.

4. A cast-iron piston $5\frac{1}{2}''$ diameter is to be turned in a lathe at 160 f.p.m. Find the r.p.m.

5. What r.p.m. should a $\frac{1}{8}''$ twist drill have that is used for cutting hard bronze?

6. A $3\frac{1}{2}''$ cast iron piston is revolving at 70 r.p.m. Get the cutting speed.

7. What cutting speed results when a 5″ cold-finished steel shaft revolves in a lathe at 50 r.p.m.? At how many r.p.m. should this work revolve to give 80 f.p.m. cutting speed?

8. What r.p.m. should a $\frac{1}{2}''$ milling tool have that is used for cutting yellow brass?

PROB. 8.

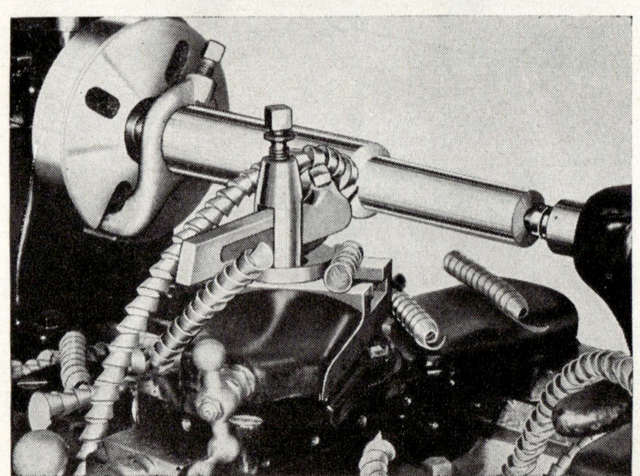

(*Courtesy of South Bend Lathe Works.*)
PROB. 9.

9. The figure shows a piece of cold-finished steel of 2″ diameter revolving in a lathe at 160 r.p.m. If the tool is high-speed steel, find the cutting speed. Is this speed too fast or too slow?

10. Find the cutting speed of the $\frac{1}{2}''$ twist drill shown on page 417 if the r.p.m. is 458.

11. What diameter of cast iron should revolve at 250 r.p.m.?

12. What is the largest brass rod that should be turned on a lathe at 346 r.p.m.?

13. A quick-change gear lathe can be run at the following r.p.m.: 8, 24, 56, 80, 120, 200, 346, 458. If the nearest r.p.m. is selected,

a. Which r.p.m. would be used to cut a 2″ brass bar?
b. Which r.p.m. would be used to cut a 2″ annealed tool steel?
c. Which r.p.m. would be used to cut a 2″ bronze?
d. Which r.p.m. would be used to cut a 2″ cast iron?
e. Which r.p.m. would be used to cut a 2″ aluminum?

14. Find the diameter of cold-finished steel that will give 80 f.p.m. cutting speed for each of the possible r.p.m. in Prob. 13.

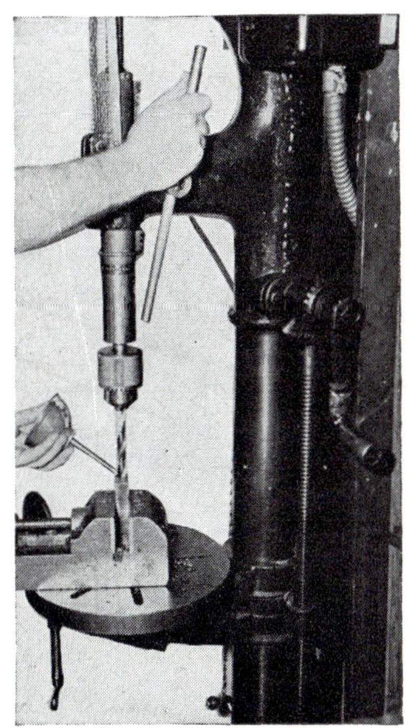

Prob. 10.

15. Find the cutting speed of 0.563″ steel if the r.p.m. is 110. Is this too fast or too slow?

16. If $2\frac{1}{4}$″ brass is running at 80 r.p.m., what is the cutting speed? Should it be more or less? How much?

17. Aluminum airplane pistons are $6\frac{1}{4}$″ in diameter. At what r.p.m. should they be turned in a lathe?

18. A 1″ diameter end mill is cutting brass. At how many r.p.m. should it run?

VARIATIONS IN CUTTING SPEEDS

There are several conditions affecting the cutting of metals that must be considered to determine the proper r.p.m. Some are the condition and set of the cutting tool, the cooling fluid (if any), the feed, grade of the material, etc.

With these factors taken into consideration, a slight variation of the speeds figured in the preceding lessons seems permissible. A shortening of the process of calculation is, therefore, advisable. The values 3.14 and 12 always appear in the formulas; for example, $C = \frac{\pi dR}{12}$ or $\frac{3.14dR}{12}$. $\frac{3.14}{12}$ is approximately $\frac{1}{4}$; and if this fraction is used, we have the following results:

Formula. $C = \frac{dR}{4}$. This formula is much shorter than the preceding one and may be used.

Example 1.—Determine the cutting speed of 3″ cast iron in a lathe if the r.p.m. is 225 and the tool is high-speed steel.

$C = \frac{dR}{4}$	Formula
$C = \frac{3 \times 225}{4}$	Substituting
$C = 168.7$ or 169 f.p.m.	Compare this with 177 as figured by the exact formula on page 387.

Example 2.—Find the r.p.m. of 2″ annealed tool steel in a lathe if the tool is high-speed steel.

$R = \frac{4C}{d}$	Formula solved for R
$R = \frac{4 \times 70}{2}$	Substituting
$R = 140$ r.p.m.	This is a little more than the 134 r.p.m. figured with the exact formula on page 388.

Example 3.—Figure the diameter of aluminum that can be cut in a lathe at 458 r.p.m. with a high-speed steel tool.

$d = \frac{4C}{R}$	Formula solved for d
$d = \frac{4 \times 300}{458}$	Substituting
$d = 2.6''$ diameter	

PROBLEMS

In the problems below, solve as indicated by the question mark (?). "Chart" refers to the chart on page 388. Tool steel can be cut only when annealed. Use the short formulas.

Number	C	R	d	Material	Tool steel used	Machine
1	Chart	?	3.625″	Cast iron	High-speed	Lathe
2	?	110	0.875″	Cold finish*	High-speed	Lathe
3	Chart	?	$2\frac{1}{4}''$	Brass	Carbon	Drill
4	Chart	300	?	Aluminum	High-speed	Mill
5	?	18	$3\frac{1}{2}''$	Cast iron*	High-speed	Drill
6	Chart	?	0.625″	Tool steel	High-speed	Drill
7	?	66	1.875″	Cast iron*	Carbon	Lathe
8	200	1920	?	Aluminum	High-speed	Drill
9	120	?	3.375″	Cast iron	High-speed	Lathe
10	Chart	?	6″	Bronze	Carbon	Mill
11	Chart	3600	?	Aluminum	Carbon	Drill
12	?	75	2.125″	Brass*	High-speed	Lathe
13	Chart	?	6.5″	Cast iron	High-speed	Lathe
14	Chart	?	15″	Cold finish	Carbon	Lathe
15	?	476	$\frac{1}{2}''$	Cold finish*	Carbon	Drill
16	?	350	0.75″	Bronze*	High-speed	Drill
17	Chart	275	?	Aluminum	High-speed	Mill
18	Chart	24	?	Cast iron	Carbon	Lathe
19	Chart	320	?	Tool steel	High-speed	Lathe
20	60	6000	?	Cold finish	Carbon	Drill
21	200	?	0.4375″	Brass	Carbon	Lathe
22	200	?	0.4375″	Brass	High-speed	Lathe
23	250	?	$\frac{1}{2}''$	Aluminum	High-speed	Drill
24	Chart	2450	?	Cold finish	Carbon	Drill

*** Is this speed too fast or too slow?**

CHAPTER 17

TAPERS

DEFINITIONS AND FORMULAS

A piece of work is said to **taper** when there is a **gradual** and **uniform increase** or **decrease** in its **diameter** or **thickness.** The machinist, wood turner, and patternmaker are often called upon to taper material. In the shop, taper turning means gradually reducing the diameter for a definite length.

Taper is shown on the automobile axle, the drill shank, the lathe center, etc.

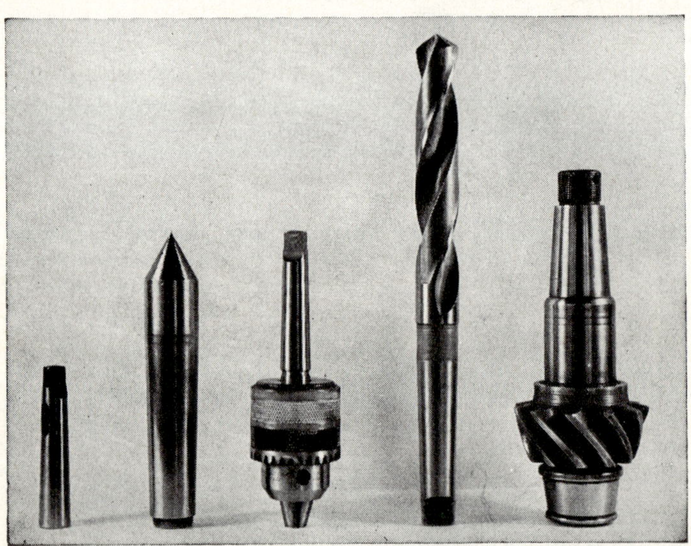

Taper is usually specified by the taper per inch or per foot. Taper per inch means the difference in diameters 1 in. apart. For example, if work tapers $\frac{1}{4}$ in. in 1-in.

length, this means that the diameters 1 in. apart differ by $\frac{1}{4}$ in. Likewise, taper per foot means the difference in diameters 1 ft. apart. For example, No. 0 Morse taper is $\frac{5}{8}$ in. per ft. This means that the diameters 1 ft. apart differ by $\frac{5}{8}$ in.

T = taper per foot.
t = taper per inch.
T_o = total taper, or difference, in diameters.
L = total length when all of piece is not tapered.
l = length of taper.
S = setover of lathe tailstock.
D = diameter of large end.
d = diameter of small end.

Note.—All measurements are in inches.

To find the total taper when the taper per inch is known, we have the following:

Rule 1.—To find the total taper, multiply the taper per inch by the length.

Formula 1. $T_o = tl$.

Example 1.—What is the total taper if a piece 9″ long is to taper $\frac{3}{32}$″ per in.?

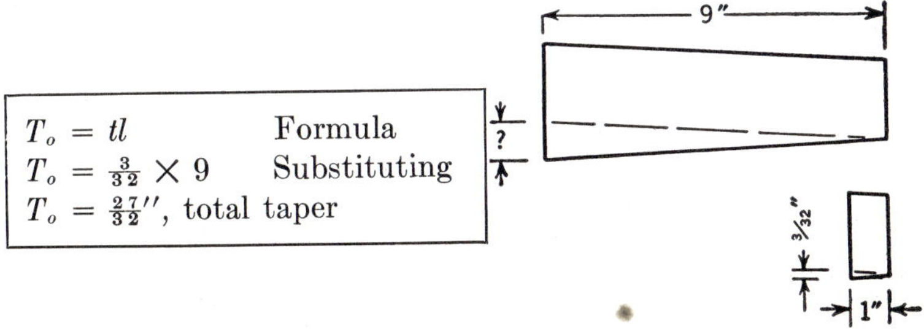

To find the taper per inch, solve Formula 1 for t.

$$t = \frac{T_o}{l}.$$

Example 2.—What is the taper per inch if a piece 9″ long tapers $\frac{27}{32}$″?

$t = \dfrac{T_o}{l}$	Formula
$t = \dfrac{.84375}{9}$	Substituting
$t = .09375$, or $\frac{3}{32}$″	

Example 2 may also be solved by the following rule:

Rule 2.—To find the taper per inch, divide the difference of the diameters by the length tapered.

Formula 2. $t = \dfrac{D - d}{l}.$

Example 3.—Find the taper per inch of a piece 9″ long that is 4″ in diameter on one end and $3\frac{5}{32}$″ on the other end.

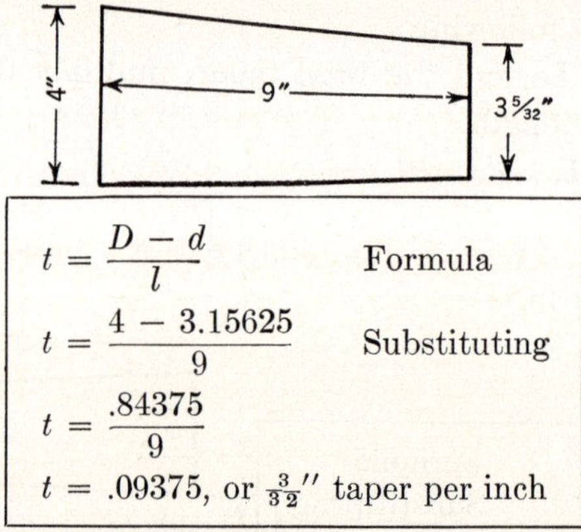

$t = \dfrac{D - d}{l}$	Formula
$t = \dfrac{4 - 3.15625}{9}$	Substituting
$t = \dfrac{.84375}{9}$	
$t = .09375$, or $\frac{3}{32}$″ taper per inch	

Note.—Obviously the solution of Example 3 is exactly the same as that of Example 2, except that in Example 3 it is necessary to find the total taper, which is .84375.

To find the total taper, T_o, when the taper per foot is known, we have the following:

Rule 3.—To find total taper, multiply the length tapered by the taper per foot, and divide by 12.

Formula 3. $T_o = \frac{lT}{12}.$

Example 4.—Find the total taper on a piece 9″ long if the taper per foot is $\frac{1}{4}$″.

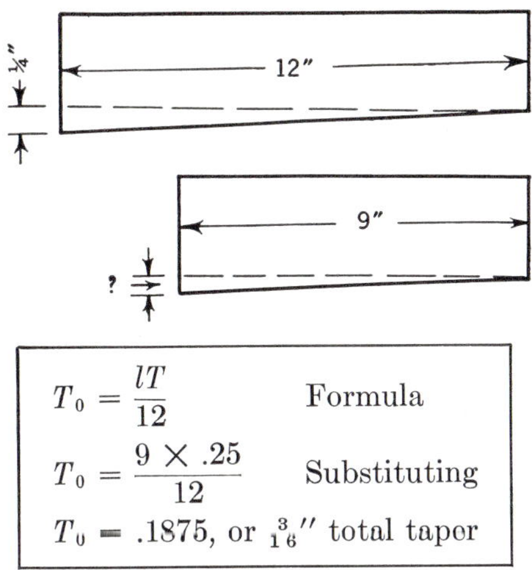

$T_0 = \frac{lT}{12}$	Formula
$T_0 = \frac{9 \times .25}{12}$	Substituting
$T_0 = .1875$, or $\frac{3}{16}$″ total taper	

The dial on the lathe taper attachment shown is graduated in units that make it necessary to know the taper per foot.

Lathe taper attachment. *(Courtesy of South Bend Lathe Works.)*

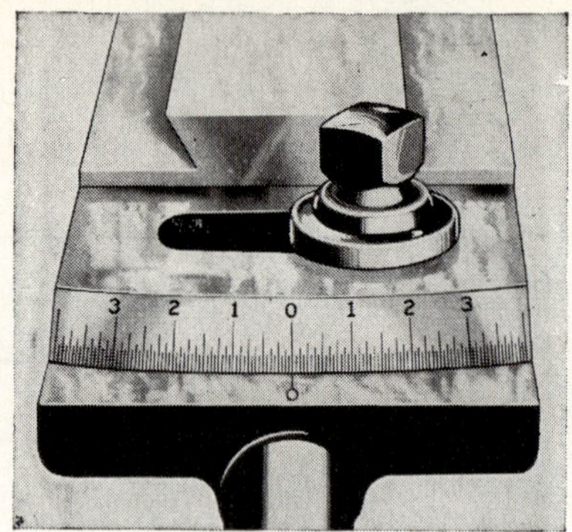

(*Courtesy of South Bend Lathe Works.*)

By solving Formula 3 for T, the taper per foot is found.

$$T = \frac{12T_o}{l}$$

Note.—This formula is twelve times as great as the formula in Example 2.

Example 5.—Find the taper per foot if a piece 9″ long tapers $\frac{3}{16}$″.

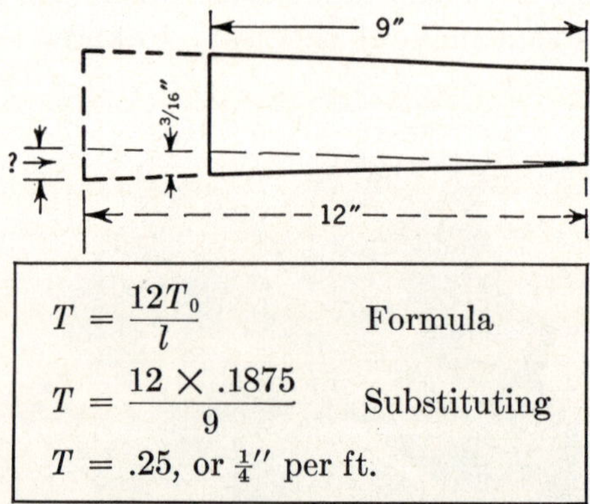

$T = \frac{12T_0}{l}$	Formula
$T = \frac{12 \times .1875}{9}$	Substituting
$T = .25$, or $\frac{1}{4}$″ per ft.	

Likewise, Formula 2 may be multiplied by 12 to give a formula for figuring the taper per foot.

Rule 4.—To find the taper per foot, divide twelve times the difference in diameters by the length tapered.

Formula 4. $T = \frac{12(D - d)}{l}$.

Example 6.—Find the taper per foot when a piece 9″ long is 3″ on the small end and $3\frac{3}{16}$″ on the large end.

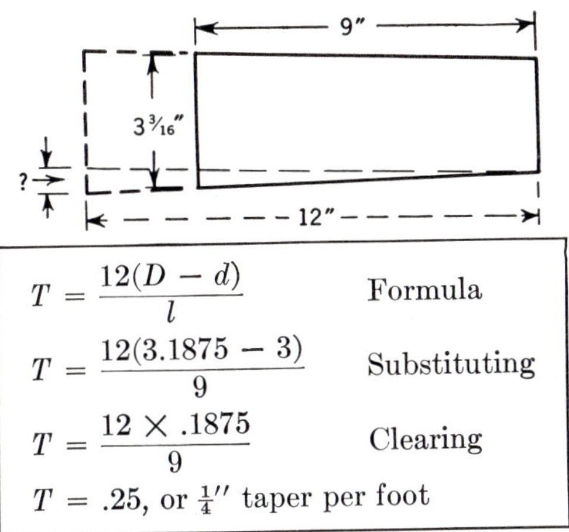

$$T = \frac{12(D - d)}{l} \qquad \text{Formula}$$

$$T = \frac{12(3.1875 - 3)}{9} \qquad \text{Substituting}$$

$$T = \frac{12 \times .1875}{9} \qquad \text{Clearing}$$

$T = .25$, or $\frac{1}{4}$″ taper per foot

Note.—Obviously the solution of Example 6 is exactly the same as that of Example 5, except that in Example 6 it is necessary to determine the total taper, which is .1875.

PROBLEMS

Give the formula used, and solve.

1. Find the total taper of a piece 6″ long that tapers at the rate of $\frac{1}{8}$″ per in.

2. What is the total taper for a 7″ long piece when the taper per inch is 0.03″?

3. How long is a piece that tapers $\frac{3}{4}$″ at the rate of $\frac{1}{8}$″ per in.?

4. What is the taper per inch on a piece 16″ long that tapers 0.04″?

5. What is the taper per foot if a piece tapers $\frac{1}{16}$″ per in.?

6. Find the total taper on a piece 6″ long at $\frac{1}{2}$″ per ft.

7. Find the taper per foot of a piece 8″ long with $\frac{5}{64}$″ taper.

8. Find the taper per inch for the figure shown.

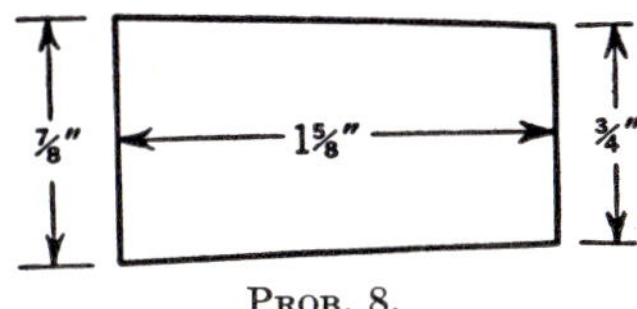

PROB. 8.

9. Find the length of a piece that tapers $\frac{3}{32}''$ per in., from 2″ to $2\frac{1}{2}''$.

10. What is the length tapered on the piece shown if it tapers $\frac{1}{4}''$ per in.?

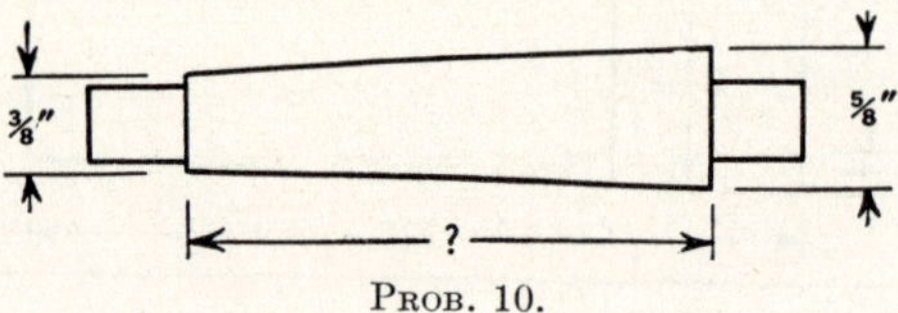

PROB. 10.

11. What is the taper per foot on a piece 10″ long that is reduced from 3″ to $2\frac{1}{2}''$?

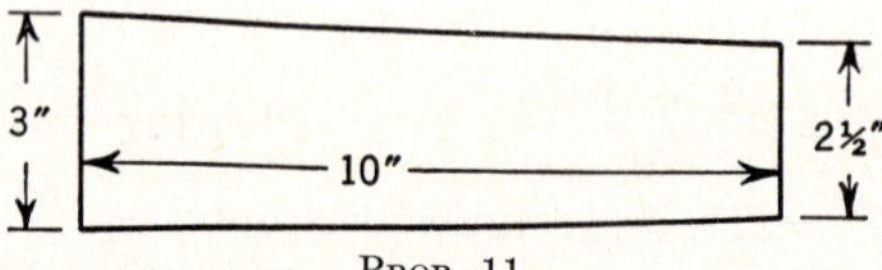

PROB. 11.

12. What is the taper per foot on a piece 8″ long when the large end is $\frac{5}{8}''$ and the small end is $\frac{1}{2}''$?

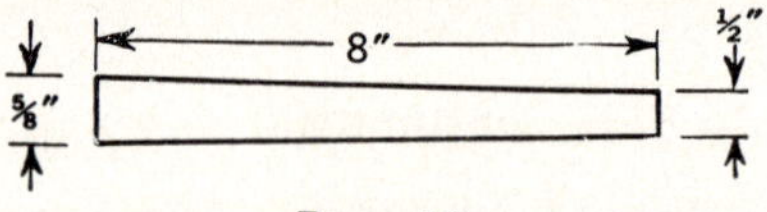

PROB. 12.

PROBLEMS

On the following illustrations, solve as indicated:

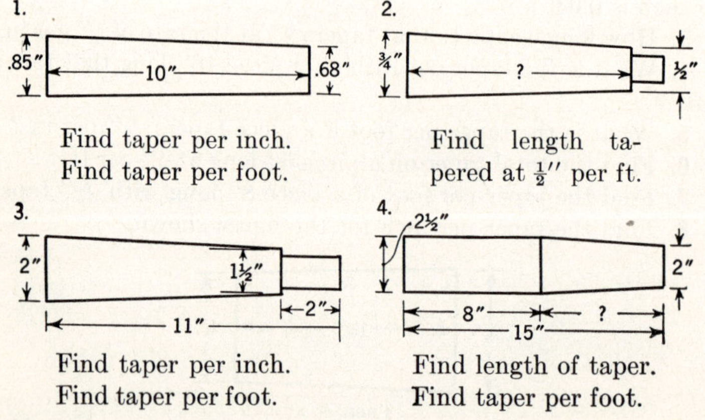

1. Find taper per inch.
Find taper per foot.

2. Find length tapered at $\frac{1}{2}''$ per ft.

3. Find taper per inch.
Find taper per foot.

4. Find length of taper.
Find taper per foot.

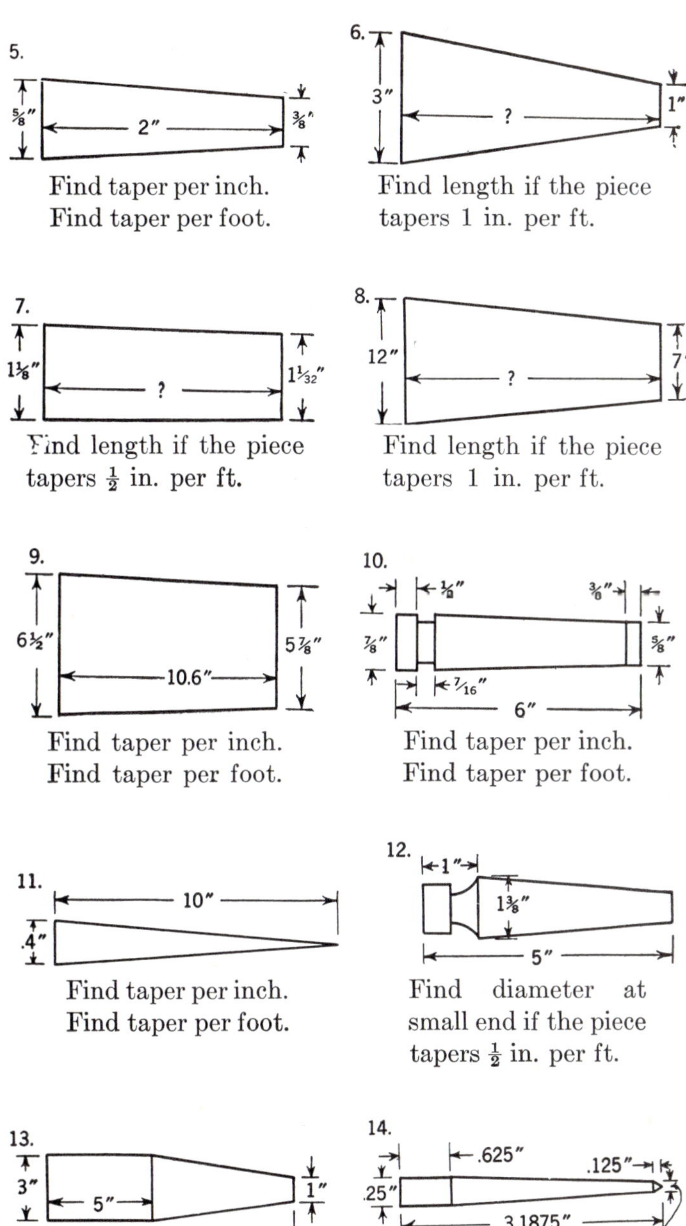

5. Find taper per inch. Find taper per foot.

6. Find length if the piece tapers 1 in. per ft.

7. Find length if the piece tapers $\frac{1}{2}$ in. per ft.

8. Find length if the piece tapers 1 in. per ft.

9. Find taper per inch. Find taper per foot.

10. Find taper per inch. Find taper per foot.

11. Find taper per inch. Find taper per foot.

12. Find diameter at small end if the piece tapers $\frac{1}{2}$ in. per ft.

13. Find taper per foot.

14. Find taper per inch.

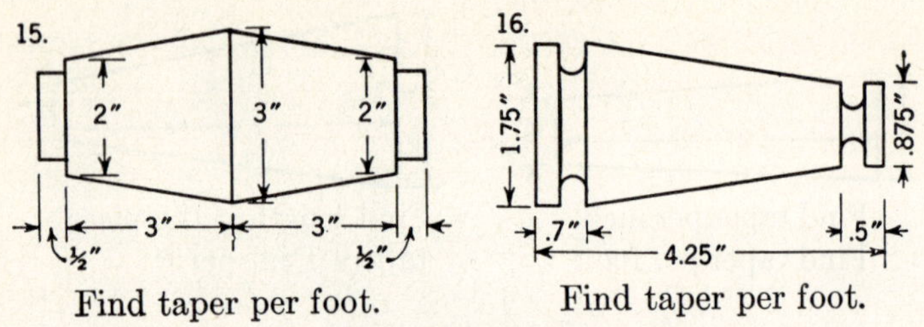

15. Find taper per foot.

16. Find taper per foot.

PROBLEMS

Solve for the missing values.

Number	l, inches	T, inches	D, inches	d, inches
1	10.6	?	$6\frac{1}{2}$	$5\frac{7}{8}$
2	8.5	?	6.6	6.38
3	25	?	$4\frac{1}{16}$	$3\frac{13}{16}$
4	?	1.6	$8\frac{1}{2}$	7
5	?	0.6	$1\frac{1}{8}$	$1\frac{1}{32}$
6	?	$\frac{3}{4}$	$\frac{15}{16}$	$\frac{5}{8}$
7	$4\frac{5}{8}$	?	$\frac{7}{8}$	$\frac{5}{8}$
8	?	0.33	1.66	1.48
9	$7\frac{3}{4}$	?	3.658	$3\frac{7}{16}$
10	?	0.458	1.88	1.55
11	?	0.010	1.10	0.975
12	$3\frac{3}{4}$	?	2.5	2.43
13	$7\frac{1}{2}$	?	2.639	2.188
14	10	?	3.61	3.52
15	?	0.01	0.75	0.21
16	15	?	$3\frac{3}{4}$	2.63
17	12.5	?	7.756	7.391
18	$15\frac{1}{2}$	?	$6\frac{3}{8}$	$6\frac{1}{4}$
19	12.32	?	1.965	1.888
20	?	0.25	2.96	2.75

TURNING TAPERS BY SETOVER OF LATHE TAILSTOCK

In straight turning on the lathe the tailstock and base are clamped at zero, as shown at *A* on the opposite page.

For taper turning, it is possible to set over the tailstock. The **setover** permits the cutting tool to reduce the end that has been set over closer to it. The **setover** is accomplished by backing off the screw *G* the distance required and screwing in the screw *F* a like distance, as shown at *B*. Setover is also called "offset."

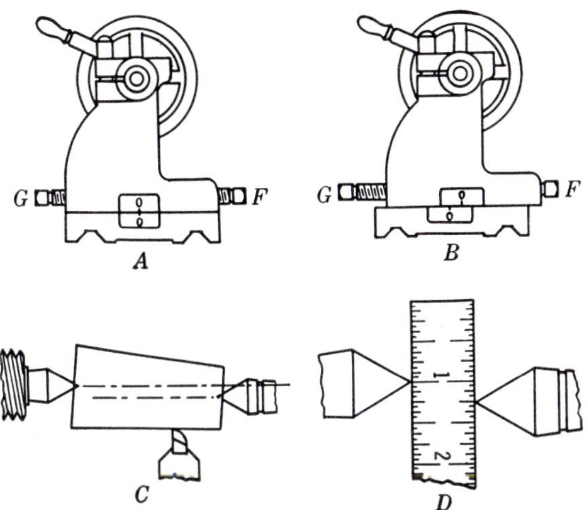

The figure at *C* shows a view of work between the centers with the tailstock set over toward the cutting tool. At *D* is shown how to measure the setover.

Rule 1.—To taper the whole length, set over the tailstock one-half the difference of the end diameters.

Formula 1. $S = \frac{D - d}{2}.$

Example 1.—Get the setover to taper a piece 5″ long from 1″ to $\frac{3}{4}$″.

1″ ¾″ 5″

$$S = \frac{D - d}{2} \quad \text{Formula}$$

$$S = \frac{1 - .75}{2} \quad \text{Substituting}$$

$S = .125$, or $\frac{1}{8}$″ setover

When only part of the piece is tapered, the setover is greater by the ratio of the whole length to the length tapered, or $\frac{L}{l}$.

Rule 2.—To find the setover to taper part of the length, multiply the ratio of the whole length to the length tapered by half the difference of the end diameters.

Formula 2. $S = \frac{L}{l} \times \frac{D - d}{2}$, or $S = \frac{L(D - d)}{2l}$

Example 2.—Find the setover to taper 5″ on a 16″ piece from $2\frac{5}{24}''$ to 2″.

2 5/24″ 2″
11″ 5″
16″

$S = \frac{L}{l} \times \frac{D - d}{2}$	Formula
$S = \frac{16}{5} \times \frac{(2\frac{5}{24} - 2)}{2}$	Substituting
$S = \frac{16}{5} \times \frac{5}{24} \times \frac{1}{2}$	
$S = \frac{1}{3}''$, setover	

PROBLEMS

Determine the setover in the following:

Number	L, inches	l, inches	D, inches	d, inches	S
1	24	5	$1\frac{1}{4}$	1	
2	33	4	$1\frac{1}{2}$	$1\frac{7}{16}$	
3	32	3	1.25	1.125	
4	12	1	1	0.875	
5	18	5	$1\frac{1}{2}$	$1\frac{5}{32}$	
6	10	1	4	3.875	
7	11	$2\frac{1}{2}$	3	$2\frac{5}{16}$	
8	36	6	6	$5\frac{43}{64}$	
9	12	$3\frac{1}{2}$	6	5.75	
10	18	5	$2\frac{1}{2}$	$2\frac{1}{8}$	
11	15	6	$1\frac{7}{8}$	$1\frac{1}{2}$	
12	$17\frac{1}{2}$	$3\frac{1}{2}$	$1\frac{5}{8}$	$1\frac{3}{8}$	
13	27	5	2.5	2.4	
14	12	10	2.4	2.36	
15	16.8	4.2	6.56	6.36	
16	15.52	3.0	$7\frac{7}{8}$	7.63	
17	10	6	5	4	
18	12	10	1	0.78	
19	3.62	3.2	1.58	1.49	
20	$8\frac{5}{8}$	5.63	$1\frac{7}{16}$	1.3	

WHEN THE WHOLE LENGTH IS TAPERED

When the whole length is tapered and the taper per inch is known, the setover is found by the following rule:

Rule 1.—To find setover, multiply the whole length by the taper per inch, and divide by 2.

Formula 1. $S = \frac{lt}{2}$.

Example 1.—Find the setover to taper a piece 16″ long when the taper per inch is $\frac{1}{48}''$.

$S = \frac{lt}{2}$ Formula

$S = \frac{16 \times 1}{2 \times 48}$ Substituting

$S = \frac{1}{6}''$ setover

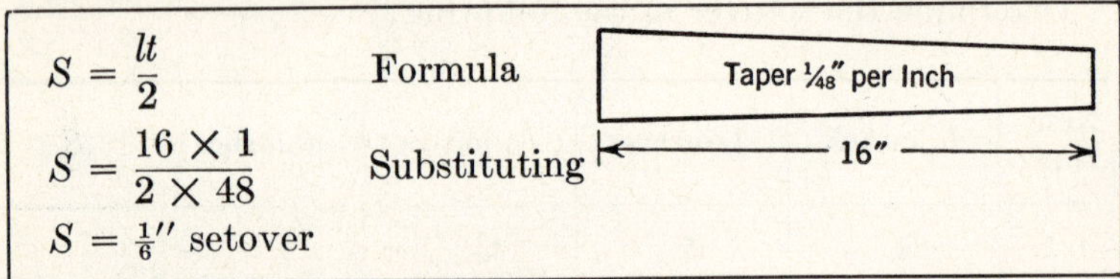

Since the taper per inch is one-twelfth of the taper per foot, when the whole length is tapered and the taper per foot is known the setover is found by the following rule:

Rule 2.—To find setover, multiply the whole length by the taper per foot, and divide by 24.

Formula 2. $S = \frac{lT}{24}.$

Example 2.—Find the setover to taper a piece 16″ long when the taper per foot is $\frac{1}{4}''$.

$S = \frac{lT}{24}$ Formula

$S = \frac{16 \times 1}{24 \times 4}$ Substituting

$S = \frac{1}{6}''$ setover The same as Example 1 above, for $\frac{1}{4}''$ per ft. is $12 \times \frac{1}{48}''$

Taper ¼″ per Foot

16″

WHEN PART OF A PIECE IS TAPERED

The taper per inch or the taper per foot is a constant factor. If only part of a piece is tapered, the setover, in using these factors, is no different than if the whole length is tapered. Hence, the part tapered may be disregarded in the calculations.

If part of the length is tapered and the taper per inch is known, the setover may be found by the following rule:

Rule 1.—To find the setover when part of the length is tapered, multiply the whole length by the taper per inch, and divide by 2.

Formula 1. $S = \frac{Lt}{2}.$

Example 1.—Find the setover to taper 5″ on a 16″ piece when taper per inch is $\frac{1}{48}$″.

$S = \frac{Lt}{2}$	Formula
$S = \frac{16 \times 1}{2 \times 48}$	Substituting. *Note.*—The part tapered is disregarded.
$S = \frac{1}{6}$″ setover	

Likewise, when the taper per foot is known, the setover may be found by the following rule:

Rule 2.—To find the setover when part of the length is tapered, multiply the whole length by the taper per foot, and divide by 24.

Formula 2. $S = \frac{LT}{24}.$

Example 2.—Find the setover to taper 5″ on a 16″ piece when the taper per foot is $\frac{1}{4}$″.

$S = \frac{LT}{24}$	Formula
$S = \frac{16 \times 1}{24 \times 4}$	Substituting. *Note.*—The part tapered is disregarded.
$S = \frac{1}{6}$″ setover	

PROBLEMS

1. Find the setover to taper a piece 8″ long when taper per inch is $\frac{1}{32}$″.

2. What is the setover to taper a piece 7″ long when the taper per inch is 0.063″?

3. What is the setover to taper a piece 8″ long when the taper per foot is $\frac{1}{8}$″?

4. What is the setover to taper a piece $7\frac{1}{2}$″ long when the taper per foot is $\frac{1}{3}$″?

5. Find the setover to taper part of a 21″ long piece when the taper per inch is $\frac{1}{16}$″.

6. Find the setover to taper 6″ on a piece 18″ long when the taper per foot is $\frac{1}{4}$″.

7. What is the setover to taper a 5″ piece from 5″ to $4\frac{1}{2}$″?

8. What is the setover to taper 5″ on a 12″ piece from $3\frac{1}{8}$″ to $2\frac{7}{8}$″?

9. What setover will result in reducing a piece from $4\frac{7}{8}$″ to $3\frac{5}{8}$″?

10. What is the taper per foot on a shank 4″ long whose small end is 0.412″ and whose large end is 0.562″?

11. What is the setover to taper 5″ on a 16″ piece from 1″ to $\frac{7}{8}$″?

12. Find the length to be tapered when the taper per foot is $1\frac{1}{2}$″ and the diameters are $1\frac{1}{2}$″ and 1″.

13. A bar is tapered 15″ of its length. The ends of taper are $2\frac{1}{2}$″ and 1″. Get the taper per foot.

14. A 2″ bar 36″ long is tapered down to 1.5″ in a length of 6″. Get the taper per foot. Get the setover of the tailstock.

In the following problems, select the proper formula, and solve it for the unknown term:

15. The taper per foot is $\frac{1}{2}$″, the length to be tapered is 5″, and the small diameter is 1″. Get the large diameter.

16. Find the small diameter when D is 2 in., l is 6 in., and T is $\frac{1}{2}$ in. per ft.

17. Find the length of taper (l) when taper per foot is $\frac{3}{8}$″, d is 2″, and D is $2\frac{1}{2}$″.

18. Find T_o when T is $\frac{1}{2}$″ and l is 7″.

19. The Morse taper is $\frac{5}{8}$ in. per ft. Find the large diameter of a shank 5 in. long if the small diameter is $\frac{13}{16}$ in.

20. The Brown and Sharpe taper is $\frac{1}{2}$ in. per ft. What is the setover to taper a piece 16 in. long with Brown and Sharpe taper?

21. The Jarno taper is 0.6 in. per ft. What is the setover for a piece $5\frac{7}{8}$ in. long with the Jarno taper?

22. The American taper is $\frac{9}{16}$ in. per ft. Calculate the total taper for a piece 10 in. long.

23. A bar of steel is tapered 1.5 in. per ft. The diameters at ends of taper are 1.5 and 1.2 in. Find the length of taper.

24. A bar is tapered 12″. The diameters are 2.68″ and 2.48″. What is the taper per foot?

25. What is the offset if a bar 11″ long is tapered 0.763″ to 0.656″?

26. What is the offset if 6″ of a 16″ bar is tapered 3.58″ to $3\frac{1}{2}$″?

27. What is the length of taper if the taper per foot is $\frac{1}{2}$″ and the diameters are 1.765″ and 1.682″?

28. Find the offset to reduce the diameter from 6.638″ to 6.521″.

29. Find the taper per foot if a bar 16″ long is tapered from $\frac{7}{8}$″ to $\frac{3}{4}$″.

30. What offset should you use to taper 3″ on a 5″ bar if diameters are 0.625″ and 0.6″?

31. A bar of steel 6″ long is to be tapered $\frac{5}{8}$″ to $\frac{13}{16}$″. What is the offset?

32. A No. 4 Morse taper of 0.623 in. per ft. is to be ground on a drill shank. If the whole length is 11 in., what is the offset?

33. What is the taper per foot if a reamer tapers $\frac{1}{8}$″ in $2\frac{1}{2}$″? In $4\frac{1}{8}$″?

34. What is the taper per foot on a pipe reamer if the diameters are $\frac{1}{4}$″ and $1\frac{1}{2}$″ in a 2″ length?

35. On a No. 1 Morse taper the diameters are 0.369″ and 0.475″. The length tapered is $2\frac{1}{8}$″. What is the taper per foot?

36. A Brown and Sharpe taper is 0.0425″ per inch. What is the taper per foot?

37. How long is the tapered part of a Brown and Sharpe taper if the taper per foot is 0.5101″ and the diameters are 0.500″ and 0.599″?

38. The Jarno taper is 0.6 in. per ft. What is the taper per inch?

39. A No. 2 taper pin is 0.193 in. in diameter at the large end and is 2 in. long. The taper is $\frac{5}{8}$ in. per ft. Find the diameter of the small end.

40. What is the taper per inch of a No. 1 Morse reamer if the small diameter is 0.369″, the large diameter is 0.475″, and the length is $2\frac{1}{8}$″?

41. Find the taper per foot of a No. 10 Brown and Sharpe taper when length is $9\frac{1}{4}''$, small diameter is 2.5″, and large diameter is 2.886″.

42. Find the taper per foot for a cone $2\frac{3}{4}''$ high if the base is $\frac{1}{2}''$ in diameter.

TURNING TAPERS BY USING THE COMPOUND REST

In turning or boring tapers greater than are possible by the tailstock setover method, on such jobs as lathe centers, tool and jig work, and diemaking, it is necessary to set the **compound rest** on the lathe at the proper angle to cut each taper.

By referring to Table XIV (page 536), the taper per foot or per inch can be converted into degrees and the desired taper may be cut by setting the compound at the specified angle. The tool is then fed along the taper with the "cross-feed." Because the compound rest is graduated in whole degrees only, the minutes and seconds given in the table can only be approximated.

Example 1.—What is the angle at which to set the compound to turn the end of a die that tapers 2 in. per ft.? (See Table XIV, page 536.)

> Under the column Angle with center line and opposite the 2 in. in column 1, the angle given is 4°45′48″.
>
> It is evident that 45′48″ is slightly more than $\frac{3}{4}°$; therefore, set the compound to approximately $4\frac{3}{4}°$.

Example 2.—The taper per inch from the center line is given as $\frac{3}{16}''$.

1. What is the included angle?
2. What is the angle with center line at which the compound should be set?
3. Approximately where should you set the compound?

$\frac{3}{16}$ = 0.1875″	
21°14′2″	1. Included angle
10°37′1″	2. Angle with center line
	3. Set compound at slightly above $10\frac{1}{2}°$.

PROBLEMS

Refer to Table XIV, page 536.

1. At what angle with center line should you set the compound for a Brown and Sharpe taper of 0.500 in. per ft.?

2. At what angle should you set the compound for a No. 0 Morse taper that is given as 0.625 in. per ft.? No. 1 Morse taper that is given as 0.600″ per ft.?

3. A No. 4 Morse taper tapers .05191 in. per in. At what approximate angle should you set the compound to cut this?

4. If the taper per inch is $\frac{1}{32}$ in., what is the included angle? What is the angle with center line?

5. When the taper is $\frac{1}{3}$ in. per in., what is the angle with center line? Approximately what is the compound setting?

6. When the taper is given as $1\frac{1}{2}$ in. per ft., what is the angle with the center line? What is the approximate compound setting?

7. If a bearing tapers $\frac{1}{16}$ in. per in., what is the exact angle with the center line at which the compound should be set?

8. A taper shank is given as $\frac{3}{8}$ in. per in. What is the included angle?

9. The taper per inch from the center line is $\frac{1}{8}$″. What is the centerline angle? What is the approximate compound setting?

10. The taper per inch from the center line is 0.1666″. What is the taper per foot? If this is too much to set over the tailstock, give the approximate angle at which to set the compound.

CHAPTER 18

SCREW THREADS

DEFINITIONS AND FORMULAS

Every mechanic is familiar with **screw threads.** There are several types of threads, and each has been designed for some definite purpose. The **American National** form of thread, which has almost entirely replaced the **sharp V** thread, is used on bolts and nuts to hold pieces of equipment together. The square thread and the **acme 29°** thread are used on machines to transmit power. The lathe shown on page 389 uses an acme thread on the lead screw. Other threads less frequently met are Brown and Sharpe worm thread, Whitworth (or British Standard) thread, British Association thread, metric (or international) thread, etc.

Sharp "V"   National Form

Square   29° Acme

The workman should learn the names applied to the thread parts, listed below:

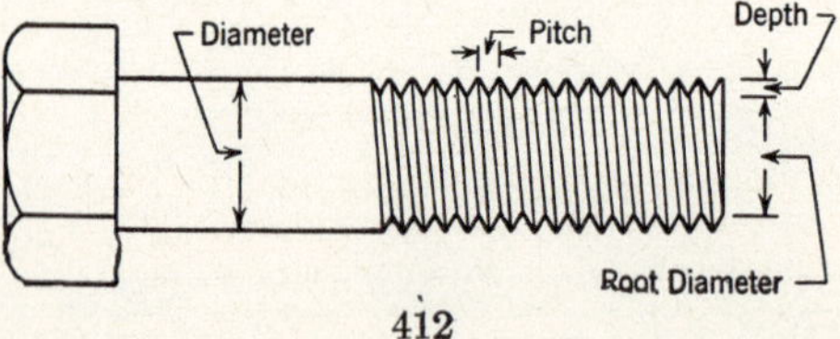

Crest, or point, is the outermost part of the thread.

Root is the bottom of the thread.

Depth (D), or height, is the vertical distance from the crest to the root.

Double depth (D_o) is an expression used to denote a depth for each side of the bolt.

Pitch (P) is the distance, in parts of an inch, between corresponding points of two adjacent threads. Pitch is often expressed as the number of threads per inch.

Lead is the distance the screw advances in one revolution.

On a bolt or tap,

Diameter, or outside diameter, is the distance perpendicular to the axis from crest to crest. It is the diameter of the bolt.

Root diameter, or bottom diameter, is the distance perpendicular to the axis from root to root. It is the bolt diameter minus the double depth.

Threads are designated by their pitch and the name of the thread. For example, an 8-pitch National Fine thread has 8 threads per inch and is the American National Fine series, or shape (commonly called N. F. or fine).

Since pitch is expressed as the number of threads per inch, it may be determined by measuring with a rule, as shown at the left, or by a screw-pitch gauge as shown at the right.

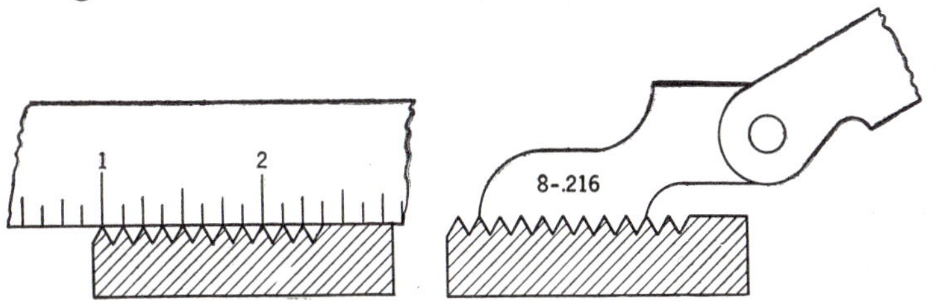

If the distance between corresponding parts on adjacent threads is $\frac{1}{8}$ in., the thread is $\frac{1}{8}$-in. pitch. In this case, there are 8 threads per inch, and the thread may be spoken of as an 8-pitch thread or a $\frac{1}{8}$-in. pitch thread. Either designation is accepted by the trade.

Rule.—To find the pitch, divide 1 in. by the number of threads per inch.

P = pitch.

N = number of threads per inch.

Formulas. $P = \frac{1}{N}$, and $N = \frac{1}{P}$.

Example 1.—Find the pitch of a thread having 13 threads per inch.

$P = \frac{1}{N}$	Formula
$P = \frac{1}{13}$	Substituting
$P = \frac{1}{13}''$, or .0769″, approximately .077″	

Example 2.—A screw has a pitch of $\frac{2}{9}''$. Find the number of threads per inch.

$N = \frac{1}{P}$	Formula solved for N
$N = \frac{1}{\frac{2}{9}} = 1 \div \frac{2}{9} = 1 \times \frac{9}{2}$	Substituting
$N = 4\frac{1}{2}$ threads per inch	

PROBLEMS

1. What is the pitch of a screw having 20 threads per inch?

2. Find the pitch of a bolt having 11 threads per inch.

3. A screw has a pitch of $\frac{2}{7}''$. Find the number of threads per inch.

4. Find the number of threads per inch on a $\frac{7}{16}''$ pitch screw.

5. The pitch is $\frac{1}{15}''$. How many threads are there per inch?

6. How many threads per inch are there on a screw of $\frac{5}{8}''$ pitch?

7. Compute the pitch of a screw with $2\frac{3}{4}$ threads per inch.

8. Find the pitch of a screw having 24 threads per inch.

9. A screw having 9 threads per inch has what pitch?

10. A $\frac{3}{4}''$ pitch screw has how many threads per inch?

11. How many threads per inch are there on a screw of $\frac{5}{13}''$ pitch?

12. What is the pitch of a screw having $2\frac{1}{2}$ threads per inch?

13. What is the pitch of a screw having 80 threads per inch?

14. How many threads per inch on a screw of 2″ pitch?

15. A screw has a pitch of $\frac{2}{45}$″. How many threads are there per inch?

16. Compute the pitch of a screw with $7\frac{1}{2}$ threads per inch.

17. Find the number of threads per inch of a $\frac{5}{16}$″ pitch screw.

18. How many threads per inch on a screw of $\frac{3}{7}$″ pitch?

19. How many crests are there on an 8-pitch thread? Sketch this thread.

20. How many roots are there on an 8-pitch thread? Sketch this thread.

LEAD

The lead of a screw is the distance the screw advances in one revolution.

In the **single-thread screw,** as considered on the preceding page, the lead is equal to the pitch. On a **double-thread** screw the lead is twice the pitch. On a **triple-thread** screw the lead is three times the pitch, etc.

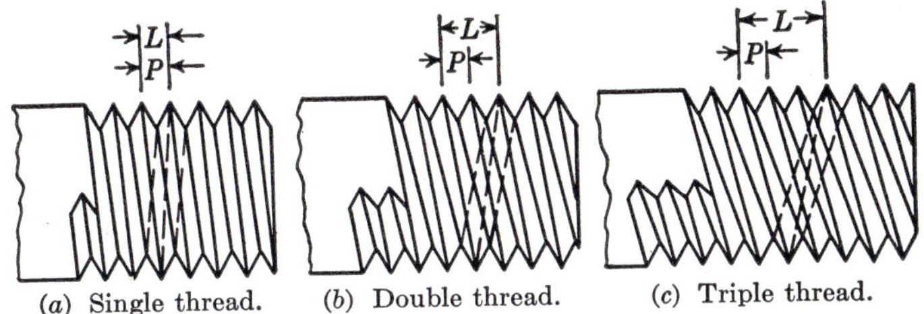

(*a*) Single thread. (*b*) Double thread. (*c*) Triple thread.

On a screw of $\frac{1}{4}$-in. pitch the lead has the following values:

On a single thread, one groove is cut spirally around the bolt. Lead = pitch = $\frac{1}{4}$ in., as shown at (*a*) above.

On a double thread, two parallel grooves are cut spirally around the bolt. Lead = 2 × pitch = $\frac{1}{2}$ in., as shown at (*b*).

On a triple thread, three parallel grooves are cut spirally around the bolt. Lead = 3 × pitch = $\frac{3}{4}$ in., as shown at (*c*).

Rule.—Lead equals the pitch multiplied by the multiple of the parallel grooves cut around the bolt.

L = lead.

m = multiple of the parallel grooves.

P = pitch.

Formula. $L = Pm.$

Example 1.—What is the lead of a double-thread screw of $\frac{1}{8}''$ pitch? (Multiple on a double thread = 2.)

$L = Pm$	Formula
$L = \frac{1}{8} \times 2$	Substituting
$L = \frac{1}{4}''$	

Example 2.—What is the pitch of a quadruple thread with $\frac{3}{4}''$ lead?

$P = \frac{L}{m}$	Formula solved for P
$P = \frac{.75}{4}$	Substituting
$P = .1875$, or $\frac{3}{16}''$	

PROBLEMS

1. A single-thread screw has pitch of $\frac{1}{8}''$. What is the lead?

2. A single-thread screw has 13 threads per inch. What is the pitch? What is the lead?

3. A double-thread screw has a lead of $\frac{1}{8}''$. What is the pitch?

4. What is the lead of a 6-pitch double-thread screw?

5. A single thread has $\frac{1}{20}''$ pitch. What is the lead?

6. What is the pitch of a triple-thread screw with a lead of $\frac{1}{4}''$?

7. Find the pitch of a quadruple-thread screw with a lead of $\frac{3}{4}''$.

8. What is the lead of a $\frac{3}{8}''$ pitch quadruple thread?

9. What is the pitch of a double thread having a lead of $\frac{1}{16}''$?

10. A single thread has $\frac{1}{14}''$ pitch. Figure the lead.

11. What are the lead and pitch of a single thread with 9 threads per inch?

12. Compute the lead of $\frac{5}{8}''$ pitch triple thread.

13. A double-thread screw has a pitch of $\frac{1}{20}''$. What is the lead?

14. What is the pitch of a quadruple-thread screw that has a 1″ lead?

15. A triple-thread screw of $\frac{1}{6}''$ pitch has what lead?

16. What is the pitch of a double-thread screw when the lead is $\frac{3}{8}''$?

17. Figure the pitch of a double-thread screw with a lead of $\frac{1}{20}''$.

18. Figure the lead of a double-thread screw with a $\frac{1}{20}''$ pitch.

19. Figure the lead and pitch of a double thread with 16 threads per inch.

20. Name the thread that has $\frac{1}{8}''$ pitch and $\frac{3}{8}''$ lead.

21. Name a thread that has 0.166″ pitch and 0.5″ lead.

22. What thread will have pitch of 0.075″ and lead of 0.225″?

DEPTH OF THREADS

Frequently the mechanic must cut threads on a screw. Also, he must cut threads into which to turn a screw. In order that the thread will possess the desired holding power, the depth of the thread, both on the screw and in the hole, is of primary importance.

The simplest thread (although not the one most commonly used) is the sharp 60° V thread shown.

The depth of a V thread is found by the use of the right triangle and square root. If one thread is taken separately,

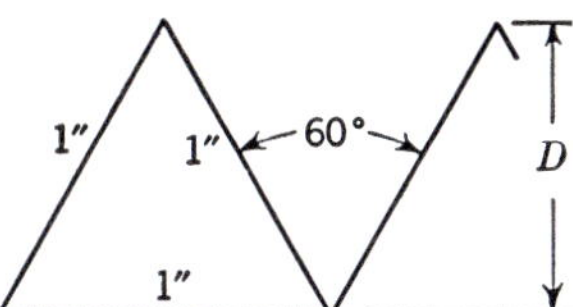

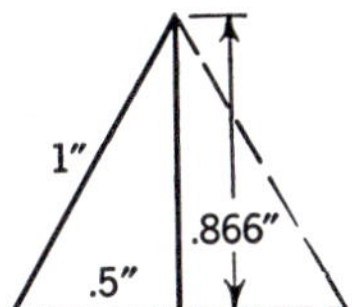

it forms an equilateral triangle. If 1 in. is taken as the pitch (distance between successive points), the height of the thread, commonly called the **depth,** may be computed by taking one-half the thread to form the right triangle shown.

$$D = \sqrt{h^2 - b^2}$$

$$D = \sqrt{1^2 - .5^2} = \sqrt{.75} = .866 \text{ in., or depth}$$

Therefore, for 1 in. pitch the depth is 0.866 in.

On a bolt the **double depth** D_o must be known.

$$D_o = .866 \text{ in.} \times 2 = 1.732 \text{ in. for 1 in. pitch}$$

If we divide D or D_o by any number, as N (number of threads per inch), we can find the D or D_o of a thread of any pitch.

Rule 1.—To find the depth of a V thread, divide .866 by the number of threads per inch.

D = depth.

N = number of threads per inch.

Formula 1. $D = \frac{.866}{N}.$

Example 1.—Get the depth of a sharp V thread when there are 3 threads per inch.

$D = \frac{.866}{N}$	Formula
$D = \frac{.866}{3}$	Substituting
$D = .289-''$	

Rule 2.—To find the double depth of a V thread, divide 1.732 by the number of threads per inch.

D_o = double depth.

Formula 2. $D_o = \frac{1.732}{N}.$

Example 2.—Find the double depth of a $\frac{1}{3}''$ pitch sharp V thread.

$D_0 = \frac{1.732}{N}$	Formula
$D_0 = \frac{1.732}{3}$	Substituting
$D_0 = .577+''$	

EXERCISES AND PROBLEMS

Write the formula each time, and solve.

1. Find the depth of threads for each of the following threads per inch: 8, 9, 10, 11, 12, 13, 14, 15, 16, 17, 18, 19, 20, 21, 22, 23, 24, 25, 26, 27, 28, 29, 30, 31, 32.

2. Find the double depth for each part of Prob. 1, above.
3. If D_o is known, derive a short formula for obtaining D.

SIZE OF DRILL FOR TAP FOR SHARP V THREAD

The size of hole to drill, into which to cut threads for a screw, is called **tap drill size** or **size of drill for tap.** The tap T is the same diameter as the stock for the screw. The drill size S is the outside diameter of the stock **minus** the double depth of the thread to be cut.

Rule.—Size of drill for a tap equals the tap diameter minus the double depth.

S = size of drill.
T = tap diameter.
N = number of threads per inch, or pitch.
$\frac{1.732}{N}$ = double depth for sharp V thread.

Formula. $S = T - \frac{1.732}{N}$.

Example.—Find the size of drill for a V-thread $\frac{1}{2}''$ tap of 13 pitch.

$S = T - \frac{1.732}{N}$	Formula
$S = \frac{1}{2} - \frac{1.732}{13}$	Substituting
$S = .500 - .133$	Change both to decimals.
$S = .367''$	The correct diameter of drill expressed in decimals. Drills commonly found in the shop are graduated in 64ths, and it is necessary to change .367″ to 64ths of an inch (see pages 71 and 72).

$S = \frac{.367}{1} \times \frac{64}{64} = \frac{23.488}{64}$ or $\frac{23''}{64}$, the nearest possible size

EXERCISES AND PROBLEMS

In the following problems, write the formula each time, and find the decimal size and the nearest 64th size of drill for each tap. Do not write in this book.

Coarse thread series					Fine thread series				
Number	Tap, inches	Pitch	Drill size		Number	Tap, inches	Pitch	Drill size	
			Decimals	64ths				Decimals	64ths
1	$\frac{1}{4}$	20			**13**	$\frac{1}{4}$	28		
2	$\frac{5}{16}$	18			**14**	$\frac{5}{16}$	24		
3	$\frac{3}{8}$	16			**15**	$\frac{3}{8}$	24		
4	$\frac{7}{16}$	14			**16**	$\frac{7}{16}$	20		
5	$\frac{1}{2}$	13			**17**	$\frac{1}{2}$	20		
6	$\frac{9}{16}$	12			**18**	$\frac{9}{16}$	18		
7	$\frac{5}{8}$	11			**19**	$\frac{5}{8}$	18		
8	$\frac{11}{16}$	11			**20**	$\frac{11}{16}$	16		
9	$\frac{3}{4}$	10			**21**	$\frac{3}{4}$	16		
10	$\frac{7}{8}$	9			**22**	$\frac{7}{8}$	14		
11	1	8			**23**	1	14		
12	$1\frac{1}{4}$	8			**24**	$1\frac{1}{4}$	12		

SIZE OF DRILL FOR TAP FOR NATIONAL FORM OF THREAD

The **National form of thread,** adopted by the National Screw Thread Commission in 1921 and since accepted by practically all manufacturers and users of threads and threading equipment, has replaced the full V thread in the United States almost entirely.

The National form of thread has two thread series, one called **Coarse** and the other called **Fine.** The National Coarse (**N.C.**) series replaces the old U.S. Standard (**U.S.S.**), or Seller's, type. The National Fine (**N.F.**) series replaces the Society of Automotive Engineers' thread (**S.A.E.**). The American Society of Mechanical Engineers' thread (**A.S.M.E.**), commonly known as machine screw thread series, has been included in the National Coarse and National Fine series with many of the sizes eliminated.

On the National form of thread, **one-eighth** the depth of a V thread has been left flat both at the crest and at the root. This makes a total of one-fourth the depth that has been deducted from the depth of a V thread. Since one-fourth off leaves three-fourths, this makes the National

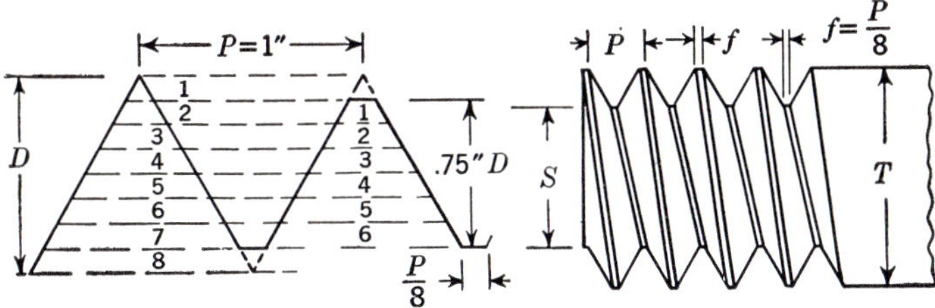

form of thread only 75 per cent (0.75) of the depth of a full V thread.

D of National threads $= .866 \times .75 = .6495$ in. for 1 in. pitch

D_o of National threads $= 1.732 \times .75 = 1.299$ in.

The last value is so near 1.3 that 1.3 is usually taken as the double depth of 1″ threads.

Rule.—Size of drill for tap equals the tap diameter minus the double depth, $\frac{1.3}{N}$.

Formula. $S = T - \frac{1.3}{N}$.

Example.—Find the size of drill for an N.C. tap ½″ diameter, 13 pitch.

$S = T - \frac{1.3}{N}$	Formula for national form
$S = \frac{1}{2} - \frac{1.3}{13}$	Substituting
$S = .500 - .100$	Changing both fractions to decimals.
$S = .400''$	The correct size expressed in decimal form
$\frac{.400}{1} \times \frac{64}{64} = \frac{25.600}{64}$, or $\frac{26}{64} = \frac{13}{32}$	This decimal changed to 64ths
$S = \frac{13}{32}''$, the nearest size in common fractions	

EXERCISES AND PROBLEMS

In the following, write the formula each time, and find the decimal size and the nearest 64th size of drill for each tap.

Full-depth National Form of Thread

National Coarse (N.C.) (Old U.S.S.)					National Fine (N.F.) (Old S.A.E.)				
Number	Tap, inches	Pitch	Drill size: Decimals	Drill size: 64ths	Number	Tap, inches	Pitch	Drill size: Decimals	Drill size: 64ths
1	$\frac{1}{4}$	20			**15**	$\frac{1}{4}$	28		
2	$\frac{5}{16}$	18			**16**	$\frac{5}{16}$	24		
3	$\frac{3}{8}$	16			**17**	$\frac{3}{8}$	24		
4	$\frac{7}{16}$	14			**18**	$\frac{7}{16}$	20		
5	$\frac{1}{2}$	13			**19**	$\frac{1}{2}$	20		
6	$\frac{9}{16}$	12			**20**	$\frac{9}{16}$	18		
7	$\frac{5}{8}$	11			**21**	$\frac{5}{8}$	18		
8	$\frac{11}{16}$	11			**22**	$\frac{11}{16}$	16		
9	$\frac{3}{4}$	10			**23**	$\frac{3}{4}$	16		
10	$\frac{7}{8}$	9			**24**	$\frac{7}{8}$	14		
11	1	8			**25**	1	14		
12	$1\frac{1}{4}$	8			**26**	$1\frac{1}{4}$	12		
13	$1\frac{1}{2}$	6			**27**	$1\frac{1}{2}$	12		
14	2	4.5			**28**	2	10		

75 PER CENT NATIONAL FORM OF THREAD

Engineers have found from tests that a common nut drilled so that it contains only 50 per cent of a full-depth thread will break the bolt before the thread will strip. A 75 per cent depth thread yields ample safety against stripping and is more economical in tapping. A full-depth thread is said to be only 5 per cent stronger than a 75 per cent depth, and yet it requires three times the power to tap. It has become accepted practice to drill holes for only a

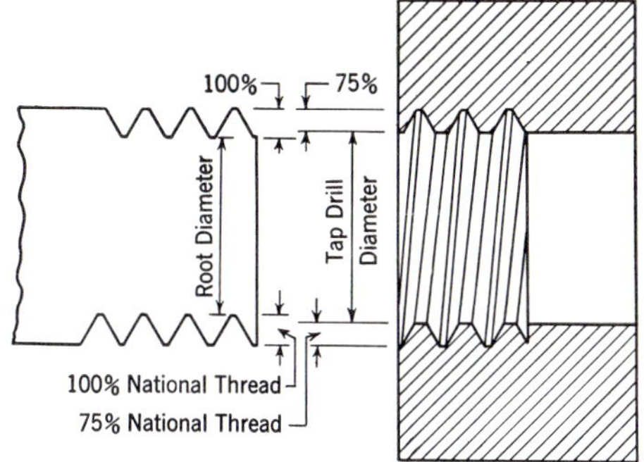

75 per cent thread. Seventy-five per cent of 1.3 gives the D_o of this thread.

$$.75 \times 1.3 = .975, \text{ the } D_o \text{ for a 1-in. thread}$$

Rule.—The size of drill for tap for a 75 per cent thread equals the tap diameter minus the double depth, $\frac{.975}{N}$.

Formula. $S = T - \frac{.975}{N}.$

Example.—Find the size of drill for a 75% depth N.C. $\frac{1}{2}''$ 13-pitch tap.

$S = T - \frac{.975}{N}$	Formula for 75% thread
$S = \frac{1}{2} - \frac{.975}{13}$	Substituting
$S = .500 - .075$	Change both fractions to decimals
$S = .425''$, the correct diameter expressed in decimals	
	This decimal changed to 64ths becomes
$\frac{.425}{1} \times \frac{64}{64} = \frac{27.200}{64}$, or $\frac{27''}{64}$, the nearest possible drill size	

EXERCISES AND PROBLEMS

In the following, write the formula each time, and find the decimal size and the nearest 64th size of drill for each tap.

National Coarse 75% thread					National Fine 75% thread				
Number	Tap, inches	Pitch	Drill size		Number	Tap, inches	Pitch	Drill size	
			Decimals	64ths				Decimals	64ths
1	$\frac{1}{4}$	20			**15**	$\frac{1}{4}$	28		
2	$\frac{5}{16}$	18			**16**	$\frac{5}{16}$	24		
3	$\frac{3}{8}$	16			**17**	$\frac{3}{8}$	24		
4	$\frac{7}{16}$	14			**18**	$\frac{7}{16}$	20		
5	$\frac{1}{2}$	13			**19**	$\frac{1}{2}$	20		
6	$\frac{9}{16}$	12			**20**	$\frac{9}{16}$	18		
7	$\frac{5}{8}$	11			**21**	$\frac{5}{8}$	18		
8	$\frac{11}{16}$	11			**22**	$\frac{11}{16}$	16		
9	$\frac{3}{4}$	10			**23**	$\frac{3}{4}$	16		
10	$\frac{7}{8}$	9			**24**	$\frac{7}{8}$	14		
11	1	8			**25**	1	14		
12	$1\frac{1}{4}$	8			**26**	$1\frac{1}{4}$	12		
13	$1\frac{1}{2}$	6			**27**	$1\frac{1}{2}$	12		
14	2	4.5			**28**	2	10		

The sizes figured in this lesson are the same as those found in most tap-drill-size charts

WIDTH OF FLAT AT ROOT AND CREST ON THE NATIONAL FORM OF THREAD

In order to sharpen a lathe tool bit to cut a thread properly, the width across the **flat** of the thread must be known.

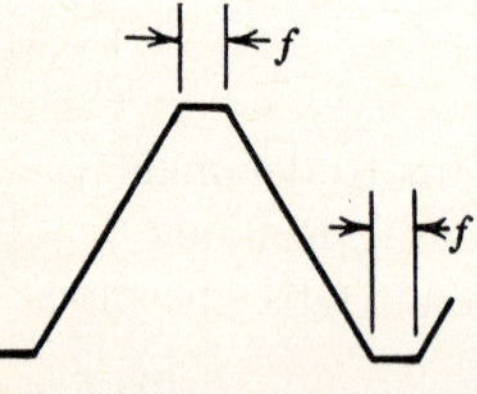

Since one-eighth of a V thread is removed at the crest and root to make the National form of thread, it follows that one-eighth the pitch will be the width of the flat and likewise the width of the point of the lathe tool bit. Let f represent the width of the flat; then, we may state the following:

Rule.—To find the flat, multiply the pitch in inches by one-eighth.

$$\frac{1}{N} = \text{pitch, inches.}$$

$$f = \text{width of flat at crest and root.}$$

Formula. $f = \frac{1}{8} \times \frac{1}{N}$, or $f = \frac{1}{8N}$.

Example.—Get the width of flat on a $\frac{7}{8}''$ 9-pitch thread. (9-pitch thread = $\frac{1}{9}''$ pitch. The $\frac{7}{8}''$ need not be considered.)

$f = \frac{1}{8} \times \frac{1}{N}$	Formula
$f = \frac{1}{8} \times \frac{1}{9}$	Substituting
$f = \frac{1}{72}''$	
$f = .0138''+$	Changed to decimal
Ans. 0.014″	the nearest thousandth

EXERCISES AND PROBLEMS

1. Find the width of flat at the crest of a $\frac{1}{8}''$ pitch thread.
2. What is the flat of a $\frac{5}{8}''$ diameter 11-pitch thread?
3. Figure the flat at the root of a 20-pitch thread.

Find the width of the point of a lathe tool bit for each of the sizes, both National Coarse and National Fine threads, given in the chart below:

National Coarse				National Fine			
Number	Tap, inches	Pitch	Width of flat, decimal inches	Number	Tap, inches	Pitch	Width of flat, decimal inches
4	$\frac{1}{4}$	20		**16**	$\frac{1}{4}$	28	
5	$\frac{5}{16}$	18		**17**	$\frac{5}{16}$	24	
6	$\frac{3}{8}$	16		**18**	$\frac{3}{8}$	24	
7	$\frac{7}{16}$	14		**19**	$\frac{7}{16}$	20	
8	$\frac{1}{2}$	13		**20**	$\frac{1}{2}$	20	
9	$\frac{9}{16}$	13		**21**	$\frac{9}{16}$	18	
10	$\frac{5}{8}$	11		**22**	$\frac{5}{8}$	18	
11	$\frac{11}{16}$	11		**23**	$\frac{11}{16}$	16	
12	$\frac{3}{4}$	10		**24**	$\frac{3}{4}$	16	
13	$\frac{13}{16}$	10					
14	$\frac{7}{8}$	9		**25**	$\frac{7}{8}$	14	
15	1	8		**26**	1	14	

THREAD CUTTING ON THE LATHE

One of the principal uses of the lathe is thread cutting. A good lathe is usually designated as a **screw cutting lathe.** The modern lathe with quick-change gears requires no computation to set it up to cut almost any desired thread. However, many lathes are in use that do not have quick-change gears. The gearing needed to give the proper r.p.m. to the work must be determined by calculation, gears selected from those supplied with the lathe and placed properly on the shafts made for the purpose.

In thread cutting the work must make the same number of revolutions as the threads per inch required while the tool feeds 1 in. along the work. The tool feed is determined by the number of threads per inch on the lead screw and the gearing between the work spindle and the lead screw.

If the spindle and the lead screw are geared 1 to 1, the spindle will make the same number of revolutions as the lead screw. With 1–1 gearing a lead screw having 6 threads per inch makes 6 revolutions while the carriage, with the cutting tool mounted on it, is feeding 1 in. along the work, and 6 threads will be cut.

If the gearing between the lead screw and spindle is 2 to 1, the spindle will make twice as many turns and 12 threads will be cut.

The ratio between the lead screw and the spindle may be changed by placing different change gears on the lead-screw shaft and the stud-gear shaft, which is in between the lead screw and the spindle.

The change gears furnished with the 13-, 15- and 16-in. South Bend lathes with 6-pitch lead screws have teeth as follows: 24, 32, 40, 44, 46, 48, 52, 56, 60, 64, 72, and 80.

SIMPLE GEARING

When the motion is transmitted from the spindle gear, through the fixed-stud gear, the change-stud gear, and the intermediate-idler gear to the lead-screw gear, the process is called **simple gearing.**

The ratio of the lead-screw thread to the thread being cut is the same as the ratio of the change-stud gear to the lead-screw gear. This may be expressed as follows:

$$\frac{\text{Threads per inch of lead screw}}{\text{Threads per inch being cut}} = \frac{\text{teeth in change-stud gear}}{\text{teeth in lead-screw gear}}$$

A rule for this may be expressed as follows:

Rule.—To find the gear ratio between the lead screw and the spindle, divide the threads per inch of the lead screw by the threads per inch being cut.

Let R = ratio.
L = threads per inch on lead screw.
T = threads per inch being cut.

Formula. $R = \frac{L}{T}.$

Example 1.—What is the gear ratio required to cut 16 threads per inch on a $\frac{3}{8}''$ bolt, and what gears should be used?

$R = \frac{L}{T} = \frac{6}{16}$	*Note.*—The diameter of the bolt does not affect the gearing calculations. Since this ratio equals the ratio of the stud gear to the lead-screw gear, gears must be selected that will give this ratio. Both terms of the ratio must be multiplied by the same number so that the results will represent gears in the set listed on page 426.
$\frac{6}{16} \times \frac{4}{4} = \frac{24}{64}$	24 teeth on the change-stud gear 64 teeth on the lead-screw gear

Example 2.—What is the gearing to cut $11\frac{1}{2}$ threads per inch?

$R = \frac{L}{T} = \frac{6}{11\frac{1}{2}} \times \frac{4}{4} = \frac{24}{46}$	24-tooth stud gear 46-tooth lead-screw gear

Example 3.—What is the gearing to cut $4\frac{1}{2}$ threads per inch?

$$R = \frac{L}{T} = \frac{6}{4\frac{1}{2}} = \frac{12}{9} = \frac{4}{3} \times \frac{8}{8} = \frac{32}{24} \quad \begin{matrix}\text{Stud gear} \\ \text{Lead-screw gear}\end{matrix}$$

EXERCISES AND PROBLEMS

Use the lathe geared 1 to 1 with the change gears listed on page 426 and a 6-pitch lead screw for the problems below.

What is the gearing necessary to cut the threads per inch in the following problems?

Number	Stock diameter, inches	Threads per inch	Number	Stock diameter, inches	Threads per inch
1	$\frac{1}{2}$	20	**7**	$\frac{3}{4}$	10
2	$\frac{5}{16}$	18	**8**	$\frac{7}{8}$	9
3	$\frac{7}{16}$	14	**9**	1	8
4	$\frac{1}{2}$	13	**10**	$1\frac{1}{8}$	7
5	$\frac{9}{16}$	12	**11**	$1\frac{1}{2}$	6
6	$\frac{5}{8}$	11	**12**	$1\frac{3}{4}$	5

13. 2 threads per inch.
14. 3 threads per inch.
15. 4 threads per inch.
16. $5\frac{1}{2}$ threads per inch.
17. $5\frac{3}{4}$ threads per inch.
18. $6\frac{1}{2}$ threads per inch.
19. $7\frac{1}{2}$ threads per inch.
20. $2\frac{1}{4}$ threads per inch.
21. 15 threads per inch.

Other threads per inch may be cut by simple gearing, simply by changing the ratio between the spindle gear and the fixed-stud gear. In the problems above this ratio was 1 to 1. The lathe described above includes gears that will make this ratio 2 to 1. Other ratios, as 3 to 1, are possible on other lathes.

In using the same set of change gears and the ratio of 2 to 1 between the spindle gear and the fixed-stud gear, it will be necessary to multiply $\frac{L}{T}$ by $\frac{2}{1}$.

Formulas. $R = \frac{L}{T} \times \frac{2}{1}$, or $R = \frac{2L}{T}$.

Example.—What change gears will be used to cut 22 threads per inch, the lathe above being used, with a fixed ratio of 2 to 1?

$$R = \frac{L}{T} \times \frac{2}{1}$$

$$R = \frac{6}{22} \times \frac{2}{1} = \frac{12}{22} \times \frac{2}{2} = \frac{24}{44} \quad \begin{array}{l}\text{Stud gear}\\ \text{Lead-screw gear}\end{array}$$

EXERCISES AND PROBLEMS

Use the lathe and change gears above and the ratio 2 to 1 to get the gearing for the following problems:

1. 24 threads per inch.
2. 26 threads per inch.
3. 28 threads per inch.
4. 30 threads per inch.
5. 32 threads per inch.
6. 36 threads per inch.
7. 40 threads per inch.

Many of the threads cut by the lathe geared 1 to 1 may also be cut with 2 to 1 gearing. Compare the following with the results of the previous section.

8. 4-pitch thread.
9. 6-pitch thread.
10. 8-pitch thread.
11. 10-pitch thread.
12. 12-pitch thread.
13. 13-pitch thread.
14. 14-pitch thread.
15. 15-pitch thread.
16. 16-pitch thread.
17. 18-pitch thread.
18. 20-pitch thread.

COMPOUND GEARING

Sometimes it is impossible to obtain the correct ratio with the simple gearing needed to cut the threads per inch desired. It is then necessary to place two additional gears in the gear train. These are put on the idler shaft, one meshing with the lead-screw gear and the other meshing with the change-stud gear.

Compound gearing requires factoring the ratio $\frac{L}{T}$, then finding gears to use with each factor.

Example 1.—Find the compound gearing to use to cut 16 threads per inch on a lathe geared 1 to 1.

$$R = \frac{L}{T} = \frac{6}{16}$$

$\frac{6}{16} = \frac{2}{4} \times \frac{3}{4}$ Factoring

$\frac{2}{4} \times \frac{20}{20} = \frac{40}{80}$ 40-tooth stud gear (driver)
80-tooth compound gear on idler shaft (driven)

$\frac{3}{4} \times \frac{8}{8} = \frac{24}{32}$ 24-tooth compound gear (driver)
32-tooth lead-screw gear (driven)

Example 2.—Find the compound gearing to cut 11 threads per inch on a lathe geared 2 to 1.

$$R = \frac{L}{T} \times \frac{2}{1} = \frac{6}{11} \times \frac{2}{1} = \frac{12}{11}$$

$\frac{12}{11} = \frac{3 \times 4}{2 \times 5\frac{1}{2}}$ Factoring

$\frac{3}{2} \times \frac{20}{20} = \frac{60}{40}$ 60-tooth stud gear (driver)
40-tooth compound gear (driven)

$\frac{4}{5\frac{1}{2}} \times \frac{8}{8} = \frac{32}{44}$ 32-tooth compound gear (driver)
44-tooth lead-screw gear (driven)

Note.—The gears could be arranged differently. The 60-tooth gear could be changed with the 32-tooth gear and the 40-tooth gear could be changed with the 44-tooth gear. The 60 and 32 must be kept as drivers and the 40 and 44 as drivens.

EXERCISES AND PROBLEMS

With the same lathe, geared 1 to 1, determine the compound gearing to cut threads of the following pitches:

1. 42 pitch. **2.** 45 pitch. **3.** 48 pitch. **4.** 10 pitch.
5. $4\frac{1}{2}$ pitch. **6.** 5 pitch. **7.** 12 pitch. **8.** 8 pitch.

With the same lathe, geared 2 to 1, determine the compound gearing to cut the following threads:

9. 40 pitch. **10.** 42 pitch. **11.** 44 pitch. **12.** 50 pitch.
13. 16 pitch. **14.** 48 pitch. **15.** 56 pitch. **16.** 64 pitch.
17. 72 pitch. **18.** 80 pitch.

CHAPTER 19

MILLING-MACHINE WORK

DIRECT INDEXING

In the machine shop, many pieces of work, like gears and reamers, must be machined very accurately on several equally spaced faces. This type of work is best done on the **milling machine.** The work is secured to a mechanism called an **index head,** dividing head, or spiral head. The most common name is index head. The index head is constructed so that the work can be rotated through any part of a revolution. The operation of rotating the work with the index head is called **indexing.**

The quickest type of indexing is called **direct indexing** or quick indexing. Direct indexing is possible only with a few divisions of the circle. For direct indexing the internal gearing of the dividing head is disengaged. An **index plate,** with 24, 30, or 36 holes around it, is fastened to the front of the work spindle. There is a plunger pin fastened to the head, which may be put into the desired hole in the plate. When the plate is turned by hand the required part of a revolution, the work also turns the same part of a revolution.

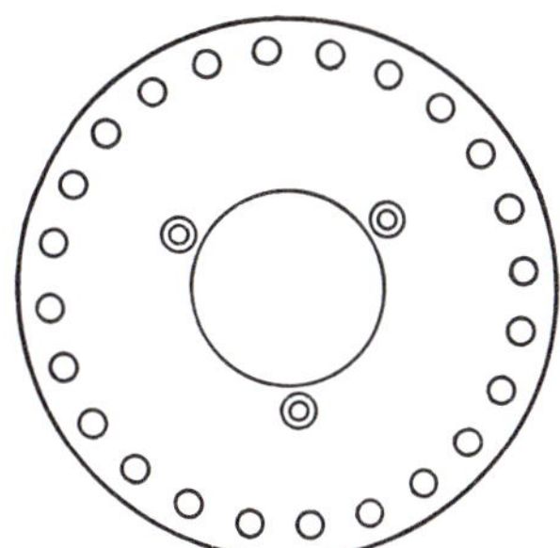

Index plate for direct indexing.

With the 24-hole plate, direct indexing is possible for all the factors of 24, as 1, 2, 3, 4, 6, 8, 12, or 24. With a 30-hole plate, 1, 2, 3, 5, 6, 10, or 15 divisions are possible. With a 36-hole plate, indexing 1, 2, 3, 4, 6, 9, 12, 18, or 36 is possible.

Example 1.—What indexing and which plate can be used to cut 6 lands (sides) on a reamer, as shown?

Milling 6 lands by direct indexing. (*Courtesy of Brown & Sharpe Manufacturing Company.*)

6 sides require $\frac{1}{6}$ revolution.
$\frac{1}{6}$ of 24 = 4 holes in the 24-hole plate
or $\frac{1}{6}$ of 30 = 5 holes in the 30-hole plate
or $\frac{1}{6}$ of 36 = 6 holes in the 36-hole plate

A few angles may be indexed by the direct method. Indexing for angles is called **angular indexing.**

Example 2.—What is the indexing to divide a circle into 30° sectors?

> There are 360° in a circle.
>
> $\frac{30}{360} = \frac{1}{12}$ revolution required for each sector
>
> $\frac{1}{12}$ of 24 = 2 holes in the 24-hole plate
>
> $\frac{1}{12}$ of 36 = 3 holes in the 36-hole plate

PROBLEMS

Use either of the plates described above.

1. Direct-index to mill 8 flutes on a hand reamer.

2. Direct-index to mill 2 flutes on a drill.

3. Direct-index to mill 3 flutes on a hand tap.

4. Direct-index to square the end of a shaft.

5. Direct-index to cut an equilateral triangle.

6. Direct-index to cut a pentagon.

7. Direct-index to cut an octagon.

8. Direct-index for milling a 10-spline shaft.

9. With the 30-hole plate, direct index for the following angles: 12°, 24°, 36°, 60°, 72°, 120°, 180°.

10. With the 24-hole plate, direct-index for the following angles: 15°, 30°, 45°, 60°, 90°, 120°, 180°.

11. With the 36-hole plate, direct-index for the following angles: 10°, 20°, 30°, 40°, 60°, 90°, 120°, 180°.

SIMPLE INDEXING

Perhaps the most common work accomplished with the index head is that called **simple indexing.**

For simple indexing the internal worm gearing of the head is engaged and the index crank is attached solidly to the end of the worm shaft, with the worm solid with the worm gear. An index plate is secured in place by means of the "back pin" or index-plate locking pin.

Interchangeable index plates, as shown, come with each index head. Each plate has several concentric rows of equally spaced holes, which makes it possible to obtain the spacing ordinarily used for gears, cutters, etc., but not possible by direct indexing.

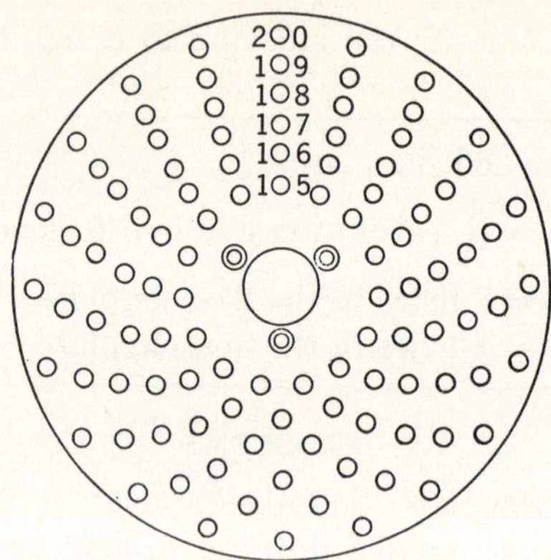

Brown and Sharpe index plate 1.

The Brown and Sharpe Manufacturing Company furnishes, with its index head, plates with circles of holes as follows:

Plate 1: 15, 16, 17, 18, 19, and 20.
Plate 2: 21, 23, 27, 29, 31, and 33.
Plate 3: 37, 39, 41, 43, 47, and 49.

Milling flutes on a reamer by simple indexing. (*Courtesy of Brown & Sharpe Manufacturing Company.*)

The Cincinnati Milling Machine Company furnishes one plate marked on both sides, which has circles of holes as follows:

Front, 24, 25, 28, 30, 34, 37, 39, 41, 42, and 43.
Back, 46, 47, 49, 51, 53, 57, 58, 59, 62, and 66.

This company makes several other plates and a "wide-range" divider unit with divisions of 2 to 400,000.

In all the examples on pages 436 to 448, the use of the term **plate 1, plate 2,** or **plate 3,** means that to index, one should use that particular index plate (described on page 434) on the Brown and Sharpe milling machine. The use of the term **front** or **back** means that to index one should use the front or back of the plate (described above) on the Cincinnati milling machine.

In the index head the reduction between the worm and the worm wheel is 40 to 1. One turn of the index crank will rotate the work one-fortieth of a turn. Forty turns of the crank will rotate the work one full turn. Thus, it is evident that if 40 is divided by the number of spaces desired it will give the revolutions of the crank needed to make the proper cuts.

Rule.—To calculate the turns of the index crank to simple-index, divide 40 by the number of equal divisions required.

Let T = turns of the crank.
N = number of divisions required.

Formula. $T = \frac{40}{N}.$

Example 1.—How many turns of the crank are required to cut 20 teeth on a gear?

$T = \frac{40}{N} = \frac{40}{20} = 2$ turns of the crank. To accomplish this, put on any index plate, pull the crank pin, and turn the crank 2 revolutions, putting the pin back in the same hole.

Example 2.—Index to mill a hexagon.

$$T = \frac{40}{N} = \frac{40}{6} = 6\tfrac{2}{3} \text{ turns of the crank}$$

To index $6\frac{2}{3}$ turns, select a plate with a circle divisible by 3. For example, with Brown and Sharpe plates, page 434, choose from among the following:

Plate 1, 6 turns and 10 holes on the 15-hole circle
6 turns and 12 holes on the 18-hole circle
Plate 2, 6 turns and 14 holes on the 21-hole circle
6 turns and 18 holes on the 27-hole circle
6 turns and 22 holes on the 33-hole circle
Plate 3, 6 turns and 26 holes on the 39-hole circle

On the Cincinnati machine, this indexing is possible on the 24-, 30-, 39-, 42-, 51-, 57-, and 66-hole circles.

Example 3.—What is the indexing to cut 230 teeth on a gear?

$$T = \frac{40}{N} = \frac{40}{230} = \frac{4}{23} \text{ revolution}$$

To index $\frac{4}{23}$ revolution, use one of the following:
Plate 2, page 434, 4 spaces on the 23-hole circle
Back, page 435, 8 spaces on the 46-hole circle

PROBLEMS

Simple-index the following, finding all the possibilities when the plates of both machines described above are used:

1. What is the simple indexing for a gear with 160 teeth?
2. What is the proper indexing for a 24-tooth gear?
3. What is the simple indexing for a 10-spline shaft?
4. Simple-index to cut an 86-tooth gear.

Simple-index for the following divisions:

5. 330.	**6.** 132.	**7.** 51.	**8.** 92.	**9.** 75.	**10.** 80.
11. 85.	**12.** 90.	**13.** 100.	**14.** 102.	**15.** 108.	**16.** 105.
17. 106.	**18.** 500.	**19.** 740.	**20.** 1000.	**21.** 1320.	**22.** 1120.

ANGULAR INDEXING

It is possible to angular-index for a great many angular divisions by the simple indexing method.

As shown on page 435, one turn of the index crank gives one-fortieth of a turn to the work. There are 360° in a circle. Therefore, $\frac{1}{40} \times 360 = 9°$ for one turn of the crank; then one-ninth of a turn equals 1°. One-ninth of a turn is possible by using one of the following:

2 spaces on the 18-hole circle.
3 spaces on the 27-hole circle.

Example 1.—What is the simple angular indexing for 19°?

> $19° = 2\frac{1}{9} \times 9°$ Hence, index as follows:
> Plate 1, page 434,
> 19° = 2 turns + 2 holes on the 18-hole circle
> or 2 turns + 3 holes on the 27-hole circle

Example 2.—Simple angular index for $\frac{1}{2}°$

> $\frac{1}{2}°$ will require $360 \times 2 = 720$ divisions.
> $T = \frac{40}{N} = \frac{40}{720} = \frac{2}{36} = \frac{1}{18}$ Hence, for $\frac{1}{2}°$, index as follows:
> Plate 1, page 434, 1 hole on the 18-hole circle

Example 3.—What is the simple angular index for 5°40′?

> $40' = \frac{2}{3}°$
> $5° = 5 \times \frac{3}{3} = \frac{15}{3}$
> $\frac{15}{3} + \frac{2}{3} = \frac{17°}{3}$ One hole on the 27-hole circle is $\frac{1}{3}°$. Then to index $\frac{17}{3}°$ (5°40′), take
> Plate 2, page 434, 17 holes on the 27-hole circle

PROBLEMS

Simple angular-index for the following angles:

1. 2.5°. **2.** 5°. **3.** 10°. **4.** 15°. **5.** 20°.
6. 30°. **7.** 40°. **8.** 60°. **9.** 6°. **10.** 4°.
11. 20′. **12.** 30′. **13.** 40′. **14.** 12°. **15.** 24°.
16. 7°40′. **17.** 3°20′. **18.** 20°20′.

COMPOUND INDEXING

Compound indexing may be used to obtain divisions that cannot be obtained by simple indexing. In compound indexing the crank is first turned as in simple indexing; then the index plate is turned in the opposite direction, carrying the crank with it. The resulting sum or difference of a turn represents the part of a circle turned by the work.

Compound indexing on a spur gear. (*Courtesy of Brown & Sharpe Manufacturing Company.*)

Compound indexing may be accomplished by applying the following rule:

Rule.—To compound-index, multiply the ratio $\frac{40}{N}$ by the product of two circles of holes, one being the outer circle, in any index plate, divided by the difference between the same two circles of holes.

Note.—The result, thus found, must be a whole number. If it is not, other circles of holes must be selected.

Example 1.—Compound-index for 99 divisions, and prove the result.

$$T = \frac{40}{N} = \frac{40}{99}.$$

That is, $\frac{40}{99}$ of a turn is required.

Try circles 33 and 27, plate 2, page 434.

33×27 = the product of the two circles of holes

$33 - 27$ = 6, the difference

$$\frac{40}{99} \times \frac{33 \times 27}{6} = 60 \text{ holes}$$

This means that 60 holes are required in each circle in opposite directions.

To index,

1. Turn crank forward 60 holes on 33-hole circle (1 turn + 27 holes).
2. Turn plate with crank backward 60 holes on the 27-hole circle (2 turns + 6 holes).

Proof:

$\frac{60}{27} - \frac{60}{33} = \frac{20}{9} - \frac{20}{11} = \frac{220}{99} - \frac{180}{99} = \frac{40}{99}$ of a turn, as required

Example 2.—Compound-index for 87 divisions, and prove the result.

$$T = \frac{40}{N} = \frac{40}{87} \text{ of a turn required}$$

Try circles 33 and 21, plate 2, page 434.

$$\frac{40}{87} \times \frac{33 \times 21}{12} = 26\frac{16}{29}, \text{ which is not a whole number}$$

Try circles 33 and 29, plate 2.

$$\frac{40}{87} \times \frac{33 \times 29}{4} = 110$$

This means that 110 holes are required in each circle in opposite directions.

To index,

1. Turn crank forward 3 turns + 11 holes on the 33-hole circle.
2. Turn plate, with crank, backward 3 turns + 23 holes on the 29-hole circle.

Proof:

$\frac{110}{29} - \frac{110}{33} = \frac{3630}{957} - \frac{3190}{957} = \frac{440}{957} = \frac{40}{87}$ of a turn, as required

PROBLEMS

Use any of the three plates on the Brown and Sharpe milling machine to compound-index the problems below. Prove all results.

1. What is the compound indexing for a 294-tooth gear?

2. What is the compound indexing to cut a gear with 147 teeth?

3. What is the compound indexing to cut 154 teeth on a gear?

4. How should you compound-index to cut 182 teeth on a gear?

5. Index by compound indexing to cut 87 divisions.

6. Index by compound indexing for 231 divisions.

7. How should you compound-index to divide a circle in 138 parts?

8. What is the compound indexing to give 91 divisions?

9. How can 174 divisions be made by compound indexing?

10. Compound-index to divide a circle in 99 parts.

11. Compound-index for 77 divisions.

12. What is the compound indexing to give 186 divisions?

DIFFERENTIAL INDEXING

The possibility of error in manually turning the index plate, as in compound indexing, is eliminated in **differential indexing.** Suitable change gears, like the change gears on a lathe, are supplied with the index head. These are installed on the index head so that turning the crank gives a differential motion to the index plate, either forward or backward, depending on the gearing. The term "differential" is used because the needed divisions are obtained by a combination of two movements, (1) turning the crank, as in simple indexing, and (2) the movement of the index plate itself, caused by the included gear train.

Differential indexing may take the place of compound indexing, but, more important, it makes possible a great many more divisions of the circle. The Brown and Sharpe index head has gears with teeth as follows: 24 (two gears), 28, 32, 40, 44, 48, 56, 64, 72, 86, and 100. With these gears, any number of divisions from 1 to 382 may be indexed by differential indexing. Many divisions from 383 to 1008 can be obtained with the gears listed above. By adding eight special gears all of the divisions from 383 to 1008 may be indexed. These gears have teeth as follows: 46, 47, 52, 58, 68, 70, 76, and 84.

Differential indexing may be accomplished by following the steps below:

1. Set up a ratio $\frac{40}{N_1}$, when N_1 is a number near the number of divisions (N) required that cannot be indexed by simple indexing.

2. Reduce the ratio $\frac{40}{N_1}$, and determine the simple indexing for it.

3. To find the differential-index number, multiply the number of divisions (N) required by the ratio of step (2) and reduce it to a mixed number.

4. Find the difference between the differential-index number and 40.

5. Resolve the difference into a simple ratio.

6*a*. If gears are available that equal this ratio, simple gearing may be used.

6*b*. If simple gearing cannot be used, compound the gearing by factoring the simple ratio and finding change gears that equal the factors.

Note.—If the differential-index number, found in step 3, is less than 40, use one idler (or the compound) in simple gearing. If it is more than 40, use two idlers (or one idler and the compound).

The ratio found in step 5 should not exceed 6 to 1, because of excessive gear stress.

If the differential-index number is less than 40, the index plate will turn in the same direction as the crank. If it is more than 40, the index plate will turn in the opposite direction from the crank.

Example 1.—Differential-index for 271 divisions, and prove the results.

Differential indexing for 271 divisions by simple gearing. See Example 1. (*Courtesy of Brown & Sharpe Manufacturing Company.*)

$$R = \frac{40}{N} = \frac{40}{271}$$ This is not possible by simple indexing with the plates furnished.

(1) Try 280 divisions as N_1.

(2) $$\frac{40}{N_1} = \frac{40}{280} = \frac{1}{7}$$ This is possible. Take 3 holes on the 21-hole circle, plate 2, page 434, or take 7 holes on the 49-hole circle, plate 3.

(3) $271 \times \frac{1}{7} = \frac{271}{7} = 38\frac{5}{7}$ Differential-index number

(4) $40 - 38\frac{5}{7} = 1\frac{2}{7}$ Difference

This means that $38\frac{5}{7}$ complete turns of the crank will be taken, which is $1\frac{2}{7}$ turns less than the 40 turns required for one complete turn of the work. By using gears with the ratio of $1\frac{2}{7}$ to 1, the index plate will make $1\frac{2}{7}$ revolutions, which with the $38\frac{5}{7}$ turns of the crank will give the 40 turns required.

(5) $1\frac{2}{7}:1 = \frac{9}{7}:1$ Difference resolved into a simple ratio

(6*a*) $\frac{9}{7} \times \frac{8}{8} = \frac{72}{56}$ By simple gearing. Gear *E* has 72 teeth. Gear *C* has 56 teeth. Use one idler, gear *D*, as shown.

To index, take 3 holes on the 21-hole circle in plate 2, page 434. Proof:

$\frac{1}{7} \times \frac{1}{40} = \frac{1}{280}$ of a turn due to the motion of the crank in the same direction as the gear motion

$\frac{9}{7} \times \frac{1}{40} \times \frac{1}{271} = \frac{9}{75,880}$ of a turn due to gear motion

$\frac{1}{280} + \frac{9}{75,880} = \frac{271}{75,880} + \frac{9}{75,880} = \frac{280}{75,880} = \frac{1}{271}$ of a turn

Example 2.—Differential index for 250 divisions, and prove the results.

Differential indexing for 250 divisions by simple gearing with two idler gears See Example 2. (*Courtesy of Brown & Sharpe Manufacturing Company.*)

$R = \frac{40}{N} = \frac{40}{250} = \frac{4}{25}$ This is not possible by simple indexing.

(1) Try 240 divisions as N_1.

(2) $\frac{40}{N_1} = \frac{40}{240} = \frac{1}{6}$ This is possible. Take 3 holes on the 18-hole circle, plate 1, page 434.

(3) $250 \times \frac{1}{6} = \frac{250}{6} = 41\frac{2}{3}$ Differential-index number

(4) $41\frac{2}{3} - 40 = 1\frac{2}{3}$ Difference

This means that $41\frac{2}{3}$ turns of the crank will be taken, or $1\frac{2}{3}$ of a turn more than the 40 turns required for one complete turn of the work. By using gears with the ratio of $1\frac{2}{3}$ to 1, together with two idlers to give a backward motion, the index plate will make $1\frac{2}{3}$ revolutions backward, which subtracted from $41\frac{2}{3}$ turns of the crank, will give the 40 turns required.

(5) $1\frac{2}{3}:1 = \frac{5}{3}:1$ Difference in a simple ratio

(6) $\frac{5}{3} \times \frac{8}{8} = \frac{40}{24}$ By simple gearing. Gear E has 40 teeth. Gear C has 24 teeth. Two idlers are used as shown. To index, take 3 holes on the 18-hole circle, plate 1, page 434.

Proof:

$$\frac{1}{6} \times \frac{1}{40} = \frac{1}{240} \text{ of a turn due to crank motion}$$

$$\frac{5}{3} \times \frac{1}{40} \times \frac{1}{250} = \frac{1}{6000} \text{ backward motion due to gears}$$

$$\frac{1}{240} - \frac{1}{6000} = \frac{25}{6000} - \frac{1}{6000} = \frac{24}{6000} = \frac{1}{250} \text{ of a turn}$$

Example 3.—Index for 319 divisions, and prove the results.

Differential indexing for 319 divisions by compound gearing and one idler gear. (*Courtesy of Brown & Sharpe Manufacturing Company.*)

$$R = \frac{40}{N} = \frac{40}{319}$$ This is not possible by simple indexing.

(1) Try 290 divisions as N_1.

(2) $\frac{40}{N_1} = \frac{40}{290} = \frac{4}{29}$ This is possible. Take 4 holes on the 29-hole circle, plate 2, page 434.

(3) $319 \times \frac{4}{29} = 44$ Differential index number.

(4) $44 - 40 = 4$ Difference

(5) $4:1$ The simple ratio

(6a) There is no simple gearing possible to give 4 to 1.

(6b) $\frac{4}{1}$ may be increased to $\frac{12}{3}$ to make factoring possible.

$$\frac{12}{3} = \frac{3 \times 4}{1 \times 3}$$ The new simple ratio factored

$$\frac{3}{1} \times \frac{24}{24} = \frac{72}{24}, \qquad \frac{4}{3} \times \frac{16}{16} = \frac{64}{48}, \qquad \frac{72 \times 64}{24 \times 48}$$

Gear E has 72 teeth, gear F has 64 teeth.

Gear G has 24 teeth, gear C has 48 teeth.

Use one idler, gear D, to reverse the direction, as shown.

To index, take 4 holes on the 29-hole circle, plate 2.

Proof:

$\frac{4}{29} \times \frac{1}{40} = \frac{1}{290}$ of a turn due to crank motion

$\frac{4}{1} \times \frac{1}{40} \times \frac{1}{319} = \frac{1}{3190}$ backward motion due to gears

$$\frac{1}{290} - \frac{1}{3190} = \frac{11 - 1}{3190} = \frac{10}{3190} = \frac{1}{319} \text{ of a turn}$$

PROBLEMS

Find the gearing and the plate to use by differential indexing for the following problems. Prove results.

1. 53 divisions (try 56).

2. 59 divisions (try 60).

3. 382 divisions.

4. 57 divisions (try 55 or 60).

5. 161 divisions (try 160).

6. Compute the differential indexing to cut a gear with 107 teeth.

7. Compute the differential indexing to cut a 173-tooth gear.

8. What is the differential indexing for 162 divisions?

9. Differential-index for 87 divisions.

10. Compute the differential indexing for cutting a gear with 201 teeth.

11. Compute the differential indexing for 227 divisions.

12. How should you differential-index to cut a gear with 149 teeth?

13. Differential-index for 111 divisions.

14. Differential-index for cutting a 71-tooth ring gear.

15. Compute the differential indexing for cutting a 51-tooth gear.

16. What is the differential indexing for a gear with 119 teeth?

17. Compute the differential indexing for a 149-tooth gear.

18. Differential-index for 352 divisions.

19. Compute the differential indexing to cut a gear with 139 teeth.

CALCULATING THE SIMPLE RATIO BY FORMULAS*

The following rule and formulas may be used to calculate the **simple ratio** of step 5 in the preceding section on differential indexing.

Rule.—Subtract the smaller product, the number of holes in the circle taken in the index plate multiplied by the ratio of gearing between index crank and spindle (usually 40), and the number of divisions required multiplied by the number of holes to be taken at each indexing, and divide this difference by the number of holes in the circle taken in the index plate.

Let s = simple ratio of the train of gearing.

R = 40, the ratio of gearing between index crank and spindle.

H = number of holes in circle taken in index plate.

N = number of divisions required.

n = number of holes taken at each indexing.

Formula 1. $s = \dfrac{RH - Nn}{H}$, when RH is greater than Nn.

Formula 2. $s = \dfrac{Nn - RH}{H}$, when RH is less than Nn.

* Adapted from *Practical Treatise on Milling*, published by the Brown and Sharpe Manufacturing Company.

Example 1.—Differential-index for 59 divisions (try $H = 33$; $n = 22$). N is 59.

(5) $$s = \frac{RH - Nn}{H}$$ Formula 1

$$s = \frac{(40 \times 33) - (59 \times 22)}{33}$$ Substituting

$$s = \frac{1320 - 1298}{33} = \frac{22}{33} = \frac{2}{3}$$ The simple ratio

(6a) $$\frac{2}{3} \times \frac{16}{16} = \frac{32}{48}$$ 32 teeth / 48 teeth

To index, take 22 holes on 33-hole circle, plate 2, page 434. By simple gearing, use one idler.

Example 2.—Index for $\frac{1}{4}°$ ($360 \div \frac{1}{4} = 1440$ divisions) (try $H = 33$; $n = 1$). N is 1440.

(5) $$s = \frac{Nn - RH}{H}$$ Formula 2

$$s = \frac{(1440 \times 1) - (40 \times 33)}{33}$$

$$s = \frac{1440 - 1320}{33} = \frac{40}{11}$$ The simple ratio

(6b) $$\frac{40}{11} = \frac{8 \times 5}{5 \times 2.2}$$ The simple ratio factored

$$\frac{8}{5} \times \frac{8}{8} = \frac{64}{40}, \quad \frac{5}{2.2} \times \frac{20}{20} = \frac{100}{44}, \quad \frac{64 \times 100}{40 \times 44}$$

In a setup similar to the illustration in Example 3, page 444, use a 64-tooth gear E, a 100-tooth gear F, a 40-tooth gear G, a 44-tooth gear C and one idler gear D.

To index, take 1 hole on the 33-hole circle, plate 2, page 434.

Note.—It is advisable to select an index circle H with a number having factors in the change gears available. If H contains a factor not in the gears, s cannot usually be obtained, unless the factor is canceled by the difference between RH and Nn or unless N contains the factor.

When RH is greater than Nn and gearing is simple, use 1 idler.

When RH is greater than Nn and gearing is compound, use no idler.

When RH is less than Nn and gearing is simple, use 2 idlers.

When RH is less than Nn and gearing is compound, use 1 idler.

Select n so that the ratio of gearing will not exceed 6 to 1. on account of excessive gear stress.

EXERCISES AND PROBLEMS

Use one of the formulas above to find the simple ratio (step 5).

1. Differential-index for 283 divisions.

2. What is the differential indexing for 179 divisions?

3. How should you differential-index to divide a circle in 157 parts?

4. Compute the differential indexing to cut a 211-tooth gear.

5. What is the differential indexing for a gear with 63 teeth?

6. Differential-index for 291 divisions.

7. How should you differential index to cut a 97-tooth gear?

8. What is the differential indexing for a gear with 109 teeth?

9. Compute the differential indexing for 313 divisions.

10. Differential-index for 381 divisions.

CHAPTER 20
MECHANICS

FORCE, WORK, AND ENERGY

In studying mechanisms and machines, one should understand the physical laws that govern motion and force.

When a force acts upon a body and produces motion (or changes its shape), **work is done.** If there is no motion, the effort expended is not considered work in the mechanical sense. Mechanical work must not be confused with physical effort. A man may attempt to lift a weight for hours; but unless he can actually raise it, no mechanical work is accomplished, although great physical effort has been expended.

The amount of work done depends not only upon the force applied but also upon the distance through which it is applied. These two factors, **force** and **distance,** determine the units of work.

In mechanical work, usually the force is expressed in **pounds.** The distance is expressed in **feet.** Work is expressed in **foot-pounds** (ft.-lb.).

A foot-pound of work results when a force of 1 lb. acts through a distance of 1 ft., or such parts and multiples as would give 1. For example, $\frac{1}{2}$ lb. could be raised 2 ft., 2 lb. could be raised $\frac{1}{2}$ ft., or 5 lb. could be raised $\frac{1}{5}$ ft., etc.

When a man lifts a 100-lb. casting from the floor to the top of a 3-ft.-high workbench, the force is 100 lb. and the distance is 3 ft. These two values multiplied together $100 \times 3 = 300$ ft.-lb. of work.

Rule.—Work is the product of the force and distance.

Let W = work.

d = distance.

f = force.

Formula 1. $W = df.$

Example.—Get the work expended to raise a 2300-lb. automobile to a height of $5\frac{1}{2}$ ft.

$W = df.$	Formula
$W = 5.5 \times 2300$	Substituting
$W = 12,650$ ft.-lb.	

Work is sometimes estimated before it is done in terms of energy. **Energy** is the capacity for doing work and is also measured in foot-pounds. To estimate the energy needed to raise the weight of 100 lb. a distance of 3 ft., the solution is the same as that given on page 449, $100 \times 3 = 300$ ft.-lb. of energy.

Mechanical energy is divided into two kinds, **potential energy** and **kinetic energy.**

Potential energy P, or possible energy, is stored up energy, due to position or some previous work. For example, a pile-driver head weighing 1000 lb. has been raised to a height of 14 ft. The work of raising the weight to its present position has resulted in storing up energy that can be used to do other work. In this case the potential energy P is $1000 \times 14 = 14,000$ ft.-lb.

Formula 2. $P = df.$

Kinetic energy K is energy in action. Thus, when the 1000-lb. head described above is released, the energy is in action and is called kinetic energy. The kinetic energy K is $1000 \times 14 = 14,000$ ft.-lb.

Formula 3. $K = df.$

PROBLEMS

State the formula used in each of the following and solve.

1. A casting weighs 450 lb. How much work will be expended to lift it to the bed of a planer 30 in. high?

2. How much work is done in raising a 10-ton beam 27 ft.?

3. How much work is used in pumping 700 cu. ft. of water into a tank 65 ft. high?

4. What amount of energy is needed to raise a 200-ton locomotive 15 ft.?

5. An elevator weighs 1680 lb. How much energy is needed to raise it 4 floors, each 12 ft. high?

6. A pile driver weighing 400 lb. is suspended from a height of 15 ft. Get the kind and amount of energy.

7. A man weighing 145 lb. climbs a ladder 14 ft. high. How many foot-pounds of work are used?

8. The 145-lb. man carried a 66-lb. weight up the ladder in addition to his own weight. Get the total foot-pounds.

9. If the man in Prob. 8 makes a trip every 5 min. for 8 hr., get the total foot-pounds of work. How much of this is pay load?

10. How many foot-pounds does the same man expend coming down the ladder each time?

11. A man with his toolbox weighing 50 lb. on his shoulder entered the elevator and rode 5 floors, each 14 ft. high. How much work did he do?

12. If he had carried the same toolbox up the stairs, how much work would he have done?

13. How much energy is needed to raise a 54-ft. beam, weighing 325 lb. per ft., to a height of 18 ft.?

14. A 165-lb. high diver plunges into the water 38 ft. below. What is his force when he strikes the water?

15. What height must a 125-lb. weight be lifted to produce 3500 ft.-lb. of work? (Solve formula for d.)

16. A baseball weighing 5 oz. was dropped from the top of the Washington Monument, 550 ft. from the ground. Compute the work.

17. 40,000 ft.-lb. of work resulted from a hammer falling 16 ft. How heavy was the hammer? (Solve the formula for f.)

18. It requires 18,000 ft.-lb. of energy to sink a pile. How heavy a head would be needed to produce this effect if the height is 16 ft.?

19. Suppose the same effect could be obtained by 5 uniform blows. What weight hammer would be used?

20. How high will a weight be raised by a pull of 88 lb. if the work is 636 ft.-lb.?

21. How heavy a load may be raised by completing 27,258 ft.-lb. of work through a distance of 66 ft.?

22. If 21,264 ft.-lb. of energy are needed to drive a pile, what weight must fall 14 ft. 3 in.?

23. What potential energy has a 1688-lb. weight, held at a height of 14 ft. 6 in.?

24. The work put into raising a 1250-lb. safe to the sixth floor of a building was 51,150 ft.-lb. Get the average height to each floor.

25. How much work was expended to fill a tank holding 3600 gal. if the lift is 43 ft.?

POWER—HORSEPOWER

In the problems on page 450 the foot-pounds of work were estimated. Obviously, one man or machine might be able to do a piece of work faster than another. The **rate,** or speed, of doing the work must be known in order to have a basis for comparing the two.

This rate of doing work is called power. Power is the result of the foot-pounds of work divided by a unit of time. Power is expressed in foot-pounds per minute, inch-pounds per minute, or inch-pounds per second, etc. For most mechanical work the foot-pound per minute is used.

If one man can raise a 100-lb. weight 10 ft. in one minute, his power equals $\frac{100 \times 10}{1} = 1000$ ft.-lb. per min. If a second man requires 5 min. to raise the 100 lb. the 10 ft., then his power is $\frac{100 \times 10}{5} = 200$ ft.-lb. per minute.

The unit of power used in mechanics has been designated as horsepower. The term horsepower (H.P.) was adopted by Sir James Watt about 1760 as a unit of expressing the power of the steam engine he had perfected. By using a very strong draft horse, pulling on a rope passed over a pulley, he raised a weight from the bottom of a well. At $2\frac{1}{2}$ m.p.h., or 220 f.p.m., his horse could raise 100 lb. 220 ft. × 100 lb. = 22,000 ft.-lbs. per minute.

So that no purchaser of one of his steam engines could complain, Watt added 50 per cent to this amount, making his **mechanical horsepower do 33,000 ft.-lb. per minute.**

This figure is probably twice what an average horse can do continuously for a 6-hr. day, but it has been accepted as a mechanical horsepower and used since.

Rule.—To find the horsepower of any machine, divide the foot-pounds of work by 33,000 multiplied by the minutes used.

$$\begin{aligned} \text{H.P.} &= \text{horsepower.} \\ d &= \text{feet.} \\ f &= \text{pounds.} \\ m &= \text{minutes.} \end{aligned}$$

Formula. $\text{H.P.} = \dfrac{df}{33{,}000m}.$

For the first man above,

$$\text{H.P.} = \frac{100 \times 10}{33{,}000 \times 1} = \frac{1}{33} \text{ or } .03 \text{ hp.}$$

For the second man,

$$\text{H.P.} = \frac{100 \times 10}{33{,}000 \times 5} = \frac{1}{165} \text{ or } .006 \text{ hp.}$$

Example.—Figure the horsepower needed to raise 11,000 lb. 50 ft. in 8 min.

$\text{H.P.} = \dfrac{df}{33{,}000m}$	Formula
$\text{H.P.} = \dfrac{11{,}000 \times 50}{33{,}000 \times 8}$	Substituting
$\text{H.P.} = 2.08+$	

PROBLEMS

Write the formula used in each of the following, and solve.

1. Solve the formula $\text{H.P.} = \dfrac{df}{33{,}000m}$ for each term.

2. Compute the horsepower needed to raise a loaded elevator weighing 6500 lb. to a height of 16 ft. in 2 min.

3. Suppose the elevator in Prob. 2 were raised in 10 sec. What horsepower would be required?

4. A force of 400 lb. acts through 5000 ft. in 5 min. Compute the horsepower.

5. The pull of an endless belt is 280 lb. The belt is traveling 5000 f.p.m. Determine the horsepower transmitted.

6. A man weighing 180 lb. climbs to the top of an 18-ft. ladder in 20 sec. What horsepower does he develop?

7. If it requires 3 hr. to raise a steel girder weighing 75 tons to a height of 64 ft., what horsepower will be needed?

8. If a man carries 6 tons of brick up a ladder 9 ft. high in 8 hr., get the foot-pounds of work. Get the horsepower.

9. Rankine, an eminent mechanical engineer, says that a man climbing a ladder does 71.5 ft.-lb. of work per second. What is his horsepower?

10. With a 2-ft. stroke a man with a hammer strikes a 15-lb. blow every 10 sec. What horsepower is he developing?

11. Rankine says that a horse drawing a cart or boat does 432 ft.-lb. of work per second. What is its mechanical horsepower?

12. What horsepower will be needed to operate a pump that raises 1 cu. ft. per sec. from a well 110 ft. deep?

13. A bullet, weighing 2 oz., leaves the muzzle of a rifle at 3000 feet per second. Find the horsepower at the muzzle.

14. Chester Horton says, "On a drive of 250 yd. the golf ball is struck with a force of 2200 lb. through a distance of 0.72 in. over a period of 0.006 sec." Get the horsepower.

15. How long would it require a 5-hp. motor to fill a tank containing 100,000 gal. of water if the lift is 28 ft.? (Use formula for minutes); (1 gal. = 8.34 lb.)

16. To what height per minute could a 12-hp. motor raise an elevator that weighs 7500 lb.? (Use formula for *d*.)

17. What weight elevator could be raised by a 22-hp. motor to a height of 115 ft. in 25 sec.? (Use formula for force.)

18. How long will it take a 5-hp. engine to raise 500 lb. 55 ft.?

19. A 40 hp. motor lifts an elevator, which has an unbalanced load of 1200 lb., to the top of a building in 30 sec. How high is the lift?

20. A pump is lifting 300 gal. of water per minute from a depth of 88 ft. Determine the horsepower needed. (8.34 lb. = 1 gal.)

21. How many gallons of water per minute can a 10-hp. gas engine raise from a well 110 ft. deep?

22. Find the horsepower needed to raise $\frac{1}{2}$ cu. ft. of water per sec. from a well 340 ft. deep.

23. How many cubic feet of water per second can be pumped from a well when the lift is 140 ft. by a 10-hp. motor?

24. How many feet will a $\frac{1}{2}$-hp. motor raise 50 gal. of water per minute?

25. How much horsepower is needed to operate a pump 10 in. in diameter making 40 f.p.m. if the lift is 32 ft.?

HORSEPOWER OF ENGINES

Indicated horsepower is the theoretical measure of the work an engine is capable of developing, not considering any losses. Indicated horsepower is found by calculation.

Brake horsepower, or effective horsepower, is **the actual amount** of power an engine is capable of transmitting. *Brake horsepower* is found by testing the engine with a **prony** brake or a **dynamometer.** Brake horsepower is horsepower remaining to do useful work after overcoming the friction and supplying the power to run the engine itself without any pay load.

Nominal horsepower is a term used in engineering as a means of **stating the dimensions of an engine.** It seldom has any relation to the horsepower an engine can develop or transmit. It is often used in figuring values for taxation purposes.

The indicated horsepower of an engine depends on the foot-pounds of work it can do per minute. The formula is

$$\text{H.P.} = \frac{df}{33{,}000m}$$

With all ordinary reciprocating engines, such as gas engines and steam engines,

d represents the product of the stroke length expressed in feet and the number of power impulses per minute.

f represents the product of the average pressure and the area of the piston expressed in square inches.

In the two-stroke cycle (also called two-cycle) gas engine and in the single-action steam engine, d equals the product of the stroke in feet multiplied by the r.p.m. There is one power impulse each r.p.m.

In the four-stroke cycle (also called four-cycle) gas engine, d equals the stroke in feet divided by 2, then multiplied by the r.p.m. The explosion, or power impulse, occurs only on every other stroke.

In the double-action steam engine, d equals twice the stroke in feet multiplied by the r.p.m. Power is applied as the piston moves both into and out of the cylinder; hence, there are two power impulses each r.p.m.

f in all engines is found by multiplying the area of the piston in square inches by the average pressure, or **mean effective pressure,** per square inch.

For multiple-cylinder engines, multiply the results obtained for one cylinder by the number of cylinders.

The standard formula for horsepower is $\text{H.P.} = \frac{PLAN}{33{,}000}$.

P = pressure per square inch. This is usually taken as the average pressure and is called the **mean effective pressure.** Mean effective pressure can be found only with a special pressure instrument.

L = total length, feet, that pressure is applied per revolution.

A = area of the piston, or cylinder bore, square inches.

N = number of revolutions of the engine crank per minute (*r.p.m.*)

HORSEPOWER OF TWO-CYCLE GAS ENGINE AND THE SINGLE-ACTION STEAM ENGINE

In the two-stroke cycle gas engine, which includes the Diesel engine and the single-action steam engine.

$$\text{H.P.} = \frac{PLAN}{33{,}000}$$

P = mean effective pressure per square inch.

L = **piston stroke, feet.**

A = area of the piston or bore, square inches ($A = \pi r^2$).

N = r.p.m.

Always multiply the horsepower of one cylinder by the number of cylinders.

Example 1.—Find the horsepower of a single-action steam engine that has a bore of 6″, a stroke of 6″, r.p.m. of 300, and mean effective pressure 110 lb.

$$\text{H.P.} = \frac{PLAN}{33,000} \qquad \text{Formula}$$

$$L = 6'' = .5', \qquad A = \pi r^2 = 3.14 \times 3 \times 3$$

$$\text{H.P.} = \frac{110 \times .5 \times 3.14 \times 3 \times 3 \times 300}{33,000} \qquad \text{Substituting}$$

$$\text{H.P.} = 14.13$$

Example 2.—Find the horsepower of a 6-cylinder Diesel engine, when the bore is 4″, stroke is 5″, mean effective pressure is 90 lb., and r.p.m. is 800.

$$\text{H.P.} = \frac{PLAN}{33,000} \qquad \text{Formula}$$

$$L = \tfrac{5}{12}\text{ ft.}, \qquad A = 3.14 \times 2 \times 2$$

$$\text{H.P.} = \frac{90 \times 5 \times 3.14 \times 2 \times 2 \times 800 \times 6}{12 \times 33,000} \qquad \text{Substituting}$$

$$\text{H.P.} = 68.5+$$

EXERCISES AND PROBLEMS

Number	Kind of engine	Inches bore	Inches stroke	R.p.m.	Pressure per sq. in.	Number of cylinders	H.P.
1	Steam	6	10	400	120	1	
2	Gas	16	18	250	100	1	
3	Diesel	12	15	220	80	1	
4	Diesel	10	12	150	90	1	
5	Gas	8	11	600	99	2	
6	Steam	10	10	200	200	2	
7	Steam	14	15	440	150	2	
8	Gas	3	5	720	88	4	
9	Diesel	8	10	600	60	6	
10	Diesel	6	7	330	100	6	
11	Diesel	18	22	450	90	1	

HORSEPOWER OF DOUBLE-ACTION STEAM ENGINES

Most steam engines, such as locomotive engines and oil-well pump engines, are double-action engines. The double-action steam engine has a power stroke both in and out for each revolution; hence, L is twice the stroke.

On the locomotive, there is always a cylinder on each side. On many locomotives the cylinders are compounded. In the case of compound engines, a high-pressure cylinder exhausts at greatly reduced pressure into a low-pressure cylinder of larger bore. Both cylinders are connected to the same crank and hence have the same r.p.m. and the same stroke. Each is figured separately, and the two results are added.

Formula. $\text{H.P.} = \dfrac{PLAN}{33{,}000}.$

P = mean effective pressure per square inch.
L = **twice the piston stroke, feet, or four times the crank throw, feet.**
A = area of the piston, or bore, square inches ($A = \pi r^2$).
N = r.p.m.

Example.—Find the horsepower of a 2-cylinder steam locomotive with 24″ bore, 26″ stroke, 220 lb. mean effective pressure, and 200 r.p.m.

$$\text{H.P.} = \frac{PLAN}{33{,}000} \qquad \text{Formula}$$

$$L = \frac{2 \times 26}{12} = \frac{52}{12}, \qquad A = \pi r^2 = 3.14 \times 12 \times 12$$

$$\text{H.P.} = \frac{220 \times 52 \times 3.14 \times 12 \times 12 \times 200 \times 2}{12 \times 33{,}000} \qquad \text{Substituting}$$

$$\text{H.P.} = 5224.96$$

PROBLEMS

Give the formula each time, and solve for horsepower of these steam engines.

Number	Inches bore	Inches stroke	R.p.m.	Pressure per sq. in.	Number of cylinders	H.P.
1	20	24	150	100	2	
2	16	20	180	110	2	
3	6	8	700	400	2	
4	3	4	600	600	6	
5	28	30	350	280	2	
6	6	6	80	100	1	

The following are values for compound double-action engines. Figure each pair of high-pressure cylinders and each pair of low-pressure cylinders, and add the horsepower.

Number	Inches bore	Inches stroke	R.p.m.	Pressure per sq. in.	Number of cylinders	H.P.
7 (high)	15	18	250	600	2	
(low)	24	18	250	50	2	
8 (high)	12	22	180	290	2	
(low)	20	22	180	66	2	
9 (high)	10	24	330	240	2	
(low)	18	24	330	35	2	

HORSEPOWER OF FOUR-CYCLE ENGINES

The **four-cycle gas engine,** as used in the automobile, **receives a power impulse for each cylinder every other revolution.** The r.p.m. divided by 2 equals the power impulses, or power strokes. Some prefer to divide the stroke by 2, leaving the r.p.m. as stated. Both methods give the same results.

Formula. $\text{H.P.} = \dfrac{PLAN}{33{,}000}.$

P = mean effective pressure per square inch.
L = **piston stroke, feet.**
A = area of the piston, or bore, square inches ($A = \pi r^2$).

N = **half the r.p.m., or the number of power impulses per cylinder.**

Multiply the horsepower of one cylinder by the number of cylinders.

Example.—Find the horsepower of a 6-cylinder automobile engine of 3″ bore, 4″ stroke, 2400 r.p.m., when mean effective pressure is 90 lb. per sq. in.

$$\text{H.P.} = \frac{PLAN}{33{,}000} \qquad \text{Formula}$$

$$L = \frac{4}{12}, \qquad N = \frac{2400}{2}$$

$$\text{H.P.} = \frac{90 \times 4 \times 3.14 \times 1.5 \times 1.5 \times 2400 \times 6}{12 \times 33{,}000 \times 2} \qquad \text{Sub.}$$

H.P. = 46.24

Or if you prefer to divide the stroke by 2, leaving the r.p.m. as given, L divided by 2 is $4 \div 2 = 2$

$$\text{H.P.} = \frac{90 \times 2 \times 3.14 \times 1.5 \times 1.5 \times 2400 \times 6}{12 \times 33{,}000} \qquad \text{Substituting}$$

H.P. = 46.24

PROBLEMS

State the formula each time, and solve for horsepower of these automobile engines.

Number	Inches bore	Inches stroke	R.p.m.	Pressure per sq. in.	Number of cylinders	H.P.
1	3	3	2000	88	4	
2	4	5	3600	90	4	
3	2.5	3	2800	110	8	
4	3.5	5	3800	120	6	
5	3	5	4400	110	8	
6	5	6	2200	70	6	
7	4.6	6	3000	80	4	
8	4	6	2700	77	4	
9	5	7	2500	88	6	
10	4	5	3200	80	8	

REVIEW PROBLEMS

Write the formula each time, and solve for horsepower in the following:

Number	Kind of engine	Inches bore	Inches stroke	R.p.m.	Pressure per sq. in.	Number of cylinders	H.P.
1	Automobile	3	5	3000	88	8	
2	2-cycle gas	6	8	600	50	1	
3	Diesel	4	6	2000	99	6	
4	Locomotive	20	30	300	240	2	
5	Single-action steam	12	15	660	120	1	
6	Automobile	$3\frac{1}{4}$	$4\frac{1}{2}$	4200	90	8	
7	Locomotive	24	24	220	240	2	
8	Diesel	$4\frac{1}{2}$	5	1500	88	4	
9	2-cycle gas	9	12	440	55	1	
10	Diesel	8	10	1800	66	8	

In the following, solve the formula H.P. $= \dfrac{PLAN}{33{,}000}$ for the missing specification, and calculate, as indicated by the question mark.

Number	Kind of engine	Inches bore	Inches stroke	R.p.m.	Pressure per sq. in.	Number of cylinders	H.P.
11	2-cycle gas	4	6	?	80	1	30
12	Automobile	3	4	3000	?	6	80
13	Diesel	6	?	1600	80	4	100
14	Double-action steam	20	24	?	220	1	1000
15	Automobile	3	?	3000	99	8	120
16	2-cycle gas	?	5	600	60	1	20
17	Locomotive	18	24	200	?	2	3000
18	Diesel	?	6	500	88	1	50
19	Automobile	?	4	3000	77	6	50
20	Diesel	5	6	?	80	4	60

CHAPTER 21

ELECTRICITY

DEFINITIONS AND FORMULAS

In almost every field of industry the workman comes in contact with electricity and electrical equipment. Though the nature of electricity is still theoretical, we do know how it acts under certain conditions and how to adapt it to practical use. The student should become familiar with its laws and their application.

Units of electrical measure are:

Volts. The **pressure** which causes the current to flow. A **volt** (also called **electromotive force**) is the unit of electrical pressure needed to push one **ampere** of current through one **ohm** of resistance.

Amperes. The **quantity** of **current** flowing. An **ampere** is the quantity of electrical current that will flow through one **ohm** of resistance when pushed by an electromotive force of one **volt.**

Ohms. The **resistance** to the flow of electrical current. An **ohm** is the resistance through which a current of one **ampere** will flow when under an electromotive force of one **volt.**

Conductors are substances, such as metals, that permit a current of electricity to flow.

Nonconductors are substances, such as glass and rubber, that do not permit a current of electricity to flow.

Ohm's law [first determined by Georg Ohm (1787–1854), a German scientist] states

Rule.—The amount of current in amperes is equal to the electrical pressure in volts divided by the resistance in ohms.

$$\text{Current} = \frac{\text{pressure}}{\text{resistance}}$$

or

$$\text{Amperes} = \frac{\text{volts}}{\text{ohms}}$$

Volts also called: Electromotive force; Electrical pressure; Pressure; Potential; Drive; Push, etc.

Amperes also called: Intensity of current; Quantity of current; Current intensity; Intensity; Current, etc.

Let I = intensity of current, amperes.
E = electromotive force, volts.
R = resistance, ohms.

Formula. $I = \frac{E}{R}$, $R = \frac{E}{I}$, $E = IR$.

Example 1.—How many amperes of current are flowing in a circuit with 22 ohms resistance, when pressure is 110 volts?

$I = \frac{E}{R}$	Formula
$I = \frac{110}{22}$	Substituting
I = 5 amp. of current flowing	

Example 2.—Find the pressure in volts when the circuit has 22 ohms resistance and the current is 5 amp.

$E = IR$	Formula
$E = 5 \times 22$	Substituting
E = 110 volts of pressure	

Example 3.—What is the resistance in ohms when a pressure of 110 volts causes a current intensity of 5 amp.?

$R = \frac{E}{I}$	Formula
$R = \frac{110}{5}$	Substituting
R = 22 ohms of resistance	

PROBLEMS

State the formula used each time, and solve.

1. Find the amperes needed when the resistance is 200 ohms and the pressure is 4 volts.

2. What voltage is needed to produce a current of 2 amp. when the resistance is 36 ohms?

3. Find the resistance of a 220-volt circuit that draws 0.5 amp.

4. If a bell has a resistance of 180 ohms on a 6-volt circuit, what amperes will flow?

5. If a motor requires 180 amp. when the resistance is 0.03 ohm, what is the voltage?

6. What is the resistance of a doorbell that uses 0.02 amp. from a 6-volt transformer?

7. How much current is needed to operate a horn on a 6-volt battery circuit that has a resistance of 4 ohms?

8. What voltage is required to push 12.5 amp. through a resistance of 75 ohms?

9. What are the ohms of resistance of a lamp that used 0.3 amp. at 12 volts?

10. How much current flows through an arc lamp on a 110-volt line if the resistance is 30 ohms?

11. What pressure is needed to drive a current of 0.5 amp. when the resistance is 200 ohms?

12. If an automobile starting motor draws 80 amp. from a 6-volt battery, what is its resistance?

13. What current is needed to start an automobile horn if the resistance is 0.05 ohm on a 6-volt battery?

14. What electromotive force is needed to force 12 amp. through a horn that shows 40 ohms resistance?

15. If the electrical pressure of a house line is 120 volts and the current flowing through a motor is 5.2 amp., what is the resistance in ohms?

16. Find the quantity of current needed to run a motor of 50 ohms resistance on a 110-volt circuit.

17. What is the electrical pressure needed to produce 0.15 amp. through a resistance of 2000 ohms?

18. What resistance permits 0.006 amp. of current to flow in a telephone receiver when pressure is 1 volt?

19. Find the intensity of current flowing through a resistance of 300 ohms on a 220-volt line.

20. If the quantity of current is 0.375 amp. and the resistance is 80 ohms, what is the electromotive force?

21. Find the quantity of current when electrical pressure is 220 volts and the resistance is 3 ohms.

22. An electric automobile heater can safely carry 2 amp. of current. Its resistance is 24 ohms. How high may the voltage be?

23. What is the resistance of a car heater that draws 2 amp. from a 6-volt battery?

24. What current is used on a light line of 120 volts, operating a lamp that has a resistance of 240 ohms?

25. A telephone receiver has a resistance of 84 ohms and requires 0.006 amp. of current. What is the electrical pressure?

26. What is the resistance when 1.11 amp. jumps a spark gap when it has 444 volts pressure?

27. What is the amperage of a relay if it operates on a 12-volt battery and has a resistance of 90 ohms?

28. If a Mazda lamp uses 1.45 amp. and has a resistance of 240 ohms, what voltage pressure must be maintained?

29. A set of door chimes is connected to a transformer that delivers 18 volts. If the mechanism must have 0.15 amp. of current, what is its resistance?

ELECTRICAL POWER

The unit of electrical work is called the **joule.** It was named after James Joule (1818–1889), an English scientist.

A joule is the amount of work performed by a current of one ampere flowing under a pressure of one volt for one second.

One joule = one volt × one ampere × one second, or **joules = volts × amperes × seconds.**

Power is the rate of doing work. If work is divided by the time, the result is power.

$$\text{Power} = \frac{\text{work}}{\text{time}}; \qquad \text{Electrical power} = \frac{\text{Electrical work}}{\text{time}}$$

The unit of electrical power is the **watt.** The watt was named after James Watt (1736–1819).

Then, power in watts $= \dfrac{\text{joules}}{\text{seconds}}$, or

$$\text{Watts} = \frac{\text{volts} \times \text{amperes} \times \text{seconds}}{\text{seconds}}$$

Since the seconds will cancel out of the equation,

$$\text{Watts} = \text{volts} \times \text{amperes}$$

Rule 1.—Power in watts is the product of volts and amperes.

P = power, watts.
E = electrical pressure, volts.
I = intensity of current, amperes.

Formula. $P = EI$.

Note.—This rule and formula hold true for all types of equipment using direct current and all equipment where the electricity is used as heat, such as lamps and toasters, with alternating current.

Example 1.—What power is used to burn a 32-volt lamp that draws 3 amp.?

$P = EI$	Formula for power in watts
$P = 32 \times 3$	Substituting
$P = 96$ watts of power	

Example 2.—What power is used by an automobile starting motor that draws 80 amp. from a 6-volt battery?

$P = EI$	Formula for power in watts
$P = 6 \times 80$	Substituting
$P = 480$ watts of power	

Example 3.—Find the watts consumed by a toaster that draws 5 amp. from a 110-volt house line.

$P = EI$	Formula for power in watts on heating devices, alternating current
$P = 110 \times 5$	Substituting
$P = 550$ watts of power	

In an alternating-current circuit, on induction equipment such as motors, the actual power is less than the theoretical power. The ratio of the actual power to the theoretical power is called **power factor.** Power factor is usually expressed as per cent or decimal. The symbol f will be used to represent the term power factor.

Rule 2.—On induction equipment, in alternating-current circuits, the power in watts is the product of the volts, the amperes, and power factor.

Formula. $P = EIf$.

Example 4.—Find the watts flowing in an alternating-current motor on a 110-volt line if amperage is 7.5 and power factor is 85 per cent.

$P = EIf$	Formula for induction equipment on alternating-current circuits
$P = 110 \times 7.5 \times 0.85$	Substituting
$P = 701.25$ watts of power	

PROBLEMS

Write the formula each time, and solve.

1. Find the rating in watts of a lamp requiring 0.5 amp. when the voltage is 110.

2. A certain soldering iron, when connected across 120 volts, takes a current of 5.5 amp. What is its rating in watts?

3. A direct-current motor runs on 200 volts and requires 30 amp. when on full load. How many watts does it require?

4. An alternating-current motor takes 26 amp. when the line voltage is 218. If the power factor is 82%, figure the watts.

5. A direct-current welding machine requires a current of 850 amp. The voltage is 8.3. What number of watts is used in this case?

6. An electrical refrigerator requires 3.5 amp. The power factor is 0.88. The house voltage is 115. How many watts are used?

7. A toaster is rated at 6 amp. on a house current of 115 volts. How many watts does it require?

8. Find the watts used by a washing-machine motor requiring 7.5 amp. when hooked on a house voltage of 110 when f is 0.83.

9. An electric fan, running on a house voltage of 110 volts, requires 0.85 amp. If the power factor is 0.85, get the watts.

10. What are the total watts required on a light circuit that has a current of 0.4 amp. and an electromotive force of 115 volts?

11. How much power is consumed by 5 lamps requiring 2.5 amp. each on a 6-volt battery?

12. An arc lamp is rated at 2.4 amp. on an 85-volt line. What power does it take?

13. How many watts are consumed by an automobile starting motor that takes 90 amp. from a 6-volt battery?

14. An electric pump, on a 220-volt circuit, requires 36 amp. What wattage is used if power factor is 90%?

15. Find the total watts used by 24 lamps, each consuming 0.5 amp. from a house line rated at 115 volts.

16. How many watts are consumed by an electric motor that draws 10 amp. when plugged in the house line of 120 volts if f is 88%?

17. A motor is rated to take 15 amp. when used on a 220-volt line. If the power factor is 0.8, what is the wattage?

18. A car heater takes 25 amp. when used on a 6-volt battery. What are the watts?

19. How many watts will be used in a house if 25 lamps drawing 0.55 amp. each on 110 volts are all turned on at one time?

20. An automobile windshield wiper requires 0.5 amp. when the pressure is 6 volts. What are the watts?

21. If 1.5 watts are required per candle power of light, what is the power in watts for a 25-candle-power tungsten lamp on 110 volts?

22. Find the watts used by a 40-candle-power lamp when the conditions are like those in Prob. 21.

23. Find the total watts used by 6 lamps of 60 candle power each when conditions are like those in Prob. 21.

24. If 1.5 watts are required for a candle power, what candle power has a 40-watt lamp? A 75-watt lamp? A 100-watt lamp?

USING THE EQUATION $P = EI$

From the equation $P = EI$, it is possible to obtain any of the quantities required when the other two are known. First solve the formula for the term wanted.

$$P = EI$$

$$I = \frac{P}{E}$$

$$E = \frac{P}{I}$$

Example 1.—Find the current taken by an electric heater operating on 120 volts if it is rated at 600 watts.

$I = \frac{P}{E}$	Formula
$I = \frac{600}{120}$	Substituting
$I = 5$ amp.	

Example 2.—Find the pressure in volts of a motor rated to take 10 amp. and to consume 800 watts.

$E = \frac{P}{I}$	Formula
$E = \frac{800}{10}$	Substituting
$E = 80$ volts	

PROBLEMS

Write the proper formula, and solve. (f not considered.)

1. An electric heater has on the name plate 1000 watts, 110 volts. What current will be drawn from the line?

2. An electric iron is rated at 110 volts and 355 watts. What amperage is used?

3. An electric toaster operating on a light circuit of 110 volts is supposed to consume 675 watts. What is the current in amperes?

4. A motor is rated to take 12 amp. and to consume 1000 watts. For what voltage is it built?

5. A toaster is rated to use 5.5 amp. and 605 watts. What voltage should be used?

6. A starting motor draws 80 amp. when consuming 480 watts. What should the battery voltage be?

7. What current in amperes flows through a circuit when the pressure is 220 volts and the watts consumed are 1220?

8. An arc lamp is rated at 150 watts on an 85-volt line. What current does it take?

9. An electric fan runs on the house line of 110 volts pressure and consumes 75 watts. How many amperes of current does it require?

10. For what pressure is a motor built that uses 3960 watts when rated at 18 amp.?

11. If the amperes of current are 5.5, on what voltage line should a washing-machine motor be used that consumes 658 watts?

12. On what voltage line should you put a refrigerator motor rated at 3.5 amp. and 420 watts?

13. How many amperes of current will flow through a 32-volt light consuming 24 watts of power?

14. An electric waffle iron uses 660 watts on a 110-volt house line. What amperage is needed to heat the iron?

15. For what voltage is a soldering iron designed that uses 6 amp. and 72 watts?

16. On what voltage line should a lamp of 40 watts be placed if the current used is $\frac{1}{2}$ amp.?

REVIEW PROBLEMS

Write the formula used, and solve. (*f* not considered.)

1. How many watts of power are going to a motor if the voltage is 500 and there are 7 amp. entering the motor?

2. If 200 watts are expended in a circuit by a current of 4 amp., what is the voltage required to drive the current?

3. An electric curling iron uses 750 watts on a 115-volt line. What amperage is needed to heat the iron?

4. On a 115-volt line, there is a device that takes 12 amp. of current. What watts are consumed?

5. How many watts are consumed by a 12-volt lamp taking 0.3 amp.?

6. A carbon-arc lamp is rated at 1000 watts on a 220-volt line. What current does it take?

7. An electric clock is rated to consume 2 watts on a 110-volt line. What is the current in amperes?

8. On what voltage line should you place a percolator that is rated to use 4.5 amp. and consumes 500 watts?

9. A cooker used 400 watts on a 115-volt line. What is its amperage?

10. A table stove using 1500 watts on a house line of 120 volts draws what current?

11. A radio is rated at 1.3 amp. on a house line of 110 volts. Figure the watts used.

12. A vacuum cleaner is supposed to use 125 watts on a 120-volt line. What amperage of current does it take?

13. If a current of 50 amp. flows through a circuit under a pressure of 220 volts, what is the power consumed?

14. How many amperes of current will flow through a 32-volt lamp that consumes 30 watts?

15. How many 50-watt lamps can be run on a 110-volt circuit with an expenditure of 48 amp. of current when each lamp requires 0.5 amp.?

16. What ampere fuses should you use on a line rated at 2000 watts if the voltage is 220 volts?

17. If a fuse must carry 50% more current than the line is rated at, what size of fuse would be used on a 2000-watt 110-volt line?

18. A 30-amp. fuse will carry what power on a 110-volt line?

19. What ampere fuse should you place on a 110-volt line that carries twelve 50-watt lamps each rated at 0.5 amp.?

20. On what voltage should you place a 15-amp. fuse to supply 1650 watts?

COST OF ELECTRIC ENERGY

Electricity used for a certain period of time constitutes electric power. It is the power which we use to perform work, and it is the power for which the "power and light" company charges.

The unit by which electrical power is measured is the **kilowatt-hour. One kilowatt (kw.) is 1000 watts.** One thousand watts used for one hour is one kilowatt-hour. Meters used to measure electricity usually register the kilowatt-hours of electrical power passing through them.

By multiplying the kilowatt-hours by the cost per kilowatt-hour the power bill may be determined.

Rule.—Cost of electricity is the product of the kilowatts, the hours, and the price per kilowatt-hour.

Example.—If the power rate is $0.08 per kilowatt-hour, how much will it cost to operate a soldering iron for 6 hr. if it draws 500 watts?

$$500 \text{ watts} = \tfrac{500}{1000} \text{ kw.} = .5 \text{ kw.}$$
$$\text{Cost} = .5 \times 6 \times .08 = \$0.24$$

PROBLEMS

1. At 5¢ per kilowatt-hour, find the cost to operate a toaster rated at 350 watts for 2 hr.

2. At 6¢ per kilowatt-hour, what will it cost to iron clothing for 4 hr. if the iron draws 700 watts?

3. A soldering iron is rated at 600 watts. Find the cost of keeping it hot 8 hr. at 4¢ per kilowatt-hour.

4. Figure the cost of burning twelve 50-watt lamps 3 hr. at 4¢ per kilowatt-hour.

5. How much will it cost to burn a heater that draws 500 watts for 24 hr. at $3\frac{1}{2}$¢ per kilowatt-hour?

6. A market has 24 lamps rated at 500 watts each. These lamps were burned $2\frac{1}{2}$ hr. a night for 6 days. Find the cost at 3¢ per kilowatt-hour.

7. A 300-watt lamp is burned 3 hr. per night for 30 days. Find the cost at 5.5¢ per kilowatt-hour.

8. A toaster is rated at 350 watts. At 5¢ per kilowatt-hour, find the cost to toast 1 slice of bread if 10 sec. are required.

9. A fruit juicer is rated at 80 watts. If it takes 3 min. to juice 1 doz. oranges, find the cost at $4\frac{1}{2}$¢ per kilowatt-hour.

10. An electric clock consumes 2 watts per hr. At 5¢ per kilowatt-hour find the cost to operate it for 30 days.

11. A waffle iron is rated at 666 watts. What will it cost to operate it for 30 min. at 5¢ per kilowatt-hour?

12. A motor is rated at 10 amp. on a 110-volt line. Find the cost to run it 1 hr. at 4¢ per kilowatt-hour.

13. A percolator is rated at 4.5 amp. on a 120-volt line. What will it cost per hour to operate it at $3\frac{3}{4}$¢ per kilowatt-hour?

14. A radio uses 140 watts. What will it cost to operate it 12 hr. at 5¢ per kilowatt-hour?

15. Find the cost of burning 12 lamps rated at 75 watts each for 2 hr. per day for 30 days at $3\frac{1}{4}$¢ per kilowatt-hour.

16. A motor is rated at 15 amp. on a 110-volt line. Find the cost to operate it 300 hr. at 2.5¢ per kilowatt-hour.

17. A fan uses 0.75 amp. on a 110-volt line. Find the cost to run it 9 hr. at 4.3¢ per kilowatt-hour.

18. Find the cost of electric energy at $0.0425 per kilowatt-hour to run a motor rated at 80 amp. for 8 hr. on a 220-volt line.

19. How much will it cost to burn a 60-watt lamp 4 hr. per night for 30 nights if the energy rate is 6¢ per kilowatt-hour?

20. A heater taking 5 amp. on 110 volts is used 6 hr. per day for 30 days. At 5¢ per kilowatt-hour, how much will it cost to operate it?

21. A motor runs on a 220-volt circuit and requires 16 amp. What will it cost to run it for 24 hr., if the rate is $0.035 per kilowatt-hour?

22. Find the cost of burning seven 60-watt lamps for 4 hr. each night for 1 month of 30 days at $0.055 per kilowatt-hour.

23. An electric refrigerator on a 110-volt line consumes 3.5 amp., and the average run per day of the motor is $7\frac{1}{2}$ hr. At 5¢ per kilowatt-hour, what will it cost to operate the box for 30 days?

24. A factory burns 1200 lamps that consume 150 watts each, 8 hr. per day, 25 days per month. At a rate of 5¢ per kilowatt-hour, what will the bill be per month?

25. A city lighting system operates 62,000 street lamps of 200 watts each from 7 P.M. to 4 A.M. daily, for 31 days. The rate for the first 200 kw.-hr. is $2\frac{1}{2}$¢, and the balance is $\frac{1}{2}$¢ per kilowatt-hour. What is the bill for the month?

26. An electric pump is rated at 10 amp. on a 220-volt line. What will it cost to operate it for 16 hr. per day for 5 months of 30 days at $0.01 per kilowatt-hour, if there is a surcharge of $4.65 per month extra to be paid?

27. Find the total electric bill for a house with 15 lamps of 60 watts each that operate an average of $1\frac{1}{2}$ hr. per day and with one refrigerator that uses 330 watts for 4 hr. per day, for 30 days, if the rate is 6¢ per kilowatt-hour.

28. The lamps on an automobile require 15 amp. on a 6-volt circuit. They are burned an average of 1 hr. per day for 365 days. Determine the cost of this energy at 5¢ per kilowatt-hour. How does this cost compare with the price of a storage battery which might be expected to last 3 years?

29. What will it cost to operate a motor running continuously for 1 year of 365 days if it requires 10 amp. on a 110-volt circuit and an industrial rate of $1\frac{1}{2}$¢ per kilowatt-hour is allowed?

30. Find the cost of running an electric heater that draws 750 watts from a 110-volt line for 30 days if the rate is 2¢ per kilowatt-hour.

31. A battery-charging outfit draws 3 amp. from a 110-volt line. Find the cost of operating it 30 days at 4¢ per kilowatt-hour.

32. A generator is required to operate a 600-light installation, each lamp being standard 55 watts. If this energy is continuous, find its value at $\frac{1}{2}$¢ per kilowatt-hour for a month of 30 days.

33. A street lighting system has 200 arc lamps that draw 10 amp. each from a 115-volt line. Determine the cost of operating these 8 hr. per night at 2¢ per kilowatt-hour.

ELECTRICAL HORSEPOWER

For comparative purposes, kilowatts of power may be converted into horsepower.

One electrical horsepower is equal to 746 watts or 0.746 kilowatts; hence, we have the following:

Rule.—To find horsepower, divide kilowatts by 0.746.

Formulas. $\text{H.P.} = \dfrac{\text{kw.}}{0.746}$, and kw. = 0.746 hp. (accurate).

Since 0.746 is almost 0.750 or $\frac{3}{4}$, one kilowatt is often taken as $\frac{3}{4}$ hp.

Formulas. Kw. = $\frac{3}{4}$ hp., and H.P. = $\frac{4}{3}$ kw. (approximate).

Example 1.—How many horsepower in 36 kw.?

$\text{H.P.} = \dfrac{\text{kw}}{.746}$.	Accurate formula
$\text{H.P.} = \dfrac{36}{.746}$	Substituting
H.P. = 48.25, accurate horsepower	
H.P. = $\frac{4}{3}$ kw.	Approximate formula
$\text{H.P.} = \dfrac{4 \times 36}{3}$	Substituting
H.P. = 48, approximate horsepower	

Example 2.—Get the kilowatts in 48 hp.

Kw. = .746 hp.	Accurate formula
Kw. = .746 × 48	Substituting
Kw. = 35.808, accurate kilowatts	
Kw. = $\frac{3}{4}$ hp.	Approximate formula
Kw. = $\frac{3 \times 48}{4}$	Substituting
Kw. = 36, approximate kilowatts	

PROBLEMS

State the formulas, and solve.

1. Find accurately how many horsepower there are in 45 kw.

2. Find approximately how many horsepower there are in 45 kw.

3. Find the accurate and the approximate horsepower in 12,346 kw.

4. Find the accurate and the approximate horsepower in 1,000,000 kw.

5. Which is greater, 1 hp. or 1 kw.?

6. Find the exact kilowatts developed by a hydroelectric plant of 85,000 hp.

7. Determine the horsepower needed to operate a generator that develops 220 volts of pressure and 125 amp.

8. A motor on a 220-volt circuit requires 36 amp. of current. What accurate horsepower is it using?

9. A generating plant maintains a pressure of 120 volts in an electric-light circuit and the ammeter indicates 66 amp. What horsepower is being generated?

10. What kilowatts are being generated by a 120-hp. gas engine?

In the following, find the watts, kilowatts, and accurate horsepower.

11. A motor requires 8 amp. on a 120-volt circuit.

12. A washing machine draws 2.5 amp. from a 110-volt line.

13. A generator produces 400 amp. at 220 volts pressure.

14. A generator maintains 110 volts, and the ammeter reads 85 amp.

15. A motor uses 36 amp. from a 220-volt circuit.

16. Electromotive force is 22,000 volts. Intensity of current is 400 amp.

READING METERS

Many of the **public utilities** sold to the consumer, such as electricity, gas, and water, are **measured by meters.** A meter face has several dials which show the cumulated total of the material that has passed through it. Some meters have dials that show line leakage. The figures below show typical meters.

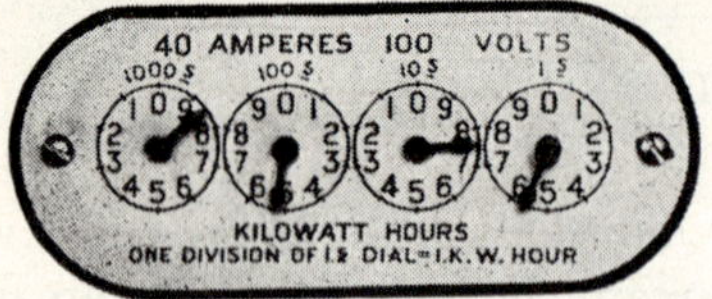

Kilowatt-hour meter. (*Courtesy of Pacific Gas and Electric Company.*)

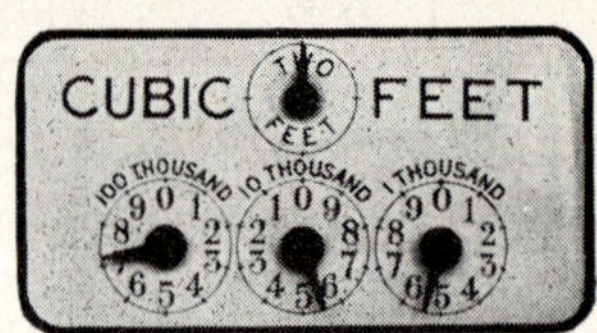

Gas meter. (*Courtesy of Pacific Gas and Electric Company.*)

Water meter. (*Courtesy of California Water Service Company.*)

On the electrical kilowatt-hour meter the right-hand dial has a 1 stamped above it. When the pointer, which turns clockwise, has made one complete turn, 1 kw.-hr. has been consumed. By then the arm on the next dial, which turns counterclockwise, will have moved from 0 to 1. After the first dial arm has made 10 turns the second dial arm will have made 1 turn and the third dial arm will have moved from 0 to 1, clockwise, etc.

Meters measuring other materials measure in a like manner.

Gas is sold by the cubic foot.

Water is sold by the cubic foot.

Rule.—The total reading is made by taking the number each hand has passed last. Start reading from the dial with the largest numbers, usually the left-hand dial.

Examples.

Kilowatt-hour meter above reads 8575 kw.-hr. Gas meter reads 755,000 cu. ft. Water meter reads 4,160,655 cu. ft.

PROBLEMS

Read the following electric meters:

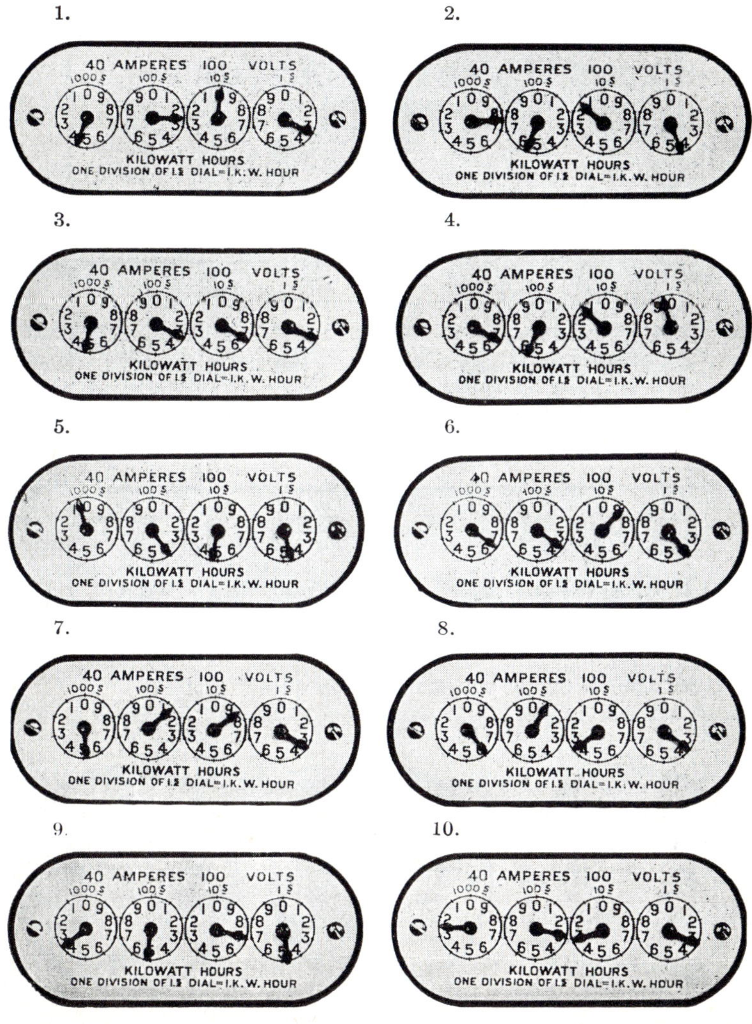

Read the following gas meters in thousand cubic feet:

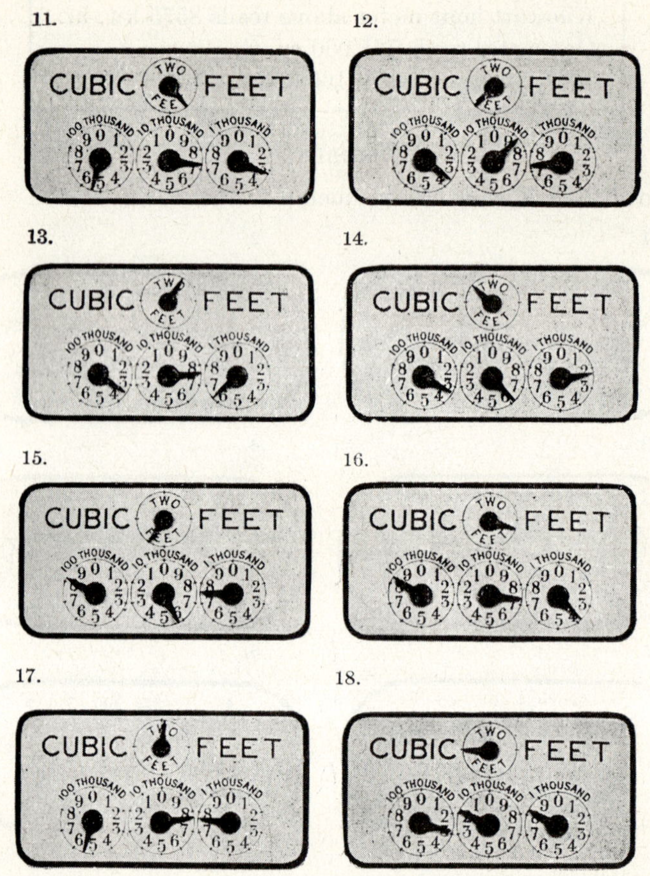

Read the following water meters in cubic feet:

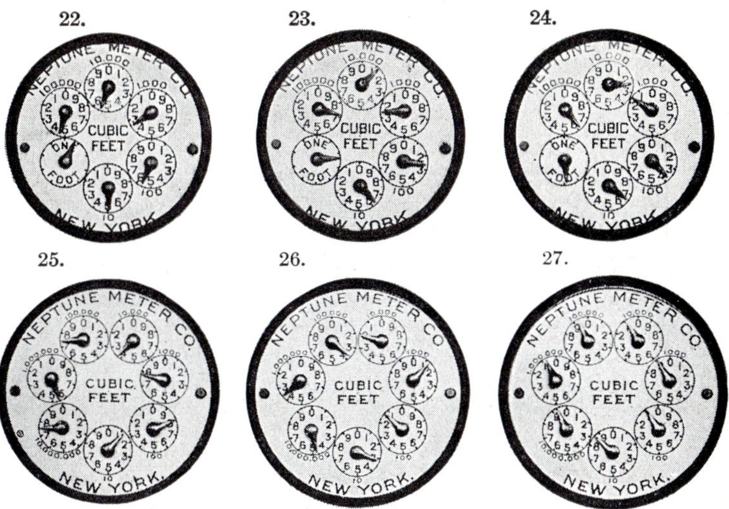

COST OF METERED UTILITIES

The cost of electricity, gas, and water varies in different localities and with the amount of each used per month. Rates are regulated by the state utility commissions. In most of the large cities of California, served by the Pacific Gas and Electric Company, electrical charges are as follows:

Service charge per month per meter..	50¢
Energy charge (to be added to service charge):	
First 35 kw.-hr. per month........	3.3¢ per kilowatt-hour
Next 165 kw.-hr. per month.......	2.2¢ per kilowatt-hour
All excess kw.-hr. per month......	1.2¢ per kilowatt-hour

The same company's rate for gas is as follows:

First 3000 cu. ft....................	$0.60 per 1000 cu. ft.
Next 7000 cu. ft....................	0.50 per 1000 cu. ft.
Next 90,000 cu. ft..................	0.35 per 1000 cu. ft.
All over 100,000 cu. ft...............	0.25 per 1000 cu. ft.
Minimum charge per meter per month	$0.70

The rate for water is as follows:

First 540 cu. ft....................	$0.1628 per 100 cu. ft.
540 to 2000 cu. ft..................	0.1320 per 100 cu. ft.
2000 to 4000 cu. ft.................	0.0792 per 100 cu. ft.
Over 4000 cu. ft....................	0.0572 per 100 cu. ft.
Minimum charge per month.........	$1.00

Example 1.—Find the cost of 63 kw.-hr. of electrical energy at the rates given above.

First 35 kw.-hr. at 3.3¢ = 1.155, or	$1.16
63 − 35 = 28 kw.-hr. at 2.2¢ = .616, or	.62
50¢ service charge per meter per month	.50
Total bill for month	$2.28

Example 2.—Find the cost of 13,200 cu. ft. of gas at the rates stated above.

3000 cu. ft. at.	$0.60 per 1000 cu. ft. =	$1.80
7000 cu. ft. at.	.50 per 1000 cu. ft. =	3.50
13,200 − 10,000 = 3,200 cu. ft. at.	.35 per 1000 cu. ft. =	1.12
Total bill for month		$6.42

Example 3.—Find the cost of 2550 cu. ft. of water at the rates stated above.

540 cu. ft. at $0.1628 per 100 cu. ft. = .879, or	$0.88
1460 cu. ft. at .1320 per 100 cu. ft. = 1.927, or	1.93
550 cu. ft. at .0792 per 100 cu. ft. = .435. or	.44
Total bill for month	$3.25

PROBLEMS

Using the rates listed on page 505, find the cost of the following utilities:

1. 3640 cu. ft. of water.
2. 9550 cu. ft. of water.
3. 12,300 cu. ft. of water.
4. 3200 cu. ft. of water.
5. 6300 cu. ft. of water.
6. 9800 cu. ft. of water.
7. 630 cu. ft. of gas.
8. 17,600 cu. ft. of gas.
9. 7800 cu. ft. of gas.
10. 25,000 cu. ft. of gas.
11. 16,800 cu. ft. of gas.
12. 800 cu. ft. of gas.
13. 77 kw.-hr. of electricity.
14. 348 kw.-hr. of electricity.
15. 123 kw.-hr. of electricity.
16. 1362 kw.-hr. of electricity.
17. 24 kw.-hr. of electricity.
18. 963 kw.-hr. of electricity.
19. 65,000 kw.-hr. of electricity.
20. 9600 kw.-hr. of electricity.

MONTHLY BILLS FOR UTILITIES

When meters are read (usually monthly), the consumption is determined by deducting the previous reading from the present reading. The cost is figured as in the preceding problems.

PROBLEMS

Determine the consumption and make out bills showing the cost of each utility and the total. Use the same rate chart as in the preceding lesson.

Number	Utility	Present reading	Previous reading
1	Electricity	1,162 kw.-hr.	1,104 kw.-hr.
	Gas	103,800 cu. ft.	99,200 cu. ft.
	Water	66,340 cu. ft.	65,110 cu. ft.
2	Electricity	9,456 kw.-hr.	9,321 kw.-hr.
	Gas	115,800 cu. ft.	103,000 cu. ft.
	Water	132,280 cu. ft.	130,100 cu. ft.
3	Electricity	8,383 kw.-hr.	8,124 kw.-hr.
	Gas	17,000 cu. ft.	3,600 cu. ft.
	Water	33,588 cu. ft.	31,006 cu. ft.
4	Electricity	4,378 kw.-hr.	3,942 kw.-hr.
	Gas	78,000 cu. ft.	76,000 cu. ft.
	Water	54,248 cu. ft.	51,162 cu. ft.
5	Electricity	2,148 kw.-hr.	1,820 kw.-hr.
	Gas	10,500 cu. ft.	9,200 cu. ft.
	Water	116,840 cu. ft.	113,324 cu. ft.
6	Electricity	1,745 kw.-hr.	1,592 kw.-hr.
	Gas	18,500 cu. ft.	0 cu. ft.
	Water	540,348 cu. ft.	538,114 cu. ft.
7	Electricity	9,345 kw.-hr.	7,846 kw.-hr.
	Gas	9,700 cu. ft.	8,900 cu. ft.
	Water	320,400 cu. ft.	316,288 cu. ft.
8	Electricity	8,842 kw.-hr.	8,320 kw.-hr.
	Gas	97,300 cu. ft.	81,100 cu. ft.
	Water	848,344 cu. ft.	844,288 cu. ft.
9	Electricity	16 kw.-hr.	0 kw.-hr.
	Gas	540 cu. ft.	0 cu. ft.
	Water	380 cu. ft.	0 cu. ft.
10	Electricity	355 kw.-hr.	315 kw.-hr.
	Gas	6,900 cu. ft.	6,600 cu. ft.
	Water	10,860 cu. ft.	10,320 cu. ft.

CHAPTER 22

MACHINES

KINDS OF MACHINES

A machine is an appliance by means of which force applied at one point, and in a definite direction, is made to produce, at another point, in some modified form, another force called work.

The work thus produced never quite equals the original force, owing to losses through friction, weights of parts, and other causes.

Some machines are simple, whereas others show great complexity. Any complicated machine is a combination of one or more of **three basic mechanisms** or some of their subdivisions. These are (1) the **lever,** (2) the **incline,** and (3) the **cord.**

The varying subdivisions of these three elements are as follows:

- 1. Lever
  - Lever..................
    - 1st class
    - 2d class
    - 3d class
    - Compound levers*
  - Pulleys
  - Gears
  - Tackle blocks
  - Differential pulleys*
  - Wheel and axle..........
    - Plain wheels
    - Compound*
    - Differential*
  - Crank*
  - Strut*
- 2. Incline
  - Inclined plane
  - Wedge
  - Screw..................
    - Plain
    - Differential*
  - Worm
  - Cam*

* These machines not discussed in this book.

3. Cord {Belts* Ropes* Cables* Chain*}

* These machines not discussed in this book.

The Lever

The simplest machine, and one of the most useful, is the **lever.** A lever is a rigid bar, turning freely on a fixed point called the **fulcrum.** The lever is used to increase the power at the expense of speed or distance moved, or vice versa, or to change the direction of motion. The bar may be straight, bent, or curved.

Levers are divided into three classes. The position of the fulcrum with reference to the weight and power determines the class to which the lever belongs.

FIRST-CLASS LEVERS

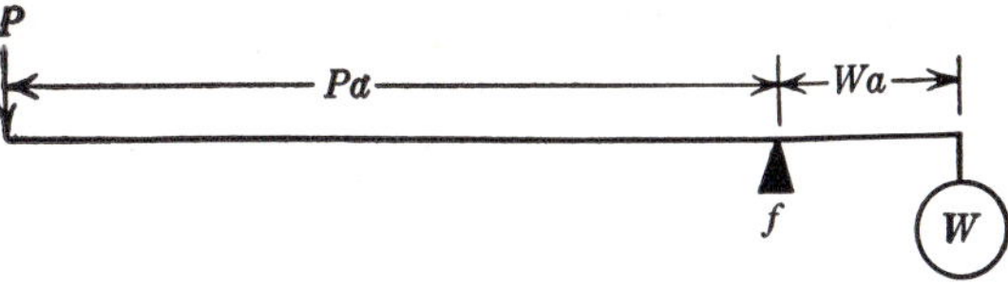

On the first-class lever as shown, *W* represents the **weight** being raised. W_a, or the **weight arm,** is the distance the weight is from the fulcrum *f* shown as a small triangle.

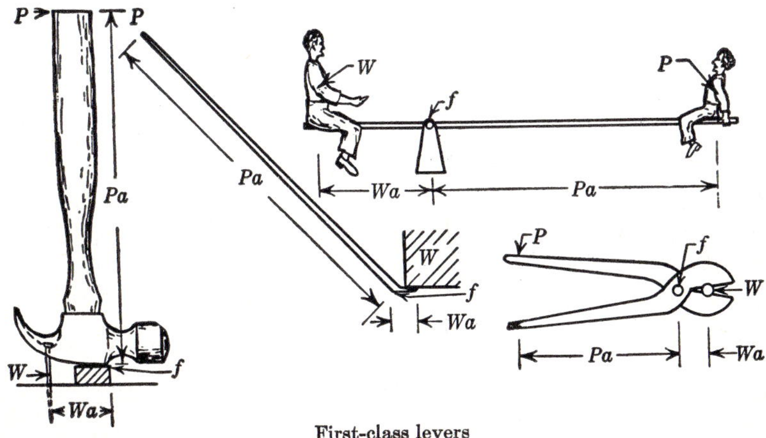

First-class levers

The sharp edge of the fulcrum gives only a little friction. The fulcrum may, however, be a common bearing or even a ball or roller bearing. P represents the position at which the power is applied, and P_a is the **power arm,** the distance the power P is from the fulcrum. Examples of the first-class lever are shown on page 483.

The fulcrum is nearer to the weight than it is to the power. This position makes it possible for a small amount of power to balance or raise a larger weight. This possibility in a machine is called **mechanical advantage.** On the lever the mechanical advantage (M.A.) is the ratio of the power arm to the weight arm. If the power and weight are known, the mechanical advantage is the ratio of the weight to power. The ratios are always reduced to their lowest terms.

Rule 1.—To find the mechanical advantage of a lever, divide the length of the power arm by the length of the weight arm.

Rule 2.—To find the mechanical advantage, divide the weight by the power.

Formulas. $\text{M.A.} = \frac{P_a}{W_a}$, or $\text{M.A.} = \frac{W}{P}$.

Example 1.—If 500 lb. can be balanced by 100 lb., find the mechanical advantage.

$\text{M.A.} = \frac{W}{P}$ Formula

$\text{M.A.} = \frac{500}{100}$, which when reduced gives a ratio of 5 to 1

M.A. = 5 to 1

Example 2.—If the power arm is 25″ and the weight arm is 5″, find the mechanical advantage.

$\text{M.A.} = \frac{P_a}{W_a}$ Formula

$\text{M.A.} = \frac{25}{5}$ which when reduced gives the ratio 5 to 1

M.A. = 5 to 1

In problems of mechanics, it is often desirable to reduce methods to simple rules or formulas. With levers, the so-called "law of levers" gives us a method of deriving formulas. The law, or rule, of levers is as follows:

Rule.—Power multiplied by length of power arm equals weight multiplied by length of weight arm.

Let P = power.

P_a = power arm.

W = weight.

W_a = weight arm.

Formula. $P \times P_a = W \times W_a$, better written

$$PP_a = WW_a.$$

This equation may be solved for any term.

All problems on machines may be solved by **inverse proportion** derived from these equal products, which for levers is

$$P:W = W_a:P_a$$

All simple machine problems presented here, as seen in the example below, may be solved by any one of three methods, namely formula, inverse ratio, and mechanical advantage. The formula method is generally the best method, but the workman may select one of the others if he chooses.

With many mechanisms, the terms **power movement** and **weight movement** are used instead of the terms power arm and weight arm. Power movement is also called **power travel.** Likewise, weight movement may be referred to as **weight travel.** The terms power movement and weight movement have the same relation to power and weight as do the terms power arm and weight arm.

With the lever, the distance the weight is moved, or the distance the power must travel, often must be determined, as will be seen in some of the problems on the following pages.

Example.—What weight can be balanced by 100 lb. of power on a first-class lever when P_a = 24 in. and W_a = 3 in.?

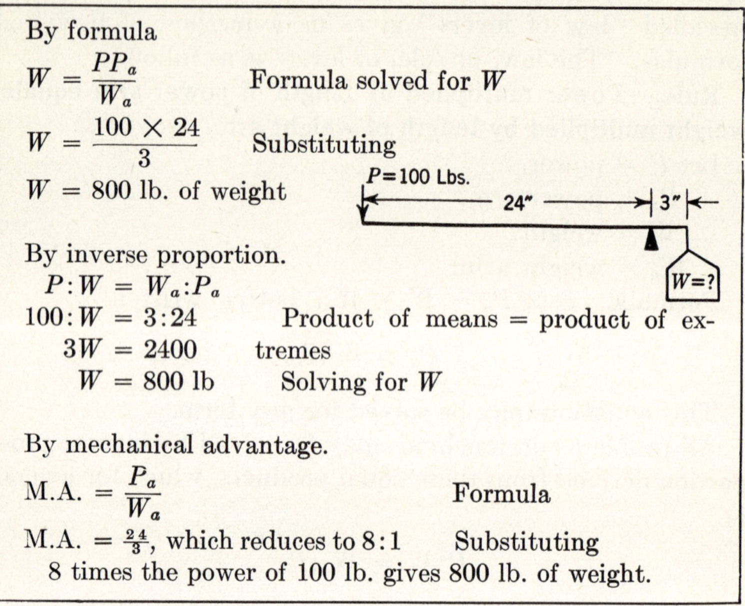

By formula

$W = \frac{PP_a}{W_a}$ Formula solved for W

$W = \frac{100 \times 24}{3}$ Substituting

W = 800 lb. of weight

By inverse proportion.

$P:W = W_a:P_a$

$100:W = 3:24$ Product of means = product of extremes

$3W = 2400$

W = 800 lb Solving for W

By mechanical advantage.

$\text{M.A.} = \frac{P_a}{W_a}$ Formula

M.A. = $\frac{24}{3}$, which reduces to 8:1 Substituting

8 times the power of 100 lb. gives 800 lb. of weight.

(*Note.*—In all machine problems, efficiency must be considered. Unless otherwise stated, the efficiency is assumed as 100 per cent and the solution is for a theoretical machine.)

PROBLEMS

Illustrate, give formula, solve, and find the mechanical advantage.

1. Solve the law of levers for each term separately. P = ?; P_a = ?; W = ?; W_a = ?.

2. What force 24 in. from the fulcrum will balance a weight of 1000 lb. that is 6 in. from the fulcrum?

3. What power is needed 10 ft. from the fulcrum to raise 6 tons that is 2 in. from the fulcrum?

4. A man grips a pair of pliers with a force of 74 lb. His grip centers 7 in. from the pivot bolt. What is the pressure on a wire being cut if it is placed only $\frac{3}{8}$ in. from the bolt?

5. A man is pulling on a hammer handle 12 in. from the point of rest. A nail being pulled is $1\frac{1}{4}$ in. from the resting point and has a resistance of 1050 lb. How much power must the man exert?

6. A man pulls the hand brake on his car with a force of 75 lb. The upper lever arm is 20 in., and the lower is $1\frac{1}{2}$ in. How much pull is exerted on the brake rod?

7. Two boys on a seesaw weigh 100 and 60 lb. The 60-lb. boy is 8 ft. 6 in. from the rest. How far out is the 100-lb. boy?

8. A stump puller is made with a handle 16 ft. long. The stump end is 4 in. long. What force can two men exerting their weight of 180 lb. each apply on the stump?

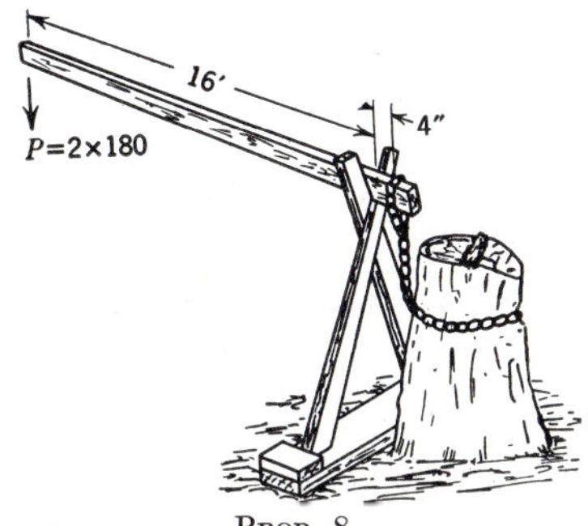

PROB. 8.

9. A weight of 240 lb. is balanced by a 20-lb. weight. The small weight is $2\frac{1}{2}$ ft. from f. Where is the larger weight?

10. What is the length of the weight arm if the power of 66 lb. $4\frac{1}{2}$ ft. from the fulcrum just balances 1 ton?

11. Weight is 729 lb. Power is 49 lb. Weight arm is 7 in. Find power arm.

12. List and illustrate as many other examples of first-class levers as you can.

13. A simple lathe test indicator is shown. B is the fulcrum. The short arm AB is $\frac{1}{2}$ in. From B to C is 8 in. If point C moves $\frac{1}{4}$ in., what is the error at A?

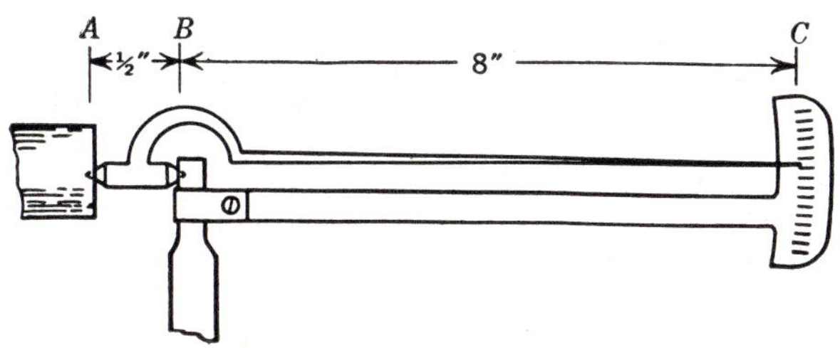

PROB. 13.

14. How far can the weight be moved when the power moves 27 in. if the weight is 3 in. from f and the power is 36 in. from f?

15. Two boys on a seesaw weigh 100 and 60 lb., respectively. The plank is 15 ft. long. Where is the fulcrum?

16. Describe or illustrate two other mechanisms using the first-class lever, but not as weight movers.

17. Describe or illustrate two mechanisms using the first-class lever with the power being applied at the short end.

SECOND-CLASS LEVERS

On the second-class lever **the weight is between the fulcrum and the power.**

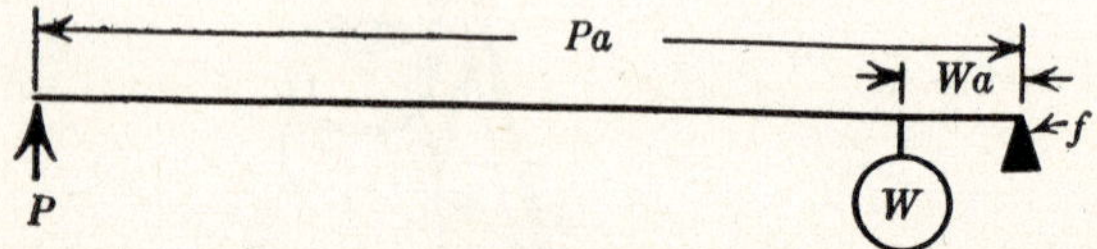

Examples of second-class levers are shown.

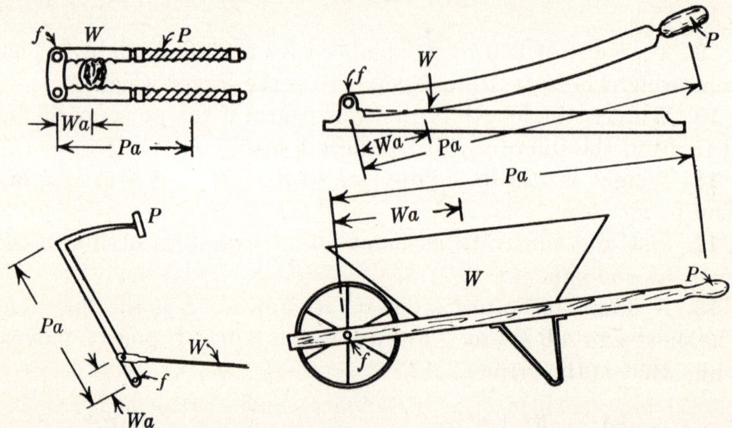

The same rules and formulas used for the first-class lever apply to the second-class lever.

$PP_a = WW_a$	Formula
$P:W = W_a:P_a$	Inverse proportion
$\text{M.A.} = \frac{P_a}{W_a}$, or $\text{M.A.} = \frac{W}{P}$	Mechanical advantage

Example.—What weight, placed 2 in. from the fulcrum, can be raised by a man exerting a lift of 150 lb. at the end of a 50-in. plank?

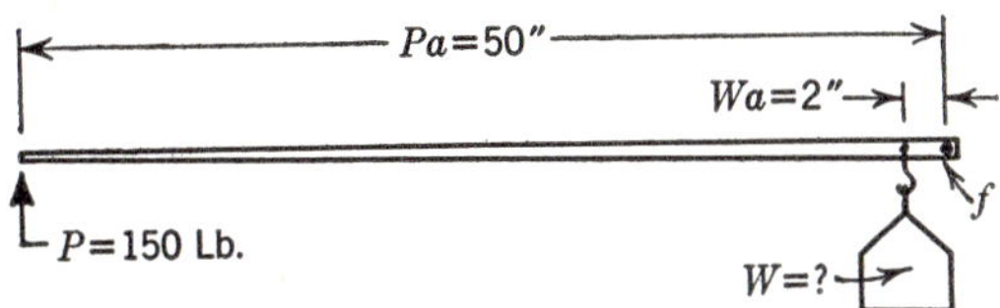

1. By formula.

$$W = \frac{PP_a}{W_a} \quad \text{Formula solved for } W$$

$$W = \frac{150 \times 50}{2} \quad \text{Substituting}$$

$$W = 3750 \text{ lb.}$$

2. By inverse proportion.

$$P:W = W_a:P_a$$

$$150:W = 2:50 \quad \text{Substituting}$$

$$2W = 7500 \quad \text{Means = extremes}$$

$$W = 3750 \text{ lb.}$$

3. By mechanical advantage.

$$\text{M.A.} = \frac{P_a}{W_a} \quad \text{Formula}$$

$$\text{M.A.} = \tfrac{50}{2} \quad \text{Substituting}$$

$$\text{M.A.} = 25 \text{ to } 1$$

$$25 \times 150 = 3750 \text{ lb.}$$

PROBLEMS

Illustrate, give formula, solve, and find mechanical advantage.

1. On a lever of the second class the weight arm is 9 in. The bar is 36 in. over all, the power arm being thus 36 in. long. What weight can a man raise if he exerts a lift of 125 lb.?

2. A man has a wheelbarrow loaded with 500 lb. of scrap iron. His grip is 48 in. from the axle. The average distance of the load is 22 in. from the axle. How much power must he exert on each handle?

3. A clutch is held in place with 100 lb. of spring. If the connection is 2 in. above the pin, what pressure is needed to open the clutch if the pedal arm is 14 in. long?

4. A nutcracker is 6 in. long. The nut is placed 1 in. from the joint. The center of the hand is 4 in. from joint. If the grip is 50 lb., what pressure is exerted on the nut?

5. A man is attempting to raise a building by putting the end of an 18-ft. timber under the wall 1 ft. If he exerts a lift of 200 lb., what lift is imparted to the wall?

6. On the foot brake of an automobile the lever is 15 in. long. A pressure of 110 lb. is exerted. The brake rod is attached $2\frac{1}{2}$ in. from the pivot. What tension is placed on the brake rod?

7. Where should you place the weight to raise 1000 lb. with a 10-ft. bar with 100 lb. of power?

8. How long a bar would be needed to raise 1 ton placed 8 in. from the fulcrum if 150 lb. of power is available?

9. Where should you place the weight of 1800 lb. to be raised with a force of 50 lb. if the bar is 6 ft. long?

10. How far from the fulcrum should you place a weight of 750 lb. if you intended to lift only 96 lb. on the end of a 6-ft.bar?

11. Can a man lift more with a first-class lever or a second-class lever if the entire bar is the same for both? Prove your answer.

12. List as many other examples of second-class levers as you can.

13. Without additional power, explain how a greater load could be handled on the wheelbarrow in Prob. 2 on page 489.

14. Why does a paper cutter shear easiest near the joint?

15. On a 6-ft. bar the weight is placed 6 in. from *f*. If the weight is raised 3 in. how far does the power move?

16. Describe or illustrate two mechanisms where power movement and weight movement are of prime importance.

17. How does the mechanical advantage of first- and second-class levers compare?

THIRD-CLASS LEVERS

On the third-class lever, the power is between the weight and the fulcrum.

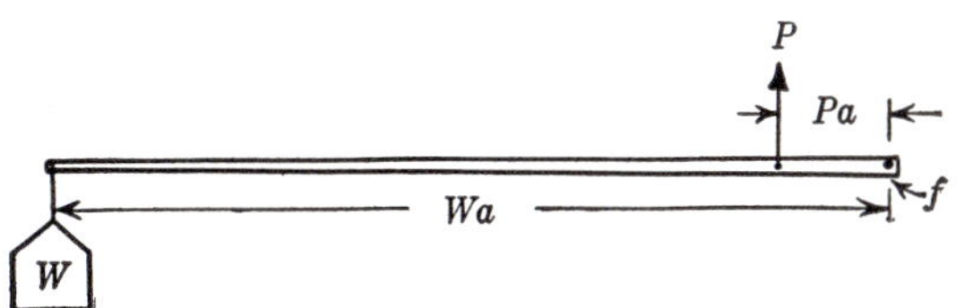

Examples of third-class levers are shown.

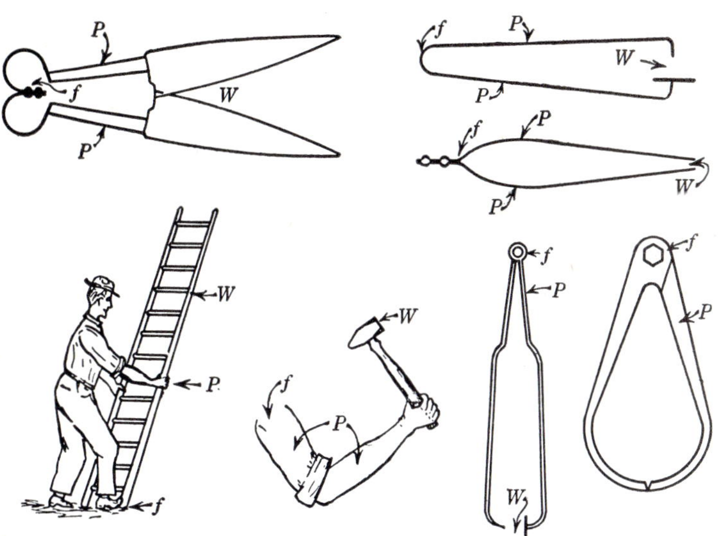

The rules and formulas used for first- and second-class levers apply to third-class levers. However, third-class levers are usually devised to sacrifice power at the expense of distance.

$PP_a = WW_a$	Formula
$P:W = W_a:P_a$	Inverse proportion
$\text{M.A.} = \frac{P_a}{W_a}$, or $\text{M.A.} = \frac{W}{P}$	Mechanical advantage

Example.—What force must be applied on the third-class lever shown below to raise a 10-lb. weight?

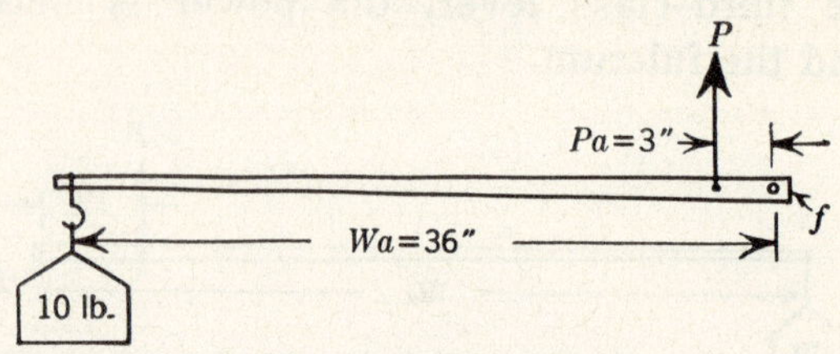

1. By formula.

$P = \frac{W W_a}{P_a}$ Formula solved for P

$P = \frac{10 \times 36}{3}$ Substituting

$P = 120$ lb.

2. By inverse proportion.

$P:W = W_a:P_a$

$P:10 = 36:3$

$3P = 360$

$P = 120$ lb.

3. By mechanical advantage.

$\text{M.A.} = \frac{P_a}{W_a}$

$\text{M.A.} = \frac{3}{36} = \frac{1}{12}$ to 1

$\frac{1}{12}P = 10$

$P = 10 \times 12 = 120$ lb.

PROBLEMS

Illustrate, give formula, solve, and find the mechanical advantage.

1. What power is needed to raise a weight of 100 lb. placed at the end of a 10-ft. bar if power is applied 2 ft. from the fulcrum?

2. What weight can be raised with the third-class lever in Prob. 1, when $P_a = 18$ in.?

3. On a sheep shear a man's grip is 60 lb. What pressure is exerted 1 in. from the tip if the center of his grip is 4 in. from the spring and the shear is 12 in. long over all?

4. On the same shear as in Prob. 3, what is the pressure at the butt of the blade if the blade is 6 in. long?

5. A man grips the middle of a pair of tweezers 5 in. long with a pressure of 5 lb. What is the force at the tips?

6. On the pair of 20-in. fire tongs shown, what force is exerted at the tips if a man squeezes the jaws 6 in. from the joint with 18 lb. pressure?

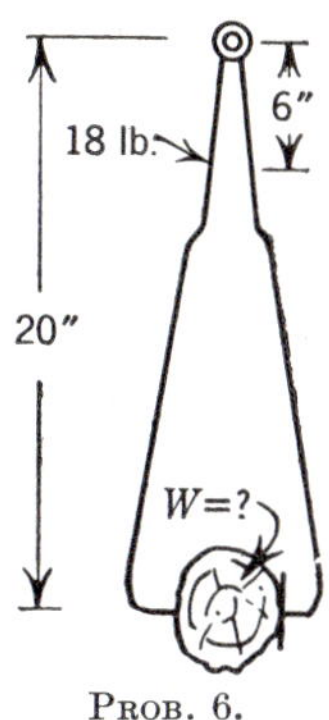

Prob. 6.

7. With the same tongs as in Prob. 6, where should you squeeze the jaws to hold a piece of fuel weighing 16 lb. if your grip is 40 lb.?

8. How does the mechanical advantage of the third-class lever compare with that of the other two classes?

9. Compare the three classes of levers on the basis of power and weight movement.

10. Illustrate or describe two mechanisms using the third-class lever not as weight movers.

11. How long a bar should you use to raise 100 lb. if your force is 1000 lb. and the power is only 1 ft. from the fulcrum?

12. List as many other examples of third-class levers as you can.

13. What weight is required to hold the valve shut on a round seat 3 in. in diameter if the steam pressure is 140 lb. per sq. in. against the valve?

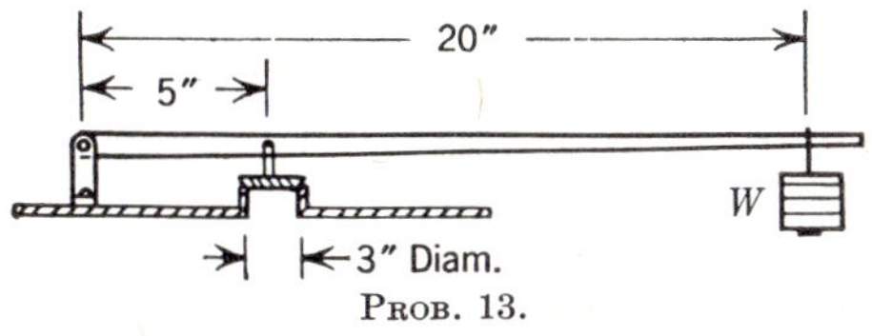

Prob. 13.

(P_a = 5 in., W_a = 20 in. $P = 0.7854 \times 3 \times 3 \times 140$).

14. What size of valve (diameter) is needed just to hold in Prob. 13 if the weight is 100 lb. and the steam pressure is 100 lb. per sq. in. (Find pressure, then area, and then diameter.)

15. In Prob. 13, what steam pressure per square inch is needed if the valve is reduced to 2 in. in diameter?

16. Where should you place the weight in Prob. 13 if the steam pressure were reduced to 100 lb. per sq. in.?

17. Where should you place the weight (Prob. 13) if the steam pressure were increased to 180 lb. per sq. in.)

THE INCLINE

The incline is a surface sloping from the horizontal. The highway climbing a hill is an incline, as are all sloping surfaces in nature. The thread on a bolt is an incline wrapped around a cylinder. A truckman loading a heavy object onto his truck finds it easier to roll the load up an inclined plane than to lift the object vertically.

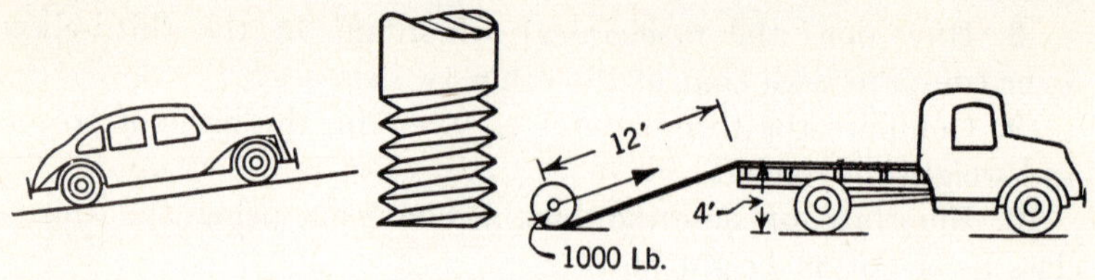

If the truckman must raise 1000 lb. to the bed of his truck 4 ft. high, he must do 1000 × 4, or 4000, ft.-lb. of work. Of course, he must produce 4000 ft.-lb. of work; but if he can spread the power over the 12-ft.-long incline shown above, he can apply less power and still raise the weight.

Rule.—Power times power movement equals weight times weight movement.

$$\text{Power} \times \text{length} = \text{weight} \times \text{height.}$$

P = power to be used.
W = weight to be raised.
L = length of plane.
H = height of lift.

Formula. $PL = WH.$

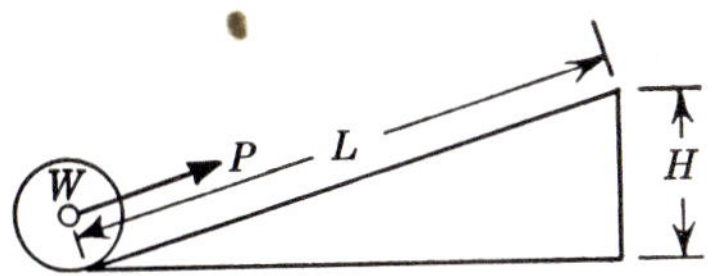

From these equal products, we can derive the **inverse proportion**

$$P:W = H:L$$

The **mechanical advantage** of an inclined plane is the ratio of the length to the height. Also, it is the ratio of the weight to the power.

$$\text{M.A.} = \frac{L}{H}, \qquad \text{and} \qquad \text{M.A.} = \frac{W}{P}$$

The power may be applied parallel to the base B instead of parallel to the incline. In that case, power moves a shorter distance; but the rule above applies and

$$\text{Power} \times \text{base} = \text{weight} \times \text{height}$$

Formula. $PB = WH.$

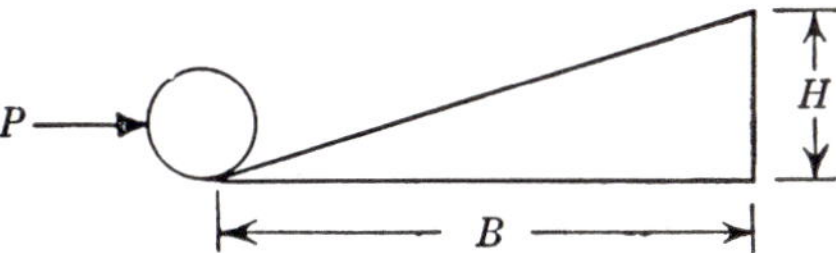

Likewise, by **inverse proportion,**

$$P:W = H:B$$

The **mechanical advantage** is

$$\text{M.A.} = \frac{B}{H}, \qquad \text{and} \qquad \text{M.A.} = \frac{W}{P}$$

The efficiency of the inclined plane depends upon the smoothness of the surfaces, the type of contact (whether sliding or rolling, etc.), in which direction power is applied, and other minor items. In the problems, these factors will be ignored and efficiency will be considered 100 per cent.

Example.—What weight can be moved up an incline 12 ft. long and 4 ft. high by a force of 100 lb., with the power applied parallel to the incline?

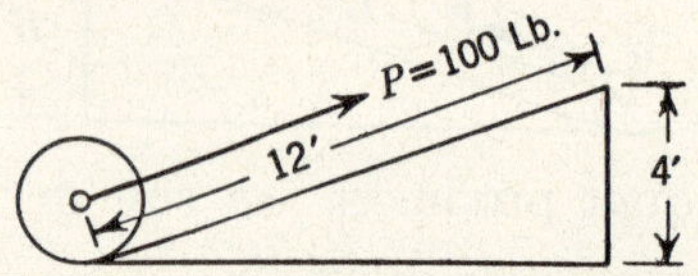

1. By formula.

$W = \frac{PL}{H}$ Formula solved for W

$W = \frac{100 \times 12}{4}$ Substituting

$W = 300$ lb.

2. By inverse proportion.

$P:L = H:W$

$100:W = 4:12$

$4W = 1200$

$W = 300$ lb.

3. By mechanical advantage.

M.A. $= \frac{L}{H}$ Formula

M.A. $= \frac{12}{4} = 3$ to 1

$3 \times 100 = 300$ lb. weight

PROBLEMS

Give formula used, illustrate, solve, and find the mechanical advantage.

1. Solve the formula for each term when pull is parallel to incline.

2. What weight can be drawn up an incline $8\frac{1}{2}$ ft. long and 3 ft. high by a force of 136 lb.?

3. Two men, each pulling 150 lb., can hold what weight on an incline 14 ft. long and 5 ft. high?

4. Solve the formula for each term when the pull is parallel to the base.

5. An incline is 6 ft. high, and the base is also 6 ft. What force is needed to hold 1 ton on the incline?

6. When the height is $5\frac{1}{2}$ ft. and the base is $27\frac{1}{2}$ ft., what weight can be moved on an incline by 144 lb. of force?

7. On an incline the base is 12 ft., the height is 9 ft., and the incline is 15 ft. Figure the force needed to hold 1000 lb. on the incline when the force is applied parallel to the base.

8. In Prob. 7, if the force is applied parallel to the incline, what force is needed?

9. Explain the reason for the difference in the answers to Probs. 7 and 8.

10. What force is needed to move a weight of 5000 lb. up an incline 7 ft. high and 24 ft. on the base?

Solve as indicated by the (?).

Number	Incline, feet	Height	Base, feet	Weight, pounds	Power, pounds
11	400	60 ft.	. . . .	3000	?
12	. . . .	4 ft.	8	1250	?
13	15	3 ft. 9 in.	. . . .	?	150
14	$82\frac{2}{3}$	$3\frac{1}{2}$ ft.	. . . .	6200	?

15. A load of 5000 lb. is to be pulled up an incline 25 ft. long and 4 ft. high. If a first-class lever with the power arm 6 ft. long and the weight arm 6 in. long is used as a pry, what power must be applied?

16. How long would an incline be if the height is 5 ft., the weight is 5000 lb., and the power is only 15 lb.?

17. List at least three applications of the incline to machinery.

THE WEDGE

The wedge is an adaptation of the inclined plane. It might be considered a pair of inclined planes with the bases joined.

Wedges are used for splitting logs, raising heavy weights short distances, and on machinery to actuate various mechanisms. Power may be supplied by a hammer blow, the pressure of jacks or levers, or by connection with a crank. Owing to friction, wedges are not very efficient. As in other mechanisms, the following may be stated.

Rule.—Power times power movement equals weights times weight movement.

$$\text{Power} \times \text{length} = \text{weight} \times \text{thickness}.$$

Let P = power.
W = weight.
L = length.
T = thickness.

Formula. $PL = WT.$

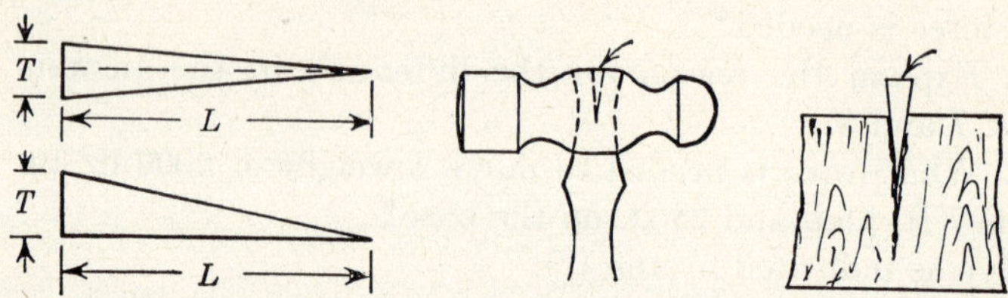

From these equal products, we can derive the **inverse proportion**

$$P:W = T:L$$

The **mechanical advantage** of a wedge is the ratio of the length to the thickness. It is also the ratio of the weight to the power.

$$\text{M.A.} = \frac{L}{T}, \qquad \text{and} \qquad \text{M.A.} = \frac{W}{P}$$

Example.—To raise a machine, pressure of 2 tons is put on the head of a wedge 12 in. long and 2 in. thick. What does the machine weigh?

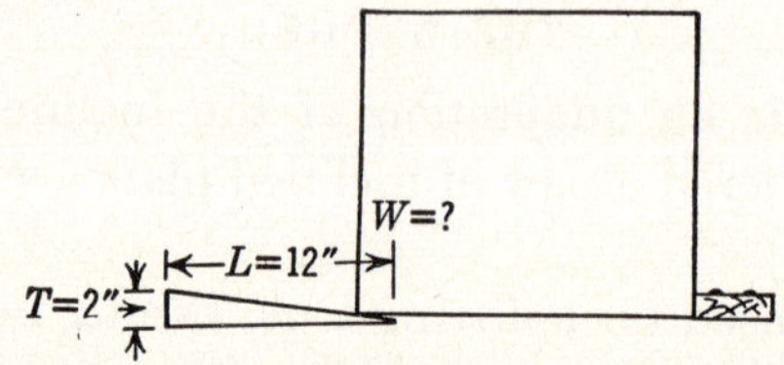

1. By formula.

$W = \frac{PL}{T}$ Formula solved for W

$W = \frac{4000 \times 12}{2}$ Substituting

$W = 24{,}000$ lb., not considering friction

2. By inverse proportion.

$$P:W = T:L$$
$$4000:W = 2:12$$
$$2W = 48{,}000$$
$$W = 24{,}000 \text{ lb.}$$

3. By mechanical advantage.

$$\text{M.A.} = \frac{L}{T}$$
$$\text{M.A.} = \tfrac{12}{2} = 6 \text{ to } 1$$
$$6 \times 4000 = 24{,}000 \text{ lb.}$$

PROBLEMS

Give formula used, illustrate, solve, and find mechanical advantage.

1. Solve the formula for each term.

Solve as indicated by (?).

Number	Power, pounds	Weight	Length, inches	Thickness, inches
2	100	?	16	2
3	746	?	21	5
4	?	10,500 lb.	15	$2\frac{1}{2}$
5	145	5 tons	?	3
6	3,000	28 tons	?	11
7	5,500	50 tons	12	?
8	45	1,000 lb.	12	?
9	166	?	$12\frac{1}{2}$	$\frac{3}{8}$
10	500	4,000 lb.	1	?
11	?	100 lb.	2	$\frac{1}{4}$
12	?	1 ton	3	$\frac{1}{2}$

13. If the loss due to friction is equal to 65%, what weight can be raised with 75 lb. of power with a wedge $\frac{3}{4}$ in. thick and 9 in. long?

14. How much power is needed to raise a 3500-lb. machine with a wedge 14 in. long and 2 in. thick if the efficiency is only 35%.

15. The available power is 3650 lb. The weight to be raised is 12 tons, and the length is 77 in. What is the thickness if the power applied is only 30% efficient?

16. With a wedge $\frac{1}{2}$ in. thick and 15 in. long to be used to raise a machine that weighs 5 tons, there is available 300 lb. of power. If the machine is only 40% efficient, how much more power must be supplied?

17. List five applications of the wedge to machinery.

18. State two reasons why the wedge might be low in mechanical efficiency.

19. Would a wedge of high mechanical advantage be higher in efficiency than one of lower mechanical advantage? Why?

20. Distinguish between the uses of wedges and inclines.

THE SCREW

The screw is an inclined plane wound around a cylinder and, when continuous, forms a common thread. The lead of the screw is the amount the thread advances for one

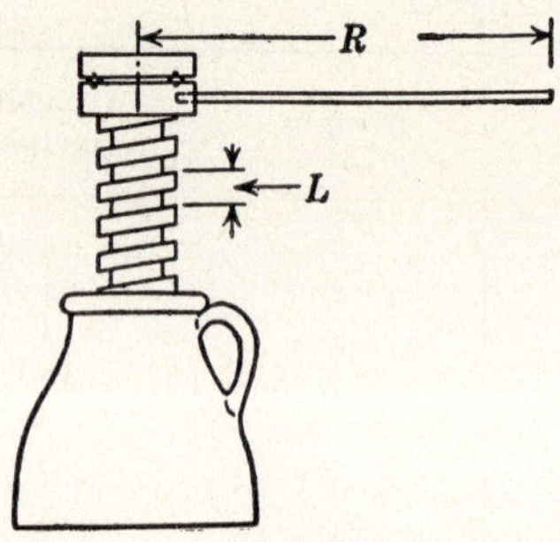

complete turn. The nut may turn, as on bolts, or the screw may turn out of a base as does the screw jack shown. If power is applied to the end of the handle R through one complete revolution, the power movement equals the circumference of the circle described by R, or $2\pi R$. During the same movement the screw will rise out of the base a distance equal to the lead L. Thus the power moves the circumference, and the weight moves the lead.

Rule.—The power multiplied by the power movement equals the weight multiplied by the weight movement.

Let P = power.
W = weight.
R = radius of handle.
L = lead of thread.

Formula. $P2\pi R = WL$, or $2\pi RP = WL$ (better arrangement).

The **inverse proportion** derived from these equal products is

$$P:W = L:2\pi R$$

The **mechanical advantage** is the ratio of the circumference ($2\pi R$) to the lead. It is also the ratio of the weight to the power.

$$\text{M.A.} = \frac{2\pi R}{L}, \qquad \text{and} \qquad \text{M.A.} = \frac{W}{P}$$

The efficiency of the screw is comparatively high, as high as 80 per cent when the angle of inclination of the screw is 45°. The angle of threads is not usually over 20°, where the efficiency reaches 74 per cent. In presses, where the mechanical advantage is required to be great, the angle is taken down to 3°, for this value the efficiency drops to about 34 per cent.

Example.—Find the weight that can be raised with a jack-screw if the handle R is 10 in. long, the lead L is $\frac{1}{4}$ in., and power P is 100 lb.

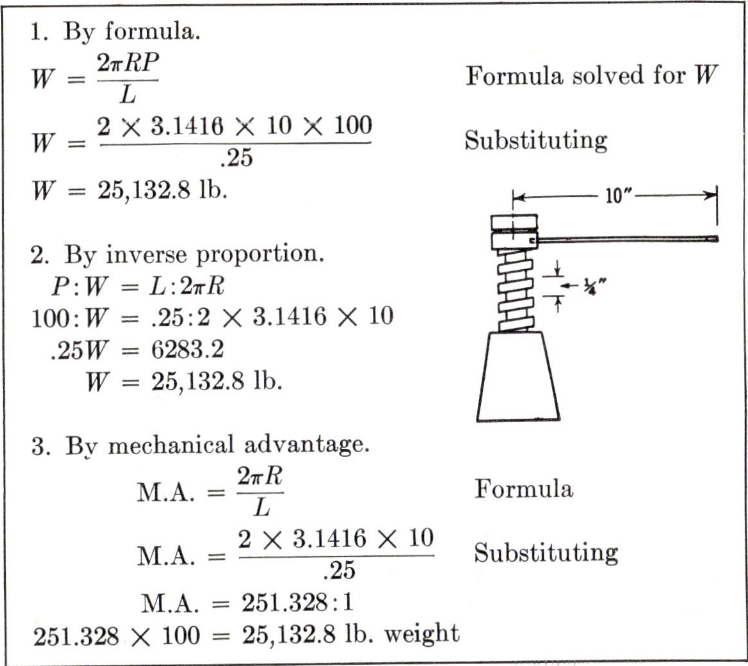

1. By formula.

$W = \dfrac{2\pi RP}{L}$ — Formula solved for W

$W = \dfrac{2 \times 3.1416 \times 10 \times 100}{.25}$ — Substituting

$W = 25{,}132.8$ lb.

2. By inverse proportion.

$$\begin{aligned} P:W &= L:2\pi R \\ 100:W &= .25:2 \times 3.1416 \times 10 \\ .25W &= 6283.2 \\ W &= 25{,}132.8 \text{ lb.} \end{aligned}$$

3. By mechanical advantage.

$\text{M.A.} = \dfrac{2\pi R}{L}$ — Formula

$\text{M.A.} = \dfrac{2 \times 3.1416 \times 10}{.25}$ — Substituting

$\text{M.A.} = 251.328:1$

$251.328 \times 100 = 25{,}132.8$ lb. weight

PROBLEMS

Write the formula, illustrate, solve, and find mechanical advantage.

1. Solve the formula $2\pi RP = WL$ for each term separately.

2. What weight can be raised with a jackscrew of $\frac{1}{4}$ in. lead if a force of 100 lb. is applied to a handle 16 in. long?

3. What weight can be raised with a 6-pitch single screw by a force of 15 lb. applied on the end of an 8-in. wrench?

4. Determine the pressure on a vise jaw if the screw has 8 threads per inch, the handle is 10 in. long, the power is 10 lb., and the efficiency is only 70%.

5. A building weighing 200 tons is to be raised with 50 jacks with $\frac{1}{2}$ in. lead on the screws. The handles are 15 in. long. Find the power needed on each jack.

6. By means of a 15-in.-diameter handwheel on a screw with $\frac{3}{8}$ in. lead, the head gate in a canal is raised. If a man applies a force of 45 lb. with each hand on the rim of the wheel, what resistance is overcome?

7. What force will be required on a wrench 10 in. long on a double-thread screw of $\frac{1}{8}''$ pitch to move a weight of 10 tons? (On the double screw, lead is twice the pitch.)

8. Find the lever required for a triple-thread screw of $\frac{1}{8}''$ pitch to give a pressure of 5 tons if the power is 75 lb., if the loss through friction is 25%.

Solve as indicated by (?).

Number	Power, pounds	Weight, pounds	Lead	Handle (R), inches	M.A.
9	12	2,000	1 in.	?	?
10	40	10,200	$\frac{1}{2}$ in.	?	?
11	210	?	1 in.	24	?
12	?	12,000	$\frac{1}{3}$ in.	20	?
13	?	60,000	$\frac{1}{4}$ in.	22	?
14	100	16,000	?	21	?
15	?	100	$\frac{1}{16}$ in.	6	?
16	312	?	$\frac{3}{8}$ in.	18	?
17	100	?	$\frac{1}{2}$ in.	15	?
18	?	20,000	3 pitch	12	?
19	90	15,500	?	10	?
20	?	2,500	4 pitch	10	?

PULLEY BLOCKS

In addition to being used to change speeds and transmit continuous power, pulleys are also used to move heavy loads at slow speeds by means of **pulley blocks.** Pulley blocks are also called **tackle blocks, block and tackle, sheave blocks,** or simply **sheaves** or **blocks.**

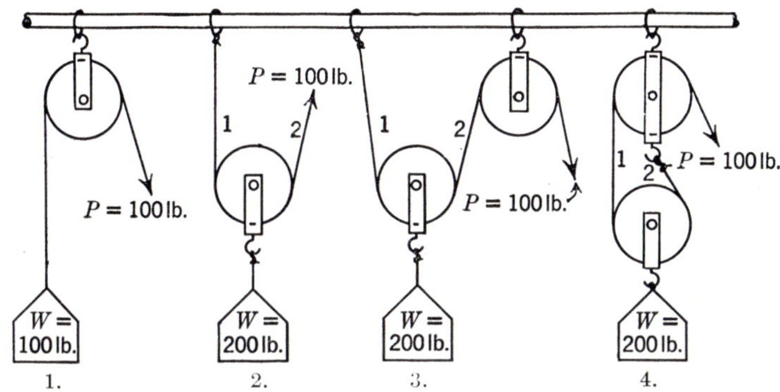

In Fig. 1, power is applied at P to raise the weight at W. The pulley serves only to change the direction of power. There is less friction with the pulley here than there would be if the rope passed over the bar above, but there is no mechanical advantage. A pull of 100 lb. at P will just support 100 lb. at W. If the rope at P is lowered 1 ft., the weight is raised 1-ft. This pulley is simply a first-class lever where $W_a = P_a$.

In Fig. 2 with the pulley movable, the power must move twice as far as the weight is lifted. It is seen that to raise the weight 1 ft., the rope on each side must be shortened 1 ft., and P would thus move 2 ft. The power, moving twice as far, gives a mechanical advantage of 2 to 1; therefore, $W = 2P$. One hundred pounds of power applied at P would raise 200 lb. of weight at W, not considering friction. It should be noted here that **the number of strands from the movable pulley is the mechanical advantage.** It is evident that one of the cords supports half the load. Thus, power equal to half the load on the other strand is enough to lift the load.

By combining Figs. 1 and 2, as shown in Fig. 3, we have the simplest form of pulley blocks. In practice, the first tie of the rope is permanently made to one pulley, as shown in Figs. 4 and 5.

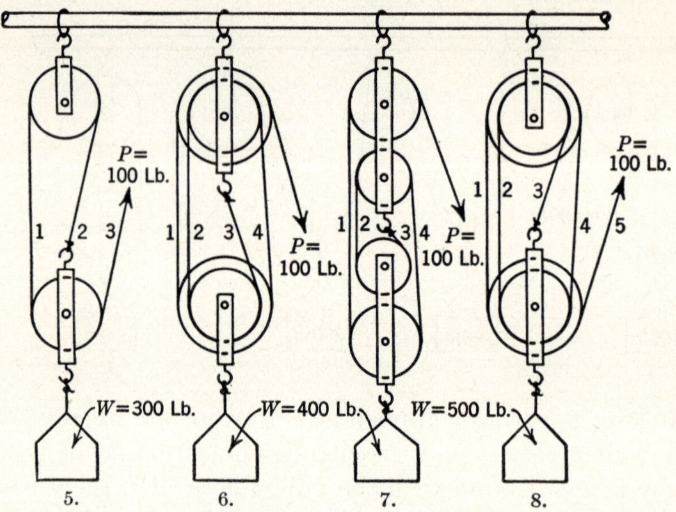

5. 6. 7. 8.

The work done by pulleys as arranged in Figs. 2, 3, and 4 is the same. The ratio of the weight to the power is 2 to 1, and the power movement is twice the weight movement in each case. This ratio is the same as the number of strands from the movable pulley.

Rule 1.—The number of strands S from the movable pulley equals the mechanical advantage of the machine.

$$S = \text{strands from movable pulley}$$

Formula 1. M.A. $= S$.

Rule 2.—The power multiplied by the number of strands from the movable pulley equals the weight.

$$P = \text{power.}$$
$$W = \text{weight.}$$

Formula 2. $W = PS.$

If the blocks of Fig. 4 are inverted, as in Fig. 5, the pull is in the direction the weight moves, just the opposite direction to the pull in Fig. 4. In Fig. 4, the pull is toward the load, and there are only two strands pulling on the load. In Fig. 5 the pull is with the load, and there are three strands pulling directly on the load.

Rule 3.—When the pull is toward the load, the strands equal the sheaves on both blocks.

Formula 3. $S =$ sheaves.

Rule 4.—When the pull is with the load, the strands equal the sheaves on both blocks, plus 1.

$$S = \text{sheaves} + 1$$

Ordinarily, several sheaves of equal size are placed in each block, or frame, on a common shaft, but because the strands could not be shown one behind the other in one-view drawings, it is common practice to illustrate the

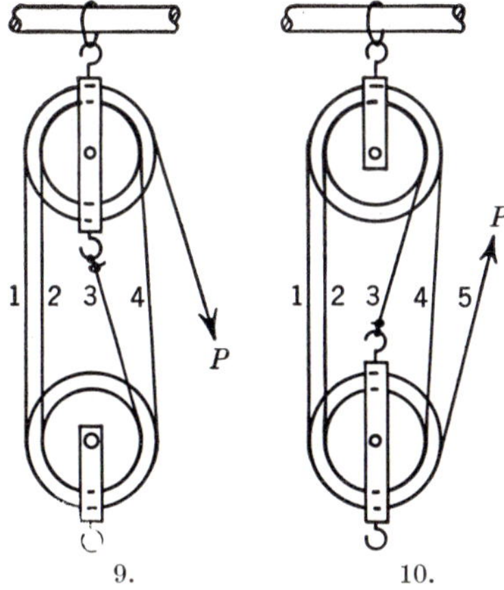

9. 10.

different sheaves by showing them different sizes, as shown in Figs. 6, 7, and 8. These figures each show two sheaves at each block. Figure 9 shows the use of two-sheave tackle blocks pulling toward the load. Figure 10 shows the use of the same two-sheave blocks pulling with the load.

Example.—With a three-sheave tackle block, power of 135 lb. is applied in the direction the weight moves. Find the strands, the mechanical advantage, and the weight, and illustrate. (Pull is with the load.)

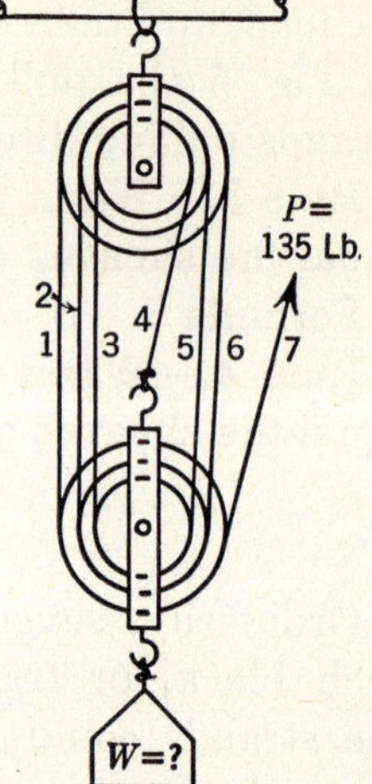

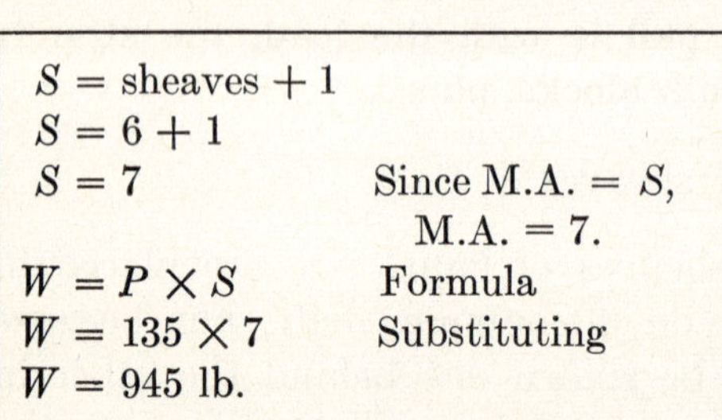

$S = \text{sheaves} + 1$	
$S = 6 + 1$	
$S = 7$	Since M.A. $= S$,
	M.A. $= 7$.
$W = P \times S$	Formula
$W = 135 \times 7$	Substituting
$W = 945$ lb.	

PROBLEMS

Illustrate, give formula, solve, and find mechanical advantage.

1. A 3-sheave block, arranged like that in Fig. 6 (to pull toward load) has 177 lb. of power. What weight can be raised?

2. With a 4-sheave tackle and downward pull, what weight can be raised by $16\frac{1}{2}$ lb. of power?

3. What weight can be raised by 75 lb. of power pulling upward with a 6-sheave tackle?

4. What power is needed to raise 1 ton with a 2-sheave tackle when the pull is downward (see Fig. 9)?

5. What power is needed in Prob. 4 if the pull is upward (see Fig. 10)?

6. The power in Prob. 5 is what per cent of that in Prob. 4?

7. What power is needed to raise 1 ton with a 3-sheave block, pulling toward the load (downward) if the net efficiency is 70%?

8. How many sheaves, with the pull upward, are needed to raise 2250 lb. with 450 lb. of power?

9. In Prob. 8, what power would be needed if the pull is downward?

10. What weight can be raised with a 2-sheave block with 66 lb. of power pulling toward the load?

11. If the efficiency of the entire mechanism is 81%, what weight could be raised with a 2-sheave tackle, pulling downward by 75 lb. of power?

12. If the efficiency is 77%, what power is needed to raise 1250 lb. with a 3-sheave block pulling in the direction the load moves?

13. As the number of sheaves in the blocks increase, likewise the friction increases as each sheave has friction on its bearing. The loss is considered as about 5% per sheave on each end. In a 2-sheave block with each pulley 95% efficient, determine the actual per cent of efficiency of the mechanism. (4 sheaves = 95% of 95% of 95% of 95%.)

14. What weight can be raised by 140 lb. of power with a 2-sheave block, pulling in the direction of the load movement, if efficiency is 95% on each pulley?

15. In addition to the advantage gained by the workman when the blocks are arranged for an upward pull, what other advantage does he have?

16. For light loads, which type of machine should you prefer? Give at least two reasons.

THE WHEEL AND AXLE

When a lever is free to rotate about its fulcrum, it becomes a wheel and the fulcrum becomes its axle. The simplest wheel-and-axle problem to consider is shown. If a cord is passed over the rim, with power applied at *P*, with the weight attached at *W*, *R* becomes the power arm and *r* becomes the weight arm. *R* and *r* are equal, and the power equals the weight. On the simple wheel, it is not the objective to raise an increased weight. The wheel simply serves to change the direction of the force. A man would pull 100 lb. at *P* to raise or balance 100 lb. at *W*.

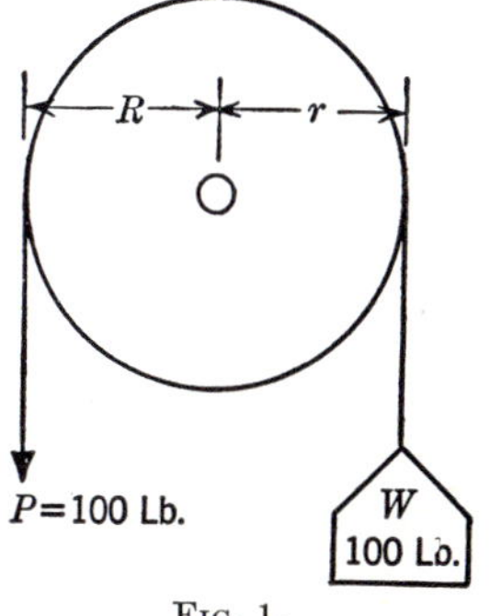

Fig. 1.

By using two pulleys or drums, rigid to each other as shown in Fig. 2, the power being applied at P over the large pulley, with the weight suspended from the rim of the small pulley, we have a mechanical advantage, because R is greater than r. The mechanism acts as a first-class lever. The large pulley, or drum, may be made into a crank as shown in Fig. 3.

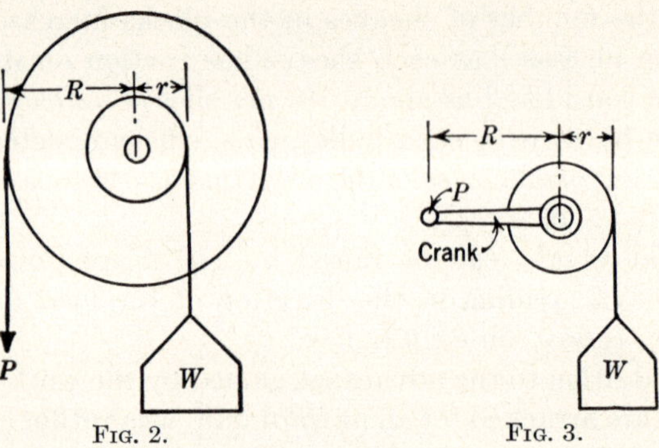

Fig. 2. Fig. 3.

By the rule for levers,

Power × power arm (R) = weight × weight arm (r)

$$PR = Wr$$

The **inverse proportion** derived from these equal products is

$$P:W = r:R$$

By **mechanical advantage**,

$$\text{M.A.} = \frac{R}{r}, \qquad \text{or} \qquad \text{M.A.} = \frac{W}{P}$$

The efficiency of the wheel and axle depends upon the load and the type of bearing. With roller bearings an efficiency as high as 98 per cent may be expected. With plain journal bearings the efficiency varies from 90 to 95 per cent.

The **windlass,** or winch, shown in Fig. 4, is an example of the wheel and axle in practical work. The handle, or

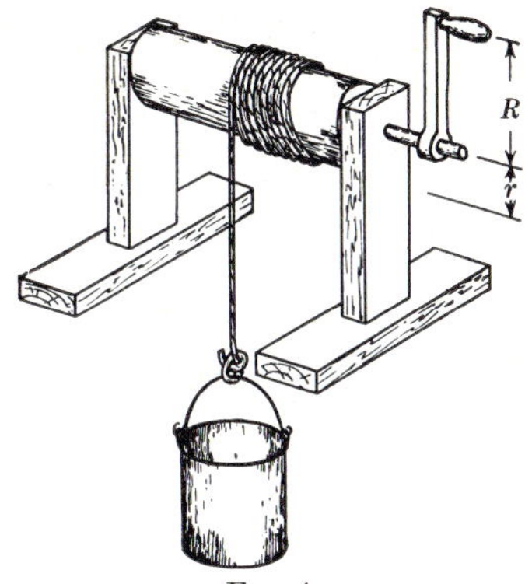

FIG. 4.

crank, is represented by R; and the radius of the drum by r.

Example 1.—What weight can be raised on a winch when the drum is 10 in. in diameter (r is 5 in.), the handle, or crank, is 15 in. long, and the power is 100 lb., efficiency not considered?

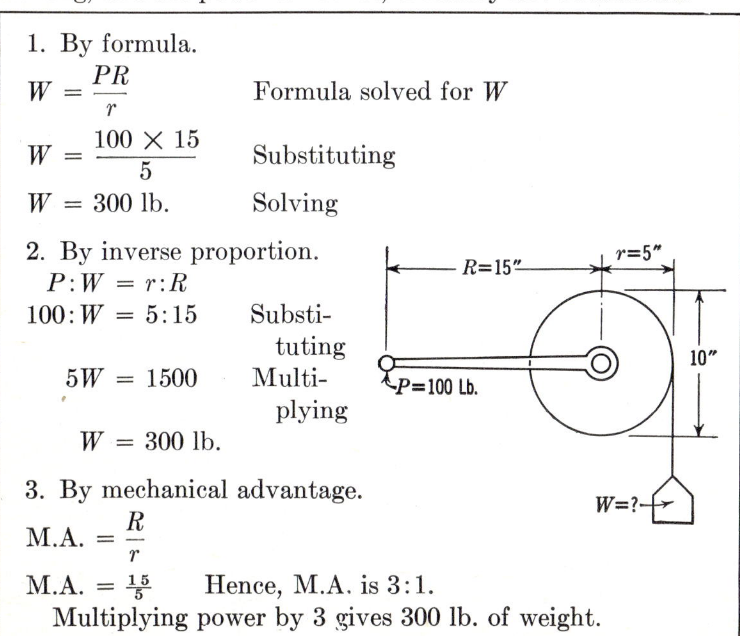

1. By formula.

$W = \frac{PR}{r}$ Formula solved for W

$W = \frac{100 \times 15}{5}$ Substituting

$W = 300$ lb. Solving

2. By inverse proportion.

$P:W = r:R$

$100:W = 5:15$ Substituting

$5W = 1500$ Multiplying

$W = 300$ lb.

3. By mechanical advantage.

$\text{M.A.} = \frac{R}{r}$

$\text{M.A.} = \frac{15}{5}$ Hence, M.A. is 3:1.

Multiplying power by 3 gives 300 lb. of weight.

Example 2.—What power is needed to raise 3000 lb. on a winch if the drum is 12 in. in diameter, the handle 16 in. long, and efficiency 90%?

$P = \frac{Wr}{R}$	Formula solved for P
$P = \frac{3000 \times 6}{16}$	Substituting
$P = 1125$ lb. theoretical power	
$P = \frac{1125}{.90}$	Theoretical power divided by the efficiency per cent changed to decimal.
$P = 1250$ lb. actual power	

Example 3.—Find the length of handle needed to raise 2000 lb. with 150 lb. of power on a winch that has a drum 8 in. in diameter and efficiency of 94%.

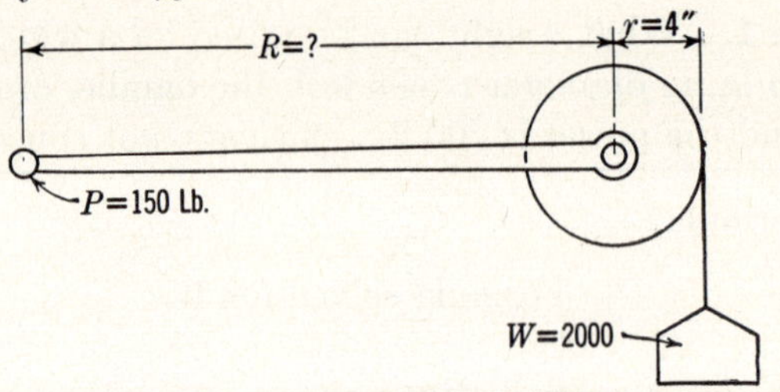

$R = \frac{Wr}{P}$	Formula solved for R
$R = \frac{2000 \times 4}{150}$	Substituting
$R = 53.33$ in. theoretical handle	
$R = \frac{53.33}{.94}$	
$R = 56.7$ in., the actual length of R needed	

PROBLEMS

Illustrate, give formula, solve, and find mechanical advantage.

1. Solve the formula for each term.

2. The diameter of the drum of a windlass is 8 in. The length of the crank is 18 in. The power at the crank is 125 lb. What weight can be raised or balanced?

3. A man exerts a pressure of 72 lb. on a crank 66 in. long. What weight can be raised on a drum 10 in. in diameter if the friction causes a loss of 18%? (Efficiency is 82%.)

4. Cranks on each end of a windlass are 20 in. long. What weight can two men raise with a force of 145 lb. each over a drum of 3 in. radius?

5. The axle of a windlass is 6 in. in diameter, and the crank is 25 in. long. If a pull of 96 lb. is applied at the end of the crank, what weight will be raised if 15% is allowed for frictional losses?

6. A windlass is to raise 1500 lb. The drum is 12 in. in diameter. What length of crank will be needed if the power is 200 lb.?

7. What diameter of drum should you provide on a winch to raise 1 ton with 100 lb. of power on a 20-in. crank?

8. The steering wheel of a ship is 6 ft. in diameter. The drum on which the ropes from the rudder are wound is 15 in. in diameter. What resistance can be overcome on the ropes by a force of 180 lb. at the wheel rim?

9. How long a crank should you provide on a windlass to raise 1200 lb. over a 3-in. axle if your power is 95 lb.?

10. In Prob. 9, what length should you make the crank if the machine is only 80% efficient?

11. What power should you provide on a windlass with an 8-in. drum to raise 1000 lb., when the crank is 20 in. long and the efficiency is only 75%?

12. What weight can be raised by a 100-lb. pull on a winch 80% efficient when the axle is 4 in. and the crank is 10 in.?

Work the following as indicated by the (?).

Number	Drum, inches	Crank	Weight	Power, pounds	Efficiency, per cent
13	6 diameter	72 in. diameter	?	120	90
14	21 circumference	20 in. crank	?	50	80
15	5 diameter	30 in. crank	750 lb.	?	85
16	14 circumference	90 in. circumference	1000 lb.	?	95
17	?	7 ft.	10 tons	160	70

THE WORM

The **worm is a screw with special-shaped threads.** The thread on a worm may be **single, double,** or **triple,** the same as threads used on bolts and machine parts.

Worm gearing.

The **worm wheel,** which meshes at right angles to the worm, appears similar to an ordinary gear wheel.

One of the most common uses of the worm is in the rear-axle reduction of automobiles, trucks, and busses. Usually, speedometers, electric meters, electric clocks, etc., incorporate worm gearing in their mechanisms. The worm is used because of the great reduction possible through its use.

When a single-thread worm makes one complete turn, the worm wheel will advance only one tooth. For example, if the worm wheel has 40 teeth, a single-thread worm will revolve forty times while the worm wheel makes only one complete revolution.

The ratio of their speeds is called their **velocity ratio.** The velocity ratio of the worm and wheel described in the preceding paragraph would be 40 to 1.

The **mechanical advantage** is the same as the velocity ratio and is also 40 to 1.

V = velocity ratio.

N = number of teeth on worm wheel.

n = number of threads on worm, as single = 1, double = 2, etc.

M.A. = mechanical advantage.

Formulas. $V = \frac{N}{n}$, and M.A. $= \frac{N}{n}$.

Example 1.—Find the velocity ratio of a triple-thread worm in mesh with a worm wheel that has 41 teeth.

$V = \frac{N}{n} = \frac{41}{3} = 13\frac{2}{3}$ to 1. This is also the M.A.

Note.—As found in this example, the velocity ratio need not be a whole number. Nor is it necessary to consider the size of the worm or worm wheel. The sizes will be determined by the strength needed and the space available.

Modern worm gearing is very efficient. Tests have shown worm gearing to be as high as $97\frac{1}{2}$ per cent efficient. This is as high as the efficiency of single-gear reductions. It would require several gears in a gear train to give the reduction of a single worm; therefore, the worm gear would be higher in efficiency.

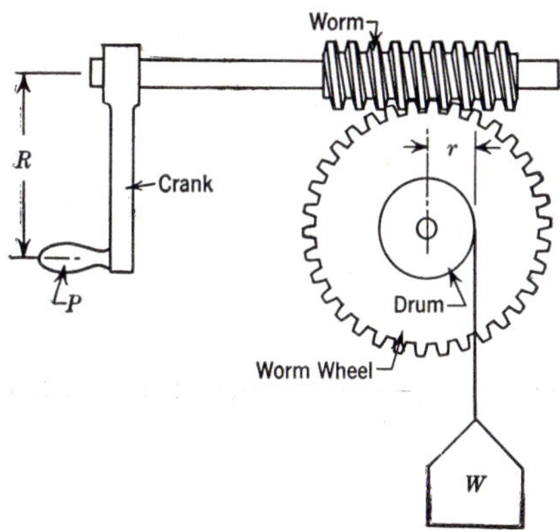

A common use of the worm as a weight-raising machine is shown. It is similar to the wheel and axle discussed on page 507, except that it has the worm and worm wheel incorporated in it. The worm is fitted with a crank R. The worm wheel is made solid to an axle, or drum, with a radius r. When the crank is turned, the cord winds around the drum, thus raising the weight. The machine has the mechanical advantage of the worm reduction $\frac{N}{n}$, multiplied by the mechanical advantage of $\frac{R}{r}$. Thus, M.A. $= \frac{N}{n} \times \frac{R}{r}$; since $V = \frac{N}{n}$, then M.A. $= \frac{RV}{r}$. However, M.A. $= \frac{NR}{nr}$ may be used if desired. The combined power arm value is $\frac{NR}{n}$.

Formulas. M.A. $= \frac{NR}{nr}$, or M.A. $= \frac{RV}{r}$.

Rule.—Power multiplied by power arm equals weight multiplied by weight arm.

P = power.
R = radius of crank.
W = weight.
r = radius of drum.
N = number of teeth on worm wheel.
n = number of threads on worm.
V = velocity ratio, which is $\frac{N}{n}$.

Formula. $\frac{PRN}{n} = Wr$, or $PRV = Wr$.

The **inverse proportion** derived from these equal products is

$$P:W = r:\frac{RN}{n}, \qquad \text{or } P:W = r:RV$$

Example 2.—Get the weight that can be raised by 100 lb. of power applied on a 15-in. crank, with a single-thread worm, with a 32-tooth worm wheel solid to a 12-in. axle.

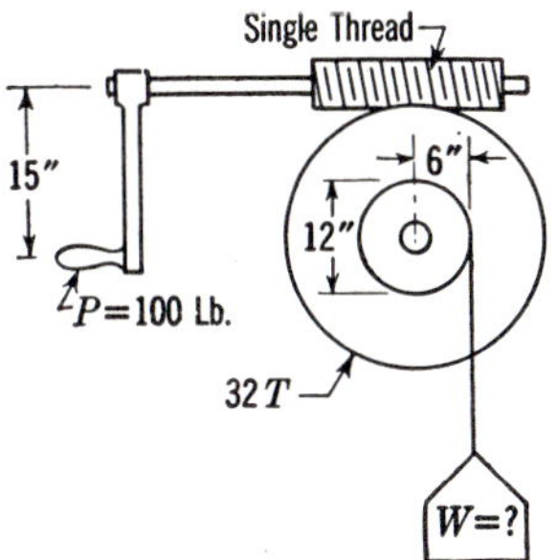

$$W = \frac{PRN}{rn} \qquad \text{Formula solved for } W$$

$$W = \frac{100 \times 15 \times 32}{6 \times 1} \qquad \text{Substituting}$$

$$= 8000 \text{ lb., not considering frictional losses}$$

By inverse proportion.

$$P:W = r:\frac{RN}{n}$$

$$100:W = 6:\frac{15 \times 32}{1}$$

$$6W = 48{,}000$$

$$W = 8000 \text{ lb.}$$

By mechanical advantage.

$$\text{M.A.} = \frac{NR}{nr}$$

$$\text{M.A.} = \frac{32 \times 15}{1 \times 6}$$

$$\text{M.A.} = \frac{480}{6} = 80 \text{ to } 1$$

$$80 \times 100 = 8000 \text{ lb.}$$

PROBLEMS

Give the formula used, illustrate, and solve.

1. Solve the formula for each letter.

2. Determine the weight it is possible to raise with a single worm in mesh with a 56-tooth wheel when the crank is 12 in. long and the drum solid to wheel is 10 in. in diameter if P is 30 lb.

3. What power will be needed at the end of a 12-in. crank to raise 5 tons over an 8-in. drum solid to a 78-tooth wheel if the worm is double? ($r = 4$ in.)

4. A triple worm with a 75-tooth worm wheel has a crank 10 in. long. The drum is 5 in. in diameter. What weight can be raised with 110 lb. of power? ($r = 2.5$ in.)

5. A single-thread worm lifts a weight of 5 tons by means of a drum 12 in. in diameter and a crank 20 in. long by means of 50 lb. of power. Get the number of teeth (N) on the worm wheel.

6. A single-thread worm with a 50-tooth worm wheel, a 9-in. crank, and an 8-in. drum lifts a weight of 4850 lb. Get the power.

7. How long a crank is needed to raise 12,000 lb. by applying 100 lb. of power on a double worm, when the worm wheel has 80 teeth and the drum is 8 in. in diameter? (Get R.)

Solve for the term indicated by (?).

Number	Power, pounds	Weight, pounds	Crank (R), inches	Worm	Worm wheel	Drum diameter, inches
8	144	?	10	Single	64 teeth	8
9	75	?	12	Double	75 teeth	9
10	?	8,000	9	Double	80 teeth	12
11	?	10,000	15	Triple	90 teeth	15
12	150	15,000	?	Single	40 teeth	12
13	25	2,000	8	Single	36 teeth	?

The figure below incorporates a set of gears between the crank and the worm. Thus the power is further reduced by the ratio of the small gear to the larger gear, $\frac{g}{G}$.

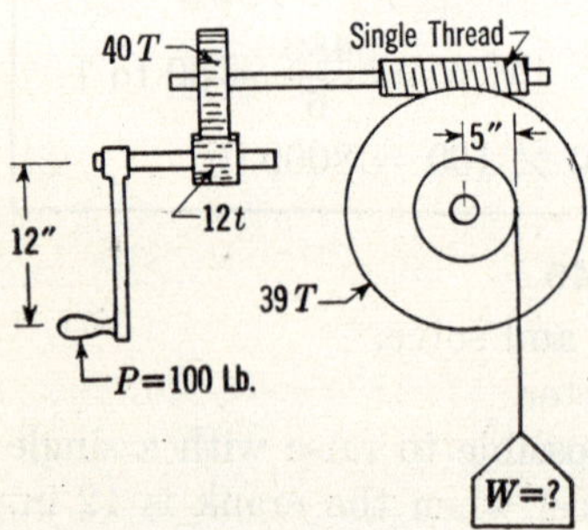

$$P = \frac{Wr}{RV} \times \frac{g}{G},$$

$$W = \frac{PRV}{r} \times \frac{G}{g}$$

14. What weight can be raised by the machine in the figure when the crank is 12 in., g has 12 teeth, G has 40 teeth, the worm is single, the worm wheel has 39 teeth, the drum is 10 in. in diameter, and the power is 100 lb.?

15. What power would be needed to raise 10 tons with the machine above?

CHAPTER 23

EFFICIENCY

RULES AND FORMULAS

In any mechanism, there is always some loss due to friction of the moving or bearing parts. In some cases, such as in sliding an object horizontally, most of the power, or force, applied must be used to overcome the resistance of friction. The force necessary to overcome friction is a wasted force and must be deducted from the input force to determine the work delivered.

Rule.—Efficiency is the ratio of output (work delivered) to input (work put in). Efficiency $= \dfrac{\textbf{output}}{\textbf{input}}$.

Example.—If a 100-lb. force is supplied to a machine and only 80 lb. results, what is the efficiency?

$$E = \frac{\text{output}}{\text{input}} = \frac{80}{100} = .80 = 80\% \text{ efficient}$$

In all the problems on the preceding pages, unless the efficiency was stated, the calculations were on the basis of 100 per cent efficiency, and the results to be obtained were for a machine theoretically frictionless.

The following table shows the average for various types of machine when properly lubricated:

	Efficiency, per Cent
Lever	90 to 97
Incline	40 to 80
Wedge	30 to 60
Screw	30 to 65
Pulley blocks	90 to 98 for each sheave
Windlass	30 to 80 depending on type of bearings
Worm	60 to 95

Other equipment is rated as follows:

	Efficiency, per Cent
Belting	96 to 98
Bicycle chains, etc	95 to 97
High-grade transmission chain	97 to 99
Half-ton chain hoists	30 to 50
Hydraulic jacks	80 to 90
Locomotive drawbar	65 to 75
Common bearings	96 to 98
Roller bearings	98
Ball bearings	99
Spur gears	93 to 97
Bevel gears	92 to 95

Marks's *Mechanical Engineers' Handbook* gives the following efficiencies for the automobile:

	Efficiency, per cent	
	Maximum	At full load
High gear	98	97
Second	95.5	95
Third (low)	94	93
Reverse	87	86

Due to frictional losses the theoretical weight that may be moved by any of the mechanical powers, such as the simple machines, the lever, wedge, or screw, is greater than the actual weight; hence the following may be stated:

Rule 1.—Theoretical weight multiplied by efficiency per cent equals the actual weight. (Change per cent to decimals before multiplying.)

Example 1.—The theoretical weight raised by a machine is 600 lb., but the efficiency is only 75 per cent. Find the actual weight.

$$600 \times .75 = 450 \text{ lb. actual weight}$$

If in a machine the weight arm has been calculated, this, we must remember, is only theoretical, an ideal situation that is never attained. It is greater than could be used; hence, we have the following:

Rule 2.—The theoretical weight arm or movement multiplied by the efficiency per cent equals the actual weight arm or movement. (Change per cent to decimals.)

Example 2.—If the theoretical weight arm of a lever has been calculated as 30 in. and the efficiency is 95 per cent, what actual weight arm must be used?

$30 \times .95 = 28.5$ in., the actual weight arm

Owing to losses the theoretical power is less than that actually needed. The theoretical power is the per cent stated of the actual power; hence, we have the following:

Rule 3.—The theoretical power divided by the efficiency per cent equals the actual power. (Change per cent to decimals before dividing.)

Example 3.—The theoretical power estimated on a wedge is 135 lb., and the efficiency is 45 per cent. Find the actual power needed.

3 00. lb. actual power needed .45)135.00

If the part of the machine to which the power is applied, such as the power arm of a lever, has been calculated, it is the theoretical power arm and is shorter than would actually be needed; hence, we have the following:

Rule 4.—The theoretical power arm, or movement, divided by the efficiency per cent equals the actual power arm, or movement. (Change per cent to decimals.)

Example 4.—The power arm of a lever has been calculated to be 25 in. long, but the lever is only 92 per cent efficient. Find the actual length of power arm needed.

27.2− in., the actual length of power arm .92)25.00

Note.—**The student should recall the types of percentage problems discussed earlier, on pages 98 to 122. A review of those pages will assist in the solution of the following.**

PROBLEMS

Refer to Chap. 22 for the rules and formulas needed to solve the problems in this section.

Find the theoretical value and actual value as indicated.

Number	Machine	Power, pounds	Power arm or power movement	Weight, pounds	Weight arm or weight movement	Theoretical value	Efficiency, per cent	Actual value
1	Lever	40	$Pa = 96''$		$Wa = 24''$		95	
2	Lever		$Pa = 42''$	480	$Wa = 2''$		97	
3	Lever	175	$Pa =$	490	$Wa = 15''$		96	
4	Lever	56	$Pa = 36''$	672	$Wa =$		90	
5	Incline	128	$L = 16'$		$H = 5'$		75	
6	Incline		$L = 75'$	2,000	$H = 12'$		60	
7	Incline	250	L	2,000	$H = 6'$		70	
8	Incline	108	$L = 32'$	1,728	H		65	
9	Incline	340	B	1,020	$H = 7'$		80	
10	Incline		$B = 10'$	4,200	$H = 3'$		70	
11	Wedge	350	$L = 18''$	10,500	T		75	
12	Wedge		$L = 10''$	500	$T = \frac{1}{2}''$		60	
13	Wedge	256	L	1,440	$T = 1''$		70	
14	Wedge	700	$L = 3\frac{1}{2}''$		$T = \frac{1}{4}''$		67	
15	Lever		$Pa = 56''$	2,000	$Wa = 7''$		90	
16	Incline	5,000	L	55,500	$H = 10'$		80	
17	Wedge	2,000	$L = 0.66''$		$T = 0.12''$		75	
18	Incline	540	$B = 81'$	18,000	$H =$		80	
19	Lever	180	$Pa = 16'$		$Wa = 6''$		98	
20	Wedge		$L = 2''$	1,000	$T = \frac{1}{8}''$		70	
21	Incline	666	$L =$	13,320	$H = 3.6''$		70	
22	Lever	140	$Pa = 45''$		$Wa = 5''$		96	

Owing to efficiency losses the mechanical advantage of a theoretical machine can never be attained; hence, we have the following:

Rule.—To find the actual mechanical advantages, multiply theoretical mechanical advantages by the efficiency per cent. (Change per cent to decimal.)

Example.—The theoretical mechanical advantage of a machine has been calculated as 25 to 1. If the efficiency is only 80%, what is the actual mechanical advantage?

$25 \times .80 = 20$	The actual mechanical advantage is 20 to 1.

PROBLEMS

In the following problems, find the theoretical value and actual value, and the theoretical and actual mechanical advantage.

Number	Machine	Power, pounds	Power arm or power movement	Weight, pounds	Weight arm or weight movement	Theoretical value	Efficiency, per cent	Actual value	Theoretical M.A.	Actual M.A.
1	Lever		$Pa = 21''$	1,600	$Wa = \frac{1}{2}''$		95			
2	Wedge	1,250	$L = 15''$		$T = 0.3''$		40			
3	Incline	500	$L = 16'$	2,400	$H =$		70			
4	Screw	120	$R = 12''$		$L = \frac{1}{8}''$		35			
5	Lever		$8'$	1,000	$3''$		90			
6	Windlass	88	R''	2,400	$r = 11''$		95			
7	Screw	95	$R = 15''$		$L = 1''$		42			
8	Pulley blocks	65	3 sheaves pull up		$w =$		75			
9	Wedge	500	L''	6,550	$T = \frac{1}{2}''$		30			
10	Incline		$L = 76'$	40,000	$H = 16'$		50			
11	Screw		$R = 18''$	50,000	$L = \frac{1}{4}''$		60			
12	Windlass	50	$R = 18''$	350	r''		90			
13	Pulley blocks	100	2 sheaves pull down		W		85			
14	Wedge		$L = 5''$	1,000	$T = \frac{1}{8}''$		40			
15	Incline	2,700	$B = 16'$		$H = 5'$		55			
16	Lever	125	Pa''	800	$Wa = 1.5''$		90			
17	Pulley blocks		4 sheaves pull up		$W = 1{,}030$		70			
18	Screw		$R = 12''$	38,000	$L = \frac{1}{2}''$		43			
19	Wedge	75	$L = 16\frac{1}{2}''$		$T = \frac{1}{3}''$		25			
20	Windlass	50	R''	375	$r = 3''$		90			

REVIEW OF MACHINES

1. State the general formula for levers, and solve it for each term. Give the mechanical advantage formulas.

2. State the general formula for inclines, and solve it for each term. Give the mechanical advantage formulas.

3. State the general formula for wedges, and solve it for each term. Give the mechanical advantage formulas.

4. State the general formula for screws, and solve it for each term. Give the mechanical advantage formulas.

5. State the general formula for windlasses, and solve it for each term. Give the mechanical advantage formulas.

6. State the general formula for pulley blocks, and solve it for each term. Give the mechanical advantage formulas.

7. State the general formula for worms, and solve it for each term. Give the mechanical advantage formulas.

In the following problems, find the theoretical value and actual value, and the theoretical and actual mechanical advantage.

Number	Machine	Power, pounds	Weight, pounds		Theoretical value	Efficiency, per cent	Actual value	Theoretical M.A.	Actual M.A.
8	Lever	45	1,395	$Wa = 31$; $Pa = ''$		95			
9	Incline	155	10,000	$L = '$; $H = 3'$		60			
10	Wedge		5,755	$L = 18''$; $T = \frac{1}{8}''$		30			
11	Screw	88		$R = 15''$; $L = \frac{1}{2}$		38			
12	Windlass		2,000	$R = 12''$; $D = 5''$		90			
13	Pulley blocks	66		2 sheaves, pull down		85			
14	Worm	77		single, 70 teeth; axle 6''; $R = 212''$		80			
15	Lever	121		$Pa = 5.5''$, $Wa = 5'$		98			
16	Incline	500		$L = 21'$, $B = 4'$		70			
17	Wedge		5,000	$L = 3''$, $T = \frac{1}{8}''$		40			
18	Windlass		650	$R = 8''$, $r = 2\frac{1}{2}''$		90			
19	Screw	21	11,000	$R = ''$, $L = \frac{1}{4}''$		50			
20	Pulley blocks		1,000	4 sheaves, pull up		70			
21	Worm		10,000	double; 80 teeth; axle, 12'' diameter, $R = 15''$		90			

APPENDIX

Table I.—Tables and Equivalents

Abbreviations are shown in parentheses.

Linear Measure

12 inches (in.) = 1 foot (ft.)
3 ft. = 1 yard (yd.)
$16\frac{1}{2}$ ft. = 1 rod (rd.)
$5\frac{1}{2}$ yd. = 1 rd.
320 rd. = 1 mile
1760 yd. = 1 mile
5280 ft. = 1 mile

Surface Measure

144 sq. in. = 1 sq. ft.
9 sq. ft. = 1 sq. yd.
$30\frac{1}{4}$ sq. yd. = 1 sq. rd.
160 sq. rd. = 1 acre
640 acres = 1 sq. mile
43,560 sq. ft. = 1 acre

Cubic Measure

1728 cu. in. = 1 cu. ft.
27 cu. ft. = 1 cu. yd.
128 cu. ft. = 1 cord

Liquid Measure

4 gills = 1 pint (pt.)
2 pt. = 1 quart (qt.)
4 qt. = 1 gallon (gal.)
231 cu. in. = 1 gal.
31.5 gal. = 1 barrel (bbl.)
42 gal. = 1 bbl. of oil
$8\frac{1}{3}$ lb. = 1 gal. water
$7\frac{1}{2}$ gal. = 1 cu. ft.

Weights of Materials

0.0963 lb. = 1 cu. in. aluminum
0.2604 lb. = 1 cu. in. cast iron
0.2833 lb. = 1 cu. in. mild steel
0.3215 lb. = 1 cu. in. copper
0.4106 lb. = 1 cu. in. lead
112 lb. = 1 cu. ft. Dowmetal
167 lb. = 1 cu. ft. aluminum
464 lb. = 1 cu. ft. cast iron
490 lb. = 1 cu. ft. mild steel
555.6 lb. = 1 cu. ft. copper
710 lb. = 1 cu. ft. lead

Avoirdupois Weight

16 ounces (oz.) = 1 pound (lb.)
100 lb. = 1 hundredweight (cwt.)
20 cwt. = 1 ton
2000 lb. = 1 ton
$8\frac{1}{3}$ lb. = 1 gal. of water
62.4 lb. = 1 cu. ft. of water
112 lb. = 1 long cwt.
2240 lb. = 1 long ton

Time Measure

60 seconds (sec.) = 1 minute (min.)
60 min. = 1 hour (hr.)
24 hr. = 1 day
7 days = 1 week
52 weeks = 1 year
365 days = 1 year
360 days = 1 year, to get interest
10 years = 1 decade

Circular, or Angular, Measure

60 sec. ($''$) = 1 min. ($'$)
$60'$ = 1 degree ($^\circ$)
90° = 1 quadrant
360° = 1 circle

Dry Measure

2 cups = 1 pt.
2 pt. = 1 qt.
4 qt. = 1 gal.
8 qt. = 1 peck (pk.)
4 pk. = 1 bushel (bu.)

Miscellaneous

12 units = 1 dozen (doz.)
12 doz. = 1 gross
144 units = 1 gross
24 sheets = 1 quire
20 quires = 1 ream
20 units = 1 score
6 ft. = 1 fathom
6080.2 = 1 nautical mile

Table II.—Decimal Equivalents of Parts of an Inch
(By 64ths)

Fraction			Decimal	Milli-meters	Fraction			Decimal	Milli-meters
		$\frac{1}{64}$	0.015625	0.3968			$\frac{33}{64}$	0.515625	13.0966
	$\frac{1}{32}$		0.03125	0.7937		$\frac{17}{32}$		0.53125	13.4934
		$\frac{3}{64}$	0.046875	1.1906			$\frac{35}{64}$	0.546875	13.8903
$\frac{1}{16}$			0.0625	1.5875	$\frac{9}{16}$			0.5625	14.2872
		$\frac{5}{64}$	0.078125	1.9843			$\frac{37}{64}$	0.578125	14.6841
	$\frac{3}{32}$		0.09375	2.3812		$\frac{19}{32}$		0.59375	15.0809
		$\frac{7}{64}$	0.109375	2.7780			$\frac{39}{64}$	0.609375	15.4778
$\frac{1}{8}$			0.125	3.1749	$\frac{5}{8}$			0.625	15.8747
		$\frac{9}{64}$	0.140625	3.5718			$\frac{41}{64}$	0.640625	16.2715
	$\frac{5}{32}$		0.15625	3.9686		$\frac{21}{32}$		0.65625	16.6684
		$\frac{11}{64}$	0.171875	4.3655			$\frac{43}{64}$	0.671875	17.0653
$\frac{3}{16}$			0.1875	4.7642	$\frac{11}{16}$			0.6875	17.4621
		$\frac{13}{64}$	0.203125	5.1592			$\frac{45}{64}$	0.703125	17.8590
	$\frac{7}{32}$		0.21875	5.5561		$\frac{23}{32}$		0.71875	18.2559
		$\frac{15}{64}$	0.234375	5.9530			$\frac{47}{64}$	0.734375	18.6527
$\frac{1}{4}$			0.25	6.3498	$\frac{3}{4}$			0.75	19.0496
		$\frac{17}{64}$	0.265625	6.7467			$\frac{49}{64}$	0.765625	19.4465
	$\frac{9}{32}$		0.28125	7.1436		$\frac{25}{32}$		0.78125	19.8433
		$\frac{19}{64}$	0.296875	7.5404			$\frac{51}{64}$	0.796875	20.2402
$\frac{5}{16}$			0.3125	7.9373	$\frac{13}{16}$			0.8125	20.6371
		$\frac{21}{64}$	0.328125	8.3342			$\frac{53}{64}$	0.828125	21.0339
	$\frac{11}{32}$		0.34375	8.7310		$\frac{27}{32}$		0.84375	21.4308
		$\frac{23}{64}$	0.359375	9.1279			$\frac{55}{64}$	0.859375	21.8277
$\frac{3}{8}$			0.375	9.5248	$\frac{7}{8}$			0.875	22.2245
		$\frac{25}{64}$	0.390625	9.9216			$\frac{57}{64}$	0.890625	22.6214
	$\frac{13}{32}$		0.40625	10.3185		$\frac{29}{32}$		0.90625	23.0183
		$\frac{27}{64}$	0.421875	10.7154			$\frac{59}{64}$	0.921875	23.4151
$\frac{7}{16}$			0.4375	11.1122	$\frac{15}{16}$			0.9375	23.8120
		$\frac{29}{64}$	0.453125	11.5091			$\frac{61}{64}$	0.953125	24.2089
	$\frac{15}{32}$		0.46875	11.9060		$\frac{31}{32}$		0.96875	24.6057
		$\frac{31}{64}$	0.484375	12.3029			$\frac{63}{64}$	0.984375	25.0026
$\frac{1}{2}$			0.5	12.6997	1			1.	25.3995

Table III.—Metric Measures

LENGTH

Unit	Abbreviation	Value, meters
Micron	μ	0.000001
Millimeter	mm.	0.001
Centimeter	cm.	0.01
Decimeter	dm.	0.1
Meter (unit)	m.	1.0
Dekameter	dkm.	10.0
Hectometer	hm.	100.0
Kilometer	km.	1,000.00
Myriameter	Mm.	10,000.0
Megameter	......	1,000,000.0

AREA

Unit	Abbreviation	Value, square meters
Square millimeter	mm.^{2}	0.000001
Square centimeter	cm.^{2}	0.0001
Square decimeter	dm.^{2}	0.01
Square meter centiare)	m.^{2}	1.0
Square dekameter (are)	a.^{2}	100.0
Hectare	ha.^{2}	10,000.0
Square kilometer	km.^{2}	1,000,000.0

VOLUME

Unit	Abbreviation	Value, liters
Milliliter	ml.	0.001
Centiliter	cl.	0.01
Deciliter	dl.	0.1
Liter (unit)	l.	1.0
Dekaliter	dkl.	10.0
Hectoliter	hl.	100.0
Kiloliter	kl.	1,000.0

CUBIC MEASURE

Unit	Abbreviation	Value, cubic meters
Cubic micron	μ^3	10^{-18}
Cubic millimeter	mm.^{3}	10^{-9}
Cubic centimeter	cm.^{3}	10^{-6}
Cubic decimeter	dm.^{3}	10^{-3}
Cubic meter	m.^{3}	1
Cubic dekameter	dkm.^{3}	10^3
Cubic hectometer	hm.^{3}	10^6
Cubic kilometer	km.^{3}	10^9

MASS

Unit	Abbreviation	Value, grams	Unit	Abbreviation	Value, grams
Microgram	μg	0.000001	Dekagram	dkg.	10.0
Milligram	mg.	0.001	Hectogram	hg.	100.0
Centigram	cg.	0.01	Kilogram	kg.	1,000.0
Decigram	dg.	0.1	Myriagram	Mg.	10,000.0
Gram (unit)	g.	1.0	Quintal	q.	100,000.0
			Ton	....	1,000,000.0

Table IV.—Metric Conversion Tables

(a) Lengths

	Inches to millimeters	Millimeters to inches	Feet to meters	Meters to feet	Yards to meters	Meters to yards	Miles to kilometers	Kilometers to miles
1	25.40	0.03937	0.3048	3.281	0.9144	1.094	1.609	0.6214
2	50.80	0.07874	0.6096	6.562	1.829	2.187	3.219	1.243
3	76.20	0.1181	0.9144	9.842	2.743	3.281	4.828	1.864
4	101.60	0.1575	1.219	13.12	3.658	4.374	6.437	2.485
5	127.00	0.1968	1.524	16.40	4.572	5.468	8.047	3.107
6	152.40	0.2362	1.829	19.68	5.486	6.562	9.656	3.728
7	177.80	0.2756	2.134	22.97	6.401	7.655	11.27	4.350
8	203.20	0.3150	2.438	26.25	7.315	8.749	12.87	4.971
9	228.60	0.3543	2.743	29.53	8.230	9.842	14.48	5.592

Example. 1 in. = 25.40 mm.

(b) Decimals of an Inch to Millimeters (0.01 to 0.99 in.)

	0	1	2	3	4	5	6	7	8	9
0.0	......	0.254	0.508	0.762	1.016	1.270	1.524	1.778	2.032	2.286
0.1	2.540	2.794	3.048	3.302	3.556	3.810	4.064	4.318	4.572	4.826
0.2	5.080	5.334	5.588	5.842	6.096	6.350	6.604	6.858	7.112	7.366
0.3	7.620	7.874	8.128	8.392	8.636	8.890	9.144	9.398	9.652	9.906
0.4	10.160	10.414	10.688	10.922	11.176	11.430	11.684	11.938	12.192	12.446
0.5	12.700	12.954	13.208	13.462	13.716	13.970	14.224	14.478	14.732	14.986
0.6	15.240	15.494	15.748	16.022	16.256	16.510	16.764	17.018	17.272	17.526
0.7	17.780	18.034	18.288	18.542	18.796	19.050	19.304	19.558	19.812	20.066
0.8	20.320	20.574	20.828	21.082	21.336	21.590	21.844	22.098	22.352	22.606
0.9	22.860	23.114	23.368	23.622	23.876	24.130	24.384	24.638	24.892	25.146

Example. 0.77 in. = 19.558 mm.

(c) Millimeters to Decimals of an Inch (1 to 99 mm.)

	0	1	2	3	4	5	6	7	8	9
0	......	0.0394	0.0787	0.1181	0.1575	0.1968	0.2362	0.2756	0.3150	0.3543
1	0.3937	0.4331	0.4724	0.5118	0.5512	0.5906	0.6299	0.6693	0.7087	0.7480
2	0.7874	0.8268	0.8661	0.9055	0.9449	0.9842	1.0236	1.0630	1.1024	1.1417
3	1.1811	1.2205	1.2598	1.2992	1.3386	1.3780	1.4173	1.4567	1.4961	1.5354
4	1.5748	1.6142	1.6535	1.6929	1.7323	1.7716	1.8110	1.8504	1.8898	1.9291
5	1.9685	2.0079	2.0472	2.0866	2.1260	2.1654	2.2047	2.2441	2.2835	2.3228
6	2.3622	2.4016	2.4409	2.4803	2.5197	2.5590	2.5984	2.6378	2.6772	2.7165
7	2.7559	2.7953	2.8346	2.8740	2.9134	2.9528	2.9921	3.0315	3.0709	3.1102
8	3.1496	3.1890	3.2283	3.2677	3.3071	3.3464	3.3858	3.4252	3.4646	3.5039
9	3.5433	3.5827	3.6220	3.6614	3.7008	3.7402	3.7795	3.8189	3.8583	3.8976

Example. 48 mm. = 1.8898 in.

Table IV.—Metric Conversion Tables.—(*Continued*)

(*d*) Areas

	Square inches to square centimeters	Square centimeters to square inches	Square feet to square meters	Square meters to square feet	Square yards to square meters	Square meters to square yards	Acres to hectares	Hectares to acres	Square miles to square kilometers	Square kilometers to square miles
1	6.452	0.1550	0.0929	10.76	0.8361	1.196	0.4047	2.471	2.59	0.3861
2	12.90	0.31	0.1859	21.53	1.672	2.392	0.8094	4.942	5.18	0.7722
3	19.356	0.465	0.2787	32.29	2.508	3.588	1.214	7.413	7.77	1.158
4	25.81	0.62	0.3716	43.06	3.345	4.784	1.619	9.884	10.36	1.544
5	32.26	0.775	0.4645	53.82	4.181	5.98	2.023	12.355	12.95	1.931
6	38.71	0.93	0.5574	64.58	5.017	7.176	2.428	14.826	15.54	2.317
7	45.16	1.085	0.6503	75.35	5.853	8.372	2.833	17.297	18.13	2.703
8	51.61	1.24	0.7432	86.11	6.689	9.568	3.237	19.768	20.72	3.089
9	58.08	1.395	0.8361	96.87	7.525	10.764	3.642	22.239	23.31	3.475

Example. 1 sq. in. = 6.452 sq. cm.

(*e*) Cubic Measure

	Cubic inches to cubic centimeters	Cubic centimeters to cubic inches	Cubic feet to cubic meters	Cubic meters to cubic feet	Cubic yards to cubic meters	Cubic meters to cubic yards	Gallons to cubic feet	Cubic feet to gallons
1	16.39	0.06102	0.02832	35.31	0.7646	1.308	0.1337	7.481
2	32.77	0.1220	0.05663	70.63	1.529	2.616	0.2674	14.96
3	49.16	0.1831	0.08495	105.9	2.294	3.924	0.4010	22.44
4	65.55	0.2441	0.1133	141.3	3.058	5.232	0.5347	29.92
5	81.94	0.3051	0.1416	176.6	3.823	6.540	0.6684	37.40
6	98.32	0.3661	0.1699	211.9	4.587	7.848	0.8021	44.88
7	114.7	0.4272	0.1982	247.2	5.352	9.156	0.9358	52.36
8	131.1	0.4882	0.2265	282.5	6.116	10.46	1.069	59.84
9	147.5	0.5492	0.2549	371.8	6.881	11.77	1.203	67.32

Example. 5 cu. m. = 176.6 cu. ft.

(*f*) Volume or Capacity

	Liquid ounces to cubic centimeters	Cubic centimeters to liquid ounces	Pints to liters	Liters to pints	Quarts to liters	Liters to quarts	Gallons to liters	Liters to gallons	Bushels to hectoliters	Hectoliters to bushels
1	29.57	0.03381	0.4732	2.113	0.9463	1.057	3.785	0.2642	0.3524	2.838
2	59.15	0.06763	0.9463	4.227	1.893	2.113	7.571	0.5284	0.7048	5.676
3	88.72	0.1014	1.420	6.340	2.839	3.170	11.36	0.7925	1.057	8.513
4	118.3	0.1353	1.893	8.454	3.785	4.227	15.14	1.057	1.410	11.35
5	147.9	0.1691	2.366	10.57	4.732	5.284	18.93	1.321	1.762	14.19
6	177.4	0.2029	2.839	12.68	5.678	6.340	22.71	1.585	2.114	17.03
7	207.0	0.2367	3.312	14.79	6.624	7.397	26.50	1.849	2.467	19.86
8	236.6	0.2705	3.785	16.91	7.571	8.454	30.23	2.113	2.819	22.70
9	266.2	0.3043	4.259	19.02	8.517	9.510	34.07	2.378	3.171	25.54

Example. 6 gal. = 22.71 l.

(*g*) Conversion Masses

	Grains to grams	Grams to grains	Ounces (avoirdupois) to grams	Grams to ounces	Pounds (avoirdupois) to kilograms	Kilograms to pounds (avoirdupois)	Short tons (2000 lb.) metric tons	Metric tons (1000 kg.) to short tons	Long tons (2240 lb.) to metric tons	Metric tons to long tons
1	0.0648	15.43	28.35	0.03527	0.4536	2.205	0.907	1.102	1.016	0.984
2	0.1296	30.86	56.70	0.07055	0.9072	4.409	1.814	2.205	2.032	1.968
3	0.1944	46.30	85.05	0.1058	1.361	6.614	2.722	3.307	3.048	2.953
4	0.2592	61.73	113.40	0.1411	1.814	8.818	3.629	4.409	4.064	3.937
5	0.3240	77.16	141.75	0.1764	2.268	11.02	4.536	5.512	5.080	4.921
6	0.3888	92.59	170.10	0.2116	2.722	13.23	5.443	6.614	6.096	5.905
7	0.4536	108.03	198.45	0.2469	3.175	15.43	6.350	7.716	7.112	6.889
8	0.5184	123.46	226.80	0.2822	3.629	17.64	7.257	8.818	8.128	7.874
9	0.5832	138.89	255.15	0.3175	4.082	19.84	8.165	9.921	9.144	8.858

Example. 8 kg. = 17.64 lb.

Table V.—Powers and Roots of Numbers 1–100

Number	Square	Cube	Square root	Cube root	Number	Square	Cube	Square root	Cube root
1	1	1	1.000	1.000	51	2,601	132,651	7.141	3.708
2	4	8	1.414	1.260	52	2,704	140,608	7.211	3.733
3	9	27	1.732	1.442	53	2,809	148,877	7.280	3.756
4	16	64	2.000	1.587	54	2,916	157,464	7.348	3.780
5	25	125	2.236	1.710	55	3,025	166,375	7.416	3.803
6	36	216	2.449	1.817	56	3,136	175,616	7.483	3.826
7	49	343	2.646	1.913	57	3,249	185,193	7.550	3.849
8	64	512	2.828	2.000	58	3,364	195,112	7.616	3.871
9	81	729	3.000	2.080	59	3,481	205,379	7.681	3.893
10	100	1,000	3.162	2.154	60	3,600	216,000	7.746	3.915
11	121	1,331	3.317	2.224	61	3,721	226,981	7.810	3.936
12	144	1,728	3.464	2.289	62	3,844	238,328	7.874	3.958
13	169	2,197	3.606	2.351	63	3,969	250,047	7.937	3.979
14	196	2,744	3.742	2.410	64	4,096	262,144	8.000	4.000
15	225	3,375	3.873	2.466	65	4,225	274,625	8.062	4.021
16	256	4,096	4.000	2.520	66	4,356	287,496	8.124	4.041
17	289	4,913	4.123	2.571	67	4,489	300,763	8.185	4.062
18	324	5,832	4.243	2.621	68	4,624	314,432	8.246	4.082
19	361	6,859	4.359	2.668	69	4,761	328,509	8.307	4.102
20	400	8,000	4.472	2.714	70	4,900	343,000	8.367	4.121
21	441	9,261	4.583	2.759	71	5,041	357,911	8.426	4.141
22	484	10,648	4.690	2.802	72	5,184	373,248	8.485	4.160
23	529	12,167	4.796	2.844	73	5,329	389,017	8.544	4.179
24	576	13,824	4.899	2.884	74	5,476	405,224	8.602	4.198
25	625	15,625	5.000	2.924	75	5,625	421,875	8.660	4.217
26	676	17,576	5.099	2.962	76	5,776	438,976	8.718	4.236
27	729	19,683	5.196	3.000	77	5,929	456,533	8.775	4.254
28	784	21,952	5.292	3.037	78	6,084	474,552	8.832	4.273
29	841	24,389	5.385	3.072	79	6,241	493,039	8.888	4.291
30	900	27,000	5.477	3.107	80	6,400	512,000	8.944	4.309
31	961	29,791	5.568	3.141	81	6,561	531,441	9.000	4.327
32	1,024	32,798	5.657	3.175	82	6,724	551,368	9.055	4.344
33	1,089	35,937	5.745	3.208	83	6,889	571,787	9.110	4.362
34	1,156	39,304	5.831	3.240	84	7,056	592,704	9.165	4.380
35	1,225	42,875	5.916	3.271	85	7,225	614,125	9.220	4.397
36	1,296	46,656	6.000	3.302	86	7,396	636,056	9.274	4.414
37	1,369	50,653	6.083	3.332	87	7,569	658,503	9.327	4.481
38	1,444	54,872	6.164	3.362	88	7,744	681,472	9.381	4.448
39	1,521	59,319	6.245	3.391	89	7,921	704,969	9.434	4.465
40	1,600	64,000	6.325	3.420	90	8,100	729,000	9.487	4.481
41	1,681	68,921	6.403	3.448	91	8,281	753,571	9.539	4.498
42	1,764	74,088	6.481	3.476	92	8,464	778,688	9.592	4.514
43	1,849	79,507	6.557	3.503	93	8,649	804,357	9.644	4.531
44	1,936	85,184	6.633	3.530	94	8,836	830,584	9.695	4.547
45	2,025	91,125	6.708	3.557	95	9,025	857,375	9.747	4.563
46	2,116	97,336	6.782	3.583	96	9,216	884,736	9.798	4.579
47	2,209	103,823	6.856	3.609	97	9,409	912,673	9.849	4.595
48	2,304	110,592	6.928	3.634	98	9,604	941,192	9.900	4.610
49	2,401	117,649	7.000	3.659	99	9,801	970,299	9.950	4.626
50	2,500	125,000	7.071	3.684	100	10,000	1,000,000	10.000	4.642

Table VI.—Size of Drills for Taps

By classification of the National Screw Thread Commission as American National Coarse and American National Fine

Tap diameter or stock size	American National Coarse threads		American National Fine threads	
	Pitch, or threads per inch	Tap-drill size, inch, number, or letter*	Pitch, or threads per inch	Tap-drill size, inch, number, or letter*
0			80	$\frac{3}{64}$ in. or 56
1	64	53	72	53
2	56	50 or 51	64	50
3	48	$\frac{5}{64}$ in.	56	45
4	40	43	48	42
5	40	38 or 39	44	47
6	32	$\frac{7}{64}$ in.	40	33
8	32	29	36	29
10	24	25	32	21
12	20	16 or 17	28	14
$\frac{1}{4}$ in.	20	7 or $\frac{3}{16}$ in.	28	3
$\frac{5}{16}$ in.	18	F or G	24	I (eye)
$\frac{3}{8}$ in.	16	$\frac{5}{16}$ in. or Q	24	Q
$\frac{7}{16}$ in.	14	U	20	$\frac{25}{64}$ in.
$\frac{1}{2}$ in.	13	$\frac{27}{64}$ in.	20	$\frac{29}{64}$ in.
$\frac{9}{16}$ in.	12	$\frac{31}{64}$ in.	18	$\frac{33}{64}$ in.
$\frac{5}{8}$ in.	11	$\frac{17}{32}$ in.	18	$\frac{37}{64}$ in.
$\frac{3}{4}$ in.	10	$\frac{21}{32}$ in.	16	$\frac{11}{16}$ in.
$\frac{7}{8}$ in.	9	$\frac{49}{64}$ in.	14	$\frac{13}{16}$ in.
1 in.	8	$\frac{7}{8}$ in.	14	$\frac{15}{16}$ in.
$1\frac{1}{8}$ in.	7	$\frac{63}{64}$ in.	12	$1\frac{3}{64}$ in.
$1\frac{1}{4}$ in.	7	$1\frac{7}{64}$ in.	12	$1\frac{11}{64}$ in.
$1\frac{1}{2}$ in.	6	$1\frac{21}{64}$ in.	12	$1\frac{27}{64}$ in.
$1\frac{3}{4}$ in.	5	$1\frac{35}{64}$ in.	10	
2 in.	$4\frac{1}{2}$	$1\frac{25}{32}$ in.	10	
$2\frac{1}{4}$ in.	$4\frac{1}{2}$	$2\frac{1}{32}$ in.	8	$2\frac{1}{8}$ in.
$2\frac{1}{2}$ in.	4	$2\frac{1}{4}$ in.	8	$2\frac{3}{8}$ in.
$2\frac{3}{4}$ in.	4	$2\frac{1}{2}$ in.	8	$2\frac{5}{8}$ in.
3 in.	4	$2\frac{3}{4}$ in.	8	$2\frac{7}{8}$ in.

* All drill sizes are to give approximately a 75 per cent depth thread.

Table VII.—American National Pipe Threads

Nominal size	Inside diam.	Outside diam.	Threads per in.	Tap drill taper	Tap drill straight
1/8	0.269	0.405	27	21/64	11/32
1/4	0.364	0.54	18	7/16	7/16
3/8	0.493	0.675	18	9/16	37/64
1/2	0.622	0.84	14	45/64	23/32
3/4	0.824	1.05	14	29/32	15/16
1	1.049	1.315	11 1/2	1 9/64	1 5/32
1 1/4	1.38	1.66	11 1/2	1 31/64	1 33/64
1 1/2	1.61	1.9	11 1/2	1 23/32	1 3/4
2	2.067	2.375	11 1/2	2 3/16	2 7/32
2 1/2	2.469	2.875	8	2 5/8	2 21/32
3	3.068	3.5	8	3 1/4	3 9/32
3 1/2	3.548	4.0	8	3 3/4	3 25/32
4	4.026	4.5	8	4 1/4	4 9/32
4 1/2	4.506	5.0	8	4 3/4	4 25/32
5	5.047	5.563	8	5 9/32	5 11/32
6	6.065	6.625	8	6 11/32	6 13/32

Table VIII.—Inside Diameter of Standard, Extra Strong, and Double Extra-strong Pipe

Nominal size, in.	Outside diam.	Inside diam., standard	Thickness	Inside diam., extra strong	Thickness	Inside diam., double-extra strong	Thickness
1/8	0.405	0.269	0.068	0.215	0.095		
1/4	0.54	0.364	0.088	0.302	0.119		
3/8	0.675	0.493	0.091	0.423	0.126		
1/2	0.84	0.622	0.109	0.546	0.147	0.252	0.294
3/4	1.05	0.824	0.113	0.742	0.154	0.434	0.308
1	1.315	1.049	0.133	0.957	0.179	0.599	0.358
1 1/4	1.66	1.38	0.14	1.278	0.191	0.896	0.382
1 1/2	1.9	1.61	0.145	1.5	0.2	1.1	0.4
2	2.375	2.067	0.154	1.939	0.218	1.503	0.436
2 1/2	2.875	2.469	0.203	2.323	0.276	1.771	0.552
3	3.5	3.068	0.216	2.9	0.3	2.3	0.6
3 1/2	4.	3.548	0.226	3.364	0.318	2.728	0.636
4	4.5	4.026	0.237	3.826	0.337	3.152	0.674
4 1/2	5.	4.506	0.247	4.29	0.355	3.58	0.71
5	5.563	5.047	0.258	4.813	0.375	4.063	0.75
6	6.625	6.065	0.28	5.761	0.432	4.897	0.864

It should be noted that the outside diameter for wrought-iron pipe is the same for all three strengths, and as the weight increases the inside diameter decreases

Table IX.—Sizes of Drills, Z to 80

Size	Decimal of inch	Numbers	Decimal of inch
Letters			
Z	0.413	26	0.147
Y	0.404	27	0.144
X	0.397	28	0.1405
W	0.386	29	0.136
V	0.377	30	0.1285
U	0.368	31	0.12
T	0.358	32	0.116
S	0.348	33	0.113
R	0.339	34	0.111
Q	0.332	35	0.11
P	0.323	36	0.1065
O	0.316	37	0.104
N	0.302	38	0.1015
M	0.295	39	0.0995
L	0.290	40	0.098
K	0.281	41	0.096
J	0.277	42	0.0935
I	0.272	43	0.089
H	0.266	44	0.086
G	0.261	45	0.082
F	0.257	46	0.081
E	0.25	47	0.0785
D	0.246	48	0.076
C	0.242	49	0.073
B	0.238	50	0.07
A	0.234	51	0.067
		52	0.0635
Numbers		53	0.0595
		54	0.055
1	0.228	55	0.052
2	0.221	56	0.0465
3	0.213	57	0.043
4	0.209	58	0.042
5	0.2055	59	0.041
6	0.204	60	0.04
7	0.201	61	0.039
8	0.199	62	0.038
9	0.196	63	0.037
10	0.1935	64	0.036
11	0.191	65	0.035
12	0.189	66	0.033
13	0.185	67	0.032
14	0.182	68	0.031
15	0.18	69	0.0292
16	0.177	70	0.028
17	0.173	71	0.026
18	0.1695	72	0.025
19	0.166	73	0.024
20	0.161	74	0.0225
21	0.159	75	0.021
22	0.157	76	0.02
23	0.154	77	0.018
24	0.152	78	0.016
25	0.1495	79	0.0145
		80	0.0135

Table X.—Continuous Drill Sizes from No. 80 to *Z*

Size	Decimal	Size	Decimal	Size	Decimal
80	0.0135	3/32	0.0938	5	0.206
79	0.0145	41	0.096	4	0.209
1/64	0.0156	40	0.098	3	0.213
78	0.016	39	0.0995	7/32	0.219
77	0.018	38	0.1015	2	0.221
76	0.02	37	0.104	1	0.228
75	0.021	36	0.1065	A	0.234
74	0.0225	7/64	0.1094	15/64	0.234
73	0.024	35	0.11	B	0.238
72	0.025	34	0.111	C	0.242
71	0.026	33	0.113	D	0.246
70	0.028	32	0.116	1/4	0.250
69	0.0292	31	0.12	E	0.250
68	0.031	1/8	0.125	F	0.257
1/32	0.0313	30	0.129	G	0.261
67	0.032	29	0.136	17/64	0.266
66	0.033	9/64	0.14	H	0.266
65	0.035	28	0.141	I	0.272
64	0.036	27	0.144	J	0.277
63	0.037	26	0.147	9/32	0.281
62	0.038	25	0.15	K	0.281
61	0.039	24	0.152	L	0.290
60	0.04	23	0.154	M	0.295
59	0.041	5/32	0.156	19/64	0.297
58	0.042	22	0.157	N	0.302
57	0.043	21	0.159	5/16	0.313
56	0.0465	20	0.161	O	0.316
3/64	0.0469	19	0.166	P	0.323
55	0.052	18	0.17	21/64	0.328
54	0.055	11/64	0.172	Q	0.332
53	0.0595	17	0.173	R	0.339
1/16	0.0625	16	0.177	11/32	0.344
52	0.0635	15	0.18	S	0.348
51	0.067	14	0.182	T	0.358
50	0.07	13	0.185	23/64	0.359
49	0.073	3/16	0.188	U	0.368
48	0.076	12	0.189	3/8	0.375
5/64	0.0781	11	0.191	V	0.377
47	0.0785	10	0.194	W	0.386
46	0.081	9	0.196	25/64	0.391
45	0.082	8	0.199	X	0.397
44	0.086	7	0.201	Y	0.404
43	0.089	13/64	0.203	13/32	0.406
42	0.0935	6	0.204	Z	0.413

Table XI.—The Cutting Speed of Twist Drills in Revolutions per Minute

Diam.	Feet per minute at the periphery of the drill											
	25′	30′	35′	40′	45′	50′	60′	70′	80′	90′	100′	110′
	R.p.m.	R.p.m.	R.p.m.	R.p.m.	R.p.m.	R.p.m.	R.p.m.	R.p.m.	R.p.m.	R.p.m.	R.p.m.	R.p.m.
1/16	1,528	1,834	2,140	2,445	2,751	3,056	3,667	4,278	4,889	5,500	6,111	6,722
1/8	764	917	1,070	1,222	1,375	1,528	1,833	2,139	2,445	2,750	3,056	3,361
3/16	507	611	713	815	917	1,019	1,222	1,446	1,630	1,833	2,037	2,241
1/4	382	458	535	611	688	764	917	1,070	1,222	1,375	1,528	1,681
5/16	306	367	428	489	550	611	733	856	978	1,100	1,222	1,345
3/8	255	306	357	408	458	509	611	713	815	917	1,019	1,120
7/16	218	262	306	349	393	437	524	611	698	768	873	960
1/2	191	229	268	306	344	382	459	535	611	688	764	840
5/8	153	184	214	245	276	306	367	428	489	550	611	672
3/4	127	153	178	203	229	254	306	357	408	458	509	560
7/8	109	131	153	175	196	219	262	306	349	393	436	480
1	96	115	134	153	172	191	229	267	306	344	382	420
1 1/8	85	102	119	136	153	170	204	238	272	306	340	373
1 1/4	76	92	107	123	137	153	183	214	245	275	305	336
1 3/8	69	83	97	111	125	139	167	195	222	250	278	306
1 1/2	64	76	89	102	115	127	153	178	204	229	255	280
1 5/8	59	71	82	94	106	117	141	165	188	212	235	259
1 3/4	55	66	76	87	98	109	131	153	175	196	218	240
1 7/8	51	61	71	82	92	102	122	143	163	183	204	224
2	47	57	67	77	86	96	115	134	153	172	191	210
2 1/4	42	51	59	68	76	85	102	119	136	153	170	187
2 1/2	38	46	54	61	69	76	92	107	122	137	153	168
2 3/4	35	42	49	56	63	70	83	97	111	125	139	153
3	32	38	54	51	57	64	76	89	102	115	127	140

Table XII.—Wire Gauges Used in the United States
(Diameters given in decimal parts of an inch)

Gauge or No.	American or Brown and Sharpe	Stubs or Birmingham	Washburn and Moen or Steel Wire	London or old English	A. S. & W. Co. Music Wire	American Screw Wire
7–0			0.49			
6–0	0.58		0.4615		0.004	
5–0	0.5165		0.4305		0.005	
4–0	0.46	0.454	0.3938	0.454	0.006	
3–0	0.4096	0.425	0.3625	0.425	0.007	0.03152
2–0	0.3648	0.38	0.331	0.38	0.008	0.04468
0	0.3249	0.34	0.3065	0.34	0.009	0.05784
1	0.2893	0.3	0.283	0.3	0.01	0.071
2	0.2576	0.284	0.2625	0.284	0.011	0.08416
3	0.2294	0.259	0.2437	0.259	0.012	0.09732
4	0.2043	0.238	0.2253	0.238	0.013	0.11048
5	0.1819	0.22	0.207	0.22	0.014	0.12364
6	0.162	0.203	0.192	0.203	0.016	0.1368
7	0.1443	0.18	0.177	0.18	0.018	0.14996
8	0.1285	0.165	0.162	0.165	0.02	0.16312
9	0.1144	0.148	0.1483	0.148	0.022	0.17628
10	0.1019	0.134	0.135	0.134	0.024	0.18944
11	0.09074	0.12	0.1205	0.12	0.026	0.2026
12	0.08081	0.109	0.1055	0.109	0.029	0.21576
13	0.07196	0.095	0.0915	0.095	0.031	0.22892
14	0.06408	0.083	0.08	0.083	0.033	0.24208
15	0.05707	0.072	0.072	0.072	0.035	0.25524
16	0.05082	0.065	0.0625	0.065	0.037	0.2684
17	0.04526	0.058	0.054	0.058	0.039	0.28156
18	0.0403	0.049	0.0475	0.049	0.041	0.29472
19	0.03589	0.042	0.041	0.04	0.043	0.30788
20	0.03196	0.035	0.0348	0.035	0.045	0.32104
21	0.02846	0.032	0.03175	0.0315	0.047	0.3342
22	0.02535	0.028	0.0286	0.0295	0.049	0.34736
23	0.02257	0.025	0.0258	0.027	0.051	0.36052
24	0.0201	0.022	0.023	0.025	0.055	0.37368
25	0.0179	0.02	0.0204	0.023	0.059	0.38684
26	0.01594	0.018	0.0181	0.0205	0.063	0.4
27	0.0142	0.016	0.0173	0.01875	0.067	0.41316
28	0.01264	0.014	0.0162	0.0165	0.071	0.42632
29	0.01126	0.013	0.015	0.0155	0.075	0.43948
30	0.01003	0.012	0.014	0.01375	0.08	0.45264
31	0.008928	0.01	0.0132	0.01225	0.085	0.4658
32	0.00795	0.009	0.0128	0.01125	0.09	0.47896
33	0.00708	0.008	0.0118	0.01025	0.095	0 49212
34	0.006304	0.007	0.0104	0.095	0.1	0.50528
35	0.005614	0.005	0.0095	0.009	0.105	0.51844
36	0.005	0.004	0.009	0.0075	0.112	0.5316
37	0.004453		0.0085	0.0065	0.118	0.54476
38	0.003965		0.008	0.00575	0.124	0.55792
39	0.003531		0.0075	0.005	0.13	0.57108
40	0.003144		0.007	0.0045	0.138	0.58424
41	0.0028		0.0066		0.146	0.5974
42	0.002494		0.0062		0.154	0.61056
43	0.002221		0.006		0.162	0.62372
44	0.001978		0.0058		0.17	0.63688
45	0.001761		0.0055		0.18	0.65004
46	0.001568		0.0052			0.6632
47	0.001397		0.005			0.67636
48	0.001244		0.0048			0.68952
49	0.001018		0.0046			0.70268
50	0.0009863		0.0044			

Table XIII.—Sheet Metal Gauges Used in the United States

Gauge No.	U.S. Standard	Brown and Sharpe[1]	Birmingham or Stubs	M. & H. zinc
7–0	0.5			
6–0	0.46875	0.58		
5–0	0.4375	0.5165		
4–0	0.40625	0.46	0.454	
3–0	0.375	0.4096	0.425	
2–0	0.34375	0.3648	0.380	
0	0.3125	0.3249	0.34	
1	0.28125	0.2893	0.3	
2	0.265625	0.2576	0.284	
3	0.25	0.2294	0.259	0.006
4	0.234375	0.2043	0.238	0.008
5	0.21875	0.1819	0.22	0.01
6	0.203125	0.162	0.203	0.012
7	0.1875	0.1443	0.18	0.014
8	0.171875	0.1285	0.165	0.016
9	0.15625	0.1144	0.148	0.018
10	0.140625	0.1019	0.134	0.02
11	0.125	0.07094	0.12	0.024
12	0.109375	0.08081	0.109	0.028
13	0.09375	0.07196	0.095	0.032
14	0.078125	0.06408	0.083	0.036
15	0.0703125	0.05707	0.072	0.04
16	0.0625	0.05082	0.065	0.045
17	0.05625	0.04526	0.058	0.05
18	0.05	0.0403	0.049	0.055
19	0.04375	0.03589	0.042	0.06
20	0.0375	0.03196	0.035	0.07
21	0.034375	0.02846	0.032	0.08
22	0.03125	0.02535	0.028	0.09
23	0.028125	0.02257	0.025	0.1
24	0.025	0.0201	0.022	0.125
25	0.021875	0.0179	0.02	0.25
26	0.01875	0.01594	0.018	0.375
27	0.0171875	0.0142	0.016	0.5
28	0.015625	0.01264	0.014	
29	0.0140625	0.01126	0.013	
30	0.0125	0.01003	0.012	
31	0.0109375	0.008928	0.01	
32	0.01015625	0.00795	0.009	
33	0.009375	0.00708	0.008	
34	0.00859375	0.006304	0.007	
35	0.0078125	0.005614	0.005	
36	0.00703125	0.005		
37	0.006640625	0.004453		
38	0.00625	0.003965		
39		0.003531		
40		0.003134		

[1] The Brown and Sharpe gauge runs downward to 50 gauge, which is 0.0009863 in. thick.

Table XIV.—Tapers and Angles

Taper per foot	Included angle			Angle with center line			Taper per inch	Taper per inch from center line
	Deg.	Min.	Sec.	Deg.	Min.	Sec.		
$\frac{1}{8}$	0	35	48	0	17	54	0.010416	0.005203
$\frac{3}{16}$	0	53	44	0	26	52	0.015625	0.007812
$\frac{1}{4}$	1	11	36	0	35	48	0.020833	0.010416
$\frac{5}{16}$	1	29	30	0	44	45	0.026042	0.013021
$\frac{3}{8}$	1	47	24	0	53	42	0.031250	0.015625
$\frac{7}{16}$	2	5	18	1	2	39	0.036458	0.018229
$\frac{1}{2}$	2	23	10	1	11	35	0.014667	0.020833
$\frac{9}{16}$	2	41	4	1	20	32	0.046875	0.023438
$\frac{5}{8}$	2	59	42	1	29	51	0.052084	0.026042
$\frac{11}{16}$	3	16	54	1	38	27	0.057292	0.028646
$\frac{3}{4}$	3	34	44	1	47	22	0.06250	0.031250
$\frac{13}{16}$	3	52	38	1	56	19	0.067708	0.033854
$\frac{7}{8}$	4	10	32	2	5	16	0.072917	0.036456
$\frac{15}{16}$	4	28	24	2	14	12	0.078125	0.039063
1	4	46	18	2	23	9	0.083330	0.041667
$1\frac{1}{4}$	5	57	48	2	58	54	0.104666	0.052084
$1\frac{1}{2}$	7	9	10	3	34	35	0.125000	0.062500
$1\frac{3}{4}$	8	20	26	4	10	13	0.145833	0.072917
2	9	31	36	4	45	48	0.166666	0.083332
$2\frac{1}{2}$	11	53	36	5	56	48	0.208333	0.104166
3	14	15	0	7	7	30	0.250000	0.125000
$3\frac{1}{2}$	16	35	40	8	17	50	0.291666	0.145833
4	18	55	28	9	27	44	0.333333	0.166666
$4\frac{1}{2}$	21	14	2	10	37	1	0.375000	0.187500
5	23	32	12	11	46	6	0.416666	0.208333
6	28	4	2	14	2	1	0.500000	0.250000

Table XV.—Circumferences, Areas, Squares, Cubes, Square and Cube Roots
(Advancing by eighths)

Diam. or no.	Circle		Square	Cube	Square root	Cube root
	Circum.	Area				
⅛	0.393	0.0123	0.0156	0.0019	0.353	0.5
¼	0.786	0.0491	0.0625	0.0156	0.5	0.623
⅜	1.179	0.1104	0.1506	0.0565	0.612	0.72
½	1.572	0.196	0.25	0.125	0.707	0.79
⅝	1.965	0.307	0.3906	0.244	0.79	0.855
¾	2.358	0.442	0.5625	0.422	0.866	0.91
⅞	2.751	0.602	0.7656	0.67	0.935	0.955
1	3.14	0.785	1	1	1	1
⅛	3.53	0.994	1.27	1.42	1.06	1.04
¼	3.93	1.227	1.56	1.95	1.118	1.077
⅜	4.32	1.485	1.89	2.6	1.173	1.112
½	4.71	1.767	2.25	3.38	1.225	1.145
⅝	5.11	2.074	2.64	4.29	1.275	1.176
¾	5.5	2.405	3.06	5.36	1.323	1.205
⅞	5.89	2.761	3.52	6.59	1.369	1.233
2	6.28	3.142	4	8	1.414	1.26
⅛	6.68	3.547	4.52	9.59	1.458	1.286
¼	7.07	3.976	5.06	11.39	1.5	1.31
⅜	7.46	4.43	5.64	13.4	1.541	1.334
½	7.85	4.909	6.25	15.63	1.581	1.358
⅝	8.25	5.412	6.89	18.08	1.62	1.38
¾	8.04	5.94	7.50	20.79	1.058	1.402
⅞	9.03	6.492	8.27	23.76	1.695	1.422
3	9.42	7.07	9	27	1.732	1.442
⅛	9.82	7.67	9.77	30.52	1.768	1.462
¼	10.21	8.3	10.56	34.32	1.803	1.482
⅜	10.6	8.95	11.39	38.44	1.837	1.5
½	11	9.62	12.25	42.88	1.871	1.518
⅝	11.39	10.32	13.14	47.63	1.904	1.535
¾	11.78	11.05	14.06	52.73	1.936	1.553
⅞	12.17	11.79	15.02	58.17	1.968	1.57
4	12.57	12.57	16	64	2	1.587
¼	13.35	14.19	18.06	76.78	2.061	1.619
½	14.14	15.9	20.25	91.13	2.121	1.651
¾	14.92	17.72	22.56	107.16	2.179	1.681
5	15.71	19.63	25.00	125	2.236	1.71
¼	16.49	21.64	27.56	144.7	2.291	1.738
½	17.28	23.76	30.25	166.37	2.345	1.765
¾	18.06	25.97	33.06	190.11	2.398	1.792
6	18.85	28.29	36	216	2.449	1.817
¼	19.64	30.68	39.06	244.14	2.5	1.832
½	20.42	33.18	42.25	274.63	2.55	1.866
¾	21.21	35.78	45.56	307.55	2.599	1.89
7	21.99	38.48	49	343	2.646	1.913
¼	22.78	41.28	52.56	381.08	2.692	1.935
½	23.56	44.18	56.25	421.88	2.739	1.957
¾	24.35	47.14	60.06	465.48	2.784	1.979
8	25.13	50.26	64	512	2.828	2.000
¼	25.92	53.46	68.06	561.52	2.872	2.021
½	26.7	56.75	72.25	614.12	2.915	2.041
¾	27.49	60.13	76.56	669.92	2.958	2.061
9	28.27	63.62	81	729	3	2.08
¼	29.06	67.2	85.56	791.45	3.041	2.098
½	29.85	70.88	90.25	857.37	3.082	2.118
¾	30.63	74.66	95.06	926.86	3.112	2.136

Table XV.—Circumferences, Areas, Squares, Cubes, Square and Cube Roots.—(*Continued*)

(Advancing by decimals)

Diam. or no.	Circle		Square	Cube	Square root	Cube root
	Circum.	Area				
0.2	0.628	0.0314	0.04	0.008	0.447	0.585
0.4	1.26	0.1256	0.16	0.064	0.633	0.737
0.6	1.88	0.2827	0.36	0.216	0.775	0.843
0.8	2.51	0.5026	0.64	0.512	0.894	0.928
1	3.14	0.7854	1	1	1	1
1.2	3.77	1.131	1.44	1.73	1.095	1.063
1.4	4.39	1.539	1.96	2.74	1.813	1.119
1.6	5.02	2.011	2.56	4.1	1.265	1.17
1.8	5.65	2.545	3.24	5.83	1.342	1.216
2	6.28	3.142	4	8	1.414	1.26
2.2	6.91	3.801	4.84	10.65	1.483	1.301
2.4	7.53	4.524	5.76	13.82	1.549	1.339
2.6	8.16	5.309	6.76	17.58	1.612	1.375
2.8	8.79	6.158	7.84	21.95	1.673	1.409
3	9.42	7.069	9	27	1.732	1.442
3.2	10.05	7.548	10.24	32.77	1.789	1.474
3.4	10.68	8.553	11.56	39.3	1.844	1.504
3.6	11.3	10.18	12.96	46.66	1.897	1.533
3.8	11.93	11.34	14.44	54.87	1.949	1.56
4	12.57	12.57	16	64	2	1.587
4.2	13.19	13.85	17.64	74.09	2.049	1.613
4.4	13.82	15.21	19.36	85.18	2.098	1.639
4.6	14.45	16.62	21.16	87.34	2.145	1.663
4.8	15.08	18.1	23.04	110.6	2.191	1.687
5	15.7	19.63	25	125	2.236	1.71
5.2	16.33	21.24	27.04	140.6	2.28	1.732
5.4	16.96	22.9	29.16	157.5	2.324	1.754
5.6	17.59	24.63	31.36	175.6	2.366	1.776
5.8	18.22	26.42	33.64	195.1	2.408	1.797
6	18.84	28.27	36	216	2.449	1.817
6.2	19.47	30.19	38.44	238.3	2.49	1.837
6.4	20.1	32.17	40.96	262.1	2.53	1.857
6.6	20.73	34.21	43.56	287.5	2.569	1.876
6.8	21.36	36.32	46.24	314.4	2.608	1.895
7	21.99	38.48	49	343	2.646	1.913
7.2	22.61	40.72	51.84	373.2	2.683	1.931
7.4	23.24	43.01	54.76	405.2	2.72	1.949
7.6	23.87	45.36	57.76	438.9	2.757	1.966
7.8	24.5	47.78	60.84	474.6	2.793	1.983
8	25.13	50.27	64	512	2.828	2
8.2	25.76	52.81	67.24	551.4	2.864	2.017
8.4	26.38	55.42	70.56	592.7	2.898	2.033
8.6	27.01	58.09	73.96	636.1	2.933	2.049
8.8	27.64	60.82	77.44	681.5	2.966	2.065
9	28.27	63.62	81	729	3	2.08
9.2	28.9	66.48	84.64	778.7	3.033	2.095
9.4	29.53	69.4	88.36	830.6	3.066	2.11
9.6	30.15	72.38	92.16	884.7	3.098	2.125
9.8	30.78	75.43	96.04	941.2	3.13	2.14
10	31.41	78.54	100	1,000	3.162	2.154
11	34.55	95.03	121	1,331	3.317	2.224
12	37.69	113	144	1,728	3.464	2.289
13	40.84	132.7	169	2,197	3.606	2.351
14	43.98	153.9	196	2,744	3.742	2.41
15	47.12	176.7	225	3,375	3.873	2.446
16	50.26	201	256	4,096	4	2.52
17	53.4	226.9	289	4,913	4.123	2.571
18	56.54	254.4	324	5,832	4.243	2.621
19	59.69	283.5	361	6,859	4.359	2.668
20	62.83	314.1	400	8,000	4.472	2.714
21	65.97	346.3	441	9,261	4.583	2.759
22	69.11	380.1	484	10,648	4.69	2.802
23	72.25	415.4	529	12,167	4.796	2.844
24	75.39	452.3	576	13,824	4.899	2.885
25	78.54	490.8	625	15,625	5	2.924
26	81.68	530.9	676	17,576	5.099	2.963
27	84.82	572.5	729	19,683	5.196	3
28	87.96	615.7	784	21,952	5.292	3.037
29	91.1	660.5	841	24,389	5.385	3.072
30	94.24	706.8	900	27,000	5.477	3.107

Table XV.—Circumferences, Areas, Squares, Cubes, Square and Cube Roots.—(*Continued*)

Diam. or no.	Circle		Square	Cube	Square root	Cube root
	Circum.	Area				
31	97.39	754.8	961	29,791	5.568	3.143
32	100.5	804.2	1,024	32,768	5,657	3.175
33	103.7	855.3	1,089	35,937	5.745	3.208
34	106.8	907.9	1,156	39,304	5,831	3.24
35	110	962.1	1,225	42,875	5.916	3.271
36	113.1	1,017.9	1,296	46,656	6	3.302
37	116.2	1,075.2	1,369	50,653	6.083	3.332
38	119.4	1,134.1	1,444	54,872	6.164	3.362
39	122.5	1,194.6	1,521	59,319	6.245	3.391
40	125.7	1,256.6	1,600	64,000	6.325	3.42
41	128.8	1,320.3	1,681	68,921	6.403	3.448
42	131.9	1,385.4	1,764	74,088	6.481	3.476
43	135.1	1,452.2	1,849	79,507	6.557	3.503
44	138.2	1,520.5	1,936	85,184	6.633	3.53
45	141.4	1,590.4	2,025	91,125	6.708	3.557
46	144.5	1,661.9	2,116	97,336	6.782	3.583
47	147.7	1,734.9	2,209	103,823	6.856	3.609
48	150.8	1,809.6	2,304	110,592	6.928	3.634
49	153.9	1,885.7	2,401	117,649	7	3.659
50	157.1	1,963.5	2,500	125,000	7.071	3.684
51	160.2	2,042.8	2,601	132,651	7.141	3.708
52	163.4	2,123.7	2,704	140,608	7.211	3.733
53	166.5	2,206.2	2,809	148,877	7.28	3.756
54	169.6	2,290.2	2,916	157,464	7.348	3.78
55	172.8	2,375.8	3,025	166,375	7.416	3.803
56	175.9	2,463	3,136	175,616	7.483	3.826
57	179.1	2,551.8	3,249	185,193	7.55	3.849
58	182.2	2,642.1	3,364	195,112	7.616	3.871
59	185.4	2,734	3,481	205,379	7.681	3.893
60	188.5	2,827.4	3,600	216,000	7.746	3.915
61	191.6	2,922.5	3,721	226,981	7.81	3.937
62	194.8	3,019.1	3,844	238,328	7.874	3.958
63	197.9	3,117.3	3,969	250,047	7.937	3.979
64	201.1	3,217	4,096	262,144	8	4
65	204.2	3,318.3	4,225	274,625	8.062	4.021
66	207.3	3,421.2	4,356	287,496	8.124	4.041
67	210.5	3,525.7	4,489	300,763	8.185	4.061
68	213.6	3,631.7	4,624	314,432	8.246	4.082
69	216.8	3,739.3	4,761	328,509	8.307	4.102
70	219.9	3,848.5	4,900	343,000	8.376	4.121
71	223.1	3,959.2	5,041	357,911	8.426	4.141
72	226.2	4,071.5	5,184	373,248	8.485	4.16
73	229.3	4,185.4	5,329	389,017	8.544	4.179
74	232.5	4,300.8	5,476	405,224	8.602	4.198
75	235.6	4,417.9	5,625	421,875	8.66	4.217
76	238.8	4,536.5	5,776	438,976	8.718	4.236
77	241.9	4,656.6	5,929	465,533	8.775	4.254
78	245	4,778.4	6,984	474,552	8.832	4.273
79	248.2	4,901.7	6,241	493,039	8.888	4.291
80	251.3	5,026.6	6,400	512,000	8.944	4.309
81	254.5	5,153	6,561	531,441	9	4.327
82	257.6	5,281	6,724	551,368	9.056	4.345
83	260.8	5,410.6	6,889	571,787	9.11	4.362
84	263.9	5,541.8	7,056	592,704	9.165	4.379
85	267.0	5,674.5	7,225	614,125	9.22	4.397
86	270.2	5,808.8	7,396	636,056	9.274	4.414
87	273.3	5,944.7	7,569	658,503	9.327	4.431
88	276.5	6,082.1	7,744	681,472	9.381	4.448
89	279.6	6,221.2	7,921	704,969	9.434	4.465
90	282.7	6,361.7	8,100	729,000	9.487	4.481
91	285.9	6,503.9	8,281	753,751	9.539	4.498
92	289.0	6,647.6	8,464	778,688	9.592	4.514
93	292.2	6,792.9	8,649	804,357	9.644	4.531
94	295.2	6,939.8	8,836	830,584	9.695	4.547
95	298.5	7,088.2	9,025	857,375	9.747	4.563
96	301.6	7,238.2	9,216	884,736	9.798	4.579
97	304.7	7,389.8	9,409	912,673	9.849	4.595
98	307.9	7,543.0	9,604	941,192	9.899	4.616
99	311.9	7,697.7	9,801	970,299	9.95	4.62

Table XVI.—Properties of Various Materials

Nonmetallic Construction Materials

Material	Average specific gravity	Average weight, pounds per cubic foot	Compression strength, pounds per square inch
Brick:			
Common	1.4 to 2.0	103	4,000
Hard	1.8 to 2.3	128	8,000
Cement, Portland	1.5	94	
Clay	1.1	63	
Concrete:			
Cinder	1.5 to 1.7	100	400 to 800
Sand and gravel	2.2 to 2.4	144	1,300 to 3,000
Slag	1.9 to 2.3	130	1,000 to 2,200
Gravel, dry loose	1.4 to 1.7	100	
Stone:			
Granite	2.7	170	19,400
Limestone	2.53	158	9,500
Marble	2.72	173	12,700
Sandstone	2.22	139	9,300
Slate	2.77	175	14,000
Trap	2.96	185	20,000
Wood, dry:			
Ash	0.55	34	
Hickory	0.74 to 0.8	48	
Mahogany	0.56 to 0.85	44	
Oak	0.64 to 0.87	48	12,000
Pine	0.43 to 0.55	36	8,000
Redwood	0.39	24	7,000
Spruce	0.45	28	8,000
Walnut	0.59 to 0.63	37	

Metals and Alloys

Name	Specific gravity	Weight: Pounds per cubic foot	Weight: Pounds per cubic inch	Cubic inches per pound	Melts, Fahrenheit	Tensile strength, pounds per square inch
Aluminum:						
Cast	2.569	160.0	0.093	10.8	1,220	15,000
Wrought	2.681	167.0	0.097	10.35	1,220	22,000 to 26,000
Brass:						
Cast	8.109	505.0	0.296	3.42	1,886	18,000 to 23,000
Rods	8.509	531.2	0.3074	3.25	1,886	35,000 to 46,000
Sheets	8.469	528.7	0.306	3.27	1,886	35,000 to 46,000
Wire	8.437	526.7	0.3048	3.28	1,886	49,000 annealed
Wire	8.437	526.7	0.3048	3.28	1,886	150,000 hard
Bronze:						
Cast	8.735	544.0	0.315	3.18	1,688	45,000
Phosphor	8.81	554.4	0.3208	3.11	1,688	63,000 annealed
Phosphor	8.81	554.4	0.3208	3.11	1,688	150,000 hard
Tobin rods	8.404	524.6	0.3036	3.29	1,688	60,000 to 80,000
Copper:						
Cast	8.622	537.0	0.311	3.22	1,981	18,000 to 30,000
Rods, bars	8.9	555.6	0.3215	3.11	1,981	33,000
Sheets	8.9	555.6	0.3215	3.11	1,981	30,000
Wire	8.9	555.6	0.3215	3.11	1,981	32,000 annealed
Wire	8.9	555.6	0.3215	3.11	1,981	60,000 hard
Iron:						
Gray cast	7.209	464.0	0.2607	3.85	2,507	18,000 to 24,000
Wrought	7.707	480.0	0.2783	3.6	2,822	45,000
Steel:						
Mild	7.868	490.0	0.2833	3.52	2,687	60,000 to 95,000
Carbon tool	7.87	491.0	0.294	3.51	2,467	80,000 to 240,000
Gold	19.316	1,203.0	0.696	1.44	1,915	20,000 to 27,500
Lead	11.4	710.0	0.411	2.43	620	1,800 to 3,300
Platinum	21.516	1,340.0	0.755	1.29	3,133	32,000 annealed
Platinum	21.516	1,340.0	0.755	1.29	3,133	56,000 hard
Silver	10.517	655.0	0.379	2.64	1,760	40,000 to 95,000
Tin	7.418	462.0	0.267	3.74	449	3,000 to 7,000
Zinc	7.14	438	0.252		787	4,000 to 6,000

Table XVII.—Square Roots of Numbers

N.	0	1	2	3	4	5	6	7	8	9	1	2	3	4	5	6	7	8	9
1.0	1.000	1.005	1.010	1.015	1.020	1.025	1.030	1.034	1.039	1.044	0	1	1	2	2	3	3	4	4
1.1	1.049	1.054	1.058	1.063	1.068	1.072	1.077	1.082	1.086	1.091	0	1	1	2	2	3	3	4	4
1.2	1.095	1.100	1.105	1.109	1.114	1.118	1.122	1.127	1.131	1.136	0	1	1	2	2	3	3	4	4
1.3	1.140	1.145	1.149	1.153	1.158	1.162	1.166	1.170	1.175	1.179	0	1	1	2	2	3	3	3	4
1.4	1.183	1.187	1.192	1.196	1.200	1.204	1.208	1.212	1.217	1.221	0	1	1	2	2	2	3	3	4
1.5	1.225	1.229	1.233	1.237	1.241	1.245	1.249	1.253	1.257	1.261	0	1	1	2	2	2	3	3	4
1.6	1.265	1.269	1.273	1.277	1.281	1.285	1.288	1.292	1.296	1.300	0	1	1	2	2	2	3	3	4
1.7	1.304	1.308	1.311	1.315	1.319	1.323	1.327	1.330	1.334	1.338	0	1	1	2	2	2	3	3	3
1.8	1.342	1.345	1.349	1.353	1.356	1.360	1.364	1.367	1.371	1.375	0	1	1	1	2	2	3	3	3
1.9	1.378	1.382	1.386	1.389	1.393	1.396	1.400	1.404	1.407	1.411	0	1	1	1	2	2	3	3	3
2.0	1.414	1.418	1.421	1.425	1.428	1.432	1.435	1.439	1.442	1.446	0	1	1	1	2	2	2	3	3
2.1	1.449	1.453	1.456	1.459	1.463	1.466	1.470	1.473	1.476	1.480	0	1	1	1	2	2	2	3	3
2.2	1.483	1.487	1.490	1.493	1.497	1.500	1.503	1.507	1.510	1.513	0	1	1	1	2	2	2	3	3
2.3	1.517	1.520	1.523	1.526	1.530	1.533	1.536	1.539	1.543	1.546	0	1	1	1	2	2	2	3	3
2.4	1.549	1.552	1.556	1.559	1.562	1.565	1.568	1.572	1.575	1.578	0	1	1	1	2	2	2	3	3
2.5	1.581	1.584	1.587	1.591	1.594	1.597	1.600	1.603	1.606	1.609	0	1	1	1	2	2	2	3	3
2.6	1.612	1.616	1.619	1.622	1.625	1.628	1.631	1.634	1.637	1.640	0	1	1	1	2	2	2	2	3
2.7	1.643	1.646	1.649	1.652	1.655	1.658	1.661	1.664	1.667	1.670	0	1	1	1	2	2	2	2	3
2.8	1.673	1.676	1.679	1.682	1.685	1.688	1.691	1.694	1.697	1.700	0	1	1	1	1	2	2	2	3
2.9	1.703	1.706	1.709	1.712	1.715	1.718	1.720	1.723	1.726	1.729	0	1	1	1	1	2	2	2	3
3.0	1.732	1.735	1.738	1.741	1.744	1.746	1.749	1.752	1.755	1.758	0	1	1	1	1	2	2	2	3
3.1	1.761	1.764	1.766	1.769	1.772	1.775	1.778	1.780	1.783	1.786	0	1	1	1	1	2	2	2	3
3.2	1.789	1.792	1.794	1.797	1.800	1.803	1.806	1.808	1.811	1.814	0	1	1	1	1	2	2	2	2
3.3	1.817	1.819	1.822	1.825	1.828	1.830	1.833	1.836	1.838	1.841	0	1	1	1	1	2	2	2	2
3.4	1.844	1.847	1.849	1.852	1.855	1.857	1.860	1.863	1.865	1.868	0	1	1	1	1	2	2	2	2
3.5	1.871	1.873	1.876	1.879	1.881	1.884	1.887	1.889	1.892	1.895	0	1	1	1	1	2	2	2	2
3.6	1.897	1.900	1.903	1.905	1.908	1.910	1.913	1.916	1.918	1.921	0	1	1	1	1	2	2	2	2
3.7	1.924	1.926	1.929	1.931	1.934	1.936	1.939	1.942	1.944	1.947	0	1	1	1	1	2	2	2	2
3.8	1.949	1.952	1.954	1.957	1.960	1.962	1.965	1.967	1.970	1.972	0	1	1	1	1	2	2	2	2
3.9	1.975	1.977	1.980	1.982	1.985	1.987	1.990	1.992	1.995	1.997	0	1	1	1	1	2	2	2	2
4.0	2.000	2.002	2.005	2.007	2.010	2.012	2.015	2.017	2.020	2.022	0	0	1	1	1	1	2	2	2
4.1	2.025	2.027	2.030	2.032	2.035	2.037	2.040	2.042	2.045	2.047	0	0	1	1	1	1	2	2	2
4.2	2.049	2.052	2.054	2.057	2.059	2.062	2.064	2.066	2.069	2.071	0	0	1	1	1	1	2	2	2
4.3	2.074	2.076	2.078	2.081	2.083	2.086	2.088	2.090	2.093	2.095	0	0	1	1	1	1	2	2	2
4.4	2.098	2.100	2.102	2.105	2.107	2.110	2.112	2.114	2.117	2.119	0	0	1	1	1	1	2	2	2
4.5	2.121	2.124	2.126	2.128	2.131	2.133	2.135	2.138	2.140	2.142	0	0	1	1	1	1	2	2	2
4.6	2.145	2.147	2.149	2.152	2.154	2.156	2.159	2.161	2.163	2.166	0	0	1	1	1	1	2	2	2
4.7	2.168	2.170	2.173	2.175	2.177	2.179	2.182	2.184	2.186	2.189	0	0	1	1	1	1	2	2	2
4.8	2.191	2.193	2.195	2.198	2.200	2.202	2.205	2.207	2.209	2.211	0	0	1	1	1	1	2	2	2
4.9	2.214	2.216	2.218	2.220	2.223	2.225	2.227	2.229	2.232	2.234	0	0	1	1	1	1	2	2	2
5.0	2.236	2.238	2.241	2.243	2.245	2.247	2.249	2.252	2.254	2.256	0	0	1	1	1	1	2	2	2
5.1	2.258	2.261	2.263	2.265	2.267	2.269	2.272	2.274	2.276	2.278	0	0	1	1	1	1	2	2	2
5.2	2.280	2.283	2.285	2.287	2.289	2.291	2.293	2.296	2.298	2.300	0	0	1	1	1	1	2	2	2
5.3	2.302	2.304	2.307	2.309	2.311	2.313	2.315	2.317	2.319	2.322	0	0	1	1	1	1	2	2	2
5.4	2.324	2.326	2.328	2.330	2.332	2.335	2.337	2.339	2.341	2.343	0	0	1	1	1	1	1	2	2
N.	**0**	**1**	**2**	**3**	**4**	**5**	**6**	**7**	**8**	**9**	**1**	**2**	**3**	**4**	**5**	**6**	**7**	**8**	**9**

Table XVII.—Square Roots of Numbers.—(*Continued*)

N.	0	1	2	3	4	5	6	7	8	9	1	2	3	4	5	6	7	8	9
5.5	2.345	2.347	2.349	2.352	2.354	2.356	2.358	2.360	2.362	2.364	0	0	1	1	1	1	1	2	2
5.6	2.366	2.369	2.371	2.373	2.375	2.377	2.379	2.381	2.383	2.385	0	0	1	1	1	1	1	2	2
5.7	2.387	2.390	2.392	2.394	2.396	2.398	2.400	2.402	2.404	2.406	0	0	1	1	1	1	1	2	2
5.8	2.408	2.410	2.412	2.415	2.417	2.419	2.421	2.423	2.425	2.427	0	0	1	1	1	1	1	2	2
5.9	2.429	2.431	2.433	2.435	2.437	2.439	2.441	2.443	2.445	2.447	0	0	1	1	1	1	1	2	2
6.0	2.449	2.452	2.454	2.456	2.458	2.460	2.462	2.464	2.466	2.468	0	0	1	1	1	1	1	2	2
6.1	2.470	2.472	2.474	2.476	2.478	2.480	2.482	2.484	2.486	2.488	0	0	1	1	1	1	1	2	2
6.2	2.490	2.492	2.494	2.496	2.498	2.500	2.502	2.504	2.506	2.508	0	0	1	1	1	1	1	2	2
6.3	2.510	2.512	2.514	2.516	2.518	2.520	2.522	2.524	2.526	2.528	0	0	1	1	1	1	1	2	2
6.4	2.530	2.532	2.534	2.536	2.538	2.540	2.542	2.544	2.546	2.548	0	0	1	1	1	1	1	2	2
6.5	2.550	2.551	2.553	2.555	2.557	2.559	2.561	2.563	2.565	2.567	0	0	1	1	1	1	1	2	2
6.6	2.569	2.571	2.573	2.575	2.577	2.579	2.581	2.583	2.585	2.587	0	0	1	1	1	1	1	2	2
6.7	2.588	2.590	2.592	2.594	2.596	2.598	2.600	2.602	2.604	2.606	0	0	1	1	1	1	1	2	2
6.8	2.608	2.610	2.612	2.613	2.615	2.617	2.619	2.621	2.623	2.625	0	0	1	1	1	1	1	2	2
6.9	2.627	2.629	2.631	2.632	2.634	2.636	2.638	2.640	2.642	2.644	0	0	1	1	1	1	1	2	2
7.0	2.646	2.648	2.650	2.651	2.653	2.655	2.657	2.659	2.661	2.663	0	0	1	1	1	1	1	2	2
7.1	2.665	2.666	2.668	2.670	2.672	2.674	2.676	2.678	2.680	2.681	0	0	1	1	1	1	1	1	2
7.2	2.683	2.685	2.687	2.689	2.691	2.693	2.694	2.696	2.698	2.700	0	0	1	1	1	1	1	1	2
7.3	2.702	2.704	2.706	2.707	2.700	2.711	2.713	2.715	2.717	2.718	0	0	1	1	1	1	1	1	2
7.4	2.720	2.722	2.724	2.726	2.728	2.729	2.731	2.733	2.735	2.737	0	0	1	1	1	1	1	1	2
7.5	2.793	2.740	2.742	2.744	2.746	2.748	2.750	2.751	2.753	2.755	0	0	1	1	1	1	1	1	2
7.6	2.757	2.759	2.760	2.762	2.764	2.766	2.768	2.769	2.771	2.773	0	0	1	1	1	1	1	1	2
7.7	2.775	2.777	2.778	2.780	2.782	2.784	2.786	2.787	2.789	2.791	0	0	1	1	1	1	1	1	2
7.8	2.793	2.795	2.796	2.798	2.800	2.802	2.804	2.805	2.807	2.809	0	0	1	1	1	1	1	1	2
7.9	2.811	2.812	2.814	2.816	2.818	2.820	2.821	2.823	2.825	2.827	0	0	1	1	1	1	1	1	2
8.0	2.828	2.830	2.832	2.834	2.835	2.837	2.839	2.841	2.843	2.844	0	0	1	1	1	1	1	1	2
8.1	2.846	2.848	2.850	2.851	2.853	2.855	2.857	2.858	2.860	2.862	0	0	1	1	1	1	1	1	2
8.2	2.864	2.865	2.867	2.869	2.871	2.872	2.874	2.876	2.877	2.879	0	0	1	1	1	1	1	1	2
8.3	2.881	2.883	2.884	2.886	2.888	2.890	2.891	2.893	2.895	2.897	0	0	1	1	1	1	1	1	2
8.4	2.898	2.900	2.902	2.903	2.905	2.907	2.909	2.910	2.912	2.914	0	0	1	1	1	1	1	1	2
8.5	2.915	2.917	2.919	2.921	2.922	2.924	2.926	2.927	2.929	2.931	0	0	1	1	1	1	1	1	2
8.6	2.933	2.934	2.936	2.938	2.939	2.941	2.943	2.944	2.946	2.948	0	0	1	1	1	1	1	1	2
8.7	2.950	2.951	2.953	2.955	2.956	2.958	2.960	2.961	2.963	2.965	0	0	1	1	1	1	1	1	2
8.8	2.966	2.968	2.970	2.972	2.973	2.975	2.977	2.978	2.980	2.982	0	0	1	1	1	1	1	1	2
8.9	2.983	2.985	2.987	2.988	2.990	2.992	2.993	2.995	2.997	2.998	0	0	1	1	1	1	1	1	2
9.0	3.000	3.002	3.003	3.005	3.007	3.008	3.010	3.012	3.013	3.015	0	0	0	1	1	1	1	1	1
9.1	3.017	3.018	3.020	3.022	3.023	3.025	3.027	3.028	3.030	3.032	0	0	0	1	1	1	1	1	1
9.2	3.033	3.035	3.036	3.038	3.040	3.041	3.043	3.045	3.046	3.048	0	0	0	1	1	1	1	1	1
9.3	3.050	3.051	3.053	3.055	3.056	3.058	3.059	3.061	3.063	3.064	0	0	0	1	1	1	1	1	1
9.4	3.066	3.068	3.069	3.071	3.072	3.074	3.076	3.077	3.079	3.081	0	0	0	1	1	1	1	1	1
9.5	3.082	3.084	3.085	3.087	3.089	3.090	3.092	3.094	3.095	3.097	0	0	0	1	1	1	1	1	1
9.6	3.098	3.100	3.102	3.103	3.105	3.106	3.108	3.110	3.111	3.113	0	0	0	1	1	1	1	1	1
9.7	3.114	3.116	3.118	3.119	3.121	3.122	3.124	3.126	3.127	3.129	0	0	0	1	1	1	1	1	1
9.8	3.130	3.132	3.134	3.135	3.137	3.138	3.140	3.142	3.143	3.145	0	0	0	1	1	1	1	1	1
9.9	3.146	3.148	3.150	3.151	3.153	3.154	3.156	3.158	3.159	3.161	0	0	0	1	1	1	1	1	1
N	**0**	**1**	**2**	**3**	**4**	**5**	**6**	**7**	**8**	**9**	**1**	**2**	**3**	**4**	**5**	**6**	**7**	**8**	**9**

Table XVII.—Square Roots of Numbers.—*(Continued)*

N.	0	1	2	3	4	5	6	7	8	9	1	2	3	4	5	6	7	8	9
10	3.162	3.178	3.194	3.209	3.225	3.240	3.256	3.271	3.286	3.302	2	3	5	6	8	9	11	12	14
11	3.317	3.332	3.347	3.362	3.376	3.391	3.406	3.421	3.435	3.450	1	3	4	6	7	9	10	12	13
12	3.464	3.479	3.493	3.507	3.521	3.536	3.550	3.564	3.578	3.592	1	3	4	6	7	8	10	11	13
13	3.606	3.619	3.633	3.647	3.661	3.674	3.688	3.701	3.715	3.728	1	3	4	5	7	8	10	11	12
14	3.742	3.755	3.768	3.782	3.795	3.808	3.821	3.834	3.847	3.860	1	3	4	5	7	8	9	11	12
15	3.873	3.886	3.899	3.912	3.924	3.937	3.950	3.962	3.975	3.987	1	3	4	5	6	8	9	10	11
16	4.000	4.012	4.025	4.037	4.050	4.062	4.074	4.087	4.099	4.111	1	2	4	5	6	7	9	10	11
17	4.123	4.135	4.147	4.159	4.171	4.183	4.195	4.207	4.219	4.231	1	2	4	5	6	7	8	10	11
18	4.243	4.254	4.266	4.278	4.290	4.301	4.313	4.324	4.336	4.347	1	2	3	5	6	7	8	9	10
19	4.359	4.370	4.382	4.393	4.405	4.416	4.427	4.438	4.450	4.461	1	2	3	5	6	7	8	9	10
20	4.472	4.483	4.494	4.506	4.517	5.528	4.539	4.550	4.561	4.572	1	2	3	4	6	7	8	9	10
21	4.583	4.593	4.604	4.615	4.626	4.637	4.648	4.658	4.669	4.680	1	2	3	4	5	6	8	9	10
22	4.690	4.701	4.712	4.722	4.733	4.743	4.754	4.764	4.775	4.785	1	2	3	4	5	6	7	8	9
23	4.796	4.806	4.817	4.827	4.837	4.848	4.858	4.868	4.879	4.889	1	2	3	4	5	6	7	8	9
24	4.899	4.909	4.919	4.930	4.940	4.950	4.960	4.970	4.980	4.990	1	2	3	4	5	6	7	8	9
25	5.000	5.010	5.020	5.030	5.040	5.050	5.060	5.070	5.079	5.089	1	2	3	4	5	6	7	8	9
26	5.099	5.109	5.119	5.128	5.138	5.148	5.158	5.167	5.177	5.187	1	2	3	4	5	6	7	8	9
27	5.196	5.206	5.215	5.225	5.235	5.244	5.254	5.263	5.273	5.282	1	2	3	4	5	6	7	8	9
28	5.292	5.301	5.310	5.320	5.329	5.339	5.348	5.357	5.367	5.376	1	2	3	4	5	6	7	7	8
29	5.385	5.394	5.404	5.413	5.422	5.431	5.441	5.450	5.459	5.468	1	2	3	4	5	5	6	7	8
30	5.477	5.486	5.495	5.505	5.514	5.523	5.532	5.541	5.550	5.559	1	2	3	4	4	5	6	7	8
31	5.568	5.577	5.586	5.595	5.604	5.612	5.621	5.630	5.639	5.648	1	2	3	3	4	5	6	7	8
32	5.657	5.666	5.675	5.683	5.692	5.701	5.710	5.718	5.727	5.736	1	2	3	3	4	5	6	7	8
33	5.745	5.753	5.762	5.771	5.779	5.788	5.797	5.805	5.814	5.822	1	2	3	3	4	5	6	7	8
34	5.831	5.840	5.848	5.857	5.865	5.874	5.882	5.891	5.899	5.908	1	2	3	3	4	5	6	7	8
35	5.916	5.925	5.933	5.941	5.950	5.958	5.967	5.975	5.983	5.992	1	2	2	3	4	5	6	7	8
36	6.000	6.008	6.017	6.025	6.033	6.042	6.050	6.058	6.066	6.075	1	2	2	3	4	5	6	7	7
37	6.083	6.091	6.099	6.107	6.116	6.124	6.132	6.140	6.148	6.156	1	2	2	3	4	5	6	7	7
38	6.164	6.173	6.181	6.189	6.197	6.205	6.213	6.221	6.229	6.237	1	2	2	3	4	5	6	6	7
39	6.245	6.253	6.261	6.269	6.277	6.285	6.293	6.301	6.309	6.317	1	2	2	3	4	5	6	6	7
40	6.325	6.332	6.340	6.348	6.356	6.364	6.372	6.380	6.387	6.395	1	2	2	3	4	5	6	6	7
41	6.403	6.411	6.419	6.427	6.434	6.442	6.450	6.458	6.465	6.473	1	2	2	3	4	5	5	6	7
42	6.481	6.488	6.496	6.504	6.512	6.519	6.527	6.535	6.542	6.550	1	2	2	3	4	5	5	6	7
43	6.557	6.565	6.573	6.580	6.588	6.595	6.603	6.611	6.618	6.626	1	2	2	3	4	5	5	6	7
44	6.633	6.641	6.648	6.656	6.663	6.671	6.678	6.686	6.693	6.701	1	2	2	3	4	5	5	6	7
45	6.708	6.716	6.723	6.731	6.738	6.745	6.753	6.760	6.768	6.775	1	1	2	3	4	4	5	6	7
46	6.782	6.790	6.797	6.804	6.812	6.819	6.826	6.834	6.841	6.848	1	1	2	3	4	4	5	6	7
47	6.856	6.863	6.870	6.877	6.885	6.892	6.899	6.907	6.914	6.921	1	1	2	3	4	4	5	6	7
48	6.928	6.935	6.943	6.950	6.957	6.964	6.971	6.979	6.986	6.993	1	1	2	3	4	4	5	6	6
49	7.000	7.007	7.014	7.021	7.029	7.036	7.043	7.050	7.057	7.064	1	1	2	3	4	4	5	6	6
50	7.071	7.078	7.085	7.092	7.099	7.106	7.113	7.120	7.127	7.134	1	1	2	3	4	4	5	6	6
51	7.141	7.148	7.155	7.162	7.169	7.176	7.183	7.190	7.197	7.204	1	1	2	3	4	4	5	6	6
52	7.211	7.218	7.225	7.232	7.239	7.246	7.253	7.259	7.266	7.273	1	1	2	3	3	4	5	6	6
53	7.280	7.287	7.294	7.301	7.308	7.314	7.321	7.328	7.335	7.342	1	1	2	3	3	4	5	5	6
54	7.348	7.355	7.362	7.369	7.376	7.382	7.389	7.396	7.403	7.409	1	1	2	3	3	4	5	5	6
N	0	1	2	3	4	5	6	7	8	9	1	2	3	4	5	6	7	8	9

Table XVII.—Square Roots of Numbers.—(*Continued*)

N	0	1	2	3	4	5	6	7	8	9	1	2	3	4	5	6	7	8	9
55	7.416	7.423	7.430	7.436	7.443	7.450	7.457	7.463	7.470	7.477	1	1	2	3	3	4	5	5	6
56	7.483	7.490	7.497	7.503	7.510	7.517	7.523	7.530	7.537	7.543	1	1	2	3	3	4	5	5	6
57	7.550	7.556	7.563	7.570	7.576	7.583	7.589	7.596	7.603	7.609	1	1	2	3	3	4	5	5	6
58	7.616	7.622	7.629	7.635	7.642	7.649	7.655	7.662	7.668	7.675	1	1	2	3	3	4	5	5	6
59	7.681	7.688	7.694	7.701	7.707	7.714	7.720	7.727	7.733	7.740	1	1	2	3	3	4	4	5	6
60	7.746	7.752	7.759	7.765	7.772	7.778	7.785	7.791	7.797	7.804	1	1	2	3	3	4	4	5	6
61	7.810	7.817	7.823	7.829	7.836	7.842	7.849	7.855	7.861	7.868	1	1	2	3	3	4	4	5	6
62	7.874	7.880	7.887	7.893	7.899	7.906	7.912	7.918	7.925	7.931	1	1	2	3	3	4	4	5	6
63	7.937	7.944	7.950	7.956	7.962	7.969	7.975	7.981	7.987	7.994	1	1	2	3	3	4	4	5	6
64	8.000	8.006	8.012	8.019	8.025	8.031	8.037	8.044	8.050	8.056	1	1	2	2	3	4	4	5	6
65	8.062	8.068	8.075	8.081	8.087	8.093	8.099	8.106	8.112	8.118	1	1	2	2	3	4	4	5	5
66	8.124	8.130	8.136	8.142	8.149	8.155	8.161	8.167	8.173	8.179	1	1	2	2	3	4	4	5	5
67	8.185	8.191	8.198	8.204	8.210	8.216	8.222	8.228	8.234	8.240	1	1	2	2	3	4	4	5	5
68	8.246	8.252	8.258	8.264	8.270	8.276	8.283	8.289	8.295	8.301	1	1	2	2	3	4	4	5	5
69	8.307	8.313	8.319	8.325	8.331	8.337	8.343	8.340	8.355	8.361	1	1	2	2	3	4	4	5	5
70	8.367	8.373	8.379	8.385	8.390	8.396	8.402	8.408	8.414	8.420	1	1	2	2	3	4	4	5	5
71	8.426	8.432	8.438	8.444	8.450	8.456	8.462	8.468	8.473	8.479	1	1	2	2	3	4	4	5	5
72	8.485	8.491	8.497	8.503	8.509	8.515	8.521	8.526	8.532	8.538	1	1	2	2	3	3	4	5	5
73	8.544	8.550	8.556	8.562	8.567	8.573	8.579	8.585	8.591	8.597	1	1	2	2	3	3	4	5	5
74	8.602	8.608	8.614	8.620	8.626	8.631	8.637	8.643	8.649	8.654	1	1	2	2	3	3	4	5	5
75	8.660	8.666	8.672	8.678	8.683	8.689	8.695	8.701	8.706	8.712	1	1	2	2	3	3	4	5	5
76	8.718	8.724	8.729	8.735	8.741	8.746	8.752	8.758	8.764	8.769	1	1	2	2	3	3	4	5	5
77	8.775	8.781	8.786	8.792	8.798	8.803	8.809	8.815	8.820	8.826	1	1	2	2	3	3	4	4	5
78	8.832	8.837	8.843	8.849	8.854	8.860	8.866	8.871	8.877	8.883	1	1	2	2	3	3	4	4	5
79	8.888	8.894	8.899	8.905	8.911	8.916	8.922	8.927	8.933	8.939	1	1	2	2	3	3	4	4	5
80	8.944	8.950	8.955	8.961	8.967	8.972	8.978	8.983	8.989	8.994	1	1	2	2	3	3	4	4	5
81	9.000	9.006	9.011	9.017	9.022	9.028	9.033	9.039	9.044	9.050	1	1	2	2	3	3	4	4	5
82	9.055	9.061	9.066	9.072	9.077	9.083	9.088	9.094	9.009	9.105	1	1	2	2	3	3	4	4	5
83	9.110	9.116	9.121	9.127	9.132	9.138	9.143	9.149	9.154	9.160	1	1	2	2	3	3	4	4	5
84	9.165	9.171	9.176	9.182	9.187	9.192	9.198	9.203	9.209	9.214	1	1	2	2	3	3	4	4	5
85	9.220	9.225	9.230	9.236	9.241	9.247	9.252	9.257	9.263	9.268	1	1	2	2	3	3	4	4	5
86	9.274	9.279	9.284	9.290	9.295	9.301	9.306	9.311	9.317	9.322	1	1	2	2	3	3	4	4	5
87	9.327	9.333	9.338	9.343	9.349	9.354	9.359	9.365	9.370	9.375	1	1	2	2	3	3	4	4	5
88	9.381	9.386	9.391	9.397	9.402	9.407	9.413	9.418	9.423	9.429	1	1	2	2	3	3	4	4	5
89	9.434	9.439	9.445	9.450	9.455	9.460	9.466	9.471	9.476	9.482	1	1	2	2	3	3	4	4	5
90	9.487	9.492	9.497	9.503	9.508	9.513	9.518	9.524	9.529	9.534	1	1	2	2	3	3	4	4	5
91	9.539	9.545	9.550	9.555	9.560	9.566	9.571	9.576	9.581	9.586	1	1	2	2	3	3	4	4	5
92	9.592	9.597	9.602	9.607	9.612	9.618	9.623	9.628	9.633	9.638	1	1	2	2	3	3	4	4	5
93	9.644	9.649	9.654	9.659	9.664	9.670	9.675	9.680	9.685	9.690	1	1	2	2	3	3	4	4	5
94	9.695	9.701	9.706	9.711	9.716	9.721	9.726	9.731	9.737	9.742	1	1	2	2	3	3	4	4	5
95	9.747	9.752	9.757	9.762	9.767	9.772	9.778	9.783	9.788	9.793	1	1	2	2	3	3	4	4	5
96	0.798	9.803	9.808	9.813	9.818	9.823	9.829	9.834	9.839	9.844	1	1	2	2	3	3	4	4	5
97	9.849	9.854	9.859	9.864	9.869	9.874	9.879	9.884	9.889	9.894	1	1	2	2	3	3	4	4	5
98	9.899	9.905	9.910	9.915	9.920	9.925	9.930	9.935	9.940	9.945	1	1	1	2	2	3	3	4	4
99	9.950	9.955	9.960	9.965	9.970	9.975	9.980	9.985	9.990	9.995	0	1	1	2	2	3	3	4	4
N	**0**	**1**	**2**	**3**	**4**	**5**	**6**	**7**	**8**	**9**	**1**	**2**	**3**	**4**	**5**	**6**	**7**	**8**	**9**

INDEX

D

G

H

I

R

S